METHODS
OF
ALGEBRAIC GEOMETRY

METHODS
OF
ALGEBRAIC GEOMETRY

by

W. V. D. HODGE, Sc.D., F.R.S.

Formerly Lowndean Professor of Astronomy and Geometry, and Fellow of Pembroke College, Cambridge

and

D. PEDOE, Ph.D.

Emeritus Professor of Mathematics University of Minnesota

VOLUME II

BOOK III: GENERAL THEORY OF ALGEBRAIC VARIETIES IN PROJECTIVE SPACE

BOOK IV: QUADRICS AND GRASSMANN VARIETIES

CAMBRIDGE
UNIVERSITY PRESS

Published by the Press Syndicate of the University of Cambridge
The Pitt Building, Trumpington Street, Cambridge CB2 1RP
40 West 20th Street, New York, NY 10011–4211, USA
10 Stamford Road, Oakleigh, Melbourne 3166, Australia

First published 1952
Reprinted 1968
First paperback edition 1968
Reissued in the Cambridge Mathematical Library 1994

ISBN 0 521 46901 5 paperback

Transferred to digital printing 2004

CONTENTS

BOOK III

GENERAL THEORY OF ALGEBRAIC VARIETIES IN PROJECTIVE SPACE

CHAPTER X: ALGEBRAIC VARIETIES

CHAPTER XI: ALGEBRAIC CORRESPONDENCES

CHAPTER XII: INTERSECTION THEORY

BOOK IV

QUADRICS AND GRASSMANN VARIETIES

CHAPTER XIII: QUADRICS

CHAPTER XIV: GRASSMANN VARIETIES

PREFACE

THIS VOLUME gives an account of the principal methods used in developing a theory of algebraic varieties in space of n dimensions. Applications of these methods are also given to some of the more important varieties which occur in projective geometry. It was originally our intention to include an account of the arithmetic theory of varieties, and of the foundations of birational geometry, but it has turned out to be more convenient to reserve these topics for a third volume. The theory of algebraic varieties developed in this volume is therefore mainly a theory of varieties in projective space.

In writing this volume we have been faced with two problems: the difficult question of what must go in and what should be left out, and the problem of the degree of generality to be aimed at. As our objective has been to give an account of the modern algebraic methods available to geometers, we have not sought generality for its own sake. There is still enough to be done in the realm of classical geometry to give these methods all the scope that could be desired, and had it been possible to confine ourselves to the classical case of geometry over the field of complex numbers, we should have been content to do so. But in order to put the classical methods on a sound basis, using algebraic methods, it is necessary to consider geometry over more general fields than the field of complex numbers. However, if the ultimate object is to provide a sound algebraic basis for classical geometry, it is only necessary to consider fields without characteristic. Since geometry over any field without characteristic conforms to the general pattern of geometry over the field of complex numbers, we have developed the theory of algebraic varieties over any field without characteristic. Thus fields with finite characteristic are not used in this book.

As for the material included, the space factor has caused us to restrict ourselves to fundamental concepts and methods. But it is our hope that a reader who has mastered the methods described in these volumes, and has seen how they can be applied to some of the standard problems in geometry, will be able to apply them to more advanced geometrical problems.

Chapter X (the first in this volume) is devoted to definitions, and to the study of the basic concepts of the theory of algebraic varieties, including irreducibility, generic points, dimension and order. The principal tool employed in this chapter is the 'Zugeordnete Form', introduced by van der Waerden and Chow, to which we have given the name 'Cayley form', for historical reasons discussed in the text. The ideas of this chapter owe much to van der Waerden, but the greater prominence given here to the Cayley form has resulted in a somewhat different, and, we believe, more complete account of the basic concepts than he has yet given.

In Chapter XI we deal with the foundations of the theory of algebraic correspondences, and apply this theory to develop the idea of multiplicities in geometry. This chapter also owes much to van der Waerden, but, following Weil, we make a sharp distinction between point-set properties of varieties and multiplicative properties. Chapter XII begins with an account of the intersections of algebraic varieties, and then goes on to develop the algebraic theory of systems of varieties. The first three chapters of this volume are therefore mainly concerned with general theory, but the principles are illustrated by applications to examples. A particularly convenient example is afforded by the Segre variety which represents an r-way space as a variety in one-way space.

Chapters XIII and XIV deal with Quadrics and Grassmann Varieties respectively. The purpose of these chapters is to show how the general methods developed in earlier chapters can be used to develop the properties of these loci on a strictly algebraic basis. The account given in each case is not intended to be complete, although a considerable range of properties is reviewed. The methods used for finding a base on a number of the varieties may appear to be rather special, but no general method is known at present.

We again have to express our thanks to Prof. T. A. A. Broadbent of the Royal Naval College, Greenwich, for valuable assistance in the preparation of the manuscript and in the correction of proofs, and to the Cambridge University Press for their assistance in technical matters.

One of us (D.P.) gratefully acknowledges the assistance he received from the Leverhulme Trustees, who granted him a Fellowship which ensured the completion of this work in a shorter time than would otherwise have been possible.

Note. The reference [II, § 4, Th. II] is to Theorem II in § 4 of Chapter II of Vol. I, which contains Chapters I–IX. This volume contains Chapters X–XIV. If a reference is to the same chapter or section, the corresponding numeral or numerals are omitted.

W. V. D. H.

D. P.

CAMBRIDGE

March 1951

BOOK III

GENERAL THEORY OF ALGEBRAIC VARIETIES IN PROJECTIVE SPACE

CHAPTER X

ALGEBRAIC VARIETIES

1. Introduction. This volume is concerned with properties of the points of projective space whose coordinates satisfy a set of homogeneous algebraic equations, these points not being treated as individuals, but as members of the aggregate of solutions of the equations.

We begin by selecting the ground field K over which our projective space is to be constructed. We shall confine ourselves to the case in which K is commutative and without characteristic, but we shall not assume that K is algebraically closed unless this requirement is specifically made. We then construct a projective space of n dimensions over K, denoting it by S_n.

In V, § 3, we saw that we could extend the ground field K to any field K^* containing K, and S_n is then extended to a space S_n^* defined over K^*. There are points $(x_0^*, \ldots, x_n^*)$ in S_n^*, where x_i^* $(i = 0, \ldots, n)$ is in K^*, which are not points of the original space S_n. When the ratios of the coordinates x_i^* are all algebraic over K we shall say that $x^* = (x_0^*, \ldots, x_n^*)$ is an algebraic point of S_n (over K). If at least one of the ratios of the coordinates x_i^* is transcendental, we say that x^* is a transcendental point of S_n. In the course of our investigations we shall have to extend the ground field many times, thus introducing points which are algebraic or transcendental over K, and we shall find it convenient to omit a reference to the extension K^* on most of these occasions. We shall therefore be considering *rational points* of S_n (that is, points the ratios of whose coordinates are in K), *algebraic points* of S_n and *transcendental points*. The term *point* without any further qualification will cover these three kinds of point.

A restriction on the fields from which the coordinates of algebraic and transcendental points are chosen is, however, necessary. If x^* is a point which is not rational, the field $K(x^*) = K(x_0^*, \ldots, x_n^*)$ is a proper extension of K, and since it is formed by adjoining a finite number of elements to K, it is an extension of a finite degree of transcendency. The geometrical content of any result which depends only on the properties of this field is not altered if we replace $K(x^*)$ by an equivalent extension of K, and x^* by the corresponding element of this new extension.

But we may have to consider a number (always finite) of extensions of K, and as we saw in III, § 4, p. 114, there may be no extension of K which contains them all. This gives rise to grave difficulties. All the extensions which arise, however, are algebraic extensions of pure transcendental extensions of finite dimension. If K_1 and K_2 are extensions of K, K_2 being of dimension d, we can construct an extension $\bar{K}_1$ of K_1 which contains the algebraic closure of an extension of K of dimension d obtained by adjoining the independent indeterminates $t_1, \ldots, t_d$, and this will contain a field K_2' isomorphic with K_2. But we cannot simply replace K_2 by K_2', for there is ambiguity in the choice of K_2'; for instance, the dimension of the field which is the intersection of K_1 and K_2' depends on the number of the t_i in K_1.

To overcome these difficulties we must ensure that the various extensions of K which we consider all belong to the same extension of K. When this condition is satisfied, the join of the various extensions considered is contained in an extension K^* of a finite degree of transcendency (the enveloping field). Results will be unaltered if K^* is replaced by an equivalent extension of K.

We can lay down a field once and for all over a given ground field K which will contain the isomorph of any enveloping field K^* which may arise, and agree that all extensions be subfields of this. Let $t_1, t_2, \ldots$ be a simple sequence of independent indeterminates over K, Σ the field consisting of rational functions of these (each element of Σ being a rational function of a finite number of t_i), and let Σ^* be the algebraic closure of Σ. Any enveloping field K^* is clearly isomorphic to a subfield of Σ^*, and hence if we replace K^* by an isomorph in Σ^*, each of the extensions in question is replaced by a subfield of Σ^*. We shall call Σ^* the 'universal field' associated with K, and in future we shall assume, without mentioning the fact explicitly, that all extensions of K are in this universal field.

We now choose an allowable coordinate system in S_n, and we consider the set of points whose coordinates satisfy the equations

$$f_i(x_0, x_1, \ldots, x_n) = 0 \quad (i = 1, 2, \ldots), \tag{1}$$

where $f_i(x_0, x_1, \ldots, x_n)$ is a homogeneous polynomial over K. In IV, § 1, we saw that the polynomials over K which are satisfied by the solutions of the set of equations (1) form an ideal, and in IV, § 2, we proved the existence of a finite set of the equations (1), say

$$f_i(x_0, x_1, \ldots, x_n) = 0 \quad (i = 1, 2, \ldots, k), \tag{2}$$

which is such that every solution of (2) satisfies the equations (1) for all values of i. Hence in considering the points of S_n which satisfy (1), we need only consider the solutions of the finite set of equations (2).

The aggregate of points defined by a set of equations (1) is called an *algebraic variety*. It may happen, of course, that there are no points satisfying the equations. While this case is of no geometrical interest, it cannot always be avoided in theoretical reasoning; but the statement of theorems will be simpler if, for the present, we assume that the varieties we are considering have at least one point. This point may, of course, be rational, algebraic or transcendental.

An algebraic variety in S_n has been defined in a particular allowable coordinate system. Let us consider what happens to the set of equations (2) when we carry out the allowable transformation of coordinates given by the equations

$$y_i = \sum_{j=0}^{n} a_{ij} x_j \quad (i = 0, \ldots, n),$$

or

$$x_i = \sum_{j=0}^{n} b_{ij} y_j \quad (i = 0, \ldots, n),$$

where (b_{ij}) is the matrix inverse to (a_{ij}).

If $(x'_0, \ldots, x'_n)$ satisfies (2), the coordinates $(y'_0, \ldots, y'_n)$ of the same point in the new coordinate system satisfy the set of equations

$$f_i(\sum_j b_{0j} y_j, \ldots, \sum_j b_{nj} y_j) = 0 \quad (i = 1, \ldots, k), \tag{3}$$

and conversely, if $(y'_0, \ldots, y'_n)$ are the coordinates of a point in the new coordinate system which satisfy (3), the coordinates $(x'_0, \ldots, x'_n)$ of the same point in the original coordinate system satisfy (2). Hence, if an aggregate of points in S_n form an algebraic variety in one coordinate system, they form an algebraic variety in any other allowable coordinate system, although the equations in the two systems may be different. In this sense the definition of an algebraic variety is independent of the coordinate system chosen.

2. Reducible and irreducible varieties. If V_1, V_2 are two algebraic varieties given by the equations

$$f_i(x_0, \ldots, x_n) = 0 \quad (i = 1, \ldots, r), \tag{1}$$

$$g_i(x_0, \ldots, x_n) = 0 \quad (i = 1, \ldots, s) \tag{2}$$

respectively, the points common to V_1 and V_2 satisfy both sets of equations simultaneously, and therefore define a third algebraic variety. We call this aggregate of points the *intersection* of V_1 and V_2, and denote it by the symbol

$$V_1 \wedge V_2.$$

This set of points is the point-set theoretic intersection of the sets V_1, V_2. Evidently

$$V_1 \wedge V_2 = V_2 \wedge V_1.$$

The points which satisfy the set of equations

$$f_i(x_0, \ldots, x_n)\, g_j(x_0, \ldots, x_n) = 0 \quad (i = 1, \ldots, r;\ j = 1, \ldots, s) \tag{3}$$

are those points, and only those, which satisfy either (1) or (2). Points which satisfy (1) or (2) evidently satisfy (3). On the other hand, let $(x'_0, \ldots, x'_n)$ be a point which is not on V_2, say, but which satisfies (3). Then for some value of j

$$g_j(x'_0, \ldots, x'_n) \neq 0.$$

If we consider the equations of the set (3) for which j has this particular value and $i = 1, \ldots, r$, we see that

$$f_i(x'_0, \ldots, x'_n) = 0 \quad (i = 1, \ldots, r).$$

Hence $(x'_0, \ldots, x'_n)$ lies on V_1. Similarly, points satisfying (3) which do not lie on V_1 lie on V_2.

We call the algebraic variety defined by (3) the *sum* of V_1 and V_2 and denote it by the symbol

$$V_1 \dotplus V_2.$$

We have shown that this symbol defines the point-set theoretic sum of the points in V_1 and the points in V_2. Evidently

$$V_1 \dotplus V_2 = V_2 \dotplus V_1.$$

If V_3 is an algebraic variety given by the equations

$$h_i(x_0, \ldots, x_n) = 0 \quad (i = 1, \ldots, t),$$

our definitions lead to the following associative and distributive laws, as in point-set theory:

$$(a) \quad V_1 \wedge (V_2 \wedge V_3) = (V_1 \wedge V_2) \wedge V_3,$$

$$(b) \quad V_1 \dotplus (V_2 \dotplus V_3) = (V_1 \dotplus V_2) \dotplus V_3,$$

$$(c) \quad V_1 \wedge (V_2 \dotplus V_3) = V_1 \wedge V_2 \dotplus V_1 \wedge V_3.$$

Now let us suppose that every solution of the equations (1) in any *algebraic* extension of K satisfies (2); that is, that every algebraic point of V_1 lies on V_2. Then by Hilbert's zero-theorem [IV, § 8]

$$[g_i(x_0, \ldots, x_n)]^{\rho_i} = \sum_{j=1}^{r} a_{ij}(x_0, \ldots, x_n) f_j(x_0, \ldots, x_n),$$

where ρ_i is a positive integer, and the $a_{ij}(x_0, \ldots, x_n)$ are forms in $K[x_0, \ldots, x_n]$. Hence $g_i(x_0, \ldots, x_n)$ *vanishes on the variety* V_1, that is, $g_i(x'_0, \ldots, x'_n) = 0$ for *all* points $(x'_0, \ldots, x'_n)$ on V_1, not only for algebraic points. Hence *every* solution of (1) satisfies (2).

We then say 'V_1 lies on V_2', or 'V_1 is contained in V_2', and write

$$V_1 \subseteq V_2,$$

or say 'V_2 contains V_1' and write

$$V_2 \supseteq V_1.$$

If $V_1 \subseteq V_2$ and $V_2 \subseteq V_1$, we must have $V_1 = V_2$. If there are points of which are not on V_1, and V_1 lies on V_2, we write

$$V_1 \subset V_2 \quad \text{or} \quad V_2 \supset V_1.$$

If $V_1 \subseteq V_2$ and $V_2 \subseteq V_3$, then $V_1 \subseteq V_3$. Similarly, if $V_1 \subset V_2$ and $V_2 \subset V_3$, then $V_1 \subset V_3$. Hence the relations $\subseteq$ and $\subset$ are transitive. Equivalently, the relations $\supseteq$ and $\supset$ are transitive.

Again, if

$$V = V_1 \dotplus V_2,$$

then $V \supseteq V_1$ and $V \supseteq V_2$. If there are points of V_1 not on V_2, so that

$$V_1 \nsubseteq V_2,$$

we must have $V \supset V_2$.

It is also clear that $$V_1 \supseteq V_1 \wedge V_2,$$

and that, if $$V_1 \nsubseteq V_2,$$

then $$V_1 \supset V_1 \wedge V_2.$$

With these preliminaries we now introduce the notion of *reducibility*. A variety V is said to be *reducible* if it can be expressed as the sum of two algebraic varieties, each distinct from V; that is, if

$$V = V_1 \dot{+} V_2,$$

where $V_1 \subset V$, $V_2 \subset V$.

If V is not reducible it is said to be *irreducible*.

Lemma. If an irreducible variety V lies in the sum of two varieties V_1 and V_2, then it is contained in one or in the other.

We have

$$V \subseteq V_1 \dot{+} V_2.$$

Then

$$\begin{aligned} V &= V_\wedge (V_1 \dot{+} V_2) \\ &= V_\wedge V_1 \dot{+} V_\wedge V_2. \end{aligned}$$

Since V is irreducible we must have either

$$V_\wedge V_1 = V$$

or

$$V_\wedge V_2 = V,$$

that is, V is contained in V_1 or in V_2.

This lemma is easily extended to the case of r varieties $V_1, V_2, \ldots, V_r$. We use it to prove

THEOREM I. *A necessary and sufficient condition for the reducibility of a variety V is the existence of a product fg of two forms $f(x_0, \ldots, x_n)$ and $g(x_0, \ldots, x_n)$ which vanishes at all points of V without either form having this property.*

We suppose in the first place that V is irreducible and that $fg = 0$ on V. If $f = 0$ defines the variety V_1, and $g = 0$ defines the variety V_2, then $fg = 0$ defines $V_1 \dot{+} V_2$, and

$$V \subseteq V_1 \dot{+} V_2,$$

since $fg = 0$ at all points of V. By the above lemma V is contained in V_1 or in V_2, and so either $f = 0$ or $g = 0$ at all points of V.

Now let us suppose that V is reducible. Then we can construct a product fg of two forms f and g which vanishes on V without either f or g doing so. In fact,

$$V = V_1 \dot{+} V_2,$$

where the varieties V_1 and V_2 are distinct and neither contains the other, by hypothesis. Hence, among the forms defining V_1 there must be at least one, f, say, which does not vanish on V_2.

Similarly, there must be a form g which vanishes on V_2 but not on V_1. The product fg vanishes for all points of V, but neither f nor g vanishes for all points of V. This proves the theorem.

We note that the question of the irreducibility of an algebraic variety depends on the choice of the ground field, and that a variety which is irreducible over K may become reducible when K is replaced by an extension K^*. We illustrate this by considering the algebraic variety V defined over K by a single equation

$$f(x) = f(x_0, \dots, x_n) = 0.$$

Let us suppose that $f(x)$ is irreducible over K. Then if $g(x)$, $h(x)$ are two forms such that their product vanishes on V, we have, by Hilbert's zero-theorem,

$$[g(x)\,h(x)]^\rho = a(x)\,f(x),$$

where ρ is a positive integer, and $a(x)$ is some form in $K[x]$. By the unique factorisation theorem [I, § 8, Th. II] it follows that $f(x)$, which is irreducible, is a factor of $g(x)$ or of $h(x)$, and hence either $g(x)$ or $h(x)$ vanishes on V. Hence, by Theorem I, V is irreducible.

On the other hand, let us suppose that $f(x)$ is reducible, and that it can be written in the form

$$f(x) = f_1(x)\,f_2(x),$$

where $f_1(x)$, $f_2(x)$ are forms in $K[x]$ *not having a common factor*. Then if V_1, V_2 are the varieties defined, respectively, by the equations

$$f_1(x) = 0$$

and

$$f_2(x) = 0,$$

neither variety contains the other. For instance, if $V_1 \subseteq V_2$, we should have, by Hilbert's zero-theorem,

$$[f_2(x)]^\sigma = a(x)\,f_1(x),$$

and hence every irreducible factor of $f_1(x)$ would be a factor of $f_2(x)$, contrary to hypothesis. Hence, since $f(x) = f_1(x)\,f_2(x)$,

$$V = V_1 \dotplus V_2,$$

where $V_1 \nsubseteq V_2$, $V_2 \nsubseteq V_1$, and so V is reducible.

Now let $f(x)$ be a form in $K[x]$ which is irreducible over K. Then the variety V defined by

$$f(x) = 0$$

is irreducible. But it may happen that there is an extension K^* of

K such that $f(x)$ is reducible in $K^*[x]$. Since K is without characteristic, all the irreducible factors of $f(x)$ over K^* are distinct, and therefore, by what has been proved above, V is reducible over the ground field K^*. An example of a form $f(x)$ with this property in the case $n = 1$ is given by the form $x_0^2 + x_1^2$. When K is the field of real numbers this is irreducible, but it is reducible over the field of complex numbers.

The criterion for reducibility given in Theorem I can be described in another way. We have seen [IV, § 1] that the polynomials in $K[x]$ which vanish on a variety V form an ideal in this ring. Every element of the ideal is a sum of homogeneous polynomials, each of which belongs to the ideal. Such an ideal is called a *homogeneous ideal*. If V is irreducible this homogeneous ideal has the property that if $f(x)$, $g(x)$ are two forms whose product belongs to the ideal, then either $f(x)$ or $g(x)$ belongs to the ideal. A homogeneous ideal with this property is said to be *prime*; thus the polynomials in $K[x_0, \ldots, x_n]$ which vanish on an irreducible algebraic variety form a prime homogeneous ideal, and conversely, a prime homogeneous ideal in $K[x_0, \ldots, x_n]$ defines an irreducible variety.

We remark that an algebraic variety V which is irreducible over K in a given coordinate system is also irreducible in any allowable coordinate system. This follows immediately from the definition of irreducibility, and the remark made at the end of § 1.

As a preliminary to the main theorem of this section we prove

THEOREM II. *A sequence of varieties $V_1, V_2, \ldots$ in S_n, where $V_1 \supset V_2 \ldots \supset V_r \supset V_{r+1} \ldots$, must terminate after a finite number of terms.*

Let the equations of V_1 be

$$f_i(x_0, \ldots, x_n) = 0 \quad (i = 1, \ldots, r_1),$$

and those of V_2

$$f_i(x_0, \ldots, x_n) = 0 \quad (i = r_1 + 1, \ldots, r_2).$$

Since $V_1 \supset V_2$ we may take the equations of V_2 to be

$$f_i(x_0, \ldots, x_n) = 0 \quad (i = 1, \ldots, r_2).$$

Similarly, let the equations of V_t be

$$f_i(x_0, \ldots, x_n) = 0 \quad (i = 1, \ldots, r_t).$$

By Hilbert's basis theorem [IV, § 2] there is a finite integer k such that there exist forms $a_{i1}(x), \ldots, a_{ik}(x)$ with the property that

$$f_i \equiv \sum_{j=1}^{k} a_{ij} f_j$$

for each value of i. Let us suppose that

$$r_{l-1} < k \leqslant r_l.$$

Then, inserting zero polynomials a_{ij} if necessary, we can write

$$f_i \equiv \sum_{j=1}^{r_l} a_{ij} f_j. \tag{4}$$

If there is a variety V_{l+1} in the given sequence, we deduce from (4) that

$$V_{l+1} \supseteq V_l.$$

But this contradicts the hypothesis

$$V_l \supset V_{l+1}.$$

The sequence therefore terminates at V_l.

We can now prove

THEOREM III. *Every algebraic variety V can be expressed as the sum of a finite number of irreducible varieties.*

Let us suppose that the theorem is not true for V. Then V must be reducible, say $V = V_1 \dotplus V_2$, where $V \supset V_1$ and $V \supset V_2$. If the theorem is true for both V_1 and V_2, it is true for V. Hence the theorem is false for either V_1 or V_2. Let us suppose that it is false for V_1. Then V_1 is reducible, and we can write $V_1 = V_1' \dotplus V_2'$, where $V_1 \supset V_1'$, $V_1 \supset V_2'$. As before, we see that the theorem must be false for either V_1' or V_2', say for V_1'. We can repeat this process indefinitely, and obtain an infinite sequence of varieties

$$V \supset V_1 \supset V_1' \supset \ldots,$$

for each of which the theorem is false. By our previous theorem a strictly descending infinite sequence of varieties does not exist, and we therefore conclude that every algebraic variety V is the sum of a finite number of irreducible varieties.

In the expression

$$V = V_1 \dotplus V_2 \dotplus \ldots \dotplus V_k$$

we may omit any component V_i which is contained in the sum

$$V_i' = V_1 \dotplus V_2 \dotplus \ldots \dotplus V_{i-1} \dotplus V_{i+1} \dotplus \ldots \dotplus V_k$$

of the remaining component varieties. For if $V_i \subseteq V_i'$, then by the lemma proved above, V_i must lie in one of the component varieties whose sum is V_i', and therefore

$$V = V_i \dotplus V_i' = V_i'.$$

A component V_i which can be omitted in this way is said to be *redundant.* When all redundant components of a sum have been omitted we say that we have a *non-contractible* representation of V as a sum of irreducible varieties.

THEOREM IV. *The representation of an algebraic variety V as a non-contractible sum of irreducible varieties is essentially unique.*

If

$$V = V_1 \dotplus V_2 \dotplus \dots \dotplus V_k = V_1' \dotplus V_2' \dotplus \dots \dotplus V_l'$$

are two non-contractible representations of V, we prove that $k = l$, and that $V_i = V_i'$ $(i = 1, \dots, k)$, after the components have been suitably arranged.

It follows from the representation that

$$V_1 \subseteq V = V_1' \dotplus V_2' \dotplus \dots \dotplus V_l',$$

and therefore, by the lemma above, there exists a value of i $(i \leqslant l)$ such that

$$V_1 \subseteq V_i'.$$

We rearrange the varieties $V_1', \dots, V_l'$ so that $i = 1$. Then

$$V_1 \subseteq V_1'.$$

A similar argument shows that for some value of j $(j \leqslant k)$

$$V_1' \subseteq V_j.$$

Hence

$$V_1 \subseteq V_1' \subseteq V_j,$$

and therefore

$$V_1 \subseteq V_j.$$

Since the representations are non-contractible it follows that $j = 1$, and since $V_1 \subseteq V_1' \subseteq V_1$, we must have $V_1 = V_1'$.

In the same way we show that

$$V_2 \subseteq V_i',$$

where $i \neq 1$, since $V_2 \nsubseteq V_1 = V_1'$. We order the varieties so that $i = 2$, and prove, as above, that $V_2 = V_2'$. The theorem follows by induction.

The varieties $V_1, V_2, \dots, V_k$ are called the *irreducible components* of V.

3. Generic points of an irreducible variety. A point $\xi = (\xi_0, \dots, \xi_n)$, where the ξ_i lie in some extension K^* of the ground field K, is said to be a generic point of a variety V (over K) if

(i) ξ lies on V, and

(ii) any form $f(x_0, \ldots, x_n)$ in $K[x_0, \ldots, x_n]$ for which $f(\xi_0, \ldots, \xi_n) = 0$ vanishes on V.

It is important to note that in defining a generic point the ground field must be specified.

As an example we show that if λ, μ are independent indeterminates over K the point $\xi = (\lambda^2, \lambda\mu, \mu^2)$ is a generic point of the variety V in S_2 whose equation is $x_1^2 - x_0 x_2 = 0$. Clearly ξ lies on V. If $f(x_0, x_1, x_2)$ is any form in $K[x_0, x_1, x_2]$ we can consider it as a polynomial in x_1, and use the division algorithm to write

$$f(x_0, x_1, x_2) \equiv (x_1^2 - x_0 x_2)\, g(x_0, x_1, x_2) + x_1 h(x_0, x_2) + k(x_0, x_2),$$

where g, h, k are homogeneous in their respective indeterminates. If $f(\lambda^2, \lambda\mu, \mu^2) = 0$, then

$$\lambda\mu\, h(\lambda^2, \mu^2) + k(\lambda^2, \mu^2) = 0. \tag{1}$$

Since λ, μ are independent indeterminates over K, the coefficient of each power-product $\lambda^r \mu^s$ in (1) must be zero. If $r+s$ is odd, the coefficient is evidently zero. If r and s are both even the coefficient is a coefficient of $k(x_0, x_2)$, and if r and s are both odd it is a coefficient of $h(x_0, x_2)$. Hence $h(x_0, x_2) \equiv 0$ and $k(x_0, x_2) \equiv 0$ and therefore

$$f(x_0, x_1, x_2) \equiv (x_1^2 - x_0 x_2)\, g(x_0, x_1, x_2).$$

It follows that f vanishes for all points of V. Hence ξ is a generic point of V.

Not every variety has a generic point. The two theorems which follow settle the question: Which varieties have generic points?

THEOREM I. *If V has a generic point ξ over K, V is irreducible over K.*

Let $f(x)$, $g(x)$ be any two homogeneous polynomials over K whose product vanishes on V. By property (i) of a generic point,

$$f(\xi)\, g(\xi) = 0.$$

Since $f(\xi)$ and $g(\xi)$ lie in the field $K(\xi)$, either $f(\xi)$ or $g(\xi)$ is zero; say $f(\xi) = 0$. Then, by property (ii) of a generic point, $f(x)$ vanishes on V. Hence, by § 2, Th. I, it follows that V is irreducible.

THEOREM II. *Every irreducible variety V has a generic point.*

We consider those elements of the field $K(x_0, \ldots, x_n)$ which can be written as the quotient $f(x)/g(x)$ of two forms $f(x)$ and $g(x)$ of equal degree, where $g(x)$ does not vanish on V. We set up an equivalence relation between these rational functions by means of V. We say that

$$\frac{f}{g} \sim \frac{h}{k}$$

if $fk - gh$ vanishes on V.

Since $fk - gh = -(hg - kf)$, and $fg - gf = 0$, the relation $\sim$ is symmetric and reflexive. We show that it is also transitive. Let us suppose that

$$\frac{f}{g} \sim \frac{h}{k}, \quad \frac{h}{k} \sim \frac{l}{m},$$

so that, on V, we have

$$fk - gh = 0, \quad hm - lk = 0.$$

Then $$fkm - ghm = 0, \quad ghm - glk = 0,$$

and therefore $$fkm - glk = 0;$$

that is, $$k(fm - gl) = 0$$

on V. But k does not vanish on V, by hypothesis, and since V is irreducible, it follows that $fm - gl = 0$ on V. That is

$$\frac{f}{g} \sim \frac{l}{m}.$$

The relation $\sim$ is therefore an equivalence relation [I, § 2, p. 10], and we can divide the rational functions we are considering into classes of equivalent functions. If we denote the class to which f/g belongs by the symbol $[f, g]$, we define addition of classes by the formula

$$[f, g] + [h, k] = [fk + gh, gk],$$

and multiplication by the formula

$$[f, g]\,[h, k] = [fh, gk].$$

It is easily verified that these definitions do not depend on the particular rational function selected in each class. We prove, by way of example, that if

$$\frac{f}{g} \sim \frac{f'}{g'}, \quad \frac{h}{k} \sim \frac{h'}{k'},$$

then $$\frac{fk + gh}{gk} \sim \frac{f'k' + g'h'}{g'k'}$$

We note in the first place that since V is irreducible and g, g', k, k' do not vanish on V, gk and $g'k'$ do not vanish on V. Since

$$(fk+gh)g'k'-(f'k'+g'h')gk = (fg'-f'g)kk'+(hk'-h'k)gg' = 0$$

on V, the result follows.

Again, addition and multiplication are commutative and associative, and multiplication is distributive over addition.

With the two laws of composition we have defined, the classes $[f, g]$ form a ring. The construction of this ring closely resembles the construction of the quotient field of an integral domain [I, § 4], with which it should be compared. The zero of this ring is $[f, g]$, where $f = 0$ on V, $g \neq 0$ on V. The ring has unity, this being represented by $[g, g]$, where $g \neq 0$ on V. Furthermore, every non-zero element $[f, g]$ of the ring has an inverse, which is represented by $[g, f]$. The ring is therefore a *field*, which we denote by K'.

K' contains a subfield isomorphic to K, this being generated by the elements $[a, 1]$, where a is in K. We may therefore construct an extension K^* of K which is isomorphic to K' and in which the element a of K corresponds to the element $[a, 1]$ of K'.

Not all the equations $x_j = 0$ $(j = 0, \ldots, n)$ can be satisfied on V. Let us suppose that x_i does not vanish on V. There are elements of K^* corresponding to the classes

$$[x_0, x_i], [x_1, x_i], \ldots, [x_i, x_i], \ldots, [x_n, x_i]$$

of K'. We denote these elements of K^* by

$$\xi_0, \xi_1, \ldots, 1, \ldots, \xi_n,$$

since in an isomorphism between two fields the unities correspond.

We now prove that $\xi = (\xi_0, \xi_1, \ldots, 1, \ldots, \xi_n)$ *is a generic point of* V.

If $f(x_0, \ldots, x_n)$, a form of degree m, vanishes on V, $[f(x_0, \ldots, x_n), x_i^m]$ is the zero class of K', and this is mapped on the zero of K^*. But if

$$f(x_0, \ldots, x_n) = \Sigma a_{\rho_0 \ldots \rho_n} x_0^{\rho_0} \ldots x_n^{\rho_n},$$

we have

$$\begin{aligned}[\Sigma a_{\rho_0 \ldots \rho_n} x_0^{\rho_0} \ldots x_n^{\rho_n}, x_i^m] &= \Sigma [a_{\rho_0 \ldots \rho_n}, 1][x_0^{\rho_0}, x_i^{\rho_0}] \ldots [x_n^{\rho_n}, x_i^{\rho_n}] \\ &= \Sigma [a_{\rho_0 \ldots \rho_n}, 1][x_0, x_i]^{\rho_0} \ldots [x_n, x_i]^{\rho_n},\end{aligned}$$

and this is mapped on the element

$$\Sigma a_{\rho_0 \ldots \rho_n} \xi_0^{\rho_0} \ldots \xi_n^{\rho_n} = f(\xi_0, \ldots, 1, \ldots, \xi_n)$$

of K^*. Hence $$f(\xi_0, \ldots, 1, \ldots, \xi_n) = 0,$$

and therefore the point $\xi = (\xi_0, \ldots, 1, \ldots, \xi_n)$ lies on V.

Again, if $f(\xi_0, \ldots, 1, \ldots, \xi_n) = 0$, a reversal of the argument just given shows that $[f(x_0, \ldots, x_n), x_i^m]$ is mapped on the zero of K^* and is therefore the zero class of K'. Hence $f(x_0, \ldots, x_n)$ vanishes on V. From these two results it follows that ξ is a generic point of V, and this concludes the proof of the theorem.

We can show that the definition of a generic point of an irreducible variety V is independent of the particular allowable coordinate system we are using, as long as the equations for the transformations of coordinates have coefficients in K. The defining properties of a generic point involve the vanishing of certain forms in $K[x_0, \ldots, x_n]$ at certain points, and these properties persist under allowable transformations of coordinates over K. The result follows.

In particular, we may assume that the coordinates are transformed so that V does not lie in $x_0 = 0$, and we therefore take the coordinates of a generic point as $\xi = (1, \xi_1, \ldots, \xi_n)$.

In the foregoing we have merely proved the existence of a generic point of an irreducible variety. We now consider the relation between two generic points of the same variety.

Let $(\xi_0, \xi_1, \ldots, \xi_n)$ and $(\eta_0, \eta_1, \ldots, \eta_n)$ be the coordinates of two generic points of V, and let us suppose that $\xi_0 \neq 0$. Then since $\eta_0 = 0$ would imply that $x_0 = 0$ on V and therefore that $\xi_0 = 0$, which is not the case, it follows that $\eta_0 \neq 0$. Hence if the coordinates of one generic point are taken as $(1, \xi_1, \ldots, \xi_n)$, the coordinates of any other may be taken as $(1, \eta_1, \ldots, \eta_n)$.

THEOREM III. *The fields $K(\xi_1, \ldots, \xi_n)$ and $K(\eta_1, \ldots, \eta_n)$ are isomorphic extensions of K in which ξ_i and η_i are corresponding elements.*

We map the ring $K[\xi_1, \ldots, \xi_n]$ on the ring $K[\eta_1, \ldots, \eta_n]$ by mapping any polynomial $f(\xi_1, \ldots, \xi_n)$ in the one on the corresponding polynomial $f(\eta_1, \ldots, \eta_n)$ in the other. If we have the non-trivial equation $f(\xi_1, \ldots, \xi_n) = g(\xi_1, \ldots, \xi_n)$ in $K[\xi_1, \ldots, \xi_n]$, we also have the equation $f(\eta_1, \ldots, \eta_n) = g(\eta_1, \ldots, \eta_n)$ in $K[\eta_1, \ldots, \eta_n]$. This follows from the fact that the first equation can be written

$$f(\xi_1/\xi_0, \ldots, \xi_n/\xi_0) - g(\xi_1/\xi_0, \ldots, \xi_n/\xi_0) = 0.$$

This equation, after multiplication by a suitable power of ξ_0, becomes a homogeneous equation which is true for ξ and therefore, by the definition of generic point, is true for η. It follows that $f(\eta_1, \ldots, \eta_n) - g(\eta_1, \ldots, \eta_n) = 0$, which is what we wished to prove. The mapping of $K[\xi]$ on $K[\eta]$ is therefore one-valued, and a similar argument shows that the inverse mapping of $K[\eta]$ on $K[\xi]$ is also one-valued.

This mapping takes sums into sums and products into products, and it is therefore an isomorphic mapping. It also maps ξ_i on η_i. The isomorphism of the rings $K[\xi]$ and $K[\eta]$ can immediately be extended to an isomorphism of the fields $K(\xi)$ and $K(\eta)$.

Since the isomorphism interchanges ξ and η, leaving the elements of K unchanged, it follows that *the algebraic properties of two generic points ξ and η are identical.*

The field $K(\xi_1, \ldots, \xi_n)$, or any one of its isomorphs, is called the *function field* of the variety V. We constructed one isomorph in proving Theorem II, namely, the field K^*.

As a converse to Theorem II we have

THEOREM IV. *Given any point ξ, which may be rational, algebraic, or transcendental over K, there exists a unique irreducible algebraic variety V, defined over K, such that ξ is a generic point of V.*

Consider the set of forms $f(x_0, \ldots, x_n)$ in $K[x_0, \ldots, x_n]$ which have the property $f(\xi) = 0$. From this set, which evidently forms a homogeneous ideal in $K[x_0, \ldots, x_n]$, we can select a finite basis

$$f_1(x_0, \ldots, x_n), \quad \ldots, \quad f_s(x_0, \ldots, x_n),$$

by Hilbert's basis theorem [IV, § 2]. The equations

$$f_i(x_0, \ldots, x_n) = 0 \quad (i = 1, \ldots, s)$$

define an algebraic variety V. The point ξ is generic on V. In fact, ξ lies on V, and if $g(x_0, \ldots, x_n)$ is such that $g(\xi_0, \ldots, \xi_n) = 0$, then since

$$g(x_0, \ldots, x_n) \equiv \sum_{i=1}^{s} a_i(x_0, \ldots, x_n) f_i(x_0, \ldots, x_n),$$

it follows that $g(x_0, \ldots, x_n)$ vanishes at all points of V. Hence ξ is a generic point of V, and V is therefore irreducible over K.

In certain types of problem which arise in algebraic geometry, an example of which is given by Theorem III above, the fact that the coordinates $(x'_0, \ldots, x'_n)$ of a point can also be written as $(\rho x'_0, \ldots, \rho x'_n)$, where ρ is not zero, is a source of inconvenience. We can overcome the objection, to a partial extent, as follows. We select a particular linear form $a_0 x_0 + \ldots + a_n x_n$, and denote by π the prime

$$a_0 x_0 + \ldots + a_n x_n = 0.$$

If x' is a point of S_n not in π, we take

$$\rho = (a_0 x_0' + \ldots + a_n x_n')^{-1},$$

and then the coordinates $(x_0'', \ldots, x_n'') = (\rho x_0', \ldots, \rho x_n')$ are such that

$$a_0 x_0'' + \ldots + a_n x_n'' = 1.$$

Conversely, if this equation is satisfied, the coordinates of a given point are uniquely determined within the class of equivalent sets of coordinates, and are said to be *normalised* with respect to the form $\sum_{i=0}^{n} a_i x_i$. The prime π is called the *prime at infinity*.

The choice of the prime at infinity is arbitrary, but in dealing with a given irreducible algebraic variety V we adopt certain conventions for the sake of simplicity. Let $(\xi_0, \ldots, \xi_n)$ be a generic point of V, and let us suppose that $\xi_i \neq 0$. If $(\eta_0, \ldots, \eta_n)$ is any other generic point of V, $\eta_i \neq 0$. We therefore normalise the coordinates with respect to x_i. Any point x' not in $x_i = 0$ has coordinates

$$x' = (x_0', \ldots, x_{i-1}', 1, x_{i+1}', \ldots, x_n')$$

which are determined by the set $(x_0', \ldots, x_{i-1}', x_{i+1}', \ldots, x_n')$. We call this set the *non-homogeneous coordinates* of x' (with respect to x_i).

Furthermore, we usually find it convenient to perform a transformation permuting the coordinates, as above, so that $x_0 \neq 0$ on V. We can then normalise coordinates so that for any point $(x_0', \ldots, x_n')$ not in $x_0 = 0$ (and the important thing to notice is that this includes the generic points) we have $x_0' = 1$. When this is done, the non-homogeneous coordinates with respect to x_0 are simply called *the non-homogeneous coordinates of the point.*

4. Generic members of systems of *k*-spaces. We now show how the notion of a generic point of an irreducible algebraic variety can be extended so as to enable us to define a generic k-space of certain systems of k-spaces in S_n. The general notion of an algebraic system of k-spaces will not be defined until § 8, p. 57, but we can consider here certain systems which frequently occur in the sequel.

At the beginning of this chapter we saw that we could generalise the notion of a point in an S_n defined over a field K by admitting points whose coordinates lie in some extension of K. We can, similarly, generalise the notion of a k-space by admitting k-spaces whose coordinates (or rather the ratios of whose coordinates) lie in an extension of K. To define these, we merely take $k+1$ points, in our generalised sense, say

$$(x_0^i, \ldots, x_n^i) \quad (i = 0, \ldots, k).$$

These points are linearly independent if and only if the equations

$$\sum_{i=0}^{k} \lambda_i x_j^i = 0 \quad (j = 0, \ldots, n)$$

have no solution, in any extension of K, other than $\lambda_0 = \ldots = \lambda_k = 0$. If the points are linearly independent the k-space defined by them is the aggregate of points

$$\left(\sum_0^k a_i x_0^i, \ldots, \sum_0^k a_i x_n^i\right),$$

where $a_0, \ldots, a_k$ belong to some extension of K. The coordinates of the k-space are $(\ldots, p_{i_0 \ldots i_k}, \ldots)$, where

$$p_{i_0 \ldots i_k} = \begin{vmatrix} x_{i_0}^0 & . & x_{i_k}^0 \\ . & . & . \\ x_{i_0}^k & . & x_{i_k}^k \end{vmatrix}.$$

The various theorems proved in Chapter VII concerning k-spaces can be immediately extended to cover these more general k-spaces. In what follows the term k-space is used in this more general sense.

We recall the two fundamental properties of a generic point ξ of a variety V:

(i) ξ is a point on V;

(ii) if $f(x)$ is any homogeneous polynomial in $K[x_0, \ldots, x_n]$ such that $f(\xi) = 0$, then $f(x') = 0$, where x' is any point of V.

For a system of k-spaces, we say that $\pi = (\ldots, \pi_{i_0 \ldots i_k}, \ldots)$ is a generic k-space of the system when the following two conditions are satisfied:

(i) π belongs to the system;

(ii) if $f(\ldots, x_{i_0 \ldots i_k}, \ldots)$ is a homogeneous polynomial in $K[\ldots, x_{i_0 \ldots i_k}, \ldots]$ such that

$$f(\ldots, \pi_{i_0 \ldots i_k}, \ldots) = 0, \quad \text{then} \quad f(\ldots, p'_{i_0 \ldots i_k}, \ldots) = 0,$$

where $(\ldots, p'_{i_0 \ldots i_k}, \ldots)$ is any k-space of the system.

We now consider certain simple systems of k-spaces, and construct generic spaces for them.

I. *The system of all k-spaces in* S_n. Let ξ_j^i $(i = 0, \ldots, k; j = 0, \ldots, n)$ be $(k+1)(n+1)$ independent indeterminates over K, and let

$$\pi_{i_0 \ldots i_k} = \begin{vmatrix} \xi_{i_0}^0 & . & \xi_{i_k}^0 \\ . & . & . \\ \xi_{i_0}^k & . & \xi_{i_k}^k \end{vmatrix}.$$

Then $\pi = (\ldots, \pi_{i_0 \ldots i_k}, \ldots)$ is a k-space, in our generalised sense. To prove that it is a generic space of the system of all k-spaces in S_n we have to show that it satisfies the condition (ii) above.

Let $f(\ldots, x_{i_0 \ldots i_k}, \ldots)$ be any homogeneous polynomial of degree m in $K[\ldots, x_{i_0 \ldots i_k}, \ldots]$ such that

$$f(\ldots, \pi_{i_0 \ldots i_k}, \ldots) = 0. \tag{1}$$

If
$$f(\ldots, |x_{i_0}^0 \ldots x_{i_k}^k|, \ldots) \equiv F(x^0, \ldots, x^k),$$
where we write

$$\begin{vmatrix} x_{i_0}^0 & . & x_{i_k}^0 \\ . & . & . \\ x_{i_0}^k & . & x_{i_k}^k \end{vmatrix} = |x_{i_0}^0 \ldots x_{i_k}^k|,$$

then it is clear that F is homogeneous of degree m in each set of indeterminates $x^i = (x_0^i, \ldots, x_n^i)$ $(i = 0, \ldots, k)$. But

$$f(\ldots, \pi_{i_0 \ldots i_k}, \ldots) \equiv F(\xi^0, \ldots, \xi^k),$$

and since the ξ_j^i are independent, (1) implies that the coefficient of each power-product of the ξ_j^i in $F(\xi^0, \ldots, \xi^k)$ is zero. This implies that

$$f(\ldots, |x_{i_0}^0 \ldots x_{i_k}^k|, \ldots) \equiv F(x^0, \ldots, x^k) \equiv 0.$$

Let $(\ldots, p'_{i_0 \ldots i_k}, \ldots)$ be any k-space in S_n. We can find elements a_j^i such that

$$p'_{i_0 \ldots i_k} = \begin{vmatrix} a_{i_0}^0 & . & a_{i_k}^0 \\ . & . & . \\ a_{i_0}^k & . & a_{i_k}^k \end{vmatrix}.$$

Then
$$f(\ldots, p'_{i_0 \ldots i_k}, \ldots) = F(a^0, \ldots, a^k) = 0.$$

Thus π satisfies the conditions for a generic k-space.

We may also work in dual Grassmann coordinates [VII, § 3]. An exactly similar argument shows that if

$$u_j^i \quad (i = 1, \ldots, n-k; j = 0, \ldots, n)$$

are $(n-k)(n+1)$ independent indeterminates over K, and if

$$\pi_{i_0 \ldots i_k} = \pi^{i_{k+1} \ldots i_n} = \begin{vmatrix} u^1_{i_{k+1}} & . & u^1_{i_n} \\ . & . & . \\ u^{n-k}_{i_{k+1}} & . & u^{n-k}_{i_n} \end{vmatrix},$$

then $(\ldots, \pi_{i_0 \ldots i_k}, \ldots)$ is a generic k-space of S_n. In particular, if $u_0, \ldots, u_n$ are independent indeterminates, $(u_0, \ldots, u_n)$ are the dual coordinates of a generic prime of S_n.

II. *The system of k-spaces through a given h-space Σ_h of S_n.* Let $A^i = (a^i_0, \ldots, a^i_n)$ $(i = 0, \ldots, h)$ be $h+1$ independent points of Σ_h, and let ξ^i_j $(i = h+1, \ldots, k; j = 0, \ldots, n)$ be $(k-h)(n+1)$ indeterminates which are independent over the field obtained by adjoining the a^i_j to K. We write

$$\pi_{i_0 \ldots i_k} = \begin{vmatrix} a^0_{i_0} & . & . & a^0_{i_k} \\ . & . & . & . \\ a^h_{i_0} & . & . & a^h_{i_k} \\ \xi^{h+1}_{i_0} & . & . & \xi^{h+1}_{i_k} \\ . & . & . & . \\ \xi^k_{i_0} & . & . & \xi^k_{i_k} \end{vmatrix}.$$

An argument exactly similar to that given in I shows that

$$\pi = (\ldots, \pi_{i_0 \ldots i_k}, \ldots)$$

is a generic member of the system of k-spaces through Σ_h.

We note that [II, § 8, Th. II]

$$\pi_{i_0 \ldots i_k} = \Sigma \pm |a^0_{j_0} \ldots a^h_{j_h}| \, |\xi^{h+1}_{j_{h+1}} \ldots \xi^k_{j_k}|$$

(summed over the derangements $j_0, \ldots, j_k$ of $i_0, \ldots, i_k$), and hence π is the join of Σ_h to the generic $(k-h-1)$-space of S_n with coordinates [VII, § 5 (8)]

$$(\ldots, |\xi^{h+1}_{i_0} \ldots \xi^k_{i_{k-h-1}}|, \ldots).$$

Conversely it is easily seen that the join of Σ_h to a generic S_{k-h-1} of S_n, regarded as a generic S_{k-h-1} over $K(a^i_j)$, is a generic S_k through Σ_h. Hence we have

THEOREM I. *If Σ_h is any h-space in S_n having coordinates in some extension K^* of the ground field K, a generic member of the system* *of k-spaces through Σ_h is obtained by joining Σ_h to any generic $(k-h-1)$-space of S_n, defined over K^*.*

It is not, however, necessary to take the $(k-h-1)$-space used above to be a generic member of the system of $(k-h-1)$-spaces of S_n. Let

$$\sum_{j=0}^{n} b_i^j x_j = 0 \quad (i = 0, \ldots, h)$$

be the equations of an $(n-h-1)$-space Σ_{n-h-1} of S_n which does not meet Σ_h. We shall denote by K^* an extension of the ground field K which contains the coordinates a_j^i of the points $A^0, \ldots, A^h$ of Σ_h and also the b_i^j which occur in the equations of Σ_{n-h-1}, and we shall regard K^* as the ground field.

If $\pi = (\ldots, \pi_{i_0 \ldots i_k}, \ldots)$ is any k-space through Σ_h, let us define a space with the coordinates $(\ldots, \bar{\pi}_{i_0 \ldots i_{k-h-1}}, \ldots)$ by the equations

$$\bar{\pi}_{i_0 \ldots i_{k-h-1}} = \sum_j | b_0^{j_0} \ldots b_h^{j_h} | \pi_{j_0 \ldots j_h i_0 \ldots i_{k-h-1}}.$$

Then $\bar{\pi} = (\ldots, \bar{\pi}_{i_0 \ldots i_{k-h-1}}, \ldots)$ is the intersection of π with Σ_{n-h-1} [VII, p. 307]. Conversely, if $\bar{\pi}$ is any $(k-h-1)$-space of Σ_{n-h-1}, and if

$$\pi_{i_0 \ldots i_k} = \Sigma \pm | a_{j_0}^0 \ldots a_{j_h}^h | \bar{\pi}_{j_{h+1} \ldots j_k},$$

where the summation is over the derangements $j_0, \ldots, j_k$ of $i_0, \ldots, i_k$, then $\pi = (\ldots, \pi_{i_0 \ldots i_k}, \ldots)$ is the join of $\bar{\pi}$ to Σ_h.

We now show that if π is a generic member (over K^*) of the system of k-spaces through Σ_h, then $\bar{\pi}$ is a generic member of the system of $(k-h-1)$-spaces which lie in Σ_{n-h-1}. In the first place, $\bar{\pi}$ lies in Σ_{n-h-1}, so that condition (i) is satisfied. Now let

$$f(\ldots, x_{i_0 \ldots i_{k-h-1}}, \ldots)$$

be a homogeneous polynomial in $K^*[\ldots, x_{i_0 \ldots i_{k-h-1}}, \ldots]$ such that

$$f(\ldots, \bar{\pi}_{i_0 \ldots i_{k-h-1}}, \ldots) = 0.$$

Then $$f\left(\ldots, \sum_j | b_0^{j_0} \ldots b_h^{j_h} | \pi_{j \ldots j_h i_0 \ldots i_{k-h-1}}, \ldots\right) = 0.$$

Let $(\ldots, \bar{p}_{i_0 \ldots i_{k-h-1}}, \ldots)$ be any $(k-h-1)$-space of Σ_{n-h-1}. If

$$p'_{i_0 \ldots i_k} = \Sigma \pm | a_{j_0}^0 \ldots a^h | \bar{p}_{j_{h+1} \ldots j_k},$$

then $(\ldots, p'_{i_0 \ldots i_k}, \ldots)$ is the k-space joining it to Σ_h. Since π is a generic k-space through Σ_h we have

$$f\left(\ldots, \sum_j | b_0^{j_0} \ldots b_h^{j_h} | p'_{j_0 \ldots j_h i_0 \ldots i_{h-k-1}}, \ldots\right) = 0.$$

But $$\sum_j |b_0^{j_0} \dots b_h^{j_h}| p'_{j_0 \dots j_h i_0 \dots i_{k-h-1}} = \sigma \bar{p}_{i \dots i_{k-h-1}},$$

where $\sigma \neq 0$. Hence
$$f(\dots, \bar{p}_{i_0 \dots i_{k-h-1}}, \dots) = 0.$$

This is condition (ii) that $\bar{\pi}$ be a generic $(k-h-1)$-space of Σ_{n-h-1}.

An exactly similar proof shows that if $\bar{\pi}$ is a generic $(k-h-1)$-space of Σ_{n-h-1}, its join to Σ_h is a generic member of the system of k-spaces of S_n through Σ_h. This result can also be deduced from the above by means of the principle of duality. Hence we have

THEOREM II. *If Σ_h and Σ_{n-h-1} are two spaces of S_n which do not meet, a necessary and sufficient condition that a k-space π through Σ_h be a generic member of the system of k-spaces through Σ_h is that π meet Σ_{n-h-1} in a $(k-h-1)$-space which is a generic member of the system of $(k-h-1)$-spaces lying in Σ_{n-h-1}.*

The results proved above lead to a useful representation of the primes of S_n which pass through a point $A = (a_0, \dots, a_n)$. If π is any prime through A, it is the join of A to an $(n-2)$-space which does not pass through A. Hence if
$$(\dots, p^{ij}, \dots)$$
are the (dual) coordinates of this $(n-2)$-space, the dual coordinates of π are
$$\left(\sum_0^n p^{0j} a_j, \dots, \sum_0^n p^{nj} a_j\right).$$

If $(\dots, p^{ij}, \dots)$ is a generic $(n-2)$-space of S_n, π is a generic prime through A. Hence the equation of any prime through A can be written in the form
$$u_0 x_0 + \dots + u_n x_n = 0, \tag{2}$$
where
$$u_i = \sum_{j=0}^n s_{ij} a_j,$$
the matrix (s_{ij}) being skew-symmetric. Conversely, it is seen at once that if u_i is of this form, (2) is the equation of a prime through A (unless, of course, each u_i is zero).

Now let t_{ij} $(i<j)$ be a set of independent indeterminates over the field obtained by adjoining $(a_0, \dots, a_n)$ to K, and let $t_{ij} = -t_{ji}$ $(i>j)$, $t_{ii} = 0$. If
$$\nu_i = \sum_{j=0}^n t_{ij} a_j,$$
then
$$\nu_0 x_0 + \dots + \nu_n x_n = 0$$

is the equation of a prime through A, and if

$$f(u_0, \dots, u_n)$$

is a form such that $$f\left(\sum_{j=0}^{n} t_{0j}a_j, \dots, \sum_{j=0}^{n} t_{nj}a_j\right) = 0, \tag{3}$$

then (3) will remain true when t_{ij} is replaced by s_{ij}, where (s_{ij}) is any skew-symmetric matrix, and therefore

$$f(u_0', \dots, u_n') = 0,$$

where $(u_0', \dots, u_n')$ is any prime through A. Hence we have

THEOREM III. *If* $\quad u_0x_0 + \dots + u_nx_n = 0$

is the equation of a prime π, a necessary and sufficient condition that it pass through the point $(a_0, \dots, a_n)$ is that

$$u_i = \sum_{j=0}^{n} s_{ij}a_j,$$

where (s_{ij}) is a skew-symmetric matrix. If the elements of (s_{ij}) are independent indeterminates over $K(a_i)$, subject only to the conditions of skew-symmetry, the prime is a generic member of the system of primes through the point.

5. The dimension of an algebraic variety. Let V be an irreducible algebraic variety in S_n defined over the ground field K. We suppose that V is not contained in $x_0 = 0$, and that $(\xi_1, \dots, \xi_n)$ are the non-homogeneous coordinates of a generic point of V. The field $K(\xi_1, \dots, \xi_n)$ is, as we have seen, defined to within an isomorphism by V, and its properties are therefore independent of the generic point chosen.

There exists an integer d, where $0 \leqslant d \leqslant n$, such that

(i) a set of d elements ξ_i, say $\xi_{i_1}, \dots, \xi_{i_d}$, is algebraically independent over K, but

(ii) any set of $d+1$ elements ξ_i is algebraically dependent over K.

Hence, if $d < n$, there exist non-zero polynomials

$$\phi_j(x_1, \dots, x_d, x_{d+1}) \quad \text{in} \quad K[x_1, \dots, x_d, x_{d+1}]$$

which are such that

$$\phi_j(\xi_{i_1}, \dots, \xi_{i_d}, \xi_{i_j}) = 0 \quad (j = d+1, \dots, n),$$

where $i_1, \dots, i_d, i_{d+1}, \dots, i_n$ is a derangement of $1, \dots, n$. In none of the polynomials $\phi_j(x_1, \dots, x_d, x_{d+1})$ can all terms containing x_{d+1} be absent, for this would imply the algebraic dependence of $\xi_{i_1}, \ldots, \xi_{i_d}$. If ζ is any element of $K(\xi_1, \ldots, \xi_n)$ it follows from III, § 3, Ths. II, III, IV that ζ is algebraically dependent on $K(\xi_{i_1}, \ldots, \xi_{i_d})$. Therefore $\xi_{i_1}, \ldots, \xi_{i_d}$ form a *minimal algebraic basis* for $K(\xi_1, \ldots, \xi_n)$ over K. The integer d was defined in Chapter III to be the dimension of $K(\xi_1, \ldots, \xi_n)$ over K. In algebraic geometry it is called *the dimension of the variety* V, $d = \dim V$.

As an immediate consequence of our definition and III, § 3, Th. V, we have the theorem:

THEOREM I. *If the dimension of an irreducible algebraic variety V is d, any $d+1$ elements of the function field $K(\xi_1, \ldots, \xi_n)$ of V are algebraically dependent over K.*

If $d = n$ the variety V is the space S_n. For if a non-zero form $f(x_0, \ldots, x_n)$ vanishes on V, then

$$f(1, \xi_1, \ldots, \xi_n) = 0,$$

and therefore there is an algebraic relation connecting $\xi_1, \ldots, \xi_n$, contrary to the definition of d. Hence any form which vanishes on V must be identically zero, and therefore every point of S_n lies on V.

THEOREM II. *If U and V are two irreducible algebraic varieties of dimensions d and e respectively, then $U \subseteq V$ implies $d \leqslant e$, and $U \subset V$ implies $d < e$.*

If we assume that $x_0 \neq 0$ on U then it follows that $x_0 \neq 0$ on V. We may therefore take generic points of U and V in non-homogeneous coordinates as $(\xi_1, \ldots, \xi_n)$ and $(\eta_1, \ldots, \eta_n)$ respectively. If $f(x_0, \ldots, x_n)$ is a form such that $f(1, \eta_1, \ldots, \eta_n) = 0$, then $f(x_0, \ldots, x_n)$ vanishes on V, and therefore on U, and therefore

$$f(1, \xi_1, \ldots, \xi_n) = 0.$$

Hence $f(1, \eta_1, \ldots, \eta_n) = 0$ implies $f(1, \xi_1, \ldots, \xi_n) = 0$.

Let us now suppose that $\xi_{i_1}, \ldots, \xi_{i_d}$ are algebraically independent. Since the *dependence* of $\eta_{i_1}, \ldots, \eta_{i_d}$ would involve the *dependence* of $\xi_{i_1}, \ldots, \xi_{i_d}$, by what we have proved above, it follows that $\eta_{i_1}, \ldots, \eta_{i_d}$ are independent. Hence $d \leqslant e$, and we have proved the first part of the theorem.

If $U \subset V$, there must be a form $g(x_0, \ldots, x_n)$ which vanishes on U but not on V; that is,

$$g(1, \xi_1, \ldots, \xi_n) = 0, \quad g(1, \eta_1, \ldots, \eta_n) \neq 0.$$

If $d = e$ and $\xi_{i_1}, \ldots, \xi_{i_d}$ are independent over K then, as above, $\eta_{i_1}, \ldots, \eta_{i_d}$ are independent over K, and since $\dim V = d$, all η_i are algebraically dependent on $\eta_{i_1}, \ldots, \eta_{i_d}$. Hence every element of $K(\eta_1, \ldots, \eta_n)$, and in particular $g(1, \eta_1, \ldots, \eta_n)$, is algebraically dependent on $\eta_{i_1}, \ldots, \eta_{i_d}$ [III, § 3, Th. I]. Therefore $g(1, \eta_1, \ldots, \eta_n)$ satisfies an irreducible equation

$$a_0(\eta) z^s + a_1(\eta) z^{s-1} + \ldots + a_s(\eta) = 0,$$

where the $a_i(\eta)$ are elements of $K[\eta_{i_1}, \ldots, \eta_{i_d}]$, and $a_s(\eta) \neq 0$, since $g(1, \eta_1, \ldots, \eta_n) \neq 0$. But since equations satisfied by the η_i are also satisfied by the ξ_i, the equation

$$a_0(\xi) z^s + a_1(\xi) z^{s-1} + \ldots + a_s(\xi) = 0$$

is satisfied by $z = g(1, \xi_1, \ldots, \xi_n)$. Hence, since $g(1, \xi_1, \ldots, \xi_n) = 0$, we must have

$$a_s(1, \xi_{i_1}, \ldots, \xi_{i_d}) = 0,$$

where $a_s(x_0, x_1, \ldots, x_d)$ is a non-zero polynomial, since

$$a_s(1, \eta_{i_1}, \ldots, \eta_{i_d}) \neq 0.$$

It follows that $\xi_{i_1}, \ldots, \xi_{i_d}$ are not algebraically independent, and from this contradiction we deduce that if $U \subset V$ then $d < e$.

We make use of this theorem in consideri mportant special cases.

THEOREM III. *If V is an irreducible variety of dimension $n-1$, it is given by a single irreducible equation*

$$f(x_0, \ldots, x_n = 0.$$

We know that V is given by a finite set of equations

$$\phi_i(x_0, \ldots, x_n) = 0 \quad (i = 1, \ldots, s).$$

Let us choose the first and write

$$\phi_1(x_0, \ldots, x_n) \equiv \prod_{j=1}^{t} f_j(x_0, \ldots, x_n),$$

where the $f_j(x_0, \ldots, x_n)$ are irreducible. Since V is irreducible and $\prod_{j=1}^{t} f_j(x_0, \ldots, x_n) = 0$ on V we know [§ 2, Th. I] that at least one of the forms $f_j(x_0, \ldots, x_n)$ must vanish on V. Hence we can find an irreducible form, $f(x_0, \ldots, x_n)$, say, which vanishes on V.

Let U be the variety defined by the equation

$$f(x_0, \ldots, x_n) = 0.$$

Then U is irreducible, by the proof given on p. 7, and since the equation of U is one of the equations satisfied by V, we must have $V \subseteq U$. Hence

$$n-1 = \dim V \leqslant \dim U \leqslant n-1,$$

since U is not the whole of S_n. It follows that

$$\dim V = \dim U,$$

and therefore, by Theorem II,

$$V = U.$$

Thus V is given by the single irreducible equation

$$f(x_0, \ldots, x_n) = 0,$$

and all forms $g(x_0, \ldots, x_n)$ which vanish on V are of the form

$$g(x_0, \ldots, x_n) = b(x_0, \ldots, x_n)\, f(x_0, \ldots, x_n).$$

Conversely we prove

THEOREM IV. *An irreducible equation*

$$f(x_0, \ldots, x_n) = 0$$

defines an irreducible variety of dimension $n-1$.

We can choose i so that $x_i \not\equiv f(x_0, \ldots, x_n)$. Since f is irreducible it is not of the form ax_i^r, and hence it must contain some x_j ($j \neq i$). Let

$$\xi_0, \ldots, \xi_{i-1}, \xi_{i+1}, \ldots, \xi_{j-1}, \xi_{j+1}, \ldots, \xi_n$$

be independent indeterminates over K. Then the equation

$$f(\xi_0, \ldots, \xi_{i-1}, 1, \xi_{i+1}, \ldots, \xi_{j-1}, x_j, \xi_{j+1}, \ldots, \xi_n) = 0$$

defines a simple algebraic extension

$$K(\xi_0, \ldots, \xi_{i-1}, \xi_{i+1}, \ldots, \xi_{j-1}, \xi_j, \xi_{j+1}, \ldots, \xi_n)$$

of

$$K(\xi_0, \ldots, \xi_{i-1}, \xi_{i+1}, \ldots, \xi_{j-1}, \xi_{j+1}, \ldots, \xi_n).$$

If any form $\phi(x_0, \ldots, x_n)$ in $K[x_0, \ldots, x_n]$ is such that

$$\phi(\xi_0, \ldots, \xi_{i-1}, 1, \xi_{i+1}, \ldots, \xi_{j-1}, \xi_j, \xi_{j+1}, \ldots, \xi_n) = 0,$$

it follows from III, § 3, Th. VI, that $\phi(x)$ must contain $f(x)$ as a factor. Hence the point

$$(\xi_0, \ldots, \xi_{i-1}, 1, \xi_{i+1}, \ldots, \xi_{j-1}, \xi_j, \xi_{j+1}, \ldots, \xi_n)$$

is a generic point of the variety given by the equation

$$f(x_0, \ldots, x_n) = 0,$$

and since $n-1$ of the ξ_i are algebraically independent over K, the variety is irreducible and of dimension $n-1$.

An irreducible variety in S_n of dimension $n-1$, given, as we have seen, by the vanishing of a single irreducible equation, is called a *primal*. When the equation is linear, so that the variety is a linear space, it is called a *prime*.

We now consider varieties of dimension zero and prove

THEOREM V. *An irreducible variety V of dimension zero consists of a finite number of points. These points are conjugate over K, and any one of them is a generic point of V.*

Let $(x'_0, \ldots, x'_n)$ be any point on V. Not all x'_i can be zero. Let us suppose that $x'_0 \neq 0$. Then we assert that *there are no points on V in the prime $x_0 = 0$*. In fact the points of V in $x_0 = 0$ would fill a variety $U \subset V$, and would therefore have dimension < 0, by Theorem II, and this is impossible.

We may therefore normalise the coordinates of the generic points with respect to x_0, and take the non-homogeneous coordinates of a generic point as $(\xi_1, \ldots, \xi_n)$. Since V is of dimension zero, each ξ_i belongs to some algebraic extension of K, and by the theorem of the primitive element [III, §6, Th. I] we may obtain the field $K(\xi_1, \ldots, \xi_n)$ by adjoining the element

$$\theta = \sum_1^n a_i \xi_i$$

to K, where $a_1, \ldots, a_n$ are suitable elements in K. Since θ is algebraic over K it satisfies an equation

$$\phi(\theta, 1) = 0,$$

where $\phi(x, y)$ is an irreducible form in $K[x, y]$ of degree m, say. Since ξ_i $(i = 1, \ldots, n)$ is in $K(\theta)$, we may write

$$\xi_i = \sum_{j=0}^{m-1} a_{ij} \theta^j \quad (a_{ij} \text{ in } K).$$

Hence a generic point of V, and therefore all points of V, satisfy the equations

$$\phi\left(\sum_1^n a_k x_k, x_0\right) = 0,$$

$$x_i x_0^{m-2} = \sum_{j=0}^{m-1} a_{ij} \left[\sum_1^n a_k x_k\right]^j x_0^{m-1-j} \quad (i = 1, \ldots, n).$$

The first equation can be factorised in a suitable extension field of K, giving m distinct values for the ratio $\left(\sum_1^n a_k x_k\right) : x_0$, and substitution in the second equation shows that V consists of at most m algebraic points which are contained in the set of m distinct points given by the whole set of equations. One of these points is $(1, \xi_1, \ldots, \xi_n)$, and if $(1, \eta_1, \ldots, \eta_n)$ is any other point of the set the fields $K(\xi_1, \ldots, \xi_n)$ and $K(\eta_1, \ldots, \eta_n)$ are evidently equivalent extensions of K. Hence any equation satisfied by ξ is also satisfied by η, and therefore V consists of m points, each of which is a generic point of V. If ξ, η are any two of them, an isomorphism can be set up between the fields $K(\xi)$ and $K(\eta)$, under which ξ is mapped on η, and the elements of K are mapped on themselves. This is what we imply when we say that the points of V are *conjugate over* K, and completes the proof of the theorem.

If the ground field K is algebraically closed we must have $m = 1$, and algebraic varieties of dimension zero consist of single points.

We conclude this section with a number of lemmas which are necessary for the following section. Later, these lemmas will be replaced by more powerful results.

Lemma I. An irreducible variety V of dimension d which becomes reducible when the ground field is extended has no component of dimension greater than d over the extended field.

Let V' be an irreducible component of V over the extension K', and let ξ' be a generic point of V'. Then ξ' satisfies the equations of V over K. If $\xi'_0 \neq 0$, not all points of V lie in $x_0 = 0$; hence $\xi_0 \neq 0$, where ξ is a generic point of V, and we may therefore suppose that ξ has coordinates $(1, \xi_1, \ldots, \xi_n)$.

Any $d+1$ of the ξ_i are algebraically dependent over K [Th. I], and therefore satisfy an equation

$$f(1, \xi_{i_1}, \ldots, \xi_{i_{d+1}}) = 0,$$

where $f(x_0, x_1, \ldots, x_{d+1})$ is a form in $K[x_0, \ldots, x_{d+1}]$. Since ξ is generic on V, all points on V satisfy this equation. In particular, the point ξ' satisfies the equation, so that

$$f(1, \xi'_{i_1}, \ldots, \xi'_{i_{d+1}}) = 0.$$

It follows that no $d+1$ of the ξ'_i are algebraically independent over K', since they are algebraically dependent over K, and therefore the dimension of V' is not greater than d.

We shall need this lemma for later applications, but at present we shall consider an extension of K over which V remains irreducible and of the same dimension.

Theorem VI. *Let $K' = K(u_1, \ldots, u_p)$, where $u_1, \ldots, u_p$ are independent indeterminates over K. Then V is irreducible and of dimension d over K'.*

We first prove this theorem for the case $p = 1$, writing $u = u_1$ and $K' = K(u)$. Let V be an irreducible variety of dimension d over K defined by the equations over K

$$f_i(x_0, \ldots, x_n) = 0 \quad (i = 1, \ldots, r),$$

and let V^* be the variety defined by these equations when the ground field is $K' = K(u)$.

If $\phi(x_0, \ldots, x_n) = 0$ is any equation over $K(u)$ satisfied by the points of V^*, we may write

$$\phi(x_0, \ldots, x_n) = \frac{\phi_1(u, x_0, \ldots, x_n)}{\phi_2(u)},$$

where $\phi_1(u, x_0, \ldots, x_n)$ is in $K[u, x_0, \ldots, x_n]$ and $\phi_2(u)$ is in $K[u]$. By Hilbert's zero-theorem there is an integer ρ such that

$$[\phi(x_0, \ldots, x_n)]^\rho \equiv \sum_1^r a_i(x_0, \ldots, x_n) f_i(x_0, \ldots, x_n),$$

where $a_i(x_0, \ldots, x_n)$ is in $K(u)[x_0, \ldots, x_n]$. Multiplying by a suitable polynomial in $K[u]$ we have the equation

$$\psi(u)[\phi_1(u, x_0, \ldots, x_n)]^\rho \equiv \sum_1^r b_i(x_0, \ldots, x_n) f_i(x_0, \ldots, x_n),$$

where $b_i(x_0, \ldots, x_n)$ is in $K[u, x_0, \ldots, x_n]$, and $\psi(u)$ is in $K[u]$. Equating the coefficients of the powers of u on both sides of this identity, we easily deduce that the coefficient of every power of u in

$$\phi_1(u, x_0, \ldots, x_n)$$

vanishes on V.

If V^* is reducible then there exist forms in $K(u)[x_0, \ldots, x_n]$,

$$\phi(x_0, \ldots, x_n) = \frac{\phi_1(u, x_0, \ldots, x_n)}{\phi_2(u)},$$

and

$$\psi(x_0, \ldots, x_n) = \frac{\psi_1(u, x_0, \ldots, x_n)}{\psi_2(u)},$$

such that $\phi\psi$ vanishes on V^* without ϕ or ψ doing so. A necessary and sufficient condition for $\phi\psi$ to vanish on V^* is that

$$\phi_1(u, x_0, \ldots, x_n)\,\psi_1(u, x_0, \ldots, x_n)$$

vanish on V^*. Writing

$$\phi_1(u, x_0, \ldots, x_n) \equiv \Sigma u^i \Phi_i(x_0, \ldots, x_n),$$

and
$$\psi_1(u, x_0, \ldots, x_n) \equiv \Sigma u^i \Psi_i(x_0, \ldots, x_n),$$

where Φ_i and Ψ_i are in $K[x_0, \ldots, x_n]$, let Φ_l and Ψ_m be, respectively, the first Φ_i and the first Ψ_i which do not vanish on V. Since

$$\phi_1(u, x_0, \ldots, x_n)\,\psi_1(u, x_0, \ldots, x_n) \equiv \Sigma u^{i+j} \Phi_i \Psi_j$$

vanishes on V^*, the result obtained above tells us that all the forms $\sum_{i+j=k} \Phi_i \Psi_j$ vanish on V. Taking $k = l+m$ we see that $\Phi_l \Psi_m$ vanishes on V. But since V is irreducible over K, either Φ_l or Ψ_m must vanish on V, and we therefore have a contradiction. It follows that V^* is irreducible over the field $K(u)$.

By Lemma I the dimension of V^* cannot exceed d. We show that it is exactly d, proving in the first place that there exists a generic point $(1, \xi_1, \ldots, \xi_n)$ of V over K such that u is transcendental over the function field $K(\xi_1, \ldots, \xi_n)$, and then that $(1, \xi_1, \ldots, \xi_n)$ is a generic point of V^* over $K(u)$.

Let η be any generic point of V over K, and let v be transcendental over $K(\eta)$. Then v is transcendental over K. The isomorphism $K(v) \simeq K(u)$ can be extended to the isomorphism $K(v, \eta) \simeq K(u, \xi)$, and this determines a generic point ξ of V over K such that u is transcendental over $K(\xi)$.

The point $(1, \xi_1, \ldots, \xi_n)$ satisfies the equations of V^*. Again, if

$$\phi(u, x_0, \ldots, x_n) \equiv \Sigma u^i \Phi_i(x_0, \ldots, x_n)$$

vanishes at the point $(1, \xi_1, \ldots, \xi_n)$, so that

$$\Sigma u^i \Phi_i(1, \xi_1, \ldots, \xi_n) = 0,$$

then, since u is transcendental over $K(\xi)$, we must have

$$\Phi_i(1, \xi_1, \ldots, \xi_n) = 0,$$

and therefore $\Phi_i(x_0, \ldots, x_n)$ vanishes on V. Hence $\phi(u, x_0, \ldots, x_n)$ vanishes on V, and since the set of points which defines V is the same as the set of points which defines V^*, it follows that $\phi(u, x_0, \ldots, x_n)$ vanishes on V^*, and therefore $(1, \xi_1, \ldots, \xi_n)$ is a generic point of V^* over $K(u)$.

If the elements $\xi_{i_1}, \ldots, \xi_{i_d}$ were algebraically dependent over $K(u)$, an argument similar to that just given would prove that the same elements were algebraically dependent over K. Hence the dimension of V^* over $K(u)$ is d.

This proves the theorem for the field $K' = K(u_1)$. A similar argument proves the theorem for the field

$$K(u_1, u_2) = K'(u_2),$$

and by a simple induction we prove the theorem for the field $K(u_1, \ldots, u_p)$.

Lemma II. If V is an irreducible variety of dimension d over K, and Σ_{n-d-1} is a generic linear space of dimension $n-d-1$, the two varieties have no common points.

Let Σ_{n-d-1} be given by the equations

$$\sum_{j=0}^{n} u_{ij} x_j = 0 \quad (i = 0, \ldots, d), \tag{1}$$

where the u_{ij} are independent indeterminates over K. By the theorem just proved, if we extend K to $K(u_{ij})$, the variety V^* over $K(u_{ij})$ which arises from V is irreducible and of dimension d.

Let $d = 0$. Then V consists of a finite number of points which are conjugate over K, and therefore V^* consists of a finite number of points which are conjugate over $K(u_{0j})$. None of these points lies in

$$\sum_{j=0}^{n} u_{0j} x_j = 0, \tag{2}$$

for if $(1, \xi_1, \ldots, \xi_n)$ are the coordinates of a point of V which lies in this prime, we should have

$$u_{00} + \sum_{j=1}^{n} u_{0j} \xi_j = 0$$

for indeterminate u_{0j}, which is absurd.

Let us assume that the theorem is true for varieties of dimension not exceeding $d-1$, and then prove that it is true for varieties of dimension d. The prime given by (2) does not contain V, and therefore meets it in a proper subvariety each of whose irreducible components over $K(u_{0j})$ has dimension $\leqslant d-1$ [Th. II]. Let V' be one such component. By the hypothesis of induction V' has no points in common with the $\Sigma_{n-(d-1)-1}$ given by the equations

$$\sum_{j=0}^{n} u_{ij} x_j = 0 \quad (i = 1, \ldots, d), \tag{3}$$

and we therefore conclude that V has no points in common with the Σ_{n-d-1} given by (1), since a point common to V and to Σ_{n-d-1} would lie on some component V' and in the linear space given by (3).

Finally, we prove

Lemma III. If V is an algebraic variety in S_n of dimension d over K, and x' is a point of S_n, a necessary and sufficient condition that x' lie on V is that a generic S_{n-d-1} through x' meets V.

We adjoin the coordinates of x' to K, and denote by K' the field thus obtained. Regarded as a variety over K', V is the sum of a finite number of irreducible varieties:

$$V = V_1 \dotplus V_2 \dotplus \dots \dotplus V_k,$$

where V_i is of dimension d_i and $d_i \leqslant d$ [Lemma I]. The necessity of the condition stated in our lemma is obvious. We now suppose that x' does not lie on V, and therefore does not lie on any component V_i, and prove that a generic S_{n-d-1} through x' does not meet any V_i, and therefore does not meet V. We proceed by induction on d. If $d = 0$ each V_i is a set of conjugate points, and hence V is a set of algebraic points. A generic S_{n-1} through x' is given by the equation

$$\sum_{i,j} s_{ij} x'_i x_j = 0,$$

where (s_{ij}) is an indeterminate skew-symmetric matrix [§ 4, Th. III]. If α is any one of the points of V, the equation

$$\sum_{i,j} s_{ij} x'_i \alpha_j = 0 \tag{4}$$

is true, for indeterminate s_{ij}, if and only if

$$x'_i \alpha_j = x'_j \alpha_i$$

for all i, j; that is, if and only if $\alpha = x'$. As we have assumed that x' does not lie on V, this is not the case for any point α of V, and therefore a generic S_{n-1} through x' does not meet V. Thus the lemma is proved when $d = 0$.

We therefore assume the truth of the lemma for varieties of dimension less than d, and consider a variety V of d dimensions. From the results of § 4 it follows that a generic S_{n-d-1} through x' is the intersection of $d+1$ independent generic primes through x'. Let $S^{(1)}_{n-1}, \dots, S^{(d+1)}_{n-1}$ be $d+1$ independent generic primes through x'. The equation of $S^{(d+1)}_{n-1}$ can be written in the form (4). If ξ is a generic point of any component V_i of V, the argument used in the case $d = 0$

is sufficient to show that, since x' does not lie on V_i, ξ, and therefore V_i, is not contained in $S_{n-1}^{(d+1)}$. Hence [Th. II] $S_{n-1}^{(d+1)}$ meets V in a subvariety V' of dimension less than d. Now $S_{n-1}^{(1)}, \ldots, S_{n-1}^{(d)}$ have in common with $S_{n-1}^{(d+1)}$ an $S_{(n-1)-(d-1)-1}$ which is a generic linear space through x' of dimension $(n-1)-(d-1)-1$. By the hypothesis of induction, this does not meet V', and since

$$(n-1)-(d-1)-1 = n-d-1,$$

it follows that a generic S_{n-d-1} of S_n through x' does not meet V.

Thus the lemma is proved. The proof can easily be generalised to show that if S_h $(h \leqslant n-d-1)$ is any h-space of S_n, a necessary and sufficient condition that S_h meet V is that a generic S_{n-d-1} through S_h meet V.

6. The Cayley form of an algebraic variety. We assume that V is an irreducible variety of dimension d over K, and that $(1, \xi_1, \ldots, \xi_n)$ is a generic point. We adjoin the elements

$$u_{ij} \quad (i = 0, \ldots, d;\ j = 1, \ldots, n),$$

which are algebraically independent over $K(\xi_1, \ldots, \xi_n)$, to K, and we define $d+1$ elements $\zeta_0, \zeta_1, \ldots, \zeta_d$ of $K(u_{ij}, \xi_1, \ldots, \xi_n)$ by the equations

$$\zeta_\sigma = -\sum_{\rho=1}^{n} u_{\sigma\rho}\xi_\rho \quad (\sigma = 0, \ldots, d). \tag{1}$$

Since V is of dimension d over $K(u_{ij})$ [§5, Th. VI], the $d+1$ elements defined in (1) are algebraically dependent over $K(u_{ij})$, by Theorem I of §5. We therefore have a relation

$$f(u_{ij};\ \zeta_0, \zeta_1, \ldots, \zeta_d) = 0, \tag{2}$$

where $f(u_{ij};\ z_0, z_1, \ldots, z_d)$ denotes a polynomial in $K[u_{ij};\ z_0, \ldots, z_d]$, which we may assume to be irreducible in this ring.

We shall subsequently replace the indeterminates $z_0, \ldots, z_d$ by indeterminates $u_{00}, \ldots, u_{d0}$, writing†

$$f(u_{ij};\ u_{00}, \ldots, u_{d0}) = F(u_0, u_1, \ldots, u_d),$$

where $$u_\sigma = (u_{\sigma 0}, \ldots, u_{\sigma n}) \quad (\sigma = 0, \ldots, d).$$

We shall justify this notation by proving in this section that $F(u_0, u_1, \ldots, u_d)$ is homogeneous of the same degree, which we

† We shall often consider the form $f(u_{ij};\ u_{00}, \ldots, u_{d0})$. When we use this notation, the suffixes i, j in u_{ij} take the values $i = 0, \ldots, d;\ j = 1, \ldots, n$.

shall denote by g, in each set of indeterminates $(u_{i0}, \ldots, u_{in})$. This form $F(u_0, u_1, \ldots, u_d)$ is the Cayley form of V.

THEOREM I. *If* $\phi(u_{ij}; z_0, \ldots, z_d)$ *is any polynomial in*

$$K[u_{ij}; z_0, \ldots, z_d]$$

such that
$$\phi(u_{ij}; \zeta_0, \ldots, \zeta_d) = 0,$$

then
$$\phi(u_{ij}; z_0, \ldots, z_d) = A(u_{ij}; z_0, \ldots, z_d) f(u_{ij}; z_0, \ldots, z_d),$$

where $A(u_{ij}; z_0, \ldots, z_d)$ *is in* $K[u_{ij}; z_0, \ldots, z_d]$, *and* $f(u_{ij}; z_0, \ldots, z_d)$ *is defined above.*

This theorem follows from III, § 3, Th. VI, provided that we can prove that d of the elements $\zeta_0, \ldots, \zeta_d$ are algebraically independent over $K(u_{ij})$. We prove that $\zeta_1, \ldots, \zeta_d$ are independent. For if they are dependent, this dependence subsists for the specialisations $\bar{\zeta}_1, \ldots, \bar{\zeta}_d$ induced by specialisations $u_{ij} \to \bar{u}_{ij}$ (cf. the proof of IV, § 9, Th. I). If we specialise the u_{ij} as follows:

$$u_{rs} \to -\delta_{i_r s} \quad (r = 1, \ldots, d;\ s = 1, \ldots, n),$$

where the $\delta_{\rho\sigma}$ are Kronecker deltas, we obtain the specialisations

$$\zeta_1 \to \xi_{i_1}, \quad \ldots, \quad \zeta_d \to \xi_{i_d},$$

and therefore if $\zeta_1, \ldots, \zeta_d$ are dependent over $K(u_{ij})$, $\xi_{i_1}, \ldots, \xi_{i_d}$ are dependent over K. Since we can choose $i_1, \ldots, i_d$ so that this is not so, it follows that $\zeta_1, \ldots, \zeta_d$ are algebraically independent over $K(u_{ij})$.

The polynomial $f(u_{ij}; z_0, \ldots, z_d)$ has certain symmetry properties. If we consider the isomorph $K(u_{ij}^*)$ of $K(u_{ij})$ in which the sets $(u_{\rho 1}, \ldots, u_{\rho n})$ and $(u_{\sigma 1}, \ldots, u_{\sigma n})$ are interchanged, the other u_{ij} being mapped on themselves, the relation (2) becomes

$$f(u_{ij}^*; \zeta_0^*, \ldots, \zeta_d^*) = 0$$

in the isomorphic field $K(u_{ij}^*)$, where $\zeta_i = \zeta_i^*$ $(i \neq \rho, \sigma)$, but $\zeta_\rho^* = \zeta_\sigma$ and $\zeta_\sigma^* = \zeta_\rho$. Instead of the polynomial

$$f(u_{ij}; z_0, \ldots, z_\rho, \ldots, z_\sigma, \ldots, z_d),$$

we obtain the polynomial

$$f(u_{ij}^*; z_0, \ldots, z_\sigma, \ldots, z_\rho, \ldots, z_d).$$

But these two polynomials can only differ in their sign. Hence, writing $z_0 = u_{00}$, $z_1 = u_{10}$, $\ldots$, $z_d = u_{d0}$, and

$$f(u_{ij}; u_{00}, \ldots, u_{d0}) = F(u_0, \ldots, u_d),$$

we can say that the interchange of the set of indeterminates u_ρ with the set u_σ in $F(u_0, \ldots, u_d)$ produces at most a change of sign in the Cayley form.

We now obtain some less evident properties of the Cayley form. From (2) we deduce that, for any particular choice of ρ, σ,

$$\frac{\partial f}{\partial u_{\rho\sigma}} + \frac{\partial f}{\partial \zeta_\rho}\frac{\partial \zeta_\rho}{\partial u_{\rho\sigma}} = 0, \tag{3}$$

where $\partial f/\partial \zeta_\rho$ is the result of substituting $\zeta_0, \ldots, \zeta_d$ for $u_{00}, \ldots, u_{d0}$ in the polynomial $\frac{\partial}{\partial u_{\rho 0}} f(u_{ij};\, u_{00}, \ldots, u_{d0})$. We prove that $\frac{\partial f}{\partial \zeta_\rho} \neq 0$.

To do this, let

$$\frac{\partial}{\partial u_{\rho 0}} f(u_{ij};\, u_{00}, \ldots, u_{d0}) = \phi(u_{ij};\, u_{00}, \ldots, u_{d0}). \tag{4}$$

Then if $\partial f/\partial \zeta_\rho = 0$, we have the equation

$$\phi(u_{ij};\, \zeta_0, \ldots, \zeta_d) = 0.$$

By Theorem I it follows that

$$\phi(u_{ij};\, u_{00}, \ldots, u_{d0}) = A(u_{ij};\, u_{00}, \ldots, u_{d0})\, f(u_{ij};\, u_{00}, \ldots, u_{d0}).$$

But ϕ is of lower degree in $u_{\rho 0}$ than is f. Hence $A = 0$ and therefore $\phi(u_{ij};\, u_{00}, \ldots, u_{d0}) = 0$. It follows from (4) that $f(u_{ij};\, u_{00}, \ldots, u_{d0})$ does not contain the indeterminate $u_{\rho 0}$. But since f is at most altered in sign if $(u_{\rho 0}, \ldots, u_{\rho n})$ is interchanged with $(u_{\sigma 0}, \ldots, u_{\sigma n})$, this implies that f does not contain $u_{00}, u_{10}, \ldots, u_{d0}$, and this is clearly absurd. Hence $\partial f/\partial \zeta_\rho \neq 0$.

Since it follows from (1) that $\partial \zeta_\rho/\partial u_{\rho\sigma} = -\xi_\sigma$, we can now write (3) in the form

$$\frac{\partial f}{\partial u_{\rho\sigma}} - \xi_\sigma \frac{\partial f}{\partial \zeta_\rho} = 0. \tag{5}$$

On multiplying (5) by $u_{\tau\sigma}$ and summing for $\sigma = 1, \ldots, n$ we obtain the equation

$$\sum_{\sigma=1}^{n} u_{\tau\sigma}\frac{\partial f}{\partial u_{\rho\sigma}} - \left(\sum_{\sigma=1}^{n} u_{\tau\sigma}\xi_\sigma\right)\frac{\partial f}{\partial \zeta_\rho} = 0.$$

Using (1), and writing $\zeta_\tau = -\sum_{\sigma=1}^{n} u_{\tau\sigma}\xi_\sigma$, this becomes

$$\sum_{\sigma=1}^{n} u_{\tau\sigma}\frac{\partial f}{\partial u_{\rho\sigma}} + \zeta_\tau \frac{\partial f}{\partial \zeta_\rho} = 0. \tag{6}$$

It follows from (6) that the polynomial

$$\sum_{\sigma=1}^{n} u_{\tau\sigma}\frac{\partial}{\partial u_{\rho\sigma}}f(u_{ij};\,u_{00},\ldots,u_{d0})+u_{\tau 0}\frac{\partial}{\partial u_{\rho 0}}f(u_{ij};\,u_{00},\ldots,u_{d0}),$$

which may be written

$$\sum_{\sigma=0}^{n} u_{\tau\sigma}\frac{\partial}{\partial u_{\rho\sigma}}f(u_{ij};\,u_{00},\ldots,u_{d0}),$$

vanishes when we substitute $\zeta_0,\ldots,\zeta_d$ for $u_{00},\ldots,u_{d0}$. If $\rho\neq\tau$ this polynomial is of lower degree than f in each of the indeterminates $u_{\rho 0},\ldots,u_{\rho n}$ and therefore, by Theorem I, it must be the zero polynomial. Hence we have

THEOREM II.

$$\sum_{\sigma=0}^{n} u_{\tau\sigma}\frac{\partial}{\partial u_{\rho\sigma}}f(u_{ij};\,u_{00},\ldots,u_{d0})=0\quad(\rho\neq\tau).$$

On the other hand, if $\rho=\tau$, it follows from Theorem I that

$$\sum_{\sigma=0}^{n} u_{\rho\sigma}\frac{\partial}{\partial u_{\rho\sigma}}f(u_{ij};\,u_{00},\ldots,u_{d0})=Af(u_{ij};\,u_{00},\ldots,u_{d0}),$$

where A is in K. Hence, by Euler's theorem, $f(u_{ij};\,u_{00},\ldots,u_{d0})$ must be homogeneous in any set $(u_{\rho 0},\ldots,u_{\rho n})$ of the indeterminates, and by our previous remark on the symmetry properties of this polynomial, the degree of homogeneity is independent of ρ. We may therefore write

$$f(u_{ij};\,u_{00},\ldots,u_{d0})=F(u_0,\ldots,u_d),$$

where $F(u_0,\ldots,u_d)$ is homogeneous, of degree g say, in each set of indeterminates $u_\rho=(u_{\rho 0},\ldots,u_{\rho n})$.

7. Properties of the Cayley form. We define an algebraic extension of the field $K(u_{ij};\,u_{10},\ldots,u_{d0})$ by means of the irreducible equation

$$f(u_{ij};\,z,u_{10},\ldots,u_{d0})=0,$$

where $f(u_{ij};\,u_{00},u_{10},\ldots,u_{d0})$ is the Cayley form of the variety V considered in the preceding section. Let the degree of this equation in z be h, where, clearly, $0<h\leqslant g$. In a suitable extension of the field the equation has h roots $z^{(1)},\ldots,z^{(h)}$.

Since f is irreducible, $\partial f/\partial z^{(i)}\neq 0$. For if $\partial f/\partial z^{(i)}=0$ the polynomials $f(u_{ij};\,z,u_{10},\ldots,u_{d0})$ and $\partial f(u_{ij};\,z,u_{10},\ldots,u_{d0})/\partial z$ have a common factor in $K(u_{ij};\,u_{10},\ldots,u_{d0})[z]$, and this contradicts the hypothesis that f is irreducible [III, § 5, Th. I].

Let
$$\xi_\rho^{(\tau)} = f_\rho^{(\tau)}/f_0^{(\tau)},$$
where
$$f_\rho^{(\tau)} = \left[\frac{\partial}{\partial u_{0\rho}} f(u_{ij};\, u_{00}, u_{10}, \ldots, u_{d0})\right]_{u_{00}=z^{(\tau)}} \qquad (\rho = 0, \ldots, n).$$

We prove that $\xi_\rho^{(\tau)}$ is algebraically independent of the indeterminates $u_{01}, \ldots, u_{0n}$. Since $\xi_\rho^{(\tau)}$ lies in the field $K(u_{ij};\, z^{(\tau)}, u_{10}, \ldots, u_{d0})$, it is algebraically dependent on the indeterminates $u_{ij}, u_{10}, \ldots, u_{d0}$, and satisfies a unique irreducible equation
$$\psi(u_{ij};\, u_{10}, \ldots, u_{d0}, \xi_\rho^{(\tau)}) = 0.$$
$\partial\xi_\rho^{(\tau)}/\partial u_{0k}$ is defined [III, § 7] by means of the equation
$$\frac{\partial\psi}{\partial\xi_\rho^{(\tau)}}\frac{\partial\xi_\rho^{(\tau)}}{\partial u_{0k}} + \frac{\partial\psi}{\partial u_{0k}} = 0,$$
and since $\partial\psi/\partial\xi_\rho^{(\tau)} \neq 0$, a necessary and sufficient condition that the polynomial ψ should not contain u_{0k} $(1 \leqslant k \leqslant n)$ is that $\partial\xi_\rho^{(\tau)}/\partial u_{0k} = 0$, for this is equivalent to $\partial\psi/\partial u_{0k} = 0$.

Now ξ_ρ is independent of u_{0k} by definition, and
$$\xi_\rho = \frac{\partial f}{\partial u_{0\rho}}\bigg/\frac{\partial f}{\partial u_{00}},$$
where $u_{00}, \ldots, u_{d0}$ are replaced by $\zeta_0, \ldots, \zeta_d$ after the differentiations have been performed. Hence, since
$$\frac{\partial\xi_\rho}{\partial u_{0k}} = 0, \left[\frac{\partial u_{i0}}{\partial u_{0k}}\right]_{u_{j0}=\zeta_j} = 0 \qquad (i \neq 0),$$
$$\left(\frac{\partial u_{00}}{\partial u_{0k}}\right)_{u_{j0}=\zeta_j} = -\,\xi_k = -\left[\frac{\partial f}{\partial u_{0k}}\bigg/\frac{\partial f}{\partial u_{00}}\right]_{u_{j0}=\zeta_j},$$
$$\frac{\partial}{\partial u_{0k}}\left[\frac{\partial f}{\partial u_{0\rho}}\bigg/\frac{\partial f}{\partial u_{00}}\right] - \left[\frac{\partial f}{\partial u_{0k}}\bigg/\frac{\partial f}{\partial u_{00}}\right]\frac{\partial}{\partial u_{00}}\left[\frac{\partial f}{\partial u_{0\rho}}\bigg/\frac{\partial f}{\partial u_{00}}\right]$$
becomes $\partial\zeta_\rho/\partial u_{0k}$ when $u_{00}, \ldots, u_{d0}$ are replaced by $\zeta_0, \ldots, \zeta_d$. From §6, Th. I, it follows that this expression is equal to
$$A(u_{ij};\, u_{00}, \ldots, u_{d0})\, f\,(u_{ij};\, u_{00}, \ldots, u_{d0})/(\partial f/\partial u_{00})^3,$$
and hence vanishes when u_{00} is replaced by $z^{(\tau)}$. It follows that if we compute $\partial\xi_\rho^{(\tau)}/\partial u_{0k}$ as we have computed $\partial\xi_\rho/\partial u_{0k}$, we obtain $\partial\xi_\rho^{(\tau)}/\partial u_{0k} = 0$. Hence $\xi_\rho^{(\tau)}$ is algebraically independent of $u_{01}, \ldots, u_{0n}$.

From the equation $f_0^{(\tau)}\xi_\rho^{(\tau)} - f_\rho^{(\tau)} = 0,$

we obtain the equation

$$f_0^{(\tau)} \sum_{\rho=1}^{n} (u_{0\rho}\xi_\rho^{(\tau)}) - \sum_{\rho=1}^{n} (u_{0\rho} f_\rho^{(\tau)}) = 0. \tag{1}$$

Since $f(u_{ij};\, u_{00}, \ldots, u_{d0})$ is homogeneous of degree g in the set of indeterminates $u_{00}, \ldots, u_{0n}$, we also have the identity

$$\sum_{\rho=1}^{n} \left(u_{0\rho} \frac{\partial f}{\partial u_{0\rho}} \right) + u_{00} \frac{\partial f}{\partial u_{00}} = g f(u_{ij};\, u_{00}, \ldots, u_{d0}).$$

Substituting $z^{(\tau)}$ for u_{00}, this becomes

$$\sum_{\rho=1}^{n} (u_{0\rho} f_\rho^{(\tau)}) + z^{(\tau)} f_0^{(\tau)} = 0.$$

Hence (1) becomes

$$f_0^{(\tau)} \sum_{\rho=1}^{n} (u_{0\rho}\xi_\rho^{(\tau)}) + z^{(\tau)} f_0^{(\tau)} = 0,$$

and since $f_0^{(\tau)} \neq 0$, we obtain the relation

$$z^{(\tau)} = - \sum_{\rho=1}^{n} u_{0\rho}\xi_\rho^{(\tau)}. \tag{2}$$

Now $\quad f(u_{ij};\, z, u_{10}, \ldots, u_{d0}) = A(u_{ij};\, u_{10}, \ldots, u_{d0}) \prod_{\tau=1}^{h} (z - z^{(\tau)}),$

and since the $\xi_\rho^{(\tau)}$ are independent of $u_{01}, \ldots, u_{0n}$, the symmetric functions of $\xi_1^{(\tau)}, \ldots, \xi_n^{(\tau)}$ are rational functions of the sets of indeterminates $u_1, \ldots, u_d$ only. Therefore

$$\prod_{\tau=1}^{h} (z - z^{(\tau)}) = \prod_{\tau=1}^{h} \left(z + \sum_{\rho=1}^{n} u_{0\rho}\xi_\rho^{(\tau)} \right) = \frac{\phi(u_{ij};\, z, u_{10}, \ldots, u_{d0})}{\psi(u_1, \ldots, u_d)},$$

where ϕ and ψ may be assumed to have no common factor. Hence

$$A(u_{ij};\, u_{10}, \ldots, u_{d0})\, \phi(u_{ij};\, z, u_{10}, \ldots, u_{d0})$$
$$= f(u_{ij};\, z, u_{10}, \ldots, u_{d0})\, \psi(u_1, \ldots, u_d),$$

and, by the unique factorisation theorem [I, § 8, Th. I], since ψ has no factor in common with ϕ, it must divide $A(u_{ij};\, u_{10}, \ldots, u_{d0})$. If $A = A'\psi$, then $f = A'\phi$. But since f is irreducible, A' must be in K, and therefore $A = A'\psi(u_1, \ldots, u_d)$ lies in $K[u_1, \ldots, u_d]$. We have therefore proved that

$$f(u_{ij};\, u_{00}, \ldots, u_{d0}) = F(u_0, \ldots, u_d) = A(u_1, \ldots, u_d) \prod_{\tau=1}^{h} \left(u_{00} + \sum_{\rho=1}^{n} u_{0\rho}\xi_\rho^{(\tau)} \right),$$

and since $F(u_0, \ldots, u_d)$ is of degree g in the indeterminates $u_{00}, \ldots u_{0n}$ it follows that $h = g$, and

$$F(u_0, \ldots, u_d) = A(u_1, \ldots, u_d) \prod_{\tau=1}^{g} \left(u_{00} + \sum_{\rho=1}^{n} u_{0\rho} \xi_\rho^{(\tau)}\right).$$

We now prove two theorems which associate the points $(1, \xi_1^{(\tau)}, \ldots, \xi_n^{(\tau)})$ with the variety V.

THEOREM I. *The points* $(1, \xi_1^{(\tau)}, \ldots, \xi_n^{(\tau)})$ *are generic points of the variety* V, *and satisfy the equations*

$$\sum_{\rho=0}^{n} u_{\sigma\rho} \xi_\rho^{(\tau)} = 0 \quad (\sigma = 1, \ldots, d).$$

Let $\phi(x_0, x_1, \ldots, x_n)$ be any form which vanishes on V. Then $\phi(1, \xi_1, \ldots, \xi_n) = 0$. Substituting $\xi_\rho = \dfrac{\partial f}{\partial u_{0\rho}} \Big/ \dfrac{\partial f}{\partial \zeta_0}$, we obtain the equation

$$\phi\left(\frac{\partial f}{\partial \zeta_0}, \frac{\partial f}{\partial u_{01}}, \ldots, \frac{\partial f}{\partial u_{0n}}\right) = 0, \tag{3}$$

where $\zeta_0, \ldots, \zeta_d$ are substituted for $u_{00}, \ldots, u_{d0}$ after the differentiations have been performed on $f(u_{ij}; u_{00}, \ldots, u_{d0})$. From (3) and § 6, Th. I, we deduce that

$$\phi\left(\frac{\partial f}{\partial u_{00}}, \frac{\partial f}{\partial u_{01}}, \ldots, \frac{\partial f}{\partial u_{0n}}\right) = A(u_{ij}; u_{00}, \ldots, u_{d0}) f(u_{ij}; u_{00}, \ldots, u_{d0}).$$

Substituting $u_{00} = z^{(\tau)}$ in this equation, it follows that

$$\phi(f_0^{(\tau)}, f_1^{(\tau)}, \ldots, f_n^{(\tau)}) = 0,$$

that is,

$$\phi(1, \xi_1^{(\tau)}, \ldots, \xi_n^{(\tau)}) = 0. \tag{4}$$

Conversely, if we are given (4) we deduce that the form

$$\phi\left(\frac{\partial f}{\partial u_{00}}, \frac{\partial f}{\partial u_{01}}, \ldots, \frac{\partial f}{\partial u_{0n}}\right)$$

vanishes when u_{00} is replaced by $z^{(\tau)}$, where $z^{(\tau)}$ is a root of the irreducible equation

$$f(u_{ij}; z, u_{10}, \ldots, u_{d0}) = 0.$$

It follows from III, § 1, that

$$\phi\left(\frac{\partial f}{\partial u_{00}}, \frac{\partial f}{\partial u_{01}}, \ldots, \frac{\partial f}{\partial u_{0n}}\right) = A(u_{ij}; u_{00}, \ldots, u_{d0}) f(u_{ij}; u_{00}, \ldots, u_{d0}),$$

and therefore, replacing $u_{00}, \ldots, u_{d0}$ by $\zeta_0, \ldots, \zeta_d$,

$$\phi\left(\frac{\partial f}{\partial \zeta_0}, \frac{\partial f}{\partial u_{01}}, \ldots, \frac{\partial f}{\partial u_{0n}}\right) = 0,$$

and therefore $$\phi(1, \xi_1, \ldots, \xi_n) = 0,$$

so that $\phi(x_0, x_1, \ldots, x_n)$ vanishes on V.

The points $(1, \xi_1^{(\tau)}, \ldots, \xi_n^{(\tau)})$ are therefore generic points on V.

Again, if $\sigma \neq 0$,

$$\sum_{\rho=0}^{n} u_{\sigma\rho} \xi_\rho^{(\tau)} = \left(\sum_{\rho=0}^{n} u_{\sigma\rho} f_\rho^{(\tau)}\right) \Big/ f_0^{(\tau)},$$

and $$\sum_{\rho=0}^{n} u_{\sigma\rho} f_\rho^{(\tau)} = \left[\sum_{\rho=0}^{n} u_{\sigma\rho} \frac{\partial}{\partial u_{0\rho}} f(u_{ij};\, u_{00}, \ldots, u_{d0})\right]_{u_{00} = z^{(\tau)}}.$$

By § 6, Th. II, the expression inside the bracket is zero. It therefore remains zero after the substitution is made. Hence the points $(1, \xi_1^{(\tau)}, \ldots, \xi_n^{(\tau)})$ satisfy the equations

$$\sum_{\rho=0}^{n} u_{\sigma\rho} \xi_\rho^{(\tau)} = 0 \quad (\sigma = 1, \ldots, d).$$

THEOREM II. *The points* $(1, \xi_1^{(\tau)}, \ldots, \xi_n^{(\tau)})$ $(\tau = 1, \ldots, g)$ *are the only solutions of the equations of* V *and the equations*

$$\sum_{\rho=0}^{n} u_{\sigma\rho} x_\rho = 0 \quad (\sigma = 1, \ldots, d).$$

We have assumed that V does not lie in $x_0 = 0$. Its intersection with $x_0 = 0$ is therefore a sum of proper subvarieties, each of which is of dimension less than d [§ 5, Th. II]. The equations

$$\left.\begin{aligned} \sum_{\rho=0}^{n} u_{\sigma\rho} x_\rho &= 0 \quad (\sigma = 1, \ldots, d),\\ x_0 &= 0, \end{aligned}\right\}$$

define a generic space of $n-d-1$ dimensions in $x_0 = 0$, and by § 5, Lemma II, this has no points in common with the intersection of V and $x_0 = 0$. Therefore V has no points for which $x_0 = 0$ in the generic S_{n-d} given by the equations

$$\sum_{\rho=0}^{n} u_{\sigma\rho} x_\rho = 0 \quad (\sigma = 1, \ldots, d). \tag{5}$$

Now let $(1, x_1', \ldots, x_n')$ be a point which satisfies the equations of V and also (5). Using the trivial identity

$$u_{\sigma 0} = \sum_{\rho=0}^{n} u_{\sigma\rho} x_\rho' - \sum_{\rho=1}^{n} u_{\sigma\rho} x_\rho' \quad (\sigma = 0, \ldots, d),$$

we have the relation

$$f(u_{ij};\, u_{00}, \ldots, u_{d0}) = f\left(u_{ij};\, -\sum_{\rho=1}^{n} u_{0\rho} x_\rho' + \sum_{\rho=0}^{n} u_{0\rho} x_\rho', \ldots, -\sum_{\rho=1}^{n} u_{d\rho} x_\rho' + \sum_{\rho=0}^{n} u_{d\rho} x_\rho'\right).$$

Hence, by Taylor's polynomial expansion [III, § 5],

$$f(u_{ij};\, u_{00}, \ldots, u_{d0}) = f\left(u_{ij};\, -\sum_{\rho=1}^{n} u_{0\rho} x_\rho', \ldots, -\sum_{\rho=1}^{n} u_{d\rho} x_\rho'\right) + \sum_{\sigma=0}^{d} A_\sigma(u_{ij};\, x')\left(\sum_{\rho=0}^{n} u_{\sigma\rho} x_\rho'\right), \quad (6)$$

where the $A_\sigma(u_{ij};\, x')$ are elements of the ring $K[u_{ij};\, x_1', \ldots, x_n']$. But since the point $(1, x_1', \ldots, x_n')$ lies on V, and

$$f(u_{ij};\, \zeta_0, \ldots, \zeta_d) = f\left(u_{ij};\, -\sum_{\rho=1}^{n} u_{0\rho} \xi_\rho, \ldots, -\sum_{\rho=1}^{n} u_{d\rho} \xi_\rho\right) = 0$$

is a relation holding for a generic point $(1, \xi_1, \ldots, \xi_n)$ of V, we also have the equation

$$f\left(u_{ij};\, -\sum_{\rho=1}^{n} u_{0\rho} x_\rho', \ldots, -\sum_{\rho=1}^{n} u_{d\rho} x_\rho'\right) = 0.$$

Furthermore $\quad \sum_{\rho=0}^{n} u_{\sigma\rho} x_\rho' = 0 \quad (\sigma = 1, \ldots, d).$

Equation (6) therefore becomes

$$f(u_{ij};\, u_{00}, \ldots, u_{d0}) = A_0(u_{ij};\, x')\left(\sum_{\rho=0}^{n} u_{0\rho} x_\rho'\right);$$

that is, $f(u_{ij};\, u_{00}, \ldots, u_{d0})$ contains the factor $\sum_{\rho=0}^{n} u_{0\rho} x_\rho'$. But we have proved that

$$f(u_{ij};\, u_{00}, \ldots, u_{d0}) = A(u_1, \ldots, u_d) \prod_{\tau=1}^{g} \left(u_{00} + \sum_{\rho=1}^{n} u_{0\rho} \xi_\rho^{(\tau)}\right).$$

Hence it follows that for some value of τ

$$(1, x_1', \ldots, x_n') = (1, \xi_1^{(\tau)}, \ldots, \xi_n^{(\tau)}).$$

From Theorems I and II we deduce

THEOREM III. *A generic S_{n-d} meets an irreducible variety V of dimension d in a finite number of points, each of which is a generic point of V over K.*

The number of points g is called the *order* of the variety. We have seen that it is the degree of the Cayley form $F(u_0, \ldots, u_d)$ in each set of the indeterminates $u_0, u_1, \ldots, u_d$.

The factorisation of the Cayley form

$$F(u_0, u_1, \ldots, u_d) = A(u_1, \ldots, u_d) \prod_{\rho=1}^{g} \left(u_{00} + \sum_{k=1}^{n} u_{0k} \xi_k^{(\rho)} \right) \tag{7}$$

will be frequently used in the sequel. It tells us that $F(u_0, \ldots, u_d)$ is the u_0-resultant [IV, § 10] of the equations of V taken together with the equations $\sum_{j=0}^{n} u_{ij} x_j = 0$ $(i = 1, \ldots, d)$ of a generic S_{n-d}.

We deduce that if u'_{ij} is any specialisation of u_{ij} $(u = 0, 1, \ldots, d;$ $j = 0, \ldots, n)$, a necessary and sufficient condition that V and the locus given by

$$\sum_{j=0}^{n} u'_{ij} x_j = 0 \quad (i = 0, 1, \ldots, d)$$

points in common is

$$F(u'_0, u'_1, \ldots, u'_d) = 0.$$

Since factorisation of the Cayley form gives a generic point of V, we also learn from (7) that no two distinct varieties can have the same Cayley form. An irreducible algebraic variety has a unique Cayley form and is uniquely determined by its Cayley form. Hence an irreducible algebraic variety V is uniquely determined by the complex of S_{n-d-1}'s which meet it. For, as we have just seen, this complex determines the Cayley form.

The idea of representing a curve in S_3 by means of the complex of lines which meet it is due to Cayley. (See Bibliographical Notes, p. 388.) It is for this reason that we have called $F(u_0, \ldots, u_d)$ the *Cayley form of V*.

We conclude this section by proving a theorem which links $F(u_0, \ldots, u_d)$, in the case $d = 1$, $n = 3$, directly with the polynomial in the Grassmann coordinates $(\ldots, p^{ij}, \ldots)$ of lines in S_3 which appears in Cayley's work.

THEOREM IV. $F(u_0, u_1, \ldots, u_d) = G(\ldots, p^{i_0 \ldots i_d}, \ldots)$, *where*

$$G(\ldots, z^{i_0 \ldots i_d}, \ldots)$$

is a polynomial of degree g in the indeterminates $z^{i_0\ldots i_d}$, and $p^{i_0\ldots i_d}$ are the dual Grassmann coordinates of the S_{n-d-1} determined by the equations

$$\sum_{j=0}^{n} u_{ij}x_j = 0 \quad (i = 0, \ldots, d).$$

To prove this theorem we make use of an operator $\Delta^{a_1\ldots a_p}_{\alpha_1\ldots\alpha_p}$ defined as follows: if ϕ is any form which is homogeneous of degree g in the indeterminates $u_i = (u_{i0}, \ldots, u_{in})$, for $i = 0, \ldots, d$, then

$$\Delta^{a_1\ldots a_p}_{\alpha_1\ldots\alpha_p}\phi = \Sigma \pm \frac{\partial^p\phi}{\partial u_{a_1\beta_1}\ldots\partial u_{a_p\beta_p}},$$

the summation being over all derangements $\beta_1, \ldots, \beta_p$ of $\alpha_1, \ldots, \alpha_p$, the positive sign being taken with even, the negative sign with odd derangements. We may express $\Delta^{a_1\ldots a_p}_{\alpha_1\ldots\alpha_p}$ in determinantal form:

$$\Delta^{a_1\ldots a_p}_{\alpha_1\ldots\alpha_p}\phi = \begin{vmatrix} \dfrac{\partial}{\partial u_{a_1\alpha_1}} & \cdot & \dfrac{\partial}{\partial u_{a_1\alpha_p}} \\ \cdot & \cdot & \cdot \\ \dfrac{\partial}{\partial u_{a_p\alpha_1}} & \cdot & \dfrac{\partial}{\partial u_{a_p\alpha_p}} \end{vmatrix}\phi.$$

If i is distinct from $a_1, \ldots, a_p$, then

$$\Delta^{ia_1\ldots a_p}_{\alpha\alpha_1\ldots\alpha_p}\phi = \begin{vmatrix} \dfrac{\partial}{\partial u_{i\alpha}} & \cdot & \dfrac{\partial}{\partial u_{i\alpha_p}} \\ \dfrac{\partial}{\partial u_{a_1\alpha}} & \cdot & \dfrac{\partial}{\partial u_{a_1\alpha_p}} \\ \cdot & \cdot & \cdot \\ \dfrac{\partial}{\partial u_{a_p\alpha}} & \cdot & \dfrac{\partial}{\partial u_{a_p\alpha_p}} \end{vmatrix}\phi,$$

and on expanding this determinantal expression in terms of the elements of the first column:

$$\Delta^{ia_1\ldots a_p}_{\alpha\alpha_1\ldots\alpha_p}\phi = \frac{\partial}{\partial u_{i\alpha}}\Delta^{a_1\ldots a_p}_{\alpha_1\ldots\alpha_p}\phi - \sum_{r=1}^{p}\frac{\partial}{\partial u_{a_r\alpha}}\Delta^{a_1\ldots a_{r-1}ia_{r+1}\ldots a_p}_{\alpha_1\ldots\ldots\ldots\ldots\ldots\alpha_p}\phi.$$

Hence

$$\sum_{\alpha=0}^{n} u_{i\alpha}\Delta^{ia_1\ldots a_p}_{\alpha\alpha_1\ldots\alpha_p}\phi$$

$$= \sum_{\alpha=0}^{n} u_{i\alpha}\frac{\partial}{\partial u_{i\alpha}}\Delta^{a_1\ldots a_p}_{\alpha_1\ldots\alpha_p}\phi - \sum_{r=1}^{p}\sum_{\alpha=0}^{n} u_{i\alpha}\frac{\partial}{\partial u_{a_r\alpha}}\Delta^{a_1\ldots a_{r-1}ia_{r+1}\ldots a_p}_{\alpha_1\ldots\ldots\ldots\ldots\ldots\alpha_p}\phi.$$

We are concerned with the right-hand side of this equation. Since i is distinct from $a_1, \ldots, a_p$, $\Delta^{a_1 \ldots a_p}_{\alpha_1 \ldots \alpha_p} \phi$ is homogeneous of degree g in $(u_{i0}, \ldots, u_{in})$. Therefore, by Euler's theorem,

$$\sum_{\alpha=0}^{n} u_{i\alpha} \frac{\partial}{\partial u_{i\alpha}} \Delta^{a_1 \ldots a_p}_{\alpha_1 \ldots \alpha_p} \phi = g \Delta^{a_1 \ldots a_p}_{\alpha_1 \ldots \alpha_p} \phi.$$

By direct computation we have

$$\Delta^{a_1 \ldots a_{r-1} i a_{r+1} \ldots a_p}_{\alpha_1 \ldots \ldots \ldots \ldots \ldots \alpha_p} \left(\sum_{\alpha=0}^{n} u_{i\alpha} \frac{\partial \phi}{\partial u_{a_r \alpha}} \right)$$

$$= \sum_{\alpha=0}^{n} u_{i\alpha} \frac{\partial}{\partial u_{a_r\alpha}} \Delta^{a_1 \ldots a_{r-1} i a_{r+1} \ldots a_p}_{\alpha_1 \ldots \ldots \ldots \ldots \ldots \alpha_p} \phi$$

$$+ \begin{vmatrix} \dfrac{\partial}{\partial u_{a_1\alpha_1}} & \cdot & \dfrac{\partial}{\partial u_{a_1\alpha_p}} \\ \cdot & \cdot & \cdot \\ \dfrac{\partial}{\partial u_{a_r\alpha_1}} & \cdot & \dfrac{\partial}{\partial u_{a_r\alpha_p}} \\ \cdot & \cdot & \cdot \\ \dfrac{\partial}{\partial u_{a_p\alpha_1}} & \cdot & \dfrac{\partial}{\partial u_{a_p\alpha_p}} \end{vmatrix} \phi$$

$$= \sum_{\alpha=0}^{n} u_{i\alpha} \frac{\partial}{\partial u_{a_r\alpha}} \Delta^{a_1 \ldots a_{r-1} i a_{r+1} \ldots a_p}_{\alpha_1 \ldots \ldots \ldots \ldots \ldots \alpha_p} \phi + \Delta^{a_1 \ldots a_p}_{\alpha_1 \ldots \alpha_p} \phi.$$

Therefore

$$\sum_{r=1}^{p} \sum_{\alpha=0}^{n} u_{i\alpha} \frac{\partial}{\partial u_{a_r\alpha}} \Delta^{a_1 \ldots a_{r-1} i a_{r+1} \ldots a_p}_{\alpha_1 \ldots \ldots \ldots \ldots \ldots \alpha_p} \phi$$

$$= \sum_{r=1}^{p} \Delta^{a_1 \ldots a_{r-1} i a_{r+1} \ldots a_p}_{\alpha_1 \ldots \ldots \ldots \ldots \ldots \alpha_p} \left(\sum_{\alpha=0}^{n} u_{i\alpha} \frac{\partial \phi}{\partial u_{a_r\alpha}} \right) - p \Delta^{a_1 \ldots a_p}_{\alpha_1 \ldots \alpha_p} \phi.$$

Hence we have proved

Lemma I. If i is distinct from $a_1, \ldots, a_p$, then

$$\sum_{\alpha=0}^{n} u_{i\alpha} \Delta^{i a_1 \ldots a_p}_{\alpha \alpha_1 \ldots \alpha_p} \phi = (g+p) \Delta^{a_1 \ldots a_p}_{\alpha_1 \ldots \alpha_p} \phi - \sum_{r=1}^{p} \Delta^{a_1 \ldots a_{r-1} i a_{r+1} \ldots a_p}_{\alpha_1 \ldots \ldots \ldots \ldots \ldots \alpha_p} \left(\sum_{\alpha=0}^{n} u_{i\alpha} \frac{\partial \phi}{\partial u_{a_r\alpha}} \right).$$

Let each of i, j be one of the set $0, 1, \ldots, d$. Then, as above,

$$\Delta^{0 \ldots d}_{\alpha_0 \ldots \alpha_d} \left(\sum_{\alpha=0}^{n} u_{i\alpha} \frac{\partial \phi}{\partial u_{j\alpha}} \right) = \sum_{\alpha=0}^{n} u_{i\alpha} \frac{\partial}{\partial u_{j\alpha}} (\Delta^{0 \ldots d}_{\alpha_0 \ldots \alpha_d} \phi) + \delta_{ij} \Delta^{0 \ldots d}_{\alpha_0 \ldots \alpha_d} \phi.$$

where δ_{ij} is a Kronecker delta. This gives us

Lemma II. If $\psi = \Delta^{0\ldots d}_{\alpha_0\ldots\alpha_d}\phi$, *then*

$$\sum_{\alpha=0}^{n} u_{i\alpha}\frac{\partial\psi}{\partial u_{j\alpha}} = \Delta^{0\ldots d}_{\alpha_0\ldots\alpha_d}\left(\sum_{\alpha=0}^{n} u_{i\alpha}\frac{\partial\phi}{\partial u_{j\alpha}}\right) - \delta_{ij}\psi.$$

We now take ϕ to be the Cayley form $F(u_0, \ldots, u_d)$. By § 6, Th. II,

$$\sum_{\alpha=0}^{n} u_{i\alpha}\frac{\partial F}{\partial u_{j\alpha}} = 0 \quad (i \neq j),$$

and therefore, by Lemma I,

$$\sum_{\alpha=0}^{n} u_{i\alpha}\Delta^{ia_1\ldots a_p}_{\alpha\alpha_1\ldots\alpha_p}F = (g+p)\Delta^{a_1\ldots a_p}_{\alpha_1\ldots\alpha_p}F.$$

Taking $i, a_1, \ldots, a_p$ to be $0, 1, \ldots, d$, and $\alpha = \alpha_0$, this becomes

$$\sum_{\alpha_0=0}^{n} u_{0\alpha_0}\Delta^{01\ldots d}_{\alpha_0\alpha_1\ldots\alpha_d}F = (g+d)\Delta^{1\ldots d}_{\alpha_1\ldots\alpha_d}F.$$

By repeated applications of this result we find that

$$\sum_{\alpha_0=0}^{n}\ldots\sum_{\alpha_d=0}^{n} u_{0\alpha_0}\ldots u_{d\alpha_d}\Delta^{0\ldots d}_{\alpha_0\ldots\alpha_d}F = (g+d)\ldots(g)F = \frac{(g+d)!}{(g-1)!}F.$$

Since $\Delta^{0\ldots d}_{\alpha_0\ldots\alpha_d}$ is skew-symmetric in its suffixes, this equation may be written

$$\sum_{\alpha_0,\ldots,\alpha_d}\begin{vmatrix} u_{0\alpha_0} & . & u_{0\alpha_d} \\ . & . & . \\ u_{d\alpha_0} & . & u_{d\alpha_d}\end{vmatrix}\Delta^{0\ldots d}_{\alpha_0\ldots\alpha_d}F = \frac{(g+d)!}{(g-1)!}F,$$

the summation on the left being over the unordered sets of $d+1$ distinct integers selected from $0, \ldots, n$. Therefore

$$F = \frac{(g-1)!}{(g+d)!}\sum_{\alpha_0,\ldots,\alpha_d}\begin{vmatrix} u_{0\alpha_0} & . & u_{0\alpha_d} \\ . & . & . \\ u_{d\alpha_0} & . & u_{d\alpha_d}\end{vmatrix}\Delta^{0\ldots d}_{\alpha_0\ldots\alpha_d}F.$$

We note that $\psi = \Delta^{0\ldots d}_{\alpha_0\ldots\alpha_d}F$ is homogeneous of degree $g-1$ in the sets of indeterminates $(u_{i0}, \ldots, u_{in})$ $(i = 0, \ldots, d)$, and by Lemma II,

$$\sum_{\alpha=0}^{n} u_{i\alpha}\frac{\partial\psi}{\partial u_{j\alpha}} = 0 \quad (i \neq j).$$

We may therefore apply the foregoing argument to $\psi = \Delta^{0\ldots d}_{\alpha_0\ldots\alpha_d}F$. At each stage the degree of the form considered is one less than at the preceding stage, and finally we operate with $\Delta^{0\ldots d}_{\alpha_0\ldots\alpha_d}$ on a form of degree one and obtain a constant. Hence we have shown that F can be expressed as a polynomial of degree g in the determinants

$$p^{\alpha_0\ldots\alpha_d} = \begin{vmatrix} u_{0\alpha_0} & . & u_{0\alpha_d} \\ . & . & . \\ u_{d\alpha_0} & . & u_{d\alpha_d} \end{vmatrix},$$

where $(\ldots, p^{i_0\ldots i_d}, \ldots)$ are the dual Grassmann coordinates of the S_{n-d-1} determined by the equations

$$\sum_{j=0}^{n} u_{ij} x_j = 0 \quad (i = 0, \ldots, d).$$

This completes the proof of Theorem IV.

Since $F(u'_0, u'_1, \ldots, u'_d) = 0$ is a necessary and sufficient condition that the S_{n-d-1} given by the equations

$$\sum_{j=0}^{n} u'_{ij} x_j = 0 \quad (i = 0, \ldots, d)$$

should meet V, it follows that if

$$F(u_0, u_1, \ldots, u_d) = G(\ldots, p^{i_0\ldots i_d}, \ldots),$$

then

$$G(\ldots, p'^{i_0\ldots i_d}, \ldots) = 0$$

is a necessary and sufficient condition that the S_{n-d-1} with dual Grassmann coordinates $(\ldots, p'^{i_0\ldots i_d}, \ldots)$ should meet V.

Let S_{n-d-2} be an $(n-d-2)$-space in S_n determined by the $n-d-1$ independent points

$$(\alpha^i_0, \ldots, \alpha^i_n) \quad (i = 0, \ldots, n-d-2).$$

The coordinates $(\ldots, p'_{i_0\ldots i_{n-d-2}}, \ldots)$ of S_{n-d-2} are given by the equations

$$p'_{i_0\ldots i_{n-d-2}} = \begin{vmatrix} \alpha^0_{i_0} & . & \alpha^0_{i_{n-d-2}} \\ . & . & . \\ \alpha^{n-d-2}_{i_0} & . & \alpha^{n-d-2}_{i_{n-d-2}} \end{vmatrix}.$$

If x' is any point of S_n not in S_{n-d-2}, the S_{n-d-1} joining x' to S_{n-d-2} has coordinates $(\ldots, q'_{i_0\ldots i_{n-d-1}}, \ldots)$, where [VII, p. 307]

$$q'_{i_0\ldots i_{n-d-1}} = \sum_{\nu=0}^{n-d-1} (-1)^{\nu} p'_{i_0\ldots i_{\nu-1} i_{\nu+1}\ldots i_{n-d-1}} x'_{i_\nu}.$$

A necessary and sufficient condition that this S_{n-d-1} meet V is

$$G(\dots, q'_{j_0\dots j_{n-d-1}}, \dots) = 0,$$

where $i_0, \dots, i_d, j_0, \dots, j_{n-d-1}$ is an even permutation of $0, \dots, n$. Now consider the form

$$G\left(\dots, \sum_{\nu=0}^{n-d-1} (-1)^\nu x_{j_\nu} p'_{j_0\dots j_{\nu-1} j_{\nu+1}\dots j_{n-d-1}}, \dots\right).$$

This vanishes identically if, and only if, a generic S_{n-d-1} through S_{n-d-2} meets V, that is [§ 5, p. 32], if and only if S_{n-d-2} meets V. If S_{n-d-2} does not meet V the equation

$$G\left(\dots, \sum_{\nu=0}^{n-d-1} (-1)^\nu x_{j_\nu} p'_{j_0\dots j_{\nu-1} j_{\nu+1}\dots j_{n-d-1}}, \dots\right) = 0 \tag{8}$$

determines a primal Π, and this primal has a simple geometrical interpretation. We note first that if x' is a point of S_{n-d-2},

$$\sum_{\nu=0}^{n-d-1} (-1)^\nu x'_{j_\nu} p'_{j_0\dots j_{\nu-1} j_{\nu+1}\dots j_{n-d-1}} = 0 \quad (\text{all } j_0, \dots, j_{n-d-1}),$$

and therefore x' lies on Π. Hence every point of S_{n-d-2} lies on Π. If x' is a point of Π not in S_{n-d-2}, the S_{n-d-1} joining x' to S_{n-d-2} meets V, and conversely any point of an S_{n-d-1} joining S_{n-d-2} to a point of V lies on Π. Hence the points of Π are the points of the S_{n-d-1} which join S_{n-d-2} to points of V, and Π is therefore *the cone projecting V from S_{n-d-2}*.

In particular, if the points $\alpha^0, \dots, \alpha^{n-d-2}$ are independent generic points of S_n, equation (8) is the equation of the cone projecting V from a generic S_{n-d-2}. Replacing the $p'_{j_0\dots j_{\nu-1} j_{\nu+1}\dots j_{n-d-1}}$ by their expression in terms of the α^i_j, we obtain the equation

$$G\left(\dots, \sum_{\nu=0}^{n-d-1} (-1)^\nu x_{j_\nu} p'_{j_0\dots j_{\nu-1} j_{\nu+1}\dots j_{n-d-1}}, \dots\right) \equiv \sum_\sigma \phi_\sigma(\alpha)\, \psi_\sigma(x) = 0,$$

where the $\phi_\sigma(\alpha)$ are the distinct power-products of the α^i_j. This equation is satisfied by every point x' of V. Hence V satisfies the equations

$$\psi_\sigma(x) = 0 \quad (\sigma = 1, 2, \dots). \tag{9}$$

Conversely if x' satisfies the equations (9), the equation (8) is satisfied, where $(\dots, p'_{i_0\dots i_{n-d-2}}, \dots)$ is a generic S_{n-d-2}. Hence a generic S_{n-d-1} through x' meets V and therefore [§ 5, Lemma III] x' lies on V. The equations (9) are therefore the equations of V.

Thus the Cayley form of V leads at once to the equation of the cone projecting V from a generic S_{n-d-2}, and V is defined uniquely by such a cone. The idea of defining a variety by means of the cone projecting it from a generic S_{n-d-2} is due to Cayley (Bibliographical Notes, p. 388).

We conclude this section by proving some results on the Cayley forms of certain elementary varieties.

(1) We first consider the Cayley form of a variety of $n-1$ dimensions in S_n. Let V be an irreducible primal of order g in S_n, and let $F(u_0, u_1, \ldots, u_{n-1})$ be its (necessarily irreducible) Cayley form. By Theorem IV,

$$F(u_0, u_1, \ldots, u_{n-1}) = G(\ldots, p^{i_0 \ldots i_{n-1}}, \ldots),$$

where

$$p^{i_0 \ldots i_{n-1}} = \begin{vmatrix} u_{0\, i_0} & \cdot & u_{0 i_{n-1}} \\ \cdot & \cdot & \cdot \\ u_{n-1\, i_0} & \cdot & u_{n-1\, i_{n-1}} \end{vmatrix}.$$

If the set $i_0, \ldots, i_{n-1}$ is a derangement of $0, 1, \ldots, i-1, i+1, \ldots, n$, we can write

$$p^{01\ldots i-1\, i+1 \ldots n} = (-1)^i x_i,$$

or

$$x_i = (-1)^i \begin{vmatrix} u_{00} & \cdot & u_{0\, i-1} & u_{0\, i+1} & \cdot & u_{0\, n} \\ \cdot & \cdot & \cdot & \cdot & \cdot & \cdot \\ u_{n-1\, 0} & \cdot & u_{n-1\, i-1} & u_{n-1\, i+1} & \cdot & u_{n-1\, n} \end{vmatrix},$$

and

$$F(u_0, u_1, \ldots, u_{n-1}) = G(\ldots, x_i, \ldots)$$

is homogeneous in $x_0, \ldots, x_n$, of degree g, and is irreducible, since $F(u_0, u_1, \ldots, u_{n-1})$ is irreducible. From the properties of the form $G(\ldots, x_i, \ldots)$ proved above, a necessary and sufficient condition that a point x' lie on V is

$$G(\ldots, x'_i, \ldots) = 0.$$

Hence

$$G(\ldots, x_i, \ldots) = 0$$

is the equation of V. Conversely if V is an irreducible primal given by the equation

$$\phi(x) = 0,$$

and the Cayley form of V is

$$F(u_0, \ldots, u_{n-1}) = G(x),$$

we see that

$$\phi(x') = 0$$

if and only if

$$G(x') = 0,$$

and therefore $$G(x) = a\phi(x),$$

where a is in K and different from zero. Hence the Cayley form of V can be taken as $F(u_0, \ldots, u_{n-1})$, where this form is obtained from $\phi(x)$ by substituting

$$x_i = (-1)^i \,|\, u_{00} \ldots u_{i-1\,i-1} u_{i\,i+1} \ldots u_{n-1\,n} \,|.$$

From this result we deduce that *the order of an irreducible primal is equal to the degree of the irreducible equation which defines it.*

(2) As a particular case of this last result we have the theorem: a prime is an irreducible variety of order one and, conversely, an irreducible variety of dimension $n-1$ and order one is a prime. We now extend this converse and prove that *an irreducible variety of dimension d and order one is a d-space.*

If V is an irreducible variety of order one, its Cayley form $F(u_0, \ldots, u_d)$ can be written in the form

$$F(u_0, \ldots, u_d) = G(\ldots, p_{i_{d+1} \ldots i_n}, \ldots),$$

where

$$p_{i_{d+1} \ldots i_n} = \epsilon_{i_0 \ldots i_n} \begin{vmatrix} u_{0 i_0} & \cdot & u_{0 i_d} \\ \cdot & \cdot & \cdot \\ u_{d i_0} & \cdot & u_{d i_d} \end{vmatrix}.$$

The equations of V are then obtained by substituting

$$p_{i_{d+1} \ldots i_n} = \sum_{\nu=d+1}^{n} (-1)^\nu x_{i_\nu} q_{i_{d+1} \ldots i_{\nu-1} i_{\nu+1} \ldots i_n},$$

where $(\ldots, q_{j_{d+2} \ldots j_n}, \ldots)$ are the coordinates of a generic S_{n-d-2} of S_n; but since $G(\ldots, p_{i_{d+1} \ldots i_n}, \ldots)$ is linear and the $q_{j_{d+2} \ldots j_n}$ are linearly independent over K [VII, § 6, Th. I] we need not substitute for the $q_{j_{d+2} \ldots j_n}$ in terms of the coordinates of a basis for the S_{n-d-2}, as above, but simply equate the coefficients of distinct $q_{j_{d+2} \ldots j_n}$ to zero. The equations of V are therefore linear, and V is therefore a linear space. Since a generic S_{n-d} meets V in one point [§ 7, Th. III], this linear space is of d dimensions.

Finally, a d-space is an irreducible variety of order one. Let S_d be a d-space in S_n given by the equations

$$\sum_{j=0}^{n} a_{ij} x_j = 0 \quad (i = d+1, \ldots, n).$$

The generic S_{n-d} given by the equations

$$\sum_{j=0}^{n} u_{ij} x_j = 0 \quad (i = 1, \ldots, d)$$

meets S_{n-d} in the unique point $(\xi_0, \ldots, \xi_n)$ obtained by solving the n equations

$$\sum_{j=0}^{n} u_{ij} x_j = 0 \quad (i = 1, \ldots, d),$$

$$\sum_{j=0}^{n} a_{ij} x_j = 0 \quad (i = d+1, \ldots, n).$$

Hence S_d is of order one, and since $(\xi_0, \ldots, \xi_n)$ is a generic point, S_d is irreducible. The Cayley form of S_d is

$$F(u_0, \ldots, u_d) = A(u_1, \ldots, u_d) \sum_{i=0}^{n} u_{0i} \xi_i,$$

for a suitable $A(u_1, \ldots, u_d)$, and since it is linear in $(u_{i0}, \ldots, u_{in})$, it can be taken as

$$F(u_0, \ldots, u_d) = \begin{vmatrix} u_{00} & \cdot & u_{0n} \\ \cdot & \cdot & \cdot \\ u_{d0} & \cdot & u_{dn} \\ a_{d+1\,0} & \cdot & a_{d+1\,n} \\ \cdot & \cdot & \cdot \\ a_{n0} & \cdot & a_{nn} \end{vmatrix}.$$

If we expand this determinant by Laplace's expansion [II, §8, Th. II] we can write

$$F(u_0, \ldots, u_d) = \sum_{i_0=0}^{n} \cdots \sum_{i_n=0}^{n} u_{0i_0} \cdots u_{di_d} p^{i_{d+1} \ldots i_n},$$

where $(i_0, \ldots, i_d, i_{d+1}, \ldots, i_n)$ is an even permutation of $(0, \ldots, n)$, and $(\ldots, p^{i_{d+1} \ldots i_n}, \ldots)$ are the dual Grassmann coordinates of S_d. This can also be written [VII, §3, Th. I] in the form

$$F(u_0, \ldots, u_d) = \sum_{i_0=0}^{n} \cdots \sum_{i_d=0}^{n} p_{i_0 \ldots i_d} u_{0i_0} \cdots u_{di_d},$$

where $(\ldots, p_{i_0 \ldots i_d}, \ldots)$ are the Grassmann coordinates of S_d.

8. Further properties of the Cayley form. In the last section we showed how to obtain the equations of the variety V from its Cayley form after we had first expressed the Cayley form as a polynomial in the determinants $|u_{0i_0} \ldots u_{di_d}|$. In this section we begin with a method for obtaining the equations of V directly from its Cayley form. We then go on to find necessary and sufficient conditions that a form $F(u_0, \ldots, u_d)$ which is homogeneous of degree g in the indeterminates $u_i = (u_{i0}, \ldots, u_{in})$ $(i = 0, \ldots, d)$ should be the Cayley form of an algebraic variety of dimension d and order g.

If $F(u_0, \ldots, u_d)$ is the Cayley form of V, it follows from § 5, Lemma III, and from the properties proved in the preceding section, that a necessary and sufficient condition for the point x' to lie on V is

$$F(v_0, \ldots, v_d) = 0, \tag{1}$$

where $v_0, \ldots, v_d$ are $d+1$ generic primes through x'. We have shown that (1) is a necessary and sufficient condition for the S_{n-d-1} given by the equations

$$\sum_{j=0}^{n} v_{ij} x_j = 0 \quad (i = 0, \ldots, d)$$

to meet V, and a generic S_{n-d-1} through x' meets V if and only if x' lies on V.

By the result proved in § 4, p. 22, a generic prime through the point x' is $\sum_{i=0}^{n} v_i x_i = 0$, where

$$v_i = \sum_{j=0}^{n} s_{ij} x'_j,$$

the elements s_{ij} being independent indeterminates $(i<j)$ and satisfying the conditions for skew-symmetry:

$$s_{ij} = -s_{ji}.$$

Hence we can take $d+1$ generic primes through x' as

$$\sum_{j=0}^{n} v_{ij} x_j = 0 \quad (i = 0, \ldots, d),$$

where
$$v_{ij} = \sum_{k=0}^{n} s^i_{jk} x'_k \quad (i = 0, \ldots, d),$$

and the matrices $\quad S^i = (s^i_{jk}) \quad (i = 0, \ldots, d)$

are skew-symmetric matrices with all elements above the principal diagonals independent indeterminates.

Hence x' lies on V if and only if

$$F(S^0 x, \ldots, S^d x) \tag{2}$$

vanishes when $x = x'$. The form (2) can be expressed as a sum of power-products of the s^i_{jk}. If we write $-s^i_{kj}$ for s^i_{jk} whenever $j<k$,

the elements s^i_{jk} $(j > k)$ are independent indeterminates, and writing (2) in the form (cf. p. 46)

$$\sum_\sigma \phi_\sigma(s)\,\psi_\sigma(x),$$

we have the result
$$\psi_\sigma(x') = 0,$$
if and only if x' lies on V. Hence *the equations*

$$\psi_\sigma(x) = 0$$

are the equations of the variety V.

If V is reducible, and its irreducible components are $V_1, \ldots, V_k$, so that
$$V = V_1 \dotplus V_2 \dotplus \ldots \dotplus V_k,$$
the dimension of V is defined to be

$$d = \max(d_1, \ldots, d_k),$$

where d_i is the dimension of V_i. If $d_1 = \ldots = d_k = d$, V is said to be a *pure* or *unmixed* variety of dimension d. In this case we define a Cayley form for V.

Let the Cayley form of V_i be $F_i(u_0, \ldots, u_d)$. Then

$$F(u_0, \ldots, u_d) = \prod_{i=1}^{k} [F_i(u_0, \ldots, u_d)]^{\rho_i},$$

where the ρ_i are any integers greater than zero, is called a Cayley form of V. Significance will be given to the integers ρ_i in a later chapter [XI], but at present we merely take them to be any positive integers. If g_i is the order of V_i, and therefore the degree of $F_i(u_0, \ldots, u_d)$ in each set of indeterminates $u_0, \ldots, u_d$, it is convenient to say that the order of the variety given by $F(u_0, \ldots, u_d)$ defined above is

$$g = \sum_{i=1}^{k} \rho_i g_i.$$

A justification of this definition will be given in Chapter XI.

We now consider the problem of determining the conditions which must be satisfied by a form $F(u_0, \ldots, u_d)$, homogeneous of degree g in each set of indeterminates $u_0, \ldots, u_d$, in order that it should be the Cayley form of a variety.

From the results of § 7 we see that the following three conditions are necessary:

1. $F(u_0, u_1, \ldots, u_d)$, regarded as a form in $(u_{00}, \ldots, u_{0n})$, can be

completely decomposed into a product of linear factors in some extension field of $K(u_1, \ldots, u_d)$, so that

$$F(u_0, u_1, \ldots, u_d) = A(u_1, \ldots, u_d) \prod_{\rho=1}^{g} \left(\sum_{i=0}^{n} u_{0i} \xi_i^{(\rho)} \right). \tag{3}$$

2. The points $\xi^{(\rho)} = (\xi_0^{(\rho)}, \ldots, \xi_n^{(\rho)})$ defined by (3) lie in each of the primes $u_1, \ldots, u_d$; that is,

$$\sum_{j=0}^{n} u_{ij} \xi_j^{(\rho)} = 0 \quad (i = 1, \ldots, d;\ \rho = 1, \ldots, g).$$

3. If $S^0, \ldots, S^d$ are skew-symmetric matrices, each having independent indeterminate elements above its principal diagonal, then

$$F(S^0 \xi^{(\rho)}, \ldots, S^d \xi^{(\rho)}) = 0 \quad (\rho = 1, \ldots, g),$$

that is, the points $\xi^{(\rho)}$ satisfy the equations of V. An equivalent condition is: if the primes $v_0, \ldots, v_d$ all pass through $\xi^{(\rho)}$, then $F(v_0, \ldots, v_d) = 0$.

We now show that these three conditions are sufficient.

Let V_ρ $(\rho = 1, \ldots, g)$ be the irreducible variety *defined over the ground field* K which has $\xi^{(\rho)}$ as generic point over K [§ 3, Th. IV]. The varieties V_ρ are not necessarily distinct. By condition 2, the points $\xi^{(1)}, \ldots, \xi^{(g)}$ lie in the S_{n-d} defined by the primes $u_1, \ldots, u_d$. We show that condition 3 implies that this S_{n-d} contains no other generic points of $V_1, \ldots, V_g$.

Let us suppose, in fact, that the S_{n-d} contains a generic point η of V_1 which is not $\xi^{(1)}$. Without any essential loss of generality we may assume that $\xi_0^{(1)} = \eta_0 = 1$. Since η and $\xi^{(1)}$ are both generic points of V_1 there is an isomorphism

$$K(\eta) \cong K(\xi^{(1)})$$

which maps η_i on $\xi_i^{(1)}$ $(i = 1, \ldots, n)$, leaving all elements of K unchanged [§ 3, Th. III]. This isomorphism may be extended [I, § 2, Th. I] to an isomorphism

$$K(\eta, u_1, \ldots, u_d) \cong K(\xi^{(1)}, w_1, \ldots, w_d).$$

The relations
$$\sum_{j=0}^{n} u_{ij} \eta_j = 0 \quad (i = 1, \ldots, d),$$

which express the fact that η lies in S_{n-d}, imply the relations

$$\sum_{j=0}^{n} w_{ij} \xi_j^{(1)} = 0 \quad (i = 1, \ldots, d).$$

If now w_0 is any further prime through $\xi^{(1)}$, so that

$$\sum_{j=0}^{n} w_{0j}\xi_j^{(1)} = 0,$$

condition 3 tells us that

$$F(w_0, w_1, \ldots, w_d) = 0.$$

Replacing w_0 by the set of indeterminates u_0, it follows that

$$F(u_0, w_1, \ldots, w_d)$$

vanishes for all specialisations $u_0 = w_0$ such that $\sum_{j=0}^{n} w_{0j}\xi_j^{(1)} = 0$. Hence [IV, § 8], $F(u_0, w_1, \ldots, w_d)$ contains $\sum_{j=0}^{n} u_{0j}\xi_j^{(1)}$ as a factor.

Reversing the isomorphism, it follows that $F(u_0, u_1, \ldots, u_d)$ contains $\sum_{j=0}^{n} u_{0j}\eta_j$ as a factor. It then follows from condition 1 that η must be one of the points $\xi^{(1)}, \ldots, \xi^{(g)}$. We conclude that the only *generic* points of any V_i which lie in the S_{n-d} defined by $u_1, \ldots, u_d$ are included in the set $\xi^{(1)}, \ldots, \xi^{(g)}$.

We can now deduce that each of the varieties $V_1, \ldots, V_g$ is of dimension d, and that $F(u_0, \ldots, u_d)$ is the Cayley form of the sum of these varieties. Since V_i is a variety defined over K, and the S_{n-d} defined by $u_1, \ldots, u_d$ is generic over this field, S_{n-d} would have no intersection with V_i if $d_i = \dim V_i$ were less than d. But $\xi^{(i)}$ is a point of V_i in this S_{n-d}. Hence $d_i \geqslant d$. Suppose that $d_i > d$. Then by § 7, Th. I, a generic S_{n-d_i} meets V_i in a set of points each of which is a generic point of V_i. The section of S_{n-d} by a generic S_{n-d_i+d} is a generic S_{n-d_i}. Hence in any generic S_{n-d_i} of S_{n-d} there is a generic point of V_i. But there is only a finite number of generic points of V_i in S_{n-d}, as we saw above. None of these lies in a generic S_{n-d_i}. Hence we cannot have $d_i > d$, and so $d_i = d$.

Each V_i is therefore of dimension d. The generic S_{n-d} determined by the primes $u_1, \ldots, u_d$ meets V_i only in generic points, and these are therefore contained in the set $\xi^{(i)}$. Hence the Cayley form of V_i is the form

$$F_i = B_i(u_1, \ldots, u_d) \prod_{\rho=1}^{g} \left(\sum_{j=0}^{n} u_{0j}\xi_j^{(\rho)}\right)^{\eta_{i\rho}},$$

where $\eta_{i\rho} = 1$ or 0 according as $\xi^{(\rho)}$ is or is not on V_i. Since each point $\xi^{(\rho)}$ is on at least one of the varieties V_i, any Cayley form of

$$V_1 \dot{+} V_2 \dot{+} \ldots \dot{+} V_g$$

is of the form

$$G(u_0, \ldots, u_d) = B(u_1, \ldots, u_d) \prod_{\rho=1}^{g} \left(\sum_{j=0}^{n} u_{0j} \xi_j^{(\rho)} \right)^{\sigma_\rho} = \prod_{i=1}^{g} F_i^{h_i},$$

where $\sigma_\rho > 0$ and $h_i > 0$. Since $F(u_0, \ldots, u_d)$ and $G(u_0, \ldots, u_d)$ have the same factors $\sum_{j=0}^{n} u_{0j} \xi_j^{(\rho)}$, it follows easily that the h_i can be so chosen that $F(u_0, \ldots, u_d) = G(u_0, \ldots, u_d)$.

As a corollary to the above, we have

THEOREM I. *The intersection of an irreducible variety V of dimension $d > 0$ over K with a generic prime $\sum_{j=0}^{n} u_j x_j = 0$ is a variety which is irreducible and of dimension $d-1$ over the field $K(u)$.*

Let $F(u_0, u_1, \ldots, u_{d-1}, u_d)$ be the Cayley form of V, and writing $u_d = u$, let us adjoin u to the ground field K, so that we now consider the variety, if there is one, defined by

$$F(u_0, \ldots, u_{d-1}, u) = G(u_0, \ldots, u_{d-1})$$

over the ground field $K' = K(u)$.

Since $F(u_0, \ldots, u_{d-1}, u)$ is irreducible in the ring $K[u_0, \ldots, u_{d-1}, u]$, it is also irreducible in $K[u][u_0, \ldots, u_{d-1}]$, and therefore in the ring $K(u)[u_0, \ldots, u_{d-1}] = K'[u_0, \ldots, u_{d-1}]$. This follows from I, § 8, Lemma 3. If $G(u_0, \ldots, u_{d-1})$ defines a variety over K', this variety will therefore be irreducible over K'.

Condition 1 is satisfied, since

$$F(u_0, \ldots, u_{d-1}, u) = G(u_0, \ldots, u_{d-1}) = A(u_1, \ldots, u_{d-1}, u) \prod_{\rho=1}^{g} \left(\sum_{i=0}^{n} u_{0i} \xi_i^{(\rho)} \right),$$

so that, over a suitable extension of $K'(u_1, \ldots, u_{d-1})$,

$$G(u_0, \ldots, u_{d-1}) = B(u_1, \ldots, u_{d-1}) \prod_{\rho=1}^{g} \left(\sum_{i=0}^{n} u_{0i} \xi_i^{(\rho)} \right).$$

Condition 2 is satisfied, since the points

$$\xi^{(\rho)} \quad (\rho = 1, \ldots, g)$$

lie in the primes $u_1, \ldots, u_{d-1}$. That is

$$\sum_{j=0}^{n} u_{ij} \xi_j^{(\rho)} = 0 \quad (i = 1, \ldots, d-1;\ \rho = 1, \ldots, g).$$

Condition 3 is satisfied, for the prime $\sum_{j=0}^{n} u_j x_j = 0$ contains $\xi^{(\rho)}$

[condition 2 for V], and if the primes $\sum_{j=0}^{n} v_{ij} x_j = 0$ $(i = 0, 1, \ldots, d-1)$ all contain $\xi^{(\rho)}$, then

$$F(v_0, v_1, \ldots, v_{d-1}, u) = G(v_0, \ldots, v_{d-1}) = 0.$$

Hence $G(u_0, \ldots, u_{d-1})$ defines an irreducible variety of dimension $d-1$ and order g over $K' = K(u)$. A generic point $\xi^{(\rho)}$ of this variety lies in $\sum_{j=0}^{n} u_j x_j = 0$ and is also generic on V. The variety is therefore contained in the intersection of V with this prime.

We now use the property that the S_{n-d-1} given by the equations

$$\left.\begin{array}{l} \sum_{j=0}^{n} u_{ij} x_j = 0 \\ \sum_{j=0}^{n} u_j x_j = 0 \end{array}\right\} \quad (i = 0, \ldots, d-1)$$

meets V if and only if

$$F(u_0, \ldots, u_{d-1}, u) = 0.$$

This implies that the S_{n-d} given by

$$\sum_{j=0}^{n} u_{ij} x_j = 0 \quad (i = 0, \ldots, d-1)$$

meets the intersection of V and the prime given by

$$\sum_{j=0}^{n} u_j x_j = 0$$

if and only if $\qquad F(u_0, \ldots, u_{d-1}, u) = 0.$

Therefore $F(u_0, \ldots, u_{d-1}, u)$ is the Cayley form of the intersection of V and the prime given by $\sum_{j=0}^{n} u_j x_j = 0$.

Another proof of this theorem will be given in the next chapter [XI, § 5, Th. II].

If we consider the intersection of V with a specialised prime π given by the equation $\sum_{j=0}^{n} v_j x_j = 0$, where $v_0, \ldots, v_n$ are either in K or in an extension of K, we obtain the

Corollary. The Cayley form of the intersection of V and the prime π given by $\sum_{j=0}^{n} v_j x_j = 0$ is $F(u_0, \ldots, u_{d-1}, v)$, where $F(u_0, \ldots, u_{d-1}, u_d)$

is the Cayley form of V, *provided that the intersection is purely* $(d-1)$*-dimensional.*

We conclude this section by proving, in the first place, that the conditions 1, 2, 3 for $F(u_0, \ldots, u_d)$ to be the Cayley form of a variety V can be expressed as a set of homogeneous algebraic equations to be satisfied by the coefficients a'_λ of F.

We suppose that $F(u_0, \ldots, u_d)$ is a form of degree g in the indeterminates $u_i = (u_{i0}, \ldots, u_{in})$ $(i = 0, \ldots, d)$ with unknown coefficients a_λ $(\lambda = 0, \ldots, D)$. Consider the equation

$$F(u_0, \ldots, u_d) = A(u_1, \ldots, u_d) \prod_{\rho=1}^{g} \left(\sum_{i=0}^{n} u_{0i} \xi_i^{(\rho)} \right),$$

and compare the coefficients on each side of the power-products of the u_{0i}. We obtain a set of equations

$$\phi_\nu(u_1, \ldots, u_d) = A(u_1, \ldots, u_d) \psi_\nu(\xi^{(1)}, \ldots, \xi^{(g)}),$$

and eliminating the $A(u_1, \ldots, u_d)$, we obtain the equations

$$(1') \qquad \phi_\mu(u_1, \ldots, u_d) \psi_\nu(\xi^{(1)}, \ldots, \xi^{(g)}) - \phi_\nu(u_1, \ldots, u_d) \psi_\mu(\xi^{(1)}, \ldots, \xi^{(g)}) = 0.$$

The conditions 2 are

$$(2') \qquad \sum_{j=0}^{n} u_{ij} \xi_j^{(\rho)} = 0 \quad (i = 1, \ldots, d;\ \rho = 1, \ldots, g).$$

The conditions 3 can be expressed by equating to zero the coefficients of the various power-products of the indeterminates s^i_{jk}. This gives a set of equations

$$(3') \qquad \chi_\mu(a_\lambda, \xi^{(\rho)}) = 0 \quad (\rho = 1, \ldots, g).$$

From the set of equations (1′), (2′), (3′), which are all homogeneous in the $\xi^{(\rho)}$, eliminate $\xi^{(1)}, \ldots, \xi^{(g)}$, forming the resultant system

$$R_\nu(a_\lambda, u_1, \ldots, u_d).$$

Let the coefficients of the power-products of the u_{ij} in the set $R_\nu(a_\lambda, u_1, \ldots, u_d)$ be the forms

$$T_\omega(a_\lambda) \quad (\omega = 1, 2, \ldots).$$

In order that $F(u_0, \ldots, u_d)$, with specialised coefficients a'_λ in K, should be the Cayley form of an algebraic variety we must have

$$T_\omega(a'_\lambda) = 0. \qquad (4)$$

Conversely, if a'_λ ($\lambda = 0, \dots, D$) is any set of elements such that the set of homogeneous equations (4) is satisfied, then

$$R_\nu(a'_\lambda, u_1, \dots, u_d) = 0,$$

and the equations (1′), (2′), (3′) can be solved for $\xi^{(1)}, \dots, \xi^{(g)}$. Hence conditions 2 and 3 are satisfied. Not all $\psi_\nu(\xi^{(1)}, \dots, \xi^{(g)})$ can be zero, since each is of the form $\xi^{(1)}_{i_1} \dots \xi^{(g)}_{i_g}$, and for some choice of $i_1, \dots, i_g$ this is not zero. We can therefore find $A'(u_1, \dots, u_d)$ so that

$$\phi_\nu(u_1, \dots, u_d) = A'(u_1, \dots, u_d)\, \psi_\nu(\xi^{(1)}, \dots, \xi^{(g)}),$$

and then $$F(u_0, \dots, u_d) = A'(u_1, \dots, u_d) \prod_{\rho=1}^{g} \left(\sum_{i=0}^{n} u_{0i} \xi_i^{(\rho)} \right),$$

and condition 1 is therefore satisfied. We have therefore proved

THEOREM II. *Necessary and sufficient conditions for a form $F(u_0, \dots, u_d)$ of degree g in each set of indeterminates u_i ($i = 0, \dots, d$) and with coefficients a'_λ to be the Cayley form of a variety of dimension d and order g are given by*

$$T_\omega(a'_\lambda) = 0 \quad (\omega = 1, 2, \dots),$$

where the $T_\omega(z_\lambda)$ are homogeneous forms in the indeterminates z_λ.

If the coefficients of $F(u_0, \dots, u_d)$ are $a'_0, \dots, a'_D$, we represent the algebraic variety of dimension d and order g whose Cayley form is $F(u_0, \dots, u_d)$ by the point $(a'_0, \dots, a'_D)$ of a projective space of D dimensions. All such points lie on the algebraic variety Ω whose equations are

$$T_\omega(x_\lambda) = 0 \quad (\omega = 1, 2, \dots).$$

By the theorem we have just proved, there is a one-to-one correspondence between the points of Ω and the algebraic varieties of dimension d and order g in S_n. For this reason we say that *the algebraic varieties of dimension d and order g in S_n form an algebraic system.*

The set of varieties in S_n which correspond to any subvariety of Ω is said to form an algebraic system. If the subvariety is irreducible, the system of corresponding varieties in S_n is said to be irreducible. If $(\xi_0, \dots, \xi_D)$ is a generic point of the subvariety, the variety in S_n whose Cayley form $F(u_0, \dots, u_d)$ has coefficients $\xi_0, \dots, \xi_D$ is called a generic variety of the irreducible system of varieties in S_n.

We can generalise the foregoing proof to show that the algebraic varieties of order g and dimension d which lie on the variety U whose equations are

$$f_i(x) = 0 \quad (i = 1, \ldots, r)$$

form an algebraic system. Indeed, necessary and sufficient conditions that $F(u_0, \ldots, u_d)$ be the Cayley form of a variety lying on U are:

$$(1) \qquad F(u_0, \ldots, u_d) = A(u_1, \ldots, u_d) \prod_{\rho=1}^{g} \left(\sum_{i=0}^{n} u_{0i} \xi_i^{(\rho)} \right).$$

$$(2) \qquad \sum_{j=0}^{n} u_{ij} \xi_j^{(\rho)} = 0 \quad (i = 1, \ldots, d;\ \rho = 1, \ldots, g).$$

$$(3) \qquad F(S^0 \xi^{(\rho)}, \ldots, S^d \xi^{(\rho)}) = 0 \quad (\rho = 1, \ldots, g).$$

$$(4) \qquad f_i(\xi^{(\rho)}) = 0 \quad (i = 1, \ldots, r;\ \rho = 1, \ldots, g).$$

The same argument as above shows that necessary and sufficient conditions are obtained by eliminating $\xi^{(1)}, \ldots, \xi^{(g)}$ from these equations. It follows that the conditions are algebraic in the coefficients of $F(u_0, \ldots, u_d)$.

9. The order of an algebraic variety; parametrisation. The order g of an irreducible variety V of dimension d over K has been defined by means of the Cayley form of V. It has the properties:

(i) the Cayley form $F(u_0, \ldots, u_d)$ of V is homogeneous of degree g in $(u_{i0}, \ldots, u_{in})$, for $i = 0, \ldots, d$;

(ii) a generic $(n-d)$-space of S_n meets V in g points.

We now study another property of the order which will prove useful later.

Let us consider the various ways we have had of writing the Cayley form:

$$\begin{aligned} f(u_{ij};\ u_{00}, \ldots, u_{d0}) &= F(u_0, \ldots, u_d) \\ &= A(u_1, \ldots, u_d) \prod_{\tau=1}^{g} \left(u_{00} + \sum_{\rho=1}^{n} u_{0\rho} \xi_\rho^{(\tau)} \right) \\ &= G(\ldots, |u_{0i_0} \ldots u_{di_d}|, \ldots). \qquad (1) \end{aligned}$$

The last of these is a homogeneous polynomial of degree g in the determinants

$$|u_{0i_0} \ldots u_{di_d}| = \begin{vmatrix} u_{0i_0} & \cdot & u_{0i_d} \\ \cdot & \cdot & \cdot \\ u_{di_0} & \cdot & u_{di_d} \end{vmatrix},$$

and hence if we compare the coefficients of u_{00}^g in the third and fourth forms we see that $A(u_1, \ldots, u_d)$ is a polynomial in $u_{11}, u_{21}, \ldots, u_{dn}$. More generally, the last expression shows that $f(u_{ij};\ u_{00}, \ldots, u_{d0})$ is of *total degree* g in $u_{00}, \ldots, u_{d0}$, that is, if $a(u_{ij})\, u_{00}^{\alpha_0} \ldots u_{d0}^{\alpha_d}$ is any term of $f(u_{ij};\ u_{00}, \ldots, u_{d0})$, then

$$\alpha_0 + \ldots + \alpha_d \leqslant g,$$

and there are terms for which the sum is exactly g (in particular, there are non-zero terms in u_{i0}^g $(i = 0, \ldots, d)$).

Now consider any specialisation $u_{ij} \to a_{ij}$, where a_{ij} is an element of K, for $j = 1, \ldots, n;\ i = 0, \ldots, d$. If we write

$$-\eta_i = a_{i1}\xi_1 + \ldots + a_{in}\xi_n \quad (i = 0, \ldots, d),$$

we can deduce, as in Chapter IV, § 9, Th. I, a relation

$$\phi(\eta_0, \ldots, \eta_d) = 0 \tag{2}$$

over K, connecting $\eta_0, \ldots, \eta_d$, from the relation [§ 6, (2)]

$$f(u_{ij};\ \zeta_0, \ldots, \zeta_d) = 0.$$

The relation (2) is of total degree g at most in $\eta_0, \ldots, \eta_d$. We consider, in particular, the case in which d of the elements $\eta_0, \ldots, \eta_d$ of the function field $K(\xi_1, \ldots, \xi_n)$ of V are algebraically independent over K. In this case there is a unique irreducible equation over K which is satisfied by $\eta_0, \ldots, \eta_d$, say

$$\psi(\eta_0, \ldots, \eta_d) = 0.$$

Then $\phi(y_0, \ldots, y_d)$ must contain $\psi(y_0, \ldots, y_d)$ as a factor [III, § 3, Th. VI]. Therefore $\psi(y_0, \ldots, y_d)$ is of total degree g at most.

We now show that there exist in K specialisations a_{ij} of u_{ij} such that the irreducible relation connecting $\eta_0, \ldots, \eta_d$ is of total degree g exactly. We require a lemma:

Lemma. If K is any field without characteristic, K^* a simple algebraic extension of K of degree n, and if u is an indeterminate over K^*, and hence over K, then $K^*(u)$ is a simple algebraic extension of $K(u)$ of degree n.

Let α be a primitive element of K^* over K [III, § 6], and let $f(x)$ be its characteristic polynomial over K. $f(x)$ is of order n. Now $K^*(u) = K(\alpha, u)$, and hence α is a primitive element of $K^*(u)$ over $K(u)$. Since α satisfies the equation

$$f(x) = 0,$$

whose coefficients are in $K(u)$, the characteristic polynomial of α over $K(u)$ is of order n at most. Suppose that it is of order m, where $m<n$. Then we have a relation

$$b_0(u)\alpha^m+\ldots+b_m(u)=0,$$

where $b_i(u)$ is in $K[u]$. Since u is indeterminate over K^*, this relation persists when we specialise u. We choose any specialisation $u\to a$ (a in K) of u such that $b_0(a)$ and $b_m(a)$ are not zero. Then we obtain an equation of degree $m<n$ over K satisfied by α. This is impossible, since the characteristic polynomial of α over K is of degree n. From this contradiction the lemma follows.

By repeated applications of the lemma we see that if $u_1, \ldots, u_r$ are independent indeterminates over K^*, $K^*(u_1, \ldots, u_r)$ is an algebraic extension of $K(u_1, \ldots, u_r)$ of degree n.

We now return to our main problem. Let K_r denote the field obtained by adjoining u_{ij} ($i=0, \ldots, r$; $j=1, \ldots, n$) to K. Repeating an argument used above, we find that the irreducible equation over K_{d-1} connecting $\zeta_0, \ldots, \zeta_{d-1}$, and $\sum_1^n v_i\xi_i$, where $v_1, \ldots, v_n$ are in K_{d-1}, is of total degree g at most. It follows that $K_{d-1}(\xi_1, \ldots, \xi_n)$ is an algebraic extension of $K_{d-1}(\zeta_0, \ldots, \zeta_{d-1})$ of degree g', where $g'\leqslant g$. By the deduction made above from the lemma, we see that $K_d(\xi_1, \ldots, \xi_n)$ is an extension of $K_d(\zeta_0, \ldots, \zeta_{d-1})$ of degree g'. But the irreducible equation over K_d connecting $\zeta_0, \ldots, \zeta_{d-1}$ with $\zeta_d=\sum_1^n u_{di}\xi_i$ is of degree g in ζ_d. Hence $g'=g$. Now $\sum_1^n v_i\xi_i$ is a primitive element of $K_{d-1}(\xi_1, \ldots, \xi_n)$ over $K_{d-1}(\zeta_0, \ldots, \zeta_{d-1})$, provided that the ratios of $v_1, \ldots, v_n$ are chosen to avoid at most a finite number of special values. This can be seen at once by a simple extension of the argument used in the theorem of the primitive element [III, §6, Th. I]. Since K contains an infinite number of elements, we can choose $v_i=a_{di}$ in K so that the irreducible equation connecting $\zeta_0, \ldots, \zeta_{d-1}$ and $\eta_d=\sum_1^n a_{di}\xi_i$ over K_{d-1} is of order g in η_d. We can at the same time ensure that the coefficient of u_{i0}^g in $f(u_{ij}; u_{00}, \ldots, u_{d0})$ does not vanish for the specialisation, for $i=0, \ldots, d-1$. For this specialisation $f(u_{ij}; z_0, \ldots, z_d)$ becomes a non-zero polynomial $f(u'_{ij}; z_0, \ldots, z_d)$ of total degree g in $z_0, \ldots, z_d$, containing terms in z_i^g ($i=0, \ldots, d$) and such that

$$f(u_{ij}; \zeta_0, \ldots, \zeta_{d-1}, \eta_d)=0.$$

Hence if $\psi(z_0, \ldots, z_d)$ is the irreducible polynomial over K_{d-1} such that $\psi(\zeta_0, \ldots, \zeta_{d-1}, \eta_d) = 0$,

$$f(u'_{ij}; z_0, \ldots, z_d) = A(z_0, \ldots, z_d)\,\psi(z_0, \ldots, z_d).$$

Since, however, ψ is of degree g in z_d, f is of total degree g, at least, in $z_0, \ldots, z_d$. But f is of total degree g at most. Hence A is in K_{d-1}. Thus the relation connecting $\zeta_0, \ldots, \zeta_{d-1}, \eta_d$ is of total degree g exactly. Since this relation contains z_i^g $(i = 1, \ldots, d-1)$, we can repeat the argument, passing from K_{d-1} to K_{d-2}, and so on. Eventually we arrive at linear expressions $\eta_0, \ldots, \eta_d$ in ξ_i with coefficients in K such that the irreducible relation over K connecting them is of total degree g. Thus we have

THEOREM I. *If V is of order g, and $(1, \xi_1, \ldots, \xi_n)$ is a normalised generic point of it, then any $d+1$ elements*

$$-\eta_i = a_{i1}\xi_1 + \ldots + a_{in}\xi_n \quad (i = 0, \ldots, d)$$

of $K(\xi_1, \ldots, \xi_n)$ satisfy an equation

$$\psi(\eta_0, \ldots, \eta_d) = 0$$

of total degree g at most, and there exist sets $\eta_0, \ldots, \eta_d$ such that the irreducible equation connecting them is of total degree g exactly.

It is clear that this property characterises the order g of V.

The argument given to establish Theorem I implies that if $-\eta_i = \sum_{j=1}^{n} a_{ij}\xi_j$ $(i = 1, \ldots, d)$ are d algebraically independent elements of $K(\xi_1, \ldots \xi_n)$, and if $u_1, \ldots, u_n$ are indeterminates over $K(\xi_1, \ldots, \xi_n)$, the unique irreducible polynomial $\psi(u, y_1, \ldots, y_d, z)$ in $K[u_1, \ldots, u_n; y_1, \ldots, y_d, z]$ connecting $\eta_1, \ldots, \eta_d$, and $\zeta = \sum_{1}^{n} u_i\xi_i$ is of total degree g at most in $y_1, \ldots, y_d, z$. ψ is sometimes called the *norm* of V with respect to $\eta_1, \ldots, \eta_d$, by analogy with the norm which is introduced in the theory of algebraic extensions. It can be regarded as a specialised form of the Cayley form, from which its properties can all be deduced. Sometimes, however, it is convenient to deal with the norm directly.

The argument leading to Theorem I shows that we can choose a_{ij} in K so that the irreducible equation connecting $\eta_i = -\sum_{j=1}^{n} a_{ij}\xi_j$ $(i = 0, \ldots, d)$ is

$$f(a_{ij}; y_0, \ldots, y_d) = 0,$$

where $f(u_{ij}; u_{00}, \ldots, u_{d0})$ is the Cayley form of V. By making a suitable change of coordinate system in S_n, we can arrange that $\xi_1, \ldots, \xi_d$ are algebraically independent over K and that the irreducible equation satisfied by $\xi_1, \ldots, \xi_{d+1}$ is

$$\psi(x_1, \ldots, x_{d+1}) \equiv f(b_{ij}; x_1, \ldots, x_{d+1}) = 0,$$

where $b_{ii+1} = 1$ $(i = 0, \ldots, d)$ and b_{ij} is zero otherwise. We suppose this done. It then follows at once that the norm of V with respect to $\xi_1, \ldots, \xi_d$ is obtained by replacing u_{ii+1} by 1 $(i = 0, \ldots, d-1)$, u_{ij} by zero $(i = 0, \ldots, d-1; j \neq i+1)$, and u_{di} by u_i in

$$f(u_{ij}; x_1, \ldots, x_d, z).$$

From the equation [§ 6, (5)]

$$\frac{\partial}{\partial u_{d\sigma}} f(u_{ij}, \zeta_0, \ldots, \zeta_d) - \xi_\sigma \frac{\partial}{\partial \zeta_d} f(u_{ij}; \zeta_0, \ldots, \zeta_d) = 0,$$

we obtain, on specialising $f(u_{ij}; x_1, \ldots, x_{d+1})$ to $\psi(x_1, \ldots, x_{d+1})$,

$$\psi_\sigma(\xi_1, \ldots, \xi_{d+1}) - \xi_\sigma \frac{\partial}{\partial \xi_{d+1}} \psi(\xi_1, \ldots, \xi_{d+1}) = 0,$$

where $\psi_\sigma(x_1, \ldots, x_{d+1})$ is some polynomial of degree g at most over K. If $\phi(x_0, \ldots, x_{d+1})$ is the homogeneous polynomial of degree g such that

$$\phi(1, x_1, \ldots, x_{d+1}) = \psi(x_1, \ldots, x_{d+1}),$$

and $\qquad \phi_\sigma(x_0, x_1, \ldots, x_{d+1}) \quad (\sigma = d+1, \ldots, n)$

are homogeneous polynomials of degree g such that

$$\phi_\sigma(1, x_1, \ldots, x_{d+1}) = \psi_\sigma(x_1, \ldots, x_{d+1}),$$

it follows that V satisfies the equations

$$\phi(x_0, \ldots, x_{d+1}) = 0, \tag{3}$$

$$x_\sigma \frac{\partial \phi}{\partial x_{d+1}} = \phi_\sigma(x_0, \ldots, x_{d+1}) \quad (\sigma = d+1, \ldots, n). \tag{4}$$

These equations therefore define a variety which contains V. We now show that if x' is any point not in $x_0 = 0$ which satisfies these equations, and if $\partial\psi/\partial x'_{d+1} \neq 0$, then x' is on V. We may suppose that $x'_0 = 1$. Let

$$h(x_0, \ldots, x_n) = 0$$

be any equation satisfied by V. Then

$$h(1, \xi_1, \ldots, \xi_n) = 0,$$

and hence

$$h\left(1, \xi_1, \ldots, \xi_d, \frac{\psi_{d+1}(\xi)}{\dfrac{\partial\psi}{\partial\xi_{d+1}}}, \ldots, \frac{\psi_n(\xi)}{\dfrac{\partial\psi}{\partial\xi_{d+1}}}\right) = 0.$$

Let t be an integer such that

$$\left[\frac{\partial\psi}{\partial x_{d+1}}\right]^t h\left(1, x_1, \ldots, x_d, \frac{\psi_{d+1}(x)}{\dfrac{\partial\psi}{\partial x_{d+1}}}, \ldots, \frac{\psi_n(x)}{\dfrac{\partial\psi}{\partial x_{d+1}}}\right) = \chi(x_1, \ldots, x_{d+1})$$

is a polynomial. Then it follows that

$$\chi(\xi_1, \ldots, \xi_{d+1}) = 0.$$

Hence $\chi(x_1, \ldots, x_{d+1})$ contains $\psi(x_1, \ldots, x_{d+1})$ as a factor, that is,

$$\chi(x_1, \ldots, x_{d+1}) \equiv B(x_1, \ldots, x_{d+1})\, \psi(x_1, \ldots, x_{d+1}).$$

Hence, since x' satisfies (3),

$$\psi(x'_1, \ldots, x'_{d+1}) = 0,$$

and therefore

$$h\left(1, x'_1, \ldots, x'_d, \frac{\psi_{d+1}(x')}{\dfrac{\partial\psi}{\partial x'_{d+1}}}, \ldots, \frac{\psi_n(x')}{\dfrac{\partial\psi}{\partial x'_{d+1}}}\right) = 0,$$

that is, $$h(x'_0, \ldots, x'_n) = 0.$$

Hence x' satisfies every equation satisfied by V, and it is therefore on V.

The equations (3) and (4) give what is called a *parametric* or *monoidal* representation of V.

10. Some algebraic lemmas. Before we deal with our next topic it will be useful to restate some of the algebraic results of Chapter III, and to add results which will be needed at a later stage. We saw that if ξ is algebraic over the ground field K, which we assume to be without characteristic, there exists an irreducible polynomial $f(x)$ in $K[x]$, unique save for a non-zero factor in K, such that $f(\xi) = 0$. If $g(x)$ is any polynomial in $K[x]$ such that $g(\xi) = 0$, then $g(x)$ is divisible by $f(x)$. This polynomial $f(x)$ is called the *characteristic polynomial* of ξ over K. If we suppose $f(x)$ normalised so that its leading coefficient is unity, we shall call it the *minimum function* of ξ over K. We shall also call the field $K(\xi)$ a *stem field.* If ξ is a root of the equation $F(x) = 0$, we say that $K(\xi)$ is *a stem* *field of* $F(x)$. In III, § 2, Th. II, we proved that the stem fields of an irreducible polynomial are all equivalent over K. We also introduced the root field of a polynomial $F(x)$. Any field which is the root field of an irreducible polynomial over K is said to be a *normal extension of* K.

If ξ and η are two elements of an extension K^* of K which are both algebraically dependent over K, they are said to be *conjugate* over K if they have the same minimum function over K. By III, § 2, Th. II, it follows that ξ and η are conjugate if and only if $K(\xi)$ and $K(\eta)$ are equivalent over K in such a way that ξ and η are corresponding elements.

If K_1 and K_2 are two algebraic extensions of K which are contained in a field K^* they are said to be conjugate over K if they are equivalent extensions of K. In particular, two stem fields of an irreducible polynomial are conjugate over K. Conversely, if K_1 is a stem fieid $K(\xi_i)$ of an irreducible polynomial $F(x)$ over K, and K_2 is conjugate to K_1 over K, then K_2 is also a stem field of $F(x)$. For, let ξ be the element of K_2 corresponding to ξ_i in the isomorphism between K_1 and K_2. Then $K_2 = K(\xi)$. Since $F(\xi_i)$ is mapped on $F(\xi)$, and $F(\xi_i) = 0$, we have $F(\xi) = 0$. Hence ξ_i and ξ are conjugate over K, and K_2 is a stem field of $F(x)$.

Let $F(x)$ be an irreducible polynomial of degree n over $K, \xi_1, \ldots, \xi_n$ the roots of $F(x)$ in a suitable extension of K, and let

$$\alpha_i = a_0 + a_1\xi_i + \ldots + a_{n-1}\xi_i^{n-1} \quad (a_i \text{ in } K) \quad (i = 1, \ldots, n).$$

Let
$$\phi(y) = (y-\alpha_1)\ldots(y-\alpha_n).$$

Then by the theory of symmetric functions [III, § 4, Th. VI], we see that $\phi(y)$ has coefficients in K. Let $\psi(y)$ be the minimum function of α_i over K. Since

$$\psi(\alpha_i) = \psi(a_0 + a_1\xi_i + \ldots + a_{n-1}\xi_i^{n-1}) = 0,$$

$\psi(a_0 + a_1 x + \ldots + a_{n-1}x^{n-1})$ must contain $F(x)$ as a factor and hence $\psi(\alpha_j) = 0$ $(j = 1, \ldots, n)$, that is, $\alpha_1, \ldots, \alpha_n$ are conjugate over K. Since $\phi(\alpha_i) = 0$, $\phi(x)$ must contain $\psi(x)$ as a factor. Suppose that $\phi(x) = [\psi(x)]^q \chi(x)$, where $q \geqslant 1$, and $\chi(x)$ is prime to $\psi(x)$. Then $\chi(\alpha_i) \neq 0$ for any i. If β is a root of $\chi(x)$ in some extension of $K(\xi_1, \ldots, \xi_n)$, $\phi(\beta) = 0$, that is

$$(\beta-\alpha_1)\ldots(\beta-\alpha_n) = 0.$$

Hence $\beta = \alpha_i$, and it follows that any root of $\chi(x)$ is a root of $\psi(x)$, and hence that $\chi(x)$ contains $\psi(x)$ as a factor, or else $\chi(x)$ has no roots. Since we have arranged that $\chi(x)$ is prime to $\psi(x)$ it follows that $\chi(x)$ is of order zero, and by equating coefficients we find that $\chi(x) = 1$. Hence we have

Lemma I. If $\xi_1, \ldots, \xi_n$ are the roots of the polynomial $F(x)$, which is irreducible over K, and

$$\alpha_i = a_0 + a_1\xi_i + \ldots + a_{n-1}\xi_i^{n-1} \quad (i = 1, \ldots, n) \quad (a_i \text{ in } K),$$

then
$$\prod_{i=1}^{n} (y - \alpha_i) = [\psi(x)]^q,$$

where $\psi(x)$ is the minimum function of one, and hence of every α_i.

We saw in III, § 3, that if $K^* = K(\alpha)$ is an algebraic extension of K, ignoring the operation of multiplication defined between the elements of K^*, we may consider K^* as a linear set over K, and the dimension of the linear set K^* over K is the degree of K^* over K. This number, which we now denote by (K^*/K), is the degree of the minimum function of α over K. If the minimum function is of the first degree, so that $(K^*/K) = 1$, we have $K^* = K$, since $\alpha \in K$.

Lemma II. If K, K', K^* are three fields such that $K \subset K' \subset K^*$, and K^* is an algebraic extension of K, then

$$(K^*/K) = (K^*/K')(K'/K).$$

Let $\xi_1, \ldots, \xi_n$ be a minimal basis for K' over K, and $\eta_1, \ldots, \eta_m$ a minimal basis for K^* over K'. Every element of K^* can be represented thus:

$$\begin{aligned}\eta^* &= \sum_{i=1}^{m} k'_i \eta_i \quad (k'_i \in K') \\ &= \sum_{i=1}^{m}\left(\sum_{j=1}^{n} k_{ij}\xi_j\right)\eta_i \quad (k_{ij} \in K) \\ &= \sum_{i=1}^{m}\sum_{j=1}^{n} k_{ij}\eta_i\xi_j.\end{aligned}$$

Every element of K^* therefore depends linearly on the mn elements $\eta_i\xi_j$ of K^*. These elements are linearly independent over K; for since the η_i form a minimal basis over K', the relation

$$\sum_{i=1}^{m}\sum_{j=1}^{n} d_{ij}\eta_i\xi_j = 0 \quad (d_{ij} \in K)$$

implies the relation
$$\sum_{j=1}^{n} d_{ij}\xi_j = 0,$$

and since the ξ_j are linearly independent over K, it follows that

$$d_{ij} = 0.$$

Hence the degree of K^* over K is mn, which proves the lemma.

Corollary. If $K \subset K' \subseteq K^*$, *and*

$$(K^*/K) = (K'/K),$$

then $K' = K^*$.

Since $(K^*/K') = 1$, we must have $K' = K^*$.

If $K \subset K^*$, any field K' such that $K \subset K' \subset K^*$ is called an *intermediate field.*

Lemma III. If K^* *is a finite algebraic extension of* K, *there is only a finite number of intermediate fields.*

By III, § 6, Th. I, we may write $K^* = K(\xi)$, since K is without characteristic. Let the minimum function of ξ over K be

$$F(x) \equiv \sum_{i=0}^{n} a_i x^i \quad (a_i \in K).$$

If K' is an intermediate field, let the minimum function of ξ over K' be

$$G(x) \equiv \sum_{i=0}^{m} a'_i x^i \quad (a'_i \in K').$$

The field $\bar{K} = K(a'_i) = K(a'_0, \ldots, a'_m)$ satisfies the relation

$$K \subset \bar{K} \subseteq K'.$$

Since $G(x)$ is irreducible in K', it is irreducible in $\bar{K}$. Again,

$$K^* = K(\xi),$$

so that
$$K^* = K^*(a'_i) = K(a'_i)(\xi) = \bar{K}(\xi).$$

Hence $(K^*/K') = (K^*/\bar{K})$, and therefore $K' = \bar{K}$. The intermediate field $K' = \bar{K} = K(a'_i)$ is uniquely determined by the polynomial $G(x)$. But $G(x)$ is a divisor of $F(x)$, and there is only a finite number of possible divisors of $F(x)$ in K^*. Hence there is only a finite number of possible fields K'.

An isomorphism of an extension K^* of a field K on itself is called an *automorphism* of K^*, and if in the automorphism the elements of K are mapped on themselves the automorphism is said to be *over* K. If ξ is any element of K^*, any automorphism of K^* which maps ξ on η maps $K(\xi)$ on the equivalent extension $K(\eta)$ of K; hence ξ and η must be conjugate over K.

We shall be mainly concerned with automorphisms over K of normal extensions of K, and it is convenient to establish a few preliminary results for this case now. We suppose as usual that K is without characteristic, and that K^* is a normal extension of K. Let θ be a primitive element of K^* over K [III, § 6, Th. I], and suppose that its minimum function $\phi(x)$ is of degree N. Since K^* is a root field and θ lies in K^*, $\phi(x)$ is completely reducible in K^* [III, § 4, Th. VII]. Further, since K is without characteristic, the roots $\theta_1 = \theta, \theta_2, \ldots, \theta_N$ of $\phi(x)$ are all distinct.

Any element α of $K^* = K(\theta)$ is of the form

$$\alpha = a_0 + a_1\theta + \ldots + a_{N-1}\theta^{N-1} \quad (a_i \in K).$$

We consider the mapping of the elements of K^* on K^* in which α maps on

$$\alpha_i = a_0 + a_1\theta_i + \ldots + a_{N-1}\theta_i^{N-1},$$

for a given i. This gives an isomorphic mapping of $K^* = K(\theta)$ on the conjugate field $K(\theta_i)$, and since $(K(\theta_i)/K) = N$, $K(\theta_i) = K^*$. Hence the mapping is an automorphism of K^* over K. There exists such an automorphism for each value of i ($i = 1, \ldots, n$), the automorphism in the case $i = 1$ mapping each element of K^* on itself. On the other hand, since each automorphism of K^* maps θ on a conjugate θ_i, every automorphism of K^* is obtained in this way. The result of following one automorphism of K^* over K by another is clearly an automorphism of K^* over K; hence the automorphisms of K^* over K form a group of order N. Thus we have

Lemma IV. If K^* is a normal extension of K of order N, the automorphisms of K^* over K form a group of order N. If θ is a primitive element of K^* over K, and $\theta_1 = \theta, \theta_2, \ldots, \theta_N$ are its conjugates over K, the automorphisms are given by mapping $a_0 + a_1\theta + \ldots + a_{N-1}\theta^{N-1}$ on $a_0 + a_1\theta_i + \ldots + a_{N-1}\theta_i^{N-1}$ (all $a_0, \ldots, a_{N-1}$ in K) for $i = 1, \ldots, N$.

An automorphism of K^* over K carries any element α of K^* into an element α' conjugate to α over K. If $\alpha_1 = \alpha, \alpha_2, \ldots, \alpha_r$ are a set of conjugate elements over K, it follows from Lemma I that $N = qr$, where q is an integer, and if the automorphism of K^* over K given by $\theta \to \theta_i$ takes α into $\alpha^{(i)}$, then

$$\prod_{i=1}^{N} (x - \alpha^{(i)}) = [\psi(x)]^q,$$

where $\psi(x)$ is the minimum function of α. Hence for each value of i there are exactly q automorphisms of K^* which take α into α_i.

In particular, if every automorphism of K^* leaves α unaltered, $q = N$, so that $r = 1$. Hence α is in K. Thus we have

Lemma V. If K^* is a normal extension of K of order N, and α is an element of K^* having conjugates $\alpha = \alpha_1, \alpha_2, \ldots, \alpha_r$, then $\alpha_1, \ldots, \alpha_r$ lie in K^*, and there are N/r automorphisms of K^* over K which take any given α_i into a given α_j. A necessary and sufficient condition that α lie in K is that every automorphism of K^* over K leaves α unaltered.

Since in any automorphism of K^* over K zero is the only element of K^* which maps on zero, it should be noted that, if $\alpha_i \neq \alpha_j$, any automorphism of K^* over K transforms α_i and α_j into distinct elements of K^*, and hence it follows that the automorphisms of K^* over K permute the set of conjugate elements $\alpha_1, \ldots, \alpha_r$.

If $K(\zeta)$ is any algebraic extension of K, and $\zeta_1 = \zeta, \zeta_2, \ldots, \zeta_n$ are the conjugates of ζ over K, the fields conjugate to $K(\zeta)$ over K are $K(\zeta_i)$ $(i = 1, \ldots, n)$. If $K(\zeta)$ is a normal extension of K, these all coincide. Conversely, if these all coincide, $K(\zeta) = K(\zeta_1, \ldots, \zeta_n)$, and hence $K(\zeta)$ is the root field of the minimum function of ζ. Hence we have

Lemma VI. A necessary and sufficient condition that a field be normal over K is that it coincide with its conjugates over K.

If $K(\zeta)$ is not a normal extension of K, its conjugates $K(\zeta_i)$ do not all coincide with $K(\zeta)$. Any normal extension of K which contains $K(\zeta)$ must contain $K(\zeta_1, \ldots, \zeta_n)$; on the other hand, $K(\zeta_1, \ldots, \zeta_n)$, being a root field, is a normal extension of K. We shall usually call $K(\zeta_1, \ldots, \zeta_n)$ *the least normal extension of K containing $K(\zeta)$*.

11. Absolutely and relatively irreducible varieties. We are now in a position to investigate in some detail the possibility of a variety which is irreducible over the ground field K becoming reducible over some extension of K. We have already pointed out in § 2 that this may happen. On the other hand, there exist varieties defined over K which remain irreducible whatever extension K^* of K is taken as ground field; for instance, a prime has this property. A variety defined over K which is irreducible when it is considered as a variety over any extension of K is said to be *absolutely irreducible*; and a variety which is irreducible over K but is reducible over a suitably chosen extension of K is said to be *relatively irreducible* (irreducible *relative to* K).

THEOREM I. *If V is an irreducible variety of dimension d over the ground field K, then, over any extension K^* of K, V is an unmixed variety of dimension d.*

Since V is irreducible over K, its Cayley form $F(u_0, \ldots, u_d)$ has coefficients in K, and is irreducible over K. Let K^* be any extension of K. Over K^*, $F(u_0, \ldots, u_d)$ may be reducible as a form in u_{ij}, say

$$F(u_0, \ldots, u_d) = \prod_{i=1}^{k} F_i^*(u_0, \ldots, u_d).$$

The factors $F_i^*(u_0, \ldots, u_d)$ are all distinct, since K is without characteristic. By the results proved in §8, $F(u_0, \ldots, u_d)$ is the Cayley form of a variety V^* of dimension d defined over K^*, and the factors $F_i^*(u_0, \ldots, u_d)$ are the Cayley forms of the distinct irreducible components V_i^* of V^*:

$$V^* = V_1^* \dotplus V_2^* \dotplus \ldots \dotplus V_k^*.$$

Thus V^* is an unmixed variety of dimension d. We have only to show that $V^* = V$, that is, that any point of V is a point of V^*, and any point of V^* is a point of V. Let $x' = (x_0', \ldots, x_n')$ be any point whose coordinates lie in any extension of K^* (and hence of K). Let $S^i = (s_{jk}^i)$ be a skew-symmetric matrix of $n+1$ rows and columns $(i = 0, \ldots, d)$ such that the elements s_{jk}^i $(j > k)$ are independent indeterminates over $K^*(x_0', \ldots, x_n')$. Then x' lies on V if and only if

$$F(S^0 x', \ldots, S^d x') = 0.$$

But in the same way x' lies on V^* if and only if this condition is satisfied. Hence x' lies on V if and only if it lies on V^*, that is, $V = V^*$.

We now consider how to find the extension of K over which a relatively irreducible variety becomes reducible. The Cayley form of an irreducible variety is determined to within a factor belonging to the ground field, but we shall find it convenient to fix this factor uniquely. Let V be of order g, and let h be any integer greater than g, which we fix once and for all. Let us also fix $(d+1)(n+1)$ integers m_{ij} such that

$$0 \leqslant m_{00} < m_{01} < \ldots < m_{0n} < m_{10} < \ldots < m_{1n} < \ldots < m_{d0} < \ldots < m_{dn}.$$

Let $\Phi(u_{ij})$ be any polynomial in the u_{ij} such that no u_{ij} appears to a power greater than g, and let $\phi(t)$ be the polynomial in t obtained by replacing u_{ij} in $\Phi(u_{ij})$ by t to the power $h^{m_{ij}}$ $(i = 0, \ldots, d;$ $j = 0, \ldots, n)$. Consider a power-product in $\Phi(u_{ij})$ in which u_{ij} has exponent ρ_{ij}. From this we obtain a term in $\phi(t)$ with the exponent

$\Sigma\rho_{ij}h^{m_{ij}}$. Another term in $\Phi(u_{ij})$ in which u_{ij} has exponent σ_{ij} leads to a term in t with exponent $\Sigma\sigma_{ij}h^{m_{ij}}$, and since $\rho_{ij} \leqslant g < h$, $\sigma_{ij} \leqslant g < h$,

$$\Sigma\rho_{ij}h^{m_{ij}} = \Sigma\sigma_{ij}h^{m_{ij}},$$

if and only if $\rho_{ij} = \sigma_{ij}$, for $i = 0, \ldots, d$; $j = 0, \ldots, n$. Hence the set of coefficients of $\Phi(u_{ij})$ must be the same as the set of coefficients of $\phi(t)$. In dealing with the Cayley form $F(u_0, \ldots, u_d)$ of any variety of order not exceeding g, we shall suppose that the multiplier in the ground field is chosen so that the coefficient of the higher power of t in the derived polynomial $f(t)$ is unity.

Now if K^* is any extension of K over which V becomes reducible,

$$F(u_0, \ldots, u_d) = \prod_{i=1}^{k} F_i^*(u_0, \ldots, u_d),$$

where $F_i^*(u_0, \ldots, u_d)$ is the Cayley form of a component of V. Passing to the corresponding polynomials in t, we have

$$f(t) = \prod_{i=1}^{k} f_i^*(t).$$

The terms of highest degree in t in $f(t)$, $f_i^*(t)$ $(i = 1, \ldots, k)$ all have unit coefficients. It follows that each coefficient of $f_i^*(t)$ lies in the root field K_1 of the polynomial $f(t)$ over K. But the set of coefficients of $f_i^*(t)$ is the same as the set of coefficients of $F_i^*(u_0, \ldots, u_d)$. Hence the coefficients of $F_i^*(u_0, \ldots, u_d)$ lie in the intersection $K^* \wedge K_1$ of the fields K^* and K_1. Therefore the decomposition of V,

$$V = V_1^* \dotplus \ldots \dotplus V_k^*,$$

can be carried out in $K^* \wedge K_1$, that is, in an algebraic extension of K. Thus any decomposition of V can be carried out in K_1.

Suppose that V can be decomposed over K_1 as follows:

$$V = \bar{V}_1 \dotplus \ldots \dotplus \bar{V}_s.$$

In any extension K^* of K_1, suppose that

$$\bar{V}_i = \sum_j V_{ij}^*.$$

Then over K^*, $\quad V = \sum_j V_{1j}^* \dotplus \ldots \dotplus \sum_j V_{sj}^*.$

This decomposition of V can, however, be carried out in K_1; hence V_{ij}^* is a variety defined over K_1, and must therefore coincide with

$\bar{V}_i$. Hence $\bar{V}_i$ must be absolutely irreducible. Thus over K_1, V decomposes into a sum of absolutely irreducible varieties.

Again, if over an extension K^* of K, V decomposes into a sum of components of which one, say V_i^*, is absolutely irreducible, V_i^* is a variety defined over $K^* \wedge K_1$, and since it is an absolutely irreducible variety it does not decompose when we extend $K^* \wedge K_1$. Hence it must be one of the components $\bar{V}_j$ into which V decomposes over K_1. Thus we have

THEOREM II. *An irreducible variety of dimension d defined over the ground field K can be decomposed, in an essentially unique way, into a sum of absolutely irreducible varieties of dimension d, and this can be done by taking the ground field to be a suitable finite algebraic extension of K.*

The field K_1 which we considered in the proof of Theorem II may not be the smallest field over which V can be decomposed into its absolutely irreducible components. Suppose that over K_1 we have

$$F(u_0, \ldots, u_d) = \prod_{i=1}^{r} \bar{F}_i(u_0, \ldots, u_d),$$

where $\bar{F}_i(u_0, \ldots, u_d)$ is the Cayley form of $\bar{V}_i$, normalised so that the coefficient of the highest power of t in the derived polynomial $\bar{f}_i(t)$ is unity. Let $\bar{K}$ be the smallest field which contains all the coefficients of the forms $\bar{F}_i(u_0, \ldots, u_d)$. Then, clearly, $\bar{K}$ is the smallest extension of K over which V can be represented as a sum of absolutely irreducible varieties. $\bar{K}$ is a finite algebraic extension of K, and is here defined by an irreducible polynomial $\phi(x)$, as in Chapter III, § 3. One zero of $\phi(x)$ is a primitive element of $\bar{K}$, and the other zeros are primitive elements of the conjugate fields of $\bar{K}$ over K. These coincide with $\bar{K}$ if and only if $\bar{K}$ is the root-field of $\phi(x)$. We now show that $\bar{K}$ is a normal extension of K.

Let $\bar{K}'$ be any conjugate of $\bar{K}$ over K. Over $\bar{K}'$ we have

$$F(u_0, \ldots, u_d) = \prod_{i=1}^{r} \bar{F}_i'(u_0, \ldots, u_d),$$

where $\bar{F}_i'(u_0, \ldots, u_d)$ is the conjugate of $\bar{F}_i(u_0, \ldots, u_d)$. Hence, over $\bar{K}'$, V decomposes into r components, and therefore, since V has only r absolutely irreducible components, each of the components of V over $\bar{K}'$ is absolutely irreducible. Hence the factors $\bar{F}_i'$ of F are just the factors $\bar{F}_i$ rearranged. It follows that $\bar{K}' = \bar{K}$, and hence $\bar{K}$ is a normal extension of K [§ 10, Lemma VI]. Thus we have

THEOREM III. *The smallest extension $\bar{K}$ of K over which V can be decomposed into its absolutely irreducible components is a normal extension of K, and V can be expressed as a sum of absolutely irreducible varieties over K^* if and only if $\bar{K} \subseteq K^*$.*

Let θ be a primitive element of $\bar{K}$ over K, and let $\theta_1 = \theta, \theta_2, \ldots, \theta_N$ be its conjugates. The coefficients of $\bar{F}_1$ can be written as polynomials

$$a_1(\theta),\ a_2(\theta),\ \ldots$$

in θ of degree $N-1$ at most. Then

$$a_1(\theta_i),\ a_2(\theta_i),\ \ldots$$

are the coefficients of $\bar{F}_j$ for some value of j. We may suppose that for $i = 1, \ldots, s$ we have $j = 1$, and for $i > s$, $j \neq 1$. We now show that if the θ_i are suitably arranged, then

$$a_1(\theta_i),\ a_2(\theta_i),\ \ldots$$

are the coefficients of $\bar{F}_j$ for $i = (j-1)s+1, \ldots, js$, and only for these values of i, and hence that $N = rs$.

We use Lemma I of § 10. Let

$$\alpha_i = \sum_j \lambda_j a_j(\theta_i),$$

where the λ_j are indeterminates, and let

$$\phi(y) = (y-\alpha_1)\ldots(y-\alpha_N).$$

By hypothesis $\alpha_1 = \alpha_2 = \ldots = \alpha_s \neq \alpha_t$ $(t > s)$, and therefore

$$\phi(y) = [\psi(y)]^s,$$

where $\psi(y)$ is the minimum function of α_i over $K(\lambda)$. Since any α_i corresponds to the coefficients of one of r Cayley forms, $\psi(y)$ must be of degree r, and $N = rs$. We may now select the sets of α_i which are equal, and our statement follows.

We now specialise the λ_i to elements p_i of K such that after specialisation $\partial\psi(\alpha_i)/\partial\alpha_i \neq 0$ $(i = 1, \ldots, N)$ [III, § 8, Th. I], and write

$$\alpha(\theta) = \Sigma p_i a_i(\theta),$$

$$\alpha(\theta_1) = \alpha(\theta_2) = \ldots = \alpha(\theta_s) = \alpha_1,$$

$$\alpha(\theta_{s+1}) = \ldots \quad \ldots = \alpha(\theta_{2s}) = \alpha_2,$$

$$\ldots\ldots\ldots\ldots\ldots\ldots\ldots\ldots\ldots\ldots$$

$$\alpha(\theta_{(r-1)s+1}) = \ldots \quad \ldots = \alpha(\theta_{rs}) = \alpha_r,$$

where $\alpha_1, \alpha_2, \ldots, \alpha_r$ are all distinct. The minimum function of α over K is

$$G(x) = \prod_{i=1}^{r} (x - \alpha_i).$$

Let us consider any coefficient $a_i(\theta)$ of $\bar{F}_1$, and let

$$a_i(\theta_1) = \ldots = a_i(\theta_s) = \beta_1,$$

$$a_i(\theta_{s+1}) = \ldots = a_i(\theta_{2s}) = \beta_2,$$

etc. Consider the equations

$$c_0 + c_1\alpha_i + \ldots + c_{r-1}\alpha_i^{r-1} = \beta_i \quad (i = 1, \ldots, r).$$

Since $\alpha_1, \ldots, \alpha_r$ are all distinct, these have solutions for $c_0, \ldots, c_{r-1}$. Moreover, it is clear that any substitution $\theta \rightarrow \theta_i$ merely permutes these equations. Hence, when we solve the equations for $c_0, \ldots, c_{r-1}$ we see that c_i is unaltered by the automorphisms of K^* over K, and hence $c_0, \ldots, c_{r-1}$ lie in K [§ 10, Lemma V]. It follows at once that the coefficients of $\bar{F}_j$ lie in the algebraic extension $K(\alpha_j)$ of K. We want to determine these fields $K(\alpha_j)$ in terms of the function field of V.

We first show that there exists a form $f_j(x_0, \ldots, x_n)$, with coefficients in $K(\alpha_j)$, which vanishes on $\bar{V}_j$, but not on $\bar{V}_i$ $(i \neq j)$. If S^i $(i = 0, \ldots, d)$ are $d+1$ independent skew-symmetric matrices of indeterminates, $\bar{F}_j(S^0x, \ldots, S^dx)$ vanishes on $\bar{V}_j$ but not on $\bar{V}_i$ $(i \neq j)$. We can therefore specialise the elements s^i_{jk} of the matrices to values in K and obtain a form $f_j(x_0, \ldots, x_n)$ which vanishes on $\bar{V}_j$ but not on any other component of V.

Now let $\xi = (\xi_0, \ldots, \xi_n)$ be a generic point of V over K. Without loss of generality, we may suppose that this is normalised so that $\xi_0 = 1$. Let K^* be any algebraic extension of K. By the convention of § 1, $K^*, \xi_1, \ldots, \xi_n$ are all in the universal field, and hence there is a smallest field $K^*(\xi_1, \ldots, \xi_n)$ containing them. $(1, \xi_1, \ldots, \xi_n)$ is then the generic point of some variety V^* in S_n over K^*. Now $K(\xi_1, \ldots, \xi_n)$ is of degree of transcendency d over K, and K^* is algebraic over K. Hence any $d+1$ of the ξ_i are algebraically dependent over K, and hence over K^*, and if $\xi_{i_1}, \ldots, \xi_{i_d}$ are algebraically dependent over K^*, they are algebraically dependent over K. Therefore there exist d of the ξ_i which are algebraically independent over K^*, and it follows that V^* is of dimension d. Now ξ satisfies the equations of V, hence $V^* \subseteq V$. But over K^*, V is a sum of d-dimensional varieties, and hence V^* is one of the components of V. The component obtained has no particular significance. We can, in fact, choose ξ so that V^* is any assigned component of V. Indeed, let V^* be any component of V over K^*, ξ^* a normalised generic point of V^*. Reversing the argument given above, we see that $K(\xi^*)$ is an extension of K of dimension d, and hence ξ^* is the generic point of a variety V_1 of dimension d defined over K. Since ξ^* is a point of V^*, and hence of V, it follows that $V_1 \subseteq V$, and since V_1 and V are of the same dimension, $V_1 = V$ [§ 5, Th. II].

Now take $K^* = \bar{K}$, and suppose that ξ is a generic point of $\bar{V}_j$ over $\dot{K}$. Let

$$f_j = \phi_0 + \alpha_j \phi_1 + \ldots + \alpha_j^{r-1} \phi_{r-1},$$

where $\phi_k(x_0, \ldots, x_n)$ is a homogeneous polynomial over K, be a form which vanishes on $\bar{V}_j$ but not on $\bar{V}_i$ $(i \neq j)$. Then

$$\phi_0(\xi) + \alpha_k \phi_1(\xi) + \ldots + \alpha_k^{r-1} \phi_{r-1}(\xi)$$

is zero if and only if $k = j$. If $G(x)$ is the characteristic polynomial of α_j over K, it follows that the equations

$$\phi_0(\xi) + x\phi_1(\xi) + \ldots + x^{r-1}\phi_{r-1}(\xi) = 0,$$

$$G(x) = 0,$$

have a single root $x = \alpha_j$ in common. This common root can therefore be found by the division algorithm, and hence $\alpha_j \in K(\xi_1, \ldots, \xi_n)$. If V is not absolutely irreducible, α_j is not in K. Hence there exist elements in $K(\xi_1, \ldots, \xi_n)$ which are algebraic over K, but which are not in K. The elements of $K(\xi_1, \ldots, \xi_n)$ which are algebraic over K form a field K' [III, § 3, Th. III]. We call this the *algebraic closure* of K in $K(\xi_1, \ldots, \xi_n)$. K is said to be *algebraically closed* in $K(\xi_1, \ldots, \xi_n)$ if $K' = K$. We have thus shown that if V is not absolutely irreducible, K is not algebraically closed in the function field of V. We also have $K(\alpha_j) \subseteq K'$.

On the other hand, if K is not algebraically closed in $K(\xi_1, \ldots, \xi_n)$, we can find an element α of $K(\xi_1, \ldots, \xi_n)$ which is not in K, but which is algebraic over K. Since α is in $K(\xi_1, \ldots, \xi_n)$, there exist two homogeneous forms $\beta(x)$ and $\gamma(x)$ with coefficients in K such that $\alpha = \beta(\xi)/\gamma(\xi)$. Let $\phi(x)$ be the characteristic polynomial of α over K, and let $\psi(x, y)$ be an irreducible form such that $\psi(x, 1) = \phi(x)$. Then

$$\psi(\beta(x), \gamma(x)) = 0$$

is satisfied by $x = \xi$, hence it is satisfied by every point of V. Let K_1 be the root field of $\phi(x)$, so that, over K_1,

$$\phi(x) = \prod_{i=1}^{r} (x-\alpha_i) \quad (r>1, \alpha_1 = \alpha).$$

Then $\prod_{i=1}^{r} (\beta(x)-\alpha_i\gamma(x))$ vanishes on V. We show that $\beta(x)-\alpha_i\gamma(x)$ does not vanish on V. Let

$$f_i(x) = 0 \quad (i = 1, \ldots, s)$$

be a basis for the equations of V over K. If $\beta(x)-\alpha_i\gamma(x)$ vanishes on V, we have, by Hilbert's zero-theorem,

$$[\beta(x)-\alpha_i\gamma(x)]^\rho = \sum_j a_j(x, \alpha_i)\, f_j(x),$$

and hence

$$[\beta(x)-y\gamma(x)]^\rho = \sum_j a_j(x, y)\, f_j(x) + a(x, y)\, \phi(y).$$

Therefore $\beta(x)-\alpha_j\gamma(x)$ must vanish on V, for $j = 1, \ldots, r$. In particular,

$$\beta(\xi)-\alpha_1\gamma(\xi) = 0,$$

and

$$\beta(\xi)-\alpha_2\gamma(\xi) = 0,$$

and therefore

$$\beta(\xi) = \gamma(\xi) = 0,$$

which is impossible, owing to the method of constructing $\beta(x)$ and $\gamma(x)$. Hence the r forms $\beta(x)-\alpha_i\gamma(x)$ do not vanish on V, but their product does. Therefore V is reducible over K_1, and hence it is only relatively irreducible over K. Thus V is absolutely irreducible if and only if K is algebraically closed in the function field of V.

Let K^* be any algebraic extension of K, and let V^* be the component of V over K^* whose generic point is ξ. Suppose, first, that K^* does not contain K', the algebraic closure of K in $K(\xi_1, \ldots, \xi_n)$. There then exists an element α in $K(\xi_1, \ldots, \xi_n) \subseteq K^*(\xi_1, \ldots, \xi_n)$, which is algebraic over K and therefore over K^* but is not in K^*. Then V^* is not absolutely irreducible. On the other hand, suppose that $K' \subseteq K^*$. We have

$$K \subseteq K' \subseteq K(\xi_1, \ldots, \xi_n),$$

and hence

$$K(\xi_1, \ldots, \xi_n) = K'(\xi_1, \ldots, \xi_n).$$

Any element of $K'(\xi_1, \ldots, \xi_n)$ which is algebraic over K' is in $K(\xi_1, \ldots, \xi_n)$ and is algebraic over K (since K' is algebraic over K).

Therefore K' is algebraically closed in $K'(\xi_1, \ldots, \xi_n)$; hence ξ is the generic point of a variety V' over K' which is absolutely irreducible. Since K^* is an extension of K', $V^* \subseteq V'$, and since V' is absolutely irreducible, $V' = V^*$. Hence V^* is absolutely irreducible if and only if $K' \subseteq K^*$.

Now let us return to the field $\bar{K}$, which is the smallest field over which V is expressible as a sum of absolutely irreducible components. The variety $\bar{V}_j$ (which has ξ as a generic point) is absolutely irreducible, and is a variety defined over $K(\alpha_j)$. Hence $K' \subseteq K(\alpha_j)$. But we have already seen that $K(\alpha_j) \subseteq K'$; hence $K' = K(\alpha_j)$. The other components $\bar{V}_i$ of V defined over $\bar{K}$ are defined over the fields conjugate to $K(\alpha_j)$ over K. Hence $\bar{K}$ is the least normal extension of K which contains K'. We sum up these results in

THEOREM IV. *An algebraic variety V which is irreducible over the field K is absolutely irreducible if and only if K is algebraically closed in the function field $K(\xi_1, \ldots, \xi_n)$ of V. If K' is the algebraic closure of K in $K(\xi_1, \ldots, \xi_n)$ and $\bar{K}$ the least normal extension of K containing K', V can be expressed as a sum of absolutely irreducible varieties over an extension K^* of K if and only if $\bar{K} \subseteq K^*$. If ξ is a generic point of V over K, and K^* is any algebraic extension of K, ξ is a generic point of some component V^* of V over K^*, and V^* is absolutely irreducible if and only if $K' \subseteq K^*$.*

Corollary. If the ground field is algebraically closed, every irreducible variety is absolutely irreducible.

12. Some properties of relatively irreducible varieties. We now consider some theorems, which will be useful later, on relatively irreducible varieties.

THEOREM I. *If V is relatively irreducible over the field K, and of dimension d, then the points of V which lie on more than one of the absolutely irreducible components of V form a variety W, defined over K, of dimension less than d.*

It should be noted that this theorem does not exclude the possibility that W is vacuous.

Let
$$V = \bar{V}_1 \dotplus \ldots \dotplus \bar{V}_r$$
be the representation of V as a sum of absolutely irreducible varieties, and let $K(\alpha_i)$ be the least extension of K over which $\bar{V}_i$ can be defined. Then $\alpha_1, \ldots, \alpha_r$ is a set of conjugate elements over K,

and $\bar{K} = K(\alpha_1, \ldots, \alpha_r)$ is the least extension of K over which V can be expressed as a sum of absolutely irreducible varieties. Let

$$g_i^{(k)} = g_{i0} + g_{i1}\alpha_k + \ldots + g_{ir-1}\alpha_k^{r-} \quad (i = 1, \ldots, t),$$

where g_{ij} is a homogeneous polynomial in $K[x_0, \ldots, x_n]$, be a basis for the equations of $\bar{V}_k$. Then

$$g_i^{(k)} = 0 \quad (i = 1, \ldots, t),$$

$$g_i^{(l)} = 0 \quad (i = 1, \ldots, t),$$

are the equations of $\bar{V}_k \wedge \bar{V}_l$. We write these in the form

$$h_i^{(kl)} = 0 \quad (i = 1, \ldots, t').$$

The coefficients of these equations are in $K(\alpha_k, \alpha_l)$. Then the equations of $W = \Sigma \bar{V}_k \wedge \bar{V}_l$ are

$$\prod_{\substack{k,l=1 \\ k \neq l}}^{r} h_{i_{kl}}^{(kl)} = 0 \quad (i_{kl} = 1, 2, \ldots).$$

The coefficients of these equations are in $K(\alpha_1, \ldots, \alpha_r) = \bar{K}$. W is certainly a variety over $\bar{K}$. Let θ be a primitive element of $\bar{K}$ over K. Any automorphism of $\bar{K}$ over K obtained by replacing θ by a conjugate over K permutes $\alpha_1, \ldots, \alpha_k$, and hence permutes among themselves the equations which we have found for W. We write the equations which we have obtained for W in the form

$$H_{i0} + H_{i1}\theta + \ldots + H_{iN-1}\theta^{N-1} = 0 \quad (i = 1, 2, \ldots),$$

where H_{ij} is in $K[x_0, \ldots, x_n]$. If $\theta = \theta_1, \theta_2, \ldots, \theta_N$ are the conjugates of θ over K,

$$H_{i0} + H_{i1}\theta_k + \ldots + H_{iN-1}\theta_k^{N-1} = 0 \quad (k = 1, \ldots, N)$$

are among the equations of W, and hence, if x' is any point of W,

$$H_{i0}(x') + H_{i1}(x')\theta_k + \ldots + H_{iN-1}(x')\theta_k^{N-1} = 0 \quad (k = 1, \ldots, N).$$

Since the Vandermonde determinant $|\theta_k^i| \neq 0$, it follows that

$$H_{ij}(x') = 0 \quad (i = 1, 2, \ldots; j = 0, \ldots, N-1).$$

Hence x' satisfies the equations

$$H_{ij}(x) = 0 \quad (i = 1, 2, \ldots; j = 0, \ldots, N-1).$$

Conversely, it is obvious that if x' satisfies these equations it satisfies the equations of W. Hence these equations define W, which is therefore defined over K.

Since $\bar{V}_k \wedge \bar{V}_l$ is of dimension less than d, $W = \Sigma \bar{V}_k \wedge \bar{V}_l$ has dimension less than d.

THEOREM II. *If U is any subvariety of V, defined over K, there are points of U on each absolutely irreducible component of V.*

U contains at least one algebraic point, say x'. Since x' lies on V, it lies on at least one of the absolutely irreducible components of V, say on $\bar{V}_1$. Let K^* be any normal extension of K which contains $\bar{K}$ (the smallest field over which V can be expressed as a sum of absolutely irreducible varieties) and also the normalised coordinates of x'. Let

$$g_i^*(x) = 0 \quad (i = 1, 2, \ldots)$$

be the equations of $\bar{V}_1$. We may suppose that the coefficients of these equations are in $K(\alpha_1) \subseteq \bar{K} \subseteq K^*$. Any automorphism of K^* over K transforms these equations into the equations of another absolutely irreducible component of V, and such an automorphism can be found which takes these equations into the equations of $\bar{V}_j$ for any selected j. The automorphism takes x' into a conjugate point y', and since

$$g_i^*(x') = 0 \quad (i = 1, 2, \ldots),$$

it follows that y' is on $\bar{V}_j$. Now the set of points conjugate to x' form an algebraic variety W of dimension zero, of which x' is a generic point [§5, Th. V]. Since x' lies on U, it follows that $W \subseteq U$, and hence y' is a point of U. Hence U has a point on each component $\bar{V}_j$ of V.

Corollary. It easily follows from this result that if K^* is any extension of K over which V is reducible, although V may not be expressible as a sum of absolutely irreducible components, there is a point of U on each component of V over K^*. If V^* is any component of V over K^*, over some extension of K^*, V^* can be expressed as a sum of absolutely irreducible components, each of which is an absolutely irreducible component of V. As there is a point of U on each of these components, there is a point of U on V^*.

13. Sections of an absolutely irreducible variety. We prove the following theorem:

THEOREM I. *If V is an absolutely irreducible variety of dimension $d > 1$, a generic prime section of V is absolutely irreducible.*

We begin by proving

Lemma I. The prime sections of V which are not absolutely irreducible form an algebraic system.

Let $F(u_0, \ldots, u_d)$ be the Cayley form of V. If

$$v_0 x_0 + \ldots + v_n x_n = 0$$

is the equation of any prime Π not containing V, the v_i lying in K or in an extension of K, the Cayley form of the intersection $V_\wedge \Pi$ is [X, § 8, Th. I, Cor.]

$$F(u_0, \ldots, u_{d-1}, v).$$

If $V_\wedge \Pi$ is not absolutely irreducible, $F(u_0, \ldots, u_{d-1}, v)$ can be factorised, in an algebraic extension of $K(v)$, into factors of order g_1 and g_2, for some g_1 and g_2 such that $g_1 + g_2 = g$ is the order of V. Conversely, if $F(u_0, \ldots, u_{d-1}, v)$ can be so factorised, the factors are the Cayley form of algebraic varieties, of orders g_1 and g_2. We shall show below that the condition that a form of order g can be factorised into factors of order g_1 and g_2 is algebraic. Assuming this, we obtain a set of algebraic equations

$$f_i^{(g_1 g_2)}(u) = 0 \quad (i = 1, 2, \ldots)$$

to be satisfied by $(v_0, \ldots, v_n)$ in order that V be expressible as the sum of varieties of orders g_1 and g_2. In the dual space S'_n with coordinate system $(u_0, \ldots, u_n)$ these equations define a variety $U_{g_1 g_2}$. We construct such a variety for every pair of numbers g_1, g_2 satisfying the equation $g_1 + g_2 = g$. Let W be the variety $\Sigma U_{g_1 g_2}$, defined over K. Then the section of V by

$$v_0 x_0 + \ldots + v_n x_n = 0$$

is not absolutely irreducible if and only if $(v_0, \ldots, v_n)$ lies on W.

If $W = S'_n$, a generic point of S'_n lies on W, and hence the generic section of V is not absolutely irreducible. If $W \neq S'_n$, a generic point of S'_n does not lie on W, and hence a generic section of V is absolutely irreducible. Since W represents the *algebraic* system of primes whose sections with V are not absolutely irreducible, the lemma follows.

It remains to show that the condition that a polynomial of order g be factorisable, over an extension of the ground field, into polynomials of orders g_1 and g_2, is algebraic in the coefficients. Let $f(a, z_1, \ldots, z_r)$ be any polynomial of order g in $z_1, \ldots, z_r$, with coefficients $a_1, a_2, \ldots$. Suppose that

$$f(a, z_1, \ldots, z_r) = f_1(z_1, \ldots, z_r) f_2(z_1, \ldots, z_r),$$

where f_1 and f_2 are polynomials of orders g_1 and g_2 with coefficients in an extension of $K(a)$. Let $\alpha_{i1}, \alpha_{i2}, \ldots$ be the coefficients of f_i $(i = 1, 2)$. Equating the coefficients of like power-products of $z_1, \ldots, z_r$, we obtain a set of algebraic relations

$$a_i = \phi_i(\alpha_{hk}) \quad (i = 1, 2, \ldots). \tag{1}$$

Introducing new indeterminates t_{ij}, the equations

$$a_i \phi_j(t_{hk}) - a_j \phi_i(t_{hk}) = 0 \quad (i, j = 1, 2, \ldots) \tag{2}$$

are homogeneous in the t_{ij}. Forming the resultant system,

$$R_\lambda(a) \quad (\lambda = 1, 2, \ldots),$$

we see that a necessary and sufficient condition that equations (2) have a solution t'_{ij} when a is replaced by a' is

$$R_\lambda(a') = 0 \quad (\lambda = 1, 2, \ldots).$$

From the form of $\phi_i(t_{hk})$ we see at once that, since not all t'_{ij} are zero, not all $\phi_i(t'_{hk})$ are zero, and hence

$$a'_i = \rho' \phi_i(t'_{hk}),$$

and since $\phi_i(t_{hk})$ is homogeneous in t_{hk} we can multiply the solution by a constant so that $\rho' = 1$. This solution of (1) leads immediately to a factorisation of f into factors of order g_1 and g_2. Hence the conditions for the factorisation of

$$f(a', z_1, \ldots, z_r)$$

into factors of orders g_1 and g_2 are

$$R_\lambda(a') = 0 \quad (\lambda = 1, 2, \ldots),$$

and are therefore algebraic.

We now apply this lemma to the proof of Theorem I. Let $\xi = (\xi_0, \ldots, \xi_n)$ be a generic point of the absolutely irreducible variety V. Without any essential loss of generality we may assume that $\xi_0 = 1$, and that $\xi_1, \ldots, \xi_d$ are algebraically independent over K. We may also suppose, without loss of generality, that we can find

$$\zeta_i = \sum_{j=1}^{n} a_{ij} \xi_j \quad (i = 1, \ldots, d),$$

where $a_{ij} \in K$, so that the irreducible equation over K,

$$f(z_0, \ldots, z_d) = 0,$$

satisfied by $\zeta_0 = \xi_1 + a\xi_2$, $\zeta_1, \ldots, \zeta_d$ is of total order g, both in $z_0, \ldots, z_d$ and in $z_1, \ldots, z_d$, for all a in K, with at most a finite number of exceptions, g being the order of V. In view of Lemma I, we have only to show that there exists some prime section of V which is absolutely irreducible. We shall show that we can choose an element c in K such that

$$x_1 + cx_2 = vx_0$$

meets V in an absolutely irreducible variety of dimension $d-1$, where v is an indeterminate. We consider the field $K(\xi_1, \xi_2)$. [This is the point at which we use the condition $d > 1$.] Let Σ' be the field of elements of $\Sigma = K(\xi_1, \ldots, \xi_n)$ which are algebraically dependent on $K(\xi_1, \xi_2)$. If a is any element of K, let Ω_a be the field formed by the elements of Σ which are algebraically dependent on $K(\xi_1 + a\xi_2)$. Then, clearly, we have

$$K(\xi_1, \xi_2) \subseteq \Omega_a(\xi_2) \subseteq \Sigma'.$$

Σ' is a simple algebraic extension of $K(\xi_1, \xi_2)$, and hence we can apply Lemma III of § 10. As a ranges over the elements of K, there is only a finite number of possible fields $\Omega_a(\xi_2)$ between $K(\xi_1, \xi_2)$ and Σ', and hence we can find two distinct elements b, c of K such that

$$\Omega_b(\xi_2) = \Omega_c(\xi_2),$$

that is,

$$\Omega_b(\xi_1 + c\xi_2) = \Omega_c(\xi_1 + b\xi_2).$$

K is algebraically closed in Σ [§ 11, Th. IV], and hence it is algebraically closed in Ω_b, which is a sub-field of Σ. We show that this implies that $K(\xi_1 + c\xi_2)$ is algebraically closed in $\Omega_b(\xi_1 + c\xi_2)$. Write $t = \xi_1 + c\xi_2$. Ω_b is of dimension one over K and contains $\xi_1 + b\xi_2$. Hence, since $b \neq c$, $\Omega_b(\xi_1 + c\xi_2)$ contains ξ_1 and ξ_2, and is therefore of dimension two over K. Hence t is transcendental over Ω_b. Let η be any element of $\Omega_b(t)$ which is algebraic over $K(t)$. If η is algebraic over Ω_b it is in Ω_b, since Ω_b is algebraically closed in Σ which contains $\Omega_b(t)$. The equation

$$a_0(t)\,\eta^m + a_1(t)\,\eta^{m-1} + \ldots + a_m(\eta) = 0,$$

which expresses the algebraic dependence of η on $K(t)$, cannot contain t, for if it did it would imply the algebraic dependence of t on Ω_b, whereas, as we have seen, t is transcendental over Ω_b. Hence η is algebraic over K, and, since K is algebraically closed in Ω_b, η must lie in K. If, on the other hand, η is transcendental over Ω_b, we write

$$\eta = \frac{f(t)}{g(t)},$$

where $f(t)$ and $g(t)$ are polynomials over Ω_b without common factor. Replacing η by $1+\eta$ or $\eta/(\eta+1)$, if necessary, we can arrange that f and g are of the same degree in t, say

$$f(t) = a_0 t^m + \ldots + a_m,$$
$$g(t) = b_0 t^m + \ldots + b_m,$$

where a_i, b_j are in Ω_b, and a_0 and b_0 are different from zero.

By hypothesis, η is algebraic over $K(t)$, but transcendental over Ω_b. Hence the algebraic relation connecting η and t over K involves t, and therefore t is algebraic over $K(\eta)$. We consider the equation

$$(b_0\eta - a_0)\, t^m + (b_1\eta - a_1)\, t^{m-1} + \ldots + (b_m\eta - a_m) = 0.$$

Since η is transcendental over Ω_b and this equation is linear in η, it follows by a simple application of Gauss's Lemma [I, p. 35] that if this equation for t over $\Omega_b(\eta)$ were reducible, one factor would have coefficients in Ω_b, and hence $f(t)$ and $g(t)$ would have a common factor, contrary to hypothesis. Hence the equation is the irreducible equation satisfied by t over $\Omega_b(\eta)$. Now t and all its conjugates over $\Omega_b(\eta)$ are algebraic over $K(\eta)$. Hence the symmetric function of t and its conjugates,

$$\zeta_i = \frac{b_i\eta - a_i}{b_0\eta - a_0},$$

is algebraic over $K(\eta)$. Let

$$c_0(\eta)\, z^r + c_1(\eta)\, z^{r-1} + \ldots + c_r(\eta)$$

be the characteristic polynomial of ζ_i over $K(\eta)$, where $c_j(\eta) \in K[\eta]$. We consider the coefficient of the highest power of η in

$$c_0(\eta)\,(b_i\eta - a_i)^r + c_1(\eta)\,(b_i\eta - a_i)^{r-1}\,(b_0\eta - a_0) + \ldots$$
$$+ c_r(\eta)\,(b_0\eta - a_0)^r = 0.$$

Since η is transcendental over Ω_b, and hence over K, this coefficient must be zero. It follows that b_i/b_0 is algebraic over K. But it is in Ω_b, and K is algebraically closed in Ω_b. It follows that b_i/b_0 is in K. Hence

$$g(t) = b_0\bar{g}(t),$$

where $\bar{g}(t)$ is in $K[t]$. Similarly, by considering η^{-1} we have

$$f(t) = a_0\bar{f}(t),$$

where $\bar{f}(t)$ is in $K[t]$. Since η and $\bar{f}(t)/\bar{g}(t)$ are algebraically dependent on $K(t)$, their quotient a_0/b_0 is algebraically dependent on this field.

But a_0/b_0 is in Ω_b. Hence, by a case already considered, a_0/b_0 is in K. Therefore η is in $K(t)$.

Any element of $\Omega_b(\xi_1+c\xi_2)$ which is algebraic over $K(\xi_1+c\xi_2)$ is therefore in $K(\xi_1+c\xi_2)$. But any element of Ω_c is algebraic over $K(\xi_1+c\xi_2)$, and

$$\Omega_c \subseteq \Omega_c(\xi_1+b\xi_2) = \Omega_b(\xi_1+c\xi_2).$$

Hence any element of Ω_c is in $K(\xi_1+c\xi_2)$. In other words,

$$\Omega_c = K(\xi_1+c\xi_2),$$

that is, $K(\xi_1+c\xi_2)$ is algebraically closed in $K(\xi_1, \ldots, \xi_n)$.

It is clear from the manner in which c was chosen that there is an infinite number of elements c in K such that $\Omega_c = K(\xi_1+c\xi_2)$. Hence we can choose c to satisfy this condition and the condition that the irreducible equation

$$f(z_0, \ldots, z_d) = 0$$

over K connecting $\zeta_0 = \xi_1+c\xi_2$, $\zeta_i = \sum_j a_{ij}\xi_j$ $(i = 1, \ldots, d)$ is of total degree g in $z_1, \ldots, z_d$.

Now take Ω_c as the new ground field, and let V^* be the variety defined over this field whose generic point is $\xi = (1, \xi_1, \ldots, \xi_n)$. Since Ω_c is of degree of transcendency one over K, and

$$\Omega_c(\xi_1, \ldots, \xi_n) = K(\xi_1, \ldots, \xi_n)$$

is of degree of transcendency d over K, $\Omega_c(\xi_1, \ldots, \xi_n)$ is of degree of transcendency $d-1$ over Ω_c. Hence V^* is of dimension $d-1$. Since ξ lies on V, $V^* \subseteq V$, and since

$$\xi_1+c\xi_2 = t,$$

V^* lies in the prime Π whose equation is

$$x_1+cx_2 = tx_0.$$

We next show that V^* is of order g. The equation

$$f(z_0, \ldots, z_d) = 0$$

connecting $\zeta_0, \ldots, \zeta_d$ over K is irreducible. Hence, by Gauss's lemma,

$$f(t, z_1, \ldots, z_d)$$

is irreducible over $K(t)$. Therefore

$$f(t, z_1, \ldots, z_d) = 0$$

is the irreducible equation connecting the elements

$$\sum_j a_{ij}\xi_j \quad (i = 1, \ldots, d)$$

of $K[\xi_1, \ldots, \xi_n]$ over $K(t)$. It follows from § 9, Theorem I, that V^* is of order g at least. Now it was shown in X, § 8, Th. I, Cor., that if $F(u_0, \ldots, u_d)$ is the Cayley form of V, $F(u_0, \ldots, u_{d-1}, v)$ is the Cayley form of $V_\wedge \Pi$, where $v = (t, -1, -c, 0, \ldots, 0)$. Hence $V_\wedge \Pi$ is the sum of varieties of dimension $d-1$ of order not exceeding g. Since $V^* \subseteq V_\wedge \Pi$, and is of order g at least, $V^* = V_\wedge \Pi$, and is of order g exactly.

Since Ω_c is algebraically closed in $\Omega_c(\xi_1, \ldots, \xi_n)$, it follows that $V^* = V_\wedge \Pi$ is an absolutely irreducible variety. Theorem I is therefore proved.

Repeating the argument of Theorem I $e \leqslant d-1$ times, we see that the section of V by a generic S_{n-e} $(e \leqslant d-1)$ is also an absolutely irreducible variety.

Another corollary of Theorem I is of importance. We may make a transformation of our coordinate system so that $c = 0$. Then we see that $K(\xi_1)$ is algebraically closed in $K(\xi_1, \ldots, \xi_n)$. Repeating this process, we have

THEOREM II. *The coordinate system in S_n can be chosen so that if $(1, \xi_1, \ldots, \xi_n)$ is a generic point of V, $\xi_1, \ldots, \xi_d$ are algebraically independent over K, and so that, if V is absolutely irreducible, $K(\xi_1, \ldots, \xi_e)$ is algebraically closed in $K(\xi_1, \ldots, \xi_n)$ $(e < d)$.*

14. Tangent spaces and simple points. Let the equations

$$f_i(x_0, \ldots, x_n) = 0 \quad (i = 1, \ldots, r) \tag{1}$$

form a basis for the equations of an irreducible variety V of dimension d over K, so that if $\phi(x)$ vanishes on V we have a relation

$$\phi(x) \equiv \sum_{i=1}^{r} a_i(x) f_i(x). \tag{2}$$

If x' is any point of V, we prove

THEOREM I. *The rank of the matrix $(\partial f_i/\partial x_j')$ does not exceed $n-d$, and there are points x' of V for which the rank is equal to $n-d$. The points at which the rank is less than $n-d$ form an algebraic subvariety on V.*

Let $(1, \xi_1, \ldots, \xi_n)$ be a generic point of V, and let us assume that $\xi_1, \ldots, \xi_d$ are algebraically independent over K. Then among the equations of V there are $n-d$ irreducible equations

$$g_i(x_0, \ldots, x_d, x_{d+i}) = 0 \quad (i = 1, \ldots, n-d).$$

If we define $\partial g_i/\partial \xi_{d+i}$ to be the result of substituting $1, \xi_1, \ldots, \xi_d, \xi_{d+i}$ for $x_0, x_1, \ldots, x_d, x_{d+i}$ in $\partial g_i/\partial x_{d+i}$, we show that

$$\partial g_i/\partial \xi_{d+i} \neq 0 \quad (i = 1, \ldots, n-d). \tag{3}$$

For if $\partial g_i/\partial x_{d+i} \equiv 0$, so that $g_i(x_0, \ldots, x_d, x_{d+i})$ does not contain x_{d+i}, the relation $g_i(1, \xi_1, \ldots, \xi_d, \xi_{d+i}) = 0$ is a relation connecting $\xi_1, \ldots, \xi_d$, and by hypothesis these are algebraically independent. On the other hand, if $\partial g_i/\partial x_{d+i} \not\equiv 0$, but $\partial g_i/\partial \xi_{d+i} = 0$, we have another relation, other than $g_i(1, \xi_1, \ldots, \xi_d, \xi_{d+i}) = 0$, connecting $\xi_1, \ldots, \xi_d, \xi_{d+i}$, and therefore, by III, § 3, Th. VI, $\partial g_i/\partial x_{d+i}$ must have a factor in common with $g_i(x_0, \ldots, x_d, x_{d+i})$, contrary to the assumption that the latter polynomial is irreducible.

From (3) we see that the $(n-d) \times (n+1)$ matrix $(\partial g_i/\partial \xi_j)$ contains a non-singular diagonal $(n-d) \times (n-d)$ submatrix, the diagonal elements being $\partial g_i/\partial \xi_{d+i}$. The rank of the matrix $(\partial g_i/\partial \xi_j)$ is therefore $n-d$.

But from (2) each $g_i(x_0, \ldots, x_d, x_{d+i})$ is expressible in terms of $f_1(x), \ldots, f_r(x)$, so that

$$g_i(x_0, \ldots, x_d, x_{d+i}) \equiv \sum_{\nu=1}^{r} b_{i\nu}(x) f_\nu(x),$$

and therefore $$\frac{\partial g_i}{\partial \xi_j} = \sum_{\nu=1}^{r} \frac{\partial b_{i\nu}}{\partial \xi_j} f_\nu(\xi) + \sum_{\nu=1}^{r} b_{i\nu}(\xi) \frac{\partial f_\nu(\xi)}{\partial \xi_j}.$$

Since $f_\nu(\xi) = 0$ it follows that

$$\frac{\partial g_i}{\partial \xi_j} = \sum_{\nu=1}^{r} b_{i\nu}(\xi) \frac{\partial f_\nu(\xi)}{\partial \xi_j} \quad (i = 1, \ldots, n-d),$$

and therefore the matrix

$$(\partial g_i/\partial \xi_j) = (b_{ij}(\xi))\,(\partial f_i(\xi)/\partial \xi_j).$$

The rank of a matrix is not increased by multiplication by another matrix [II, § 6, Th. I], and since the rank of $(\partial g_i/\partial \xi_j)$ is $n-d$, it follows that the rank of the matrix $(\partial f_i(\xi)/\partial \xi_j)$ is at least $n-d$.

We now prove that the rank of this matrix cannot exceed $n-d$, from which it follows that its rank is $n-d$. It will then follow that the rank of the matrix

$$(\partial f_i/\partial \xi_j) \quad (i = 1, \ldots, r; j = 0, \ldots, n) \tag{4}$$

is also $n-d$, and this is what we wish to prove. For Euler's theorem gives us the equations

$$\sum_{j=0}^{n} x_j \frac{\partial f_i}{\partial x_j} = m_i f_i(x_0, \ldots, x_n),$$

where m_i is the degree of $f_i(x_0, \ldots, x_n)$. Hence

$$\frac{\partial f_i}{\partial \xi_0} = -\sum_{j=1}^{n} \xi_j \frac{\partial f_i}{\partial \xi_j},$$

and the rank of the matrix (4) is therefore the same as the rank of the matrix obtained from it by omitting the first column, this column being a linear combination of the others.

We therefore return to the matrix

$$(\partial f_i/\partial \xi_j) \quad (i = 1, \ldots, r; j = 1, \ldots, n).$$

Since $f_i(1, \xi_1, \ldots, \xi_n) = 0$ $(i = 1, \ldots, r)$, we deduce that

$$\frac{\partial f_i}{\partial \xi_k} + \sum_{j=1}^{n-d} \frac{\partial f_i}{\partial \xi_{d+j}} \frac{\partial \xi_{d+j}}{\partial \xi_k} = 0 \quad (k = 1, \ldots, d), \tag{5}$$

by III, § 7, Th. I, and also

$$\frac{\partial g_j}{\partial \xi_k} + \frac{\partial g_j}{\partial \xi_{d+j}} \frac{\partial \xi_{d+j}}{\partial \xi_k} = 0 \quad (j = 1, \ldots, n-d). \tag{6}$$

Substituting from (6) in (5), we have the equations

$$\frac{\partial f_i}{\partial \xi_k} = \sum_{j=1}^{n-d} \frac{\partial f_i}{\partial \xi_{d+j}} \left(\frac{\partial g_j}{\partial \xi_k} \Big/ \frac{\partial g_j}{\partial \xi_{d+j}} \right) \quad (k = 1, \ldots, d).$$

These show that each of the first d columns of $(\partial f_i/\partial \xi_j)$ $(i = 1, \ldots, r;$ $j = 1, \ldots, n)$ is a linear combination of the last $n-d$ columns. Hence the rank cannot exceed $n-d$. We have therefore proved that if x' is a generic point of V, the rank of the matrix $(\partial f_i/\partial x'_j)$ $(i = 1, \ldots, r;$ $j = 0, \ldots, n)$ is $n-d$.

Since determinants of order $n-d+1$ extracted from the matrix $(\partial f_i/\partial x_j)$ vanish at a generic point of V, they vanish at all points of V. Hence if x' is any point of V the rank of the matrix $(\partial f_i/\partial x'_j)$ cannot exceed $n-d$. Since the rank is $n-d$ at a generic point, the points at which the rank is less than $n-d$ form a proper subvariety. This subvariety is algebraic, its equations being given by equating to zero all determinants of order $n-d$ which can be extracted from $(\partial f_i/\partial x_j)$.

A point x' of V at which the matrix $(\partial f_i/\partial x'_j)$ is of rank $n-d$ is called a *simple point* of V. Any point of V which is not simple is said to be *non-simple*, or *singular*. Any subvariety of V all of whose points are singular is called a singular subvariety of V, and the subvariety of V which consists of all the singular points of V is called the *singular locus* of V.

We have seen that a generic point of V is simple. It is convenient to prove here that there are algebraic points of V which are simple. If the singular locus of V is vacuous, this follows at once from the fact that there are points on V which are algebraic [IV, § 6, p. 162]. If V has a singular locus S, the dimension s of S satisfies the conditions $0 \leqslant s < d$. We can find a prime Π_1 defined over K not containing any component of S. Then $\Pi_1 \wedge S$ is of dimension $s-1$ at most. Similarly, we can find a prime Π_2 not containing any component of $\Pi_1 \wedge S$ (and hence distinct from Π_1). Proceeding in this way, we can find an $(n-d)$-space S_{n-d} defined over K which does not meet the singular locus of V. It will be proved in Chapter XI [§ 5, Th. III and § 6, Th. I] that any S_{n-d} has a non-vacuous intersection with V. Hence we can find an $(n-d)$-space defined over K which meets V in a variety U having no point in common with the singular locus. Any algebraic point of U is simple for V.

We obtain the same set of simple points of V if we adopt another basis

$$h_i(x_0, \ldots, x_n) = 0 \quad (i = 1, \ldots, t)$$

for the equations of V. For in the first place

$$h_\nu(x) \equiv \sum_{\rho=1}^{r} a_{\nu\rho}(x) f_\rho(x),$$

and therefore, if x' is any point on V,

$$\frac{\partial h_\nu}{\partial x'_\sigma} = \sum_{\rho=1}^{r} a_{\nu\rho}(x') \frac{\partial f_\rho}{\partial x'_\sigma}.$$

Hence the rank of $(\partial h_i/\partial x'_j)$ cannot exceed the rank of $(\partial f_i/\partial x'_j)$. Conversely, the rank of the second matrix cannot exceed that of the first, and the two matrices have the same rank for all points x' of V. Hence the definition of simple points of V is independent of the basis chosen for the equations of V.

If x' is a simple point of V the equations

$$\sum_{j=0}^{n} x_j \frac{\partial f_i}{\partial x'_j} = 0 \quad (i = 1, \ldots, r)$$

define a linear space S_d. This space contains x', for by Euler's theorem,

$$\sum_{j=0}^{n} x'_j \frac{\partial f_i}{\partial x'_j} = m_i f_i(x') = 0,$$

where m_i is the degree of the form $f_i(x)$. We call the space S_d *the tangent space to V at the point x'*. In the particular case in which V is a primal, of dimension $n-1$, given by the vanishing of a single equation

$$f(x_0, \ldots, x_n) = 0,$$

the point x' is a simple point if not all the coefficients $\partial f/\partial x'_j$ in the equation of the tangent prime vanish. We evidently have the theorem:

THEOREM II. *The tangent space to a variety V at a simple point x' is the intersection of all the tangent primes at x' to the primals through V which have a simple point at x'.*

CHAPTER XI

ALGEBRAIC CORRESPONDENCES

In previous chapters we have met, from time to time, with examples in which there is a correspondence, given by a set of algebraic equations, between the points of two spaces, or between the points of two algebraic varieties. For instance, if S_m and S_n are two spaces in which $(x_0, \ldots, x_m)$ and $(y_0, \ldots, y_n)$ are allowable coordinate systems, a projective correspondence between the spaces is given by the equations

$$y_i \sum_{k=0}^{m} a_{jk} x_k - y_j \sum_{k=0}^{m} a_{ik} x_k = 0 \quad (i, j = 0, \ldots, n).$$

Such correspondences are particular examples of *algebraic correspondences*, and the purpose of this chapter is to give a precise definition of algebraic correspondences, and to develop their fundamental properties.

The ground field K is assumed to be without characteristic, but unless specific mention is made of the fact, we shall not assume that K is algebraically closed.

1. Varieties in r-way projective space. In Chapter V, § 10, we defined an r-way projective space $S_{n_1 \ldots n_r}$ over a ground field K. Let $(x_0^{(1)}, \ldots, x_{n_1}^{(1)}; x_0^{(2)}, \ldots, x_{n_2}^{(2)}; \ldots; x_0^{(r)}, \ldots, x_{n_r}^{(r)})$ be an allowable coordinate system in this space. The theory of algebraic varieties in ordinary, or 1-way, projective space, can easily be extended to give a theory of algebraic varieties in $S_{n_1 \ldots n_r}$. It will be enough to outline the steps of the development. The proofs are exactly similar to those given in Chapter X for the case $r = 1$.

Consider the set of polynomials

$$f_i(x^{(1)}, x^{(2)}, \ldots, x^{(r)}) \quad (i = 1, 2, \ldots),$$

where each polynomial has coefficients in K, and is homogeneous in each set of indeterminates $x^{(1)} = (x_0^{(1)}, \ldots, x_{n_1}^{(1)})$, $x^{(2)} = (x_0^{(2)}, \ldots, x_{n_2}^{(2)})$, $\ldots$, $x^{(r)} = (x_0^{(r)}, \ldots, x_{n_r}^{(r)})$. The set of points $('x_0^{(1)}, \ldots, 'x_{n_1}^{(1)}; \ldots; 'x_0^{(r)}, \ldots, 'x_{n_r}^{(r)})$ in $S_{n_1 \ldots n_r}$ which satisfy the equations

$$f_i('x^{(1)}, 'x^{(2)}, \ldots, 'x^{(r)}) = 0 \quad (i = 1, 2, \ldots)$$

define an algebraic variety in $S_{n_1 \dots n_r}$. It is easily seen that this definition does not depend on the choice of an allowable coordinate system in $S_{n_1 \dots n_r}$.

By using Hilbert's basis theorem we show that any algebraic variety in r-way space can be defined by a finite number of equations. The intersection $V_1 \wedge V_2$ and sum $V_1 \dotplus V_2$ of two algebraic varieties are defined as in 1-way space. So are the relations

$$V_1 \subseteq V_2, \quad V_1 \subset V_2.$$

An algebraic variety V is said to be reducible if it can be written in the form

$$V = V_1 \dotplus V_2,$$

where $V_1 \subset V$, and $V_2 \subset V$. Defining an irreducible variety over K as one which is not reducible, we show that if an irreducible variety V lies in the sum of two varieties V_1 and V_2 then it is completely contained in one or in the other. It follows from this that a necessary and sufficient condition for the reducibility of a variety V is the existence of a product fg of two polynomials $f(x^{(1)}, x^{(2)}, \dots, x^{(r)})$ and $g(x^{(1)}, x^{(2)}, \dots, x^{(r)})$ (each homogeneous in each set of indeterminates $x^{(1)}, x^{(2)}, \dots, x^{(r)}$) which vanishes at all points of V without either polynomial having this property. The main theorem of Chapter X, § 2, that every algebraic variety V can be expressed as the sum of a finite number of irreducible varieties, is true for varieties in $S_{n_1 \dots n_r}$, as is also the theorem following, that the representation of V as a non-contractible sum of irreducible varieties is essentially unique.

A point $(\xi_0^{(1)}, \dots, \xi_{n_1}^{(1)}; \xi_0^{(2)}, \dots, \xi_{n_2}^{(2)}; \dots; \xi_0^{(r)}, \dots, \xi_{n_r}^{(r)})$ is said to be a *generic point* of a variety V if it lies on V, and if the relation

$$f(\xi^{(1)}, \xi^{(2)}, \dots, \xi^{(r)}) = 0$$

implies that $f(x^{(1)}, x^{(2)}, \dots, x^{(r)})$ is zero on V, $f(x^{(1)}, x^{(2)}, \dots, x^{(r)})$ having coefficients in K and being homogeneous in each set of indeterminates $x^{(1)}, x^{(2)}, \dots, x^{(r)}$.

As in the case of 1-way spaces, we prove that if V has a generic point it is irreducible. Conversely, if V is irreducible, it has a generic point. To show this, we consider the rational functions $f(x^{(1)}, x^{(2)}, \dots, x^{(r)})/g(x^{(1)}, x^{(2)}, \dots, x^{(r)})$ in $K(x^{(1)}, x^{(2)}, \dots, x^{(r)})$, where $f(x^{(1)}, x^{(2)}, \dots, x^{(r)})$ is homogeneous of degree m_1 in $x^{(1)}$, m_2 in $x^{(2)}$, ..., and $g(x^{(1)}, x^{(2)}, \dots, x^{(r)})$ is homogeneous of degree m_1 in $x^{(1)}$, m_2 in $x^{(2)}$.... We say that two such rational functions f/g, f'/g' are equivalent ($m_1, m_2, \dots$ are not necessarily the same for the two rational functions) if and only if the polynomial $fg'-f'g$, which is homogeneous in each set of indeterminates $x^{(1)}, x^{(2)}, \ldots, x^{(r)}$, vanishes on V. Defining addition and multiplication in the usual way, we show that the sets of equivalent functions are elements of a field K^* which contains a subfield which can be identified with K. Without loss of generality we may assume that $x_0^{(1)}, x_0^{(2)}, \ldots, x_0^{(r)}$ do not vanish on V. If then the elements of K^* which are defined by the rational functions $(1, x_1^{(1)}/x_0^{(1)}, \ldots, x_{n_1}^{(1)}/x_0^{(1)}; \ldots; 1, x_1^{(r)}/x_0^{(r)}, \ldots, x_{n_r}^{(r)}/x_0^{(r)})$ are $(1, \xi_1^{(1)}, \ldots, \xi_{n_1}^{(1)}; \ldots; 1, \xi_1^{(r)}, \ldots, \xi_{n_r}^{(r)})$ respectively, we can show that the point $(1, \xi_1^{(1)}, \ldots, \xi_{n_1}^{(1)}; \ldots; 1, \xi_1^{(r)}, \ldots, \xi_{n_r}^{(r)})$ is a generic point of V.

If $('\xi_0^{(1)}, '\xi_1^{(1)}, \ldots, '\xi_{n_1}^{(1)}; \ldots; '\xi_0^{(r)}, '\xi_1^{(r)}, \ldots, '\xi_{n_r}^{(r)})$ is any other generic point of V, none of $'\xi_0^{(1)}, \ldots, '\xi_0^{(r)}$ can be zero, and we may therefore normalise the coordinates of this point so that $'\xi_0^{(1)} = \ldots = '\xi_0^{(r)} = 1$. We can then show that the fields

$$K(\xi_1^{(1)}, \ldots, \xi_{n_1}^{(1)}; \ldots; \xi_1^{(r)}, \ldots, \xi_{n_r}^{(r)}), \quad K('\xi_1^{(1)}, \ldots, '\xi_{n_1}^{(1)}; \ldots; '\xi_1^{(r)}, \ldots, '\xi_{n_r}^{(r)})$$

are isomorphic extensions of K in which $\xi_i^{(j)}$ corresponds to $'\xi_i^{(j)}$. The field K^*, or any of its isomorphs, is called *the function field of* V.

As in X, § 3, Th. IV, we show that any point $(\xi^{(1)}, \xi^{(2)}, \ldots, \xi^{(r)})$, whose sets of coordinates

$$\xi^{(i)} = (\xi_0^{(i)}, \ldots, \xi_{n_i}^{(i)})$$

are in some extension of K, defines an irreducible algebraic variety V over K of which $(\xi^{(1)}, \xi^{(2)}, \ldots, \xi^{(r)})$ is a generic point. Let the equations of V be

$$f_i(x^{(1)}, x^{(2)}, \ldots, x^{(r)}) = 0 \quad (i = 1, \ldots, l). \tag{1}$$

Any solution $('x^{(1)}, 'x^{(2)}, \ldots, 'x^{(r)})$ of these equations is said to be a *proper specialisation* of $(\xi^{(1)}, \xi^{(2)}, \ldots, \xi^{(r)})$. Equivalently we can say that $('x^{(1)}, 'x^{(2)}, \ldots, 'x^{(r)})$ is a proper specialisation of $(\xi^{(1)}, \xi^{(2)}, \ldots, \xi^{(r)})$ if all algebraic relations

$$f(x^{(1)}, x^{(2)}, \ldots, x^{(r)}) = 0,$$

with coefficients in K, which are homogeneous in each set of indeterminates $x^{(1)}, x^{(2)}, \ldots, x^{(r)}$ and are satisfied by $(\xi^{(1)}, \xi^{(2)}, \ldots, \xi^{(r)})$ are also satisfied by $('x^{(1)}, 'x^{(2)}, \ldots, 'x^{(r)})$.

If $('x^{(1)}, 'x^{(2)}, \ldots, 'x^{(r)})$ is a proper specialisation of $(\xi^{(1)}, \xi^{(2)}, \ldots, \xi^{(r)})$ and $s < r$ then $('x^{(1)}, 'x^{(2)}, \ldots, 'x^{(s)})$ is evidently a proper specialisation of $(\xi^{(1)}, \xi^{(2)}, \ldots, \xi^{(s)})$. To prove the converse let

$$h_i(x^{(1)}, \ldots, x^{(s)}) \quad (i = 1, 2, \ldots)$$

be the resultant system of the set of polynomials

$$f_i(x^{(1)}, x^{(2)}, \ldots, x^{(r)}) \quad (i = 1, \ldots, l)$$

which occur in (1), that is, the equations of V, with respect to $x^{(s+1)}, \ldots, x^{(r)}$. Since the equations

$$f_i(\xi^{(1)}, \ldots, \xi^{(s)}, x^{(s+1)}, \ldots, x^{(r)}) = 0 \quad (i = 1, \ldots, l)$$

have a solution $\xi^{(s+1)}, \ldots, \xi^{(r)}$, it follows that

$$h_i(\xi^{(1)}, \ldots, \xi^{(s)}) = 0 \quad (i = 1, 2, \ldots).$$

If $'x^{(1)}, \ldots, 'x^{(s)}$ is a proper specialisation of $\xi^{(1)}, \ldots, \xi^{(s)}$,

$$h_i('x^{(1)}, \ldots, 'x^{(s)}) = 0 \quad (i = 1, 2, \ldots),$$

and hence the resultant system of

$$f_i('x^{(1)}, \ldots, 'x^{(s)}, x^{(s+1)}, \ldots, x^{(r)}) \quad (i = 1, \ldots, l)$$

vanishes, and therefore there exists a point

$$('x^{(1)}, \ldots, 'x^{(s)}, 'x^{(s+1)}, \ldots, 'x^{(r)})$$

on V. This proves

THEOREM I. *If $'x^{(1)}, \ldots, 'x^{(r)}$ is a proper specialisation of $\xi^{(1)}, \ldots, \xi^{(r)}$, then $'x^{(1)}, \ldots, 'x^{(s)}$ is a proper specialisation of $\xi^{(1)}, \ldots, \xi^{(s)}$, $s < r$. Conversely, if $'x^{(1)}, \ldots, 'x^{(s)}$ is a proper specialisation of $\xi^{(1)}, \ldots, \xi^{(s)}$, then there exists a proper specialisation $'x^{(1)}, \ldots, 'x^{(s)}, 'x^{(s+1)}, \ldots, 'x^{(r)}$ of $\xi^{(1)}, \ldots, \xi^{(s)}, \xi^{(s+1)}, \ldots, \xi^{(r)}$.*

The *dimension* of an irreducible variety V in r-way projective space is defined as follows. Let $(\xi_0^{(1)}, \ldots, \xi_{n_1}^{(1)}; \ldots; \xi_0^{(r)}, \ldots, \xi_{n_r}^{(r)})$ be the coordinates of a generic point, which we may suppose normalised so that $\xi_0^{(1)} = \ldots = \xi_0^{(r)} = 1$. The dimension of V is the maximum number d of the normalised coordinates $\xi_1^{(1)}, \ldots, \xi_{n_r}^{(r)}$ which are algebraically independent over K. As in III, § 3, Th. V we may show that any $d+1$ elements of the function field K^* of V are algebraically dependent over K.

The point $(1, \xi_1^{(1)}, \ldots, \xi_{n_1}^{(1)})$ is a generic point of a variety in S_{n_1}. Let us call it V_1. Similarly, the point $(1, \xi_1^{(1)}, \ldots, \xi_{n_1}^{(1)}; 1, \xi_1^{(2)}, \ldots, \xi_{n_2}^{(2)})$ is a generic point of a variety V_{12} in $S_{n_1 n_2}, \ldots$. The point

$$(1, \xi_1^{(1)}, \ldots, \xi_{n_1}^{(1)}; \ldots; 1, \xi_1^{(r)}, \ldots, \xi_{n_r}^{(r)})$$

defines $V_{12\ldots r} = V$. By Theorem I a characteristic property of these varieties is that if $('x^{(1)}, \ldots, 'x^{(s)})$ is any point of $V_{1\ldots s}$ then there is a point $('x^{(1)}, \ldots, 'x^{(s)}, 'x^{(s+1)})$ of $V_{1\ldots s+1}$, and conversely, if

$$('x^{(1)}, \ldots, 'x^{(s)}, 'x^{(s+1)})$$

is any point of $V_{1\ldots s+1}$, then $('x^{(1)}, \ldots, 'x^{(s)})$ is a point of $V_{1\ldots s}$.

The dimension d_1 of V_1 is the maximum number of the coordinates $\xi_1^{(1)}, \ldots, \xi_{n_1}^{(1)}$ which are algebraically independent over K. Let us suppose that $\xi_1^{(1)}, \ldots, \xi_{d_1}^{(1)}$ are algebraically independent. Now let d_{12} be the maximum number of the coordinates $\xi_1^{(2)}, \ldots, \xi_{n_2}^{(2)}$ which are algebraically independent over $K(\xi_1^{(1)}, \ldots, \xi_{n_1}^{(1)})$. Then if $\xi_1^{(2)}, \ldots, \xi_{d_{12}}^{(2)}$, say, are algebraically independent over $K(\xi_1^{(1)}, \ldots, \xi_{n_1}^{(1)})$, they are also algebraically independent over $K(\xi_1^{(1)}, \ldots, \xi_{d_1}^{(1)})$, and conversely, since algebraic *dependence* is a transitive property [III, § 3, Th. IV].

It follows that the maximum number of coordinates $\xi_1^{(1)}, \ldots, \xi_{n_1}^{(1)}$, $\xi_1^{(2)}, \ldots, \xi_{n_2}^{(2)}$ which are independent over K is $d_1 + d_{12}$. This is the dimension of V_{12}. Proceeding in this way we show that $d_1 + d_{12} + \ldots d_{1\ldots s}$ is the dimension of $V_{1\ldots s}$, where $d_{1\ldots j}$ is the dimension of $K(\xi^{(1)}, \ldots, \xi^{(j)})$ over $K(\xi^{(1)}, \ldots, \xi^{(j-1)})$. In particular,

$$d_1 + d_{12} + \ldots + d_{12\ldots r} = d.$$

The sets of coordinates $\xi^{(1)}, \xi^{(2)}, \ldots, \xi^{(r)}$ may be considered in a different order. Let $i, j, k, \ldots, p$ be any derangement of the integers $1, 2, \ldots, r$, and let us consider the sets of coordinates in the order $i, j, k, \ldots, p$. Defining $d_i, d_{ij}, \ldots$ as above, we obtain the equation

$$d_i + d_{ij} + \ldots + d_{ij\ldots p} = d. \tag{2}$$

We thus obtain $r!$ equations for d. The special case $r = 2$ will be of particular interest to us in § 3.

2. Segre's representation of r-way projective space. We now show, following Segre (see Bibliographical Notes), how to represent an r-way projective space $S_{n_1\ldots n_r}$ by the points of an algebraic variety Ω in 1-way projective space. The representation is such that there is a one-to-one correspondence, without exception, between the points of $S_{n_1\ldots n_r}$ and the points of Ω. To any algebraic variety in $S_{n_1\ldots n_r}$ there corresponds an algebraic variety lying on Ω, and conversely. Corresponding varieties are either both reducible or both irreducible, and when both are irreducible they have the same dimension.

Let N be the integer defined by the equation

$$N+1 = \prod_{i=1}^{r} (n_i+1).$$

In a projective space S_N over K choose an allowable coordinate system $(x_{0\dots 0}, \dots, x_{n_1\dots n_r})$. Each x-coordinate has r suffixes; the first suffix runs from 0 to n_1, the second from 0 to n_2, ..., the rth from 0 to n_r. We shall usually write the coordinates of the point

$$x_{i_1\dots i_r} = a_{i_1\dots i_r} \quad (i_j = 0, \dots, n_j;\ j = 1, \dots, r)$$

as $(\dots, a_{i_1\dots i_r}, \dots)$.

Consider the variety in S_N, which we shall call Ω, with generic point

$$(\dots, \xi_{i_1\dots i_r}, \dots) = (\dots, \xi^{(1)}_{i_1} \xi^{(2)}_{i_2} \dots \xi^{(r)}_{i_r}, \dots),$$

where $\xi^{(1)}_0, \dots, \xi^{(1)}_{n_1}; \xi^{(2)}_0, \dots, \xi^{(2)}_{n_2}; \dots; \xi^{(r)}_0, \dots, \xi^{(r)}_{n_r}$ are independent indeterminates. Points on Ω clearly satisfy the equations

$$x_{i_1\dots i_r} x_{j_1\dots j_r} - x_{i_1\dots i_{s-1} j_s i_{s+1}\dots i_r} x_{j_1\dots j_{s-1} i_s j_{s+1}\dots j_r} = 0, \qquad (1)$$

for $s = 1, \dots, r$ and for all admissible values of $i_1, \dots, i_r, j_1, \dots, j_r$. We show that there is a one-to-one correspondence between the points of $S_{n_1\dots n_r}$ and those of Ω.

Let $P = ('x^{(1)}_0, \dots, 'x^{(1)}_{n_1}; \dots; 'x^{(r)}_0, \dots, 'x^{(r)}_{n_r})$ be any point of $S_{n_1\dots n_r}$. Since not all the $x^{(1)}$-coordinates are zero, not all the $x^{(2)}$-coordinates are zero, ..., and not all the $x^{(r)}$-coordinates are zero, there is at least one product $'x^{(1)}_{i_1}\, 'x^{(2)}_{i_2} \dots 'x^{(r)}_{i_r}$ which is not zero. Hence there is a point

$$P^* = (\dots, 'x^{(1)}_{i_1}\, 'x^{(2)}_{i_2} \dots 'x^{(r)}_{i_r}, \dots)$$

on Ω corresponding to the point P of $S_{n_1\dots n_r}$. This point P^* on Ω remains unchanged if we take $P = (\lambda' x^{(1)}, \dots, \rho' x^{(r)})$ instead of $P = ('x^{(1)}, \dots, 'x^{(r)})$. In particular, if the coordinates of P are normalised so that $'x^{(1)}_0 = 'x^{(2)}_0 = \dots = 'x^{(r)}_0 = 1$, the coordinate $x'_{0\dots 0}$ of the corresponding point P^* on Ω is 1.

Conversely, let $P^* = (\dots, x'_{i_1\dots i_r}, \dots)$ be any point on Ω. It will be sufficient to consider the case in which $x'_{0\dots 0} \neq 0$. We normalise the coordinates of P^* so that $x'_{0\dots 0} = 1$, and define a point

$$P = ('x^{(1)}, 'x^{(2)}, \dots, 'x^{(r)})$$

in $S_{n_1\dots n_r}$ by the equations

$$\left.\begin{aligned} 'x^{(1)}_i &= x'_{i0\dots 0}, \\ 'x^{(2)}_i &= x'_{0i\dots 0}, \\ &\cdot\ \ \cdot\ \ \cdot\ \ \cdot \\ 'x^{(r)}_i &= x'_{00\dots i}. \end{aligned}\right\}$$

Since P^* satisfies the equations (1),

$$x'_{i_1\dots i_r} x'_{0\dots 0} = x'_{i_1\dots i_{r-1}0} x'_{0\dots 0 i_r} = x'_{i_1\dots i_{r-1}0}\, 'x^{(r)}_{i_r},$$

and proceeding by induction it is easily seen that

$$x'_{i_1\dots i_r} = 'x^{(1)}_{i_1}\, 'x^{(2)}_{i_2} \dots 'x^{(r)}_{i_r}.$$

Hence the point on Ω which corresponds to $P = ('x^{(1)}, 'x^{(2)}, \dots, 'x^{(r)})$ in $S_{n_1\dots n_r}$ is the point P^*, and the correspondence between P and P^* is one-to-one without exception.

The variety Ω is usually called a *Segre variety*. We shall not discuss its geometrical properties in any detail here, but merely draw attention to the r systems of linear spaces which lie on it.

If $'x^{(1)}, 'x^{(2)}, \dots, 'x^{(r-1)}$ are fixed points, the point

$$(\dots, 'x^{(1)}_{i_1}\, 'x^{(2)}_{i_2} \dots 'x^{(r-1)}_{i_{r-1}} x^{(r)}_{i_r}, \dots)$$

on Ω which corresponds to the point $('x^{(1)}, 'x^{(2)}, \dots, 'x^{(r)})$ of $S_{n_1\dots n_r}$ describes a linear space of dimension n_r, this being the dimension of the linear space S_{n_r} described by the corresponding point $'x^{(r)}$. Hence to every point of $S_{n_1\dots n_{r-1}}$ there corresponds an S_{n_r} on Ω, and distinct points of the one yield distinct linear spaces of the other. It is clear that no two of these linear spaces have any points in common, since the correspondence between Ω and $S_{n_1\dots n_r}$ is one-to-one. Similarly, there is a system of linear spaces, each of dimension n_{r-1}, on Ω corresponding to the points of $S_{n_1\dots n_{r-2} n_r}$. No two of these spaces have any points in common. We thus obtain r systems of linear spaces on Ω. One space of each system passes through any given point of Ω.

The simplest example of a Segre variety is obtained by taking $r = 2$, $n_1 = n_2 = 1$. We then have one equation for Ω,

$$x_{00}x_{11} - x_{01}x_{10} = 0.$$

This is the equation of a quadric surface in S_3. The linear spaces on Ω are the two systems of generators of the quadric. In this case any two generators of opposite systems meet in a point.

We now show that there corresponds to any algebraic variety in $S_{n_1\dots n_r}$ an algebraic variety lying on Ω, and conversely. Let

$$f(x^{(1)}, x^{(2)}, \dots, x^{(r)}) = 0 \tag{2}$$

be any equation satisfied by a variety V in $S_{n_1\dots n_r}$, where the polynomial $f(x^{(1)}, x^{(2)}, \dots, x^{(r)})$ is homogeneous of degree m_1 in the $x^{(1)}$-coordinates, of degree m_2 in the $x^{(2)}$-coordinates, ..., of degree

m_r in the $x^{(r)}$-coordinates. Let $m > \max[m_1, \ldots, m_r]$. Then if g_i $(i = 1, 2, \ldots)$ be the set of all power-products which are of degree $m - m_1$ in the $x^{(1)}$-coordinates, $m - m_2$ in the $x^{(2)}$-coordinates, ..., $m - m_r$ in the $x^{(r)}$-coordinates, the equations

$$g_i = 0 \quad (i = 1, 2, \ldots)$$

evidently have no solution in $S_{n_1 \ldots n_r}$. Hence the solutions of the equations

$$g_i f(x^{(1)}, x^{(2)}, \ldots, x^{(r)}) = 0 \quad (i = 1, 2, \ldots)$$

are the same as the solutions of (2) in $S_{n_1 \ldots n_r}$. If the equations of any variety in $S_{n_1 \ldots n_r}$ are

$$f_i(x^{(1)}, x^{(2)}, \ldots, x^{(r)}) = 0 \quad (i = 1, \ldots, s), \tag{3}$$

we may therefore assume that these equations are homogeneous and of the same degree m_i in the $x^{(1)}, x^{(2)}, \ldots, x^{(r)}$ coordinates.

The equations of V can therefore be written (possibly in more than one way) in the form

$$F_i(\ldots, x^{(1)}_{i_1} x^{(2)}_{i_2} \ldots x^{(r)}_{i_r}, \ldots) = 0 \quad (i = 1, \ldots, s). \tag{4}$$

If the point $P = ('x^{(1)}_0, \ldots, 'x^{(1)}_{n_1}; 'x^{(2)}_0, \ldots, 'x^{(2)}_{n_2}; \ldots; 'x^{(r)}_0, \ldots, 'x^{(r)}_{n_r})$ lies on the variety V whose equations are given by (4), the corresponding point $P^* = (\ldots, x'_{i_1 \ldots i_r}, \ldots) = (\ldots, 'x^{(1)}_{i_1} 'x^{(2)}_{i_2} \ldots 'x^{(r)}_{i_r}, \ldots)$ lies on the variety V^* on Ω whose equations are

$$F_i(\ldots, x_{i_1 \ldots i_r}, \ldots) = 0 \quad (i = 1, \ldots, s), \tag{5}$$

together with the equations of Ω. Conversely, from the equations (5) of a variety V^* on Ω we can deduce equations (4), which define a variety V in $S_{n_1 \ldots n_r}$.

The correspondence between V and V^* is one-to-one. If V is irreducible, and $(\xi^{(1)}_0, \ldots, \xi^{(1)}_{n_1}; \ldots; \xi^{(r)}_0, \ldots, \xi^{(r)}_{n_r})$ is a generic point of V, the point $(\ldots, \xi^{(1)}_{i_1} \xi^{(2)}_{i_2} \ldots \xi^{(r)}_{i_r}, \ldots)$ is clearly a generic point of V^*, which is therefore irreducible. Conversely, if $(\ldots, \xi_{i_1 \ldots i_r}, \ldots)$ is a generic point of an irreducible variety V^* on Ω, and the coordinates are normalised so that $\xi_{0 \ldots 0} = 1$, the point $(\xi^{(1)}_0, \ldots, \xi^{(1)}_{n_1}; \ldots; \xi^{(r)}_0, \ldots, \xi^{(r)}_{n_r})$, where

$$\left.\begin{aligned} \xi^{(1)}_i &= \xi_{i0 \ldots 0}, \\ \xi^{(2)}_i &= \xi_{0i \ldots 0}, \\ &\cdots \\ \xi^{(r)}_i &= \xi_{00 \ldots i} \end{aligned}\right\} \tag{6}$$

is easily shown to be a generic point of V. Thus irreducible varieties in $S_{n_1 \dots n_r}$ correspond to irreducible varieties on Ω, and conversely.

We also see that if V^* corresponds to V, *the dimension of* V^* *is equal to the dimension of* V. For (6) shows that the number of algebraically independent elements in the set of coordinates of a normalised generic point of V is equal to the number of algebraically independent elements of type $\xi_{0 \dots i \dots 0}$ in the set of coordinates of a normalised generic point of V^*. The relation

$$\xi_{i_1 \dots i_r} = \xi^{(1)}_{i_1} \xi^{(2)}_{i_2} \dots \xi^{(r)}_{i_r}$$

shows that all other elements $\xi_{i_1 \dots i_r}$ in the set of coordinates of a normalised generic point of V^* are algebraically dependent on these elements of type $\xi_{0 \dots i \dots 0}$. Therefore the dimension of V is equal to the dimension of V^*.

Finally, if $V_1, V_2, \dots$ are any algebraic varieties in $S_{n_1 \dots n_r}$, $V_1^*, V_2^*, \dots$ the corresponding varieties on Ω, then

$$V_i \dotplus V_j \quad \text{corresponds to} \quad V_i^* \dotplus V_j^*,$$
$$V_i \wedge V_j \quad \text{corresponds to} \quad V_i^* \wedge V_j^*,$$

and $V_i \subseteq V_j$ (or $V_i \subset V_j$) implies $V_i^* \subseteq V_j^*$ (or $V_i^* \subset V_j^*$), and conversely.

The representation we have just discussed suggests how we should define the Cayley form of an irreducible variety V of dimension d in $S_{n_1 \dots n_r}$. If V^* is the variety on Ω (in the projective space S_N) which corresponds to V, the Cayley form of V^*, which we denote by $F(u_0, u_1, \dots, u_d)$, is the u_0-resultant of the equations of V^* and of the d equations

$$\sum_{j_1} \dots \sum_{j_r} u_{ij_1 \dots j_r} x_{j_1 \dots j_r} = 0 \quad (i = 1, \dots, d).$$

This, however, is the same as the u_0-resultant of the equations of V and of the d equations

$$\sum_{j_1} \dots \sum_{j_r} u_{ij_1 \dots j_r} x^{(1)}_{j_1} x^{(2)}_{j_2} \dots x^{(r)}_{j_r} = 0 \quad (i = 1, \dots, d).$$

We can write

$$F(u_0, u_1, \dots, u_d) = A(u_1, \dots, u_d) \prod_{\nu=1}^{g} \left(\sum_{j_1} \dots \sum_{j_r} u_{0j_1 \dots j_r} \xi^{(\nu)}_{j_1 \dots j_r} \right),$$

where $(\dots, \xi^{(\nu)}_{j_1 \dots j_r}, \dots)$ is a generic point of V^*. Since

$$\xi^{(\nu)}_{j_1 \dots j_r} = {}^{(\nu)}\xi^{(1)}_{j_1} \dots {}^{(\nu)}\xi^{(r)}_{j_r},$$

$$F(u_0, u_1, \dots, u_d) = A(u_1, \dots, u_d) \prod_{\nu=1}^{g} \left(\sum_{j_1} \dots \sum_{j_r} u_{0j_1 \dots j_r} {}^{(\nu)}\xi^{(1)}_{j_1} \dots {}^{(\nu)}\xi^{(r)}_{j_r} \right), \quad (7)$$

where $({}^{(\nu)}\xi_0^{(1)}, \ldots, {}^{(\nu)}\xi_{n_1}^{(1)}; \ldots; {}^{(\nu)}\xi_0^{(r)}, \ldots, {}^{(\nu)}\xi_{n_r}^{(r)})$ is a generic point of V. We shall take (7) to be the Cayley form of the variety V in $S_{n_1\ldots n_r}$.

The variety Ω is, as we have seen, defined in S_N by the equations

$$x_{i_1\ldots i_r}x_{j_1\ldots j_r} - x_{i_1\ldots i_{s-1}j_s i_{s+1}\ldots i_r}x_{j_1\ldots j_{s-1}i_s j_{s+1}\ldots j_r} = 0.$$

It is of interest to show that these equations form a basis for all the equations satisfied by Ω. We prove

THEOREM I. *If $f(\ldots, x_{i_1\ldots i_r}, \ldots)$ vanishes on the Segre variety Ω, then we can express f in the form*

$$f \equiv \sum_1^q A_k f_k,$$

where A_k is a polynomial in the indeterminates $x_{i_1\ldots i_r}$, and, arranged in any convenient order,

$$f_k \equiv x_{i_1\ldots i_r}x_{j_1\ldots j_r} - x_{i_1\ldots i_{s-1}j_s i_{s+1}\ldots i_r}x_{j_1\ldots j_{s-1}i_s j_{s+1}\ldots j_r},$$

there being q of these forms f_k when $i_1, \ldots, i_r, j_1, \ldots, j_r$ assume all admissible values and $s = 1, \ldots, r$.

We note in the first place that $f(\ldots, x_{i_1\ldots i_r}, \ldots)$ cannot be linear. In fact, if the form

$$\Sigma A_{i_1\ldots i_r} x_{i_1\ldots i_r}$$

vanishes on Ω, we have the equation

$$\Sigma A_{i_1\ldots i_r} \xi_{i_1}^{(1)} \xi_{i_2}^{(2)} \ldots \xi_{i_r}^{(r)} = 0,$$

where each term in the sum represents a distinct power-product of the indeterminates $\xi_{i_1}^{(1)}, \xi_{i_2}^{(2)}, \ldots, \xi_{i_r}^{(r)}$. Hence every coefficient $A_{i_1\ldots i_r}$ must be zero. The Segre variety Ω is therefore not contained in a linear space of dimension less than N.

If $f(\ldots, x_{i_1\ldots i_r}, \ldots)$ is of degree >1 we proceed by considering *standard power-products*. Let

$$P \equiv x_{i_1\ldots i_r} x_{j_1\ldots j_r} \ldots x_{p_1\ldots p_r}$$

be a power-product of degree m. We suppose that the factors are arranged so that $i_1 \leqslant j_1 \leqslant \ldots \leqslant p_1$. This product can be represented by the tableau

$$\begin{matrix} i_1 & j_1 & . & a_1 & b_1 & . & p_1 \\ i_2 & j_2 & . & a_2 & b_2 & . & p_2 \\ . & . & . & . & . & . & . \\ i_l & j_l & . & a_l & b_l & . & p_l \\ . & . & . & . & . & . & . \\ i_r & j_r & . & a_r & b_r & . & p_r \end{matrix}$$

which consists of r rows and m columns. We say that the tableau, or the corresponding power-product, is *standard* if the set of integers in each row is non-decreasing as we proceed along the row from left to right. Let us suppose that the given power-product P is not standard, that the elements in the first $l-1$ rows are non-decreasing, but that in the lth row we have

$$i_l \leqslant j_l \leqslant \ldots \leqslant a_l, \quad a_l > b_l.$$

Let $$f_\rho \equiv x_{a_1 \ldots a_r} x_{b_1 \ldots b_r} - x_{a_1 \ldots a_{l-1} b_l a_{l+1} \ldots a_r} x_{b_1 \ldots b_{l-1} a_l b_{l+1} \ldots b_r},$$

and let c_ρ denote the power-product of degree $m-2$ corresponding to the tableau shown with the a and b columns omitted. Then

$$P - c_\rho f_\rho$$

is a power-product of degree m whose tableau is obtained from the one shown by interchanging a_l and b_l, leaving all other elements unaltered. By a finite number of interchanges any given tableau can be made standard, so that there exists a set of power-products c_k of degree $m-2$ such that

$$P - \sum_1^q c_k f_k \equiv P^*,$$

where P^* is a standard power-product. It follows that if $f(\ldots, x_{i_1 \ldots i_r}, \ldots)$ is any form of degree m, there exist forms $A_1, \ldots, A_q$ of degree $m-2$ such that

$$f - \sum_1^q A_k f_k \equiv f^*,$$

where f^* is a form of degree m which is a linear combination of standard power-products of degree m.

We observe at this point that to any power-product

$$\xi_{i_1}^{(1)} \xi_{j_1}^{(1)} \ldots \xi_{t_1}^{(1)} \xi_{i_2}^{(2)} \xi_{j_2}^{(2)} \ldots \xi_{t_2}^{(2)} \ldots \xi_{i_r}^{(r)} \xi_{j_r}^{(r)} \ldots \xi_{t_r}^{(r)}$$

which is of degree m in each set of indeterminates

$$(\xi_1^{(1)}, \ldots, \xi_{n_1}^{(1)}), \quad (\xi_1^{(2)}, \ldots, \xi_{n_2}^{(2)}), \quad \ldots, \quad (\xi_1^{(r)}, \ldots, \xi_{n_r}^{(r)}),$$

there corresponds a unique standard power-product of degree m, say

$$x_{\alpha_1 \ldots \alpha_r} x_{\beta_1 \ldots \beta_r} \ldots x_{\delta_1 \ldots \delta_r}.$$

In fact we simply arrange the factors in the product so that

$$i_1 \leqslant j_1 \leqslant \ldots \leqslant t_1; \quad i_2 \leqslant j_2 \leqslant \ldots \leqslant t_2; \quad \ldots; \quad i_r \leqslant j_r \leqslant \ldots \leqslant t_r,$$

in the notation used above, and then

$$x_{i_1 i_2 \ldots i_r} x_{j_1 j_2 \ldots j_r} \ldots x_{l_1 l_2 \ldots l_r}$$

is the standard power-product which corresponds to the given one under the transformation

$$x_{\rho_1 \rho_2 \ldots \rho_r} = \xi^{(1)}_{\rho_1} \xi^{(2)}_{\rho_2} \ldots \xi^{(r)}_{\rho_r}.$$

Since the factors can be ordered as indicated in only one way, there is only one standard power-product which corresponds to it. We deduce from this result that if $f^*(\ldots, x_{i_1 \ldots i_r}, \ldots)$ is a linear combination of *standard* power-products, and if

$$f^*(\ldots, \xi^{(1)}_{i_1} \xi^{(2)}_{i_2} \ldots \xi^{(r)}_{i_r}, \ldots) \equiv 0,$$

then
$$f^*(\ldots, x_{i_1 \ldots i_r}, \ldots) \equiv 0.$$

Now let us suppose that $f(\ldots, x_{i_1 \ldots i_r}, \ldots)$ vanishes on Ω. Then $f(\ldots, \xi^{(1)}_{i_1} \xi^{(2)}_{i_2} \ldots \xi^{(r)}_{i_r}, \ldots) \equiv 0$. But, as proved above,

$$f \equiv f^* + \sum_1^q A_k f_k,$$

and $f_k(\ldots, \xi^{(1)}_{i_1} \xi^{(2)}_{i_2} \ldots \xi^{(r)}_{i_r}, \ldots) \equiv 0$, so that $f^*(\ldots, \xi^{(1)}_{i_1} \xi^{(2)}_{i_2} \ldots \xi^{(r)}_{i_r}, \ldots) \equiv 0$. It follows that

$$f^*(\ldots, x_{i_1 \ldots i_r}, \ldots) \equiv 0,$$

and therefore
$$f \equiv \sum_1^q A_k f_k.$$

3. Two-way algebraic correspondences. Let $S_{n_1}, S_{n_2}, \ldots, S_{n_r}$ be r 1-way projective spaces, and choose allowable coordinate systems $(x^{(1)}_0, \ldots, x^{(1)}_{n_1})$, $(x^{(2)}_0, \ldots, x^{(2)}_{n_2})$, $\ldots$, $(x^{(r)}_0, \ldots, x^{(r)}_{n_r})$ in them. Construct an r-way projective space $S_{n_1 \ldots n_r}$ in which

$$(x^{(1)}_0, \ldots, x^{(1)}_{n_1}; x^{(2)}_0, \ldots, x^{(2)}_{n_2}; \ldots; x^{(r)}_0, \ldots, x^{(r)}_{n_r})$$

is an allowable coordinate system. Now consider any algebraic variety V in $S_{n_1 \ldots n_r}$ given by the equations

$$f_i(x^{(1)}, x^{(2)}, \ldots, x^{(r)}) = 0 \quad (i = 1, \ldots, s).$$

If $('x^{(1)}_0, \ldots, 'x^{(1)}_{n_1}; \ldots; 'x^{(1)}_0, \ldots, 'x^{(r)}_{n_r})$ is any point on V, we say that *the points* $('x^{(1)}_0, \ldots, 'x^{(1)}_{n_1}), ('x^{(2)}_0, \ldots, 'x^{(2)}_{n_r}), \ldots, ('x^{(r)}_0, \ldots, 'x^{(r)}_{n_r})$ *in* $S_{n_1}, S_{n_2}, \ldots, S_{n_r}$ *are related by the r-way algebraic correspondence* V. The correspondence is said to be *reducible* or *irreducible* according as V is reducible or irreducible.

In applications a 2-way correspondence, which we usually call, simply, a correspondence, is the most important, and we shall confine ourselves mainly to this case. The reader will have no difficulty in extending the results proved to r-way correspondences. In any case, an r-way correspondence between $S_{n_1}, S_{n_2}, \ldots, S_{n_r}$ may always be regarded as a correspondence between S_{n_1} and the space which contains the Segre variety of $S_{n_2 \ldots n_r}$.

Correspondences which have already been studied in Vol. 1 are *collineations*, which can be defined by the equations

$$y_i = \Sigma a_{ik} x_k \quad \text{or} \quad y_i \Sigma a_{jk} x_k - y_j \Sigma a_{ik} x_k = 0,$$

and *correlations*, which can be defined by the bilinear equation

$$x' A y = \Sigma a_{jk} x_j y_k = 0.$$

Now let S_m and S_n be two projective spaces (which may possibly coincide) in which $(x_0, \ldots, x_m)$ and $(y_0, \ldots, y_n)$ are allowable coordinate systems, and let C be a correspondence defined by the equations

$$f_i(x, y) = 0 \quad (i = 1, \ldots, r). \tag{1}$$

If we form the resultant system of these equations with respect to $(y_0, \ldots, y_n)$ we obtain a set of homogeneous equations

$$g_i(x) = 0 \quad (i = 1, \ldots, s), \tag{2}$$

with the property that a necessary and sufficient condition that the equations

$$f_i(x', y) = 0 \quad (i = 1, \ldots, r)$$

should have a solution is that

$$g_i(x') = 0 \quad (i = 1, \ldots, s).$$

The equations (2) define an algebraic variety M in S_m which contains those points and only those points x' of S_m to which there correspond points y' in S_n under the correspondence C. It may happen that M is vacuous; at the other extreme we may have $M = S_m$.

Similarly, if we form the resultant system of the equations (1) with respect to $(x_0, \ldots, x_m)$, we obtain a set of equations

$$h_i(y) = 0 \quad (i = 1, \ldots, t), \tag{3}$$

which define a variety N in S_n of points to which there correspond points in S_m under the correspondence C.

If we regard the correspondence C as *assigning* points y' to certain points x', we may call these points y' the *image-points* of x', and we then call the points x' the *original* or *object-points*. We shall sometimes write $x' \to y'$. With this terminology the variety M is called the *object-variety*, and the variety N the *image-variety*. We shall say that the correspondence C is *between* the varieties M and N. If (x', y') is a point-pair of the correspondence, x' belongs to M and y' to N, and to every point x' of M (y' of N) there corresponds at least one point y' of N (x' of M).

If x' is any point of M, the corresponding points of S_n constitute the variety whose equations are

$$f_i(x', y) = 0 \quad (i = 1, \ldots, r),$$

where the coefficients, of course, lie in $K(x')$. These equations define an algebraic variety in S_n which necessarily lies on N. We denote this subvariety by $N(x')$. Similarly, if y' is a point of N, the points x' which correspond to y' form the subvariety $M(y')$ of M whose equations are

$$f_i(x, y') = 0 \quad (i = 1, \ldots, r).$$

Let C be an irreducible correspondence between S_m and S_n. Then the variety V in S_{mn} given by (1) is irreducible. If (ξ, η) is a generic point of V, then a point x' is a point of the object-variety M if and only if there exists a point y' in S_n such that (x', y') lies on V; that is, if there exists a point y' in S_n such that (x', y') is a proper specialisation of (ξ, η). By § 1, Th. I, a necessary and sufficient condition that there exist a point y' with this property is that x' be a proper specialisation of ξ. Hence, a necessary and sufficient condition that x' lie on M is that it be a proper specialisation of ξ. We deduce that ξ is a generic point of M. Similarly, η is a generic point of the image-variety N. In particular, we conclude that if C is an irreducible correspondence the object- and image-varieties of C are irreducible.

The converse of this result is not true. We show by an example that M and N may both be irreducible while the correspondence C is reducible. Let $m = n$, and let $A = (a_{ij})$ and $B = (b_{ij})$ be two distinct non-singular $(n+1) \times (n+1)$ matrices over K. The equations

$$x_i \sum_{k=0}^{n} a_{jk} y_k - x_j \sum_{k=0}^{n} a_{ik} y_k = 0 \quad (i, j = 0, \ldots, n),$$

$$x_i \sum_{k=0}^{n} b_{jk} y_k - x_j \sum_{k=0}^{n} b_{ik} y_k = 0 \quad (i, j = 0, \ldots, n),$$

each define irreducible correspondences C_1 and C_2, with generic points $(A\eta, \eta)$ and $(B\eta, \eta)$ respectively. For each correspondence $M = S_m$ and $N = S_n$, and $N(\xi)$, $M(\eta)$ consist of single points. The correspondence

$$C = C_1 \dot{+} C_2$$

is reducible, but still has $M = S_m$ and $N = S_n$.

4. The Principle of Counting Constants. Let the correspondence C which is defined by the equations

$$f_i(x, y) = 0 \quad (i = 1, \ldots, r) \tag{1}$$

be irreducible, and let $(1, \xi_1, \ldots, \xi_m; 1, \eta_1, \ldots, \eta_n)$ be a generic point of C. We define the dimension q of C as the maximum number q of the normalised coordinates $\xi_1, \ldots, \eta_n$ which are algebraically independent over K. If a is the number of the ξ_i which are algebraically independent over K, and b is the number of the η_i which are algebraically independent over the field $K(\xi_1, \ldots, \xi_m)$, then, as we have seen [§ 1],

$$q = a + b.$$

Similarly, if c is the number of the η_i which are algebraically independent over K, and d is the number of the ξ_i which are algebraically independent over $K(\eta_1, \ldots, \eta_n)$, then

$$q = c + d.$$

Hence

$$a + b = q = c + d.$$

For a more general result on r-way correspondences, see § 1.

We can attach a geometrical meaning to the numbers a, b, c, d. In § 3 we proved that the point ξ is a generic point of the object-variety M, and therefore a is the dimension of M. Similarly, c is the dimension of N. We now prove that *the subvariety $N(\xi)$ of N which corresponds to a generic point ξ of M is irreducible over the field $K(\xi_1, \ldots, \xi_m)$ and of dimension b.*

$N(\xi)$ consists of all points y' which are such that the point-pair (ξ, y') belongs to the correspondence C. Hence, all homogeneous algebraic relations over K which hold for (ξ, η) hold for (ξ, y'). By the substitution $\xi_0 = 1$ these relations cease to be homogeneous in the ξ_i, but continue to be homogeneous in the η_i. They can therefore be considered as homogeneous relations in the η_i with coefficients in the field $K(\xi_1, \ldots, \xi_m)$. It easily follows that all homogeneous algebraic relations with coefficients in $K(\xi)$ which are valid for the point η are also valid for the point y' on $N(\xi)$. Since η satisfies the equations of $N(\xi)$, it follows that η is a generic point of $N(\xi)$ over $K(\xi)$. It also follows that $N(\xi)$ is irreducible over the field $K(\xi)$ and is of dimension b.

The variety $N(\xi)$ may become reducible over an extension of the field $K(\xi)$, but its absolutely irreducible components will all have the same dimension b [X, § 11, Th. I]. We have proved the theorem which is known as

The Principle of Counting Constants: If in a q-dimensional irreducible correspondence between the varieties M and N there corresponds to a generic point ξ of the a-dimensional variety M a b-dimensional variety $N(\xi)$ of points on N, and if, conversely, there corresponds to a generic point η of the c-dimensional variety N a d-dimensional variety $M(\eta)$ of points on M, then $N(\xi)$ and $M(\eta)$ are irreducible, and

$$a+b = q = c+d. \tag{2}$$

It is important to remember that all generic point-pairs (ξ, η) of the correspondence C are algebraically equivalent to each other; this is also true for the generic points of M and N respectively [X, § 3, Th. III]. If $\xi, \bar{\xi}$ are generic points of M, there is an isomorphism

$$K(\xi) \cong K(\bar{\xi})$$

which maps ξ on $\bar{\xi}$ and leaves all elements of K unchanged. This can be extended to an isomorphism

$$K(\xi, \eta) \cong K(\bar{\xi}, \bar{\eta}), \tag{3}$$

where η is mapped on $\bar{\eta}$, and therefore, since η is a generic point of N over K, $\bar{\eta}$ is also a generic point of N. We have proved that η is a generic point of $N(\xi)$ over $K(\xi)$. From the isomorphism (3) it follows that $\bar{\eta}$ is a generic point of $N(\bar{\xi})$. Again, if η' is a generic point of $N(\xi)$,

$$K(\xi, \eta) \cong K(\xi, \eta'),$$

and therefore η' is a generic point of N over K.

Hence we may begin with any generic point ξ of M and find a generic point $\bar{\eta}$ of the subvariety $N(\xi)$ which corresponds to ξ; or, conversely, we may begin with any generic point of N. In each case we obtain the same numbers a, b, c, d and the same properties of a generic point-pair (ξ, η) of the correspondence.

In applications the formula (2) is often used to find the dimension c of the image-variety N when a, b and d are given. If it is found that

$c = n$, we can deduce that the image-variety N fills the whole space S_n.

If $b = d = 0$, to a generic point ξ of M there corresponds a finite set of β points on N, and to a generic point η of N there corresponds a finite set of α points on M. We then say that there is *an* (α, β) *correspondence between M and N*. As we shall soon see, an infinite number of points may correspond to a *special* point of M.

We now consider the case when $\alpha = \beta = 1$. If ξ is a normalised generic point of M, the corresponding variety $N(\xi)$ consists of one point η. The normalised coordinates of η lie in an algebraic extension of $K(\xi)$ of degree 1, by the proof of Theorem V in Chapter X, § 5. Hence the normalised coordinates of η lie in $K(\xi)$. Similarly, the normalised coordinates of ξ lie in $K(\eta)$. Hence $K(\eta) \subseteq K(\xi)$ and $K(\xi) \subseteq K(\eta)$, so that

$$K(\xi) = K(\eta).$$

The function fields of M and N are therefore identical. If, conversely, we are given the weaker condition that the function fields of the irreducible varieties M and N are equivalent extensions of K, we can set up a $(1, 1)$ algebraic correspondence, which is usually called a *birational* correspondence, between M and N. If ξ, η are normalised generic points of M, N respectively,

$$K(\xi) \cong K(\eta).$$

Let $\bar{\eta}_i$ be the image of η_i in $K(\xi)$ under this isomorphism. Then $\bar{\eta} = (1, \bar{\eta}_1, \ldots, \bar{\eta}_n)$ is a generic point of N, and $K(\xi) = K(\bar{\eta})$. Hence, without loss of generality, we may assume that

$$K(\xi) = K(\eta).$$

Then $\quad \eta_i = R_i(\xi) \quad (i = 1, \ldots, n), \qquad \xi_i = S_i(\eta) \quad (i = 1, \ldots, m),$

where $R_i(x_0, x_1, \ldots, x_m)$ is a rational function of $x_0, x_1, \ldots, x_m$, and $S_i(y_0, y_1, \ldots, y_n)$ is a rational function of $y_0, y_1, \ldots, y_n$. If

$$R_i(x) = A_i(x)/B_i(x),$$

where the $A_i(x)$, $B_i(x)$ may be taken as forms of the same degree, and $S_i(y) = C_i(y)/D_i(y)$, the correspondence given by the equations

$$y_i B_i(x) - y_0 A_i(x) = 0 \quad (i = 1, \ldots, n),$$

$$x_i D_i(y) - x_0 C_i(y) = 0 \quad (i = 1, \ldots, m),$$

has (ξ, η) as a point-pair, and therefore contains an irreducible correspondence for which M and N are object- and image-varieties.

η corresponds to ξ and ξ corresponds to η, the correspondence being $(1,1)$. Two varieties between which we can establish a birational correspondence are said to be *birationally equivalent*. A necessary and sufficient condition for the birational equivalence of two varieties is that their function fields be equivalent extensions of K.

We conclude this section by giving an example of an irreducible one-to-one correspondence in which there corresponds to a special point of M a *line* of points of N. We take $m = n = 2$, and we consider the well-known quadratic transformation in the plane given by the equations

$$x_0y_0 = x_1y_1 = x_2y_2.$$

To a point (x_0', x_1', x_2') for which not more than one of the coordinates is zero there corresponds the unique point given by the equations

$$x_0'y_0 = x_1'y_1 = x_2'y_2,$$

but if $(x_0', x_1', x_2') = (1, 0, 0)$, say, any point on $y_0 = 0$ corresponds to x'. We show that the correspondence is irreducible by proving that if ξ, η, ζ are indeterminates, the point $(\xi, \eta, \zeta;\ \eta\zeta, \zeta\xi, \xi\eta)$ is a generic point of the correspondence.

In the first place this point satisfies the equations of the correspondence. We must show that if $f(x_0, x_1, x_2;\ y_0, y_1, y_2)$ is a polynomial of degree l in the x and of degree p in the y such that $f(\xi, \eta, \zeta;\ \eta\zeta, \zeta\xi, \xi\eta) = 0$, then f is zero for all points of the correspondence.

To do this we consider any power-product $x_0^a x_1^b x_2^c y_0^\alpha y_1^\beta y_2^\gamma$, where $a+b+c = l$, and $\alpha+\beta+\gamma = p$. This can be reduced, *modulo* $x_0y_0 - x_2y_2$ and $x_1y_1 - x_2y_2$, to a power-product in which either a or α and either b or β is zero. Four possible standard types of power-product result, and when $(\xi, \eta, \zeta;\ \eta\zeta, \zeta\xi, \xi\eta)$ is substituted for $(x_0, x_1, x_2;\ y_0, y_1, y_2)$,

$$x_2^l y_0^\alpha y_1^\beta y_2^\gamma \quad \text{becomes} \quad \xi^{p-\alpha}\eta^{p-\beta}\zeta^{l+p-\gamma} \quad (\alpha+\beta+\gamma = p),$$

$$x_1^b x_2^c y_0^\alpha y_2^\gamma \quad \text{becomes} \quad \xi^{p-\alpha}\eta^{p+b}\zeta^{p+c-\gamma} \quad (b+c = l;\ \alpha+\gamma = p),$$

$$x_0^a x_2^c y_1^\beta y_2^\gamma \quad \text{becomes} \quad \xi^{p+a}\eta^{p-\beta}\zeta^{p+c-\gamma} \quad (a+c = l;\ \beta+\gamma = p),$$

$$x_0^a x_1^b x_2^c y_2^p \quad \text{becomes} \quad \xi^{p+a}\eta^{p+b}\zeta^{c} \quad (a+b+c = l).$$

Hence any possible power-product in x_0, x_1, x_2, y_0, y_1, y_2 which is of degree l in the x and of degree p in the y becomes, after the substitution

$$(x_0, x_1, x_2;\ y_0, y_1, y_2) = (\xi, \eta, \zeta;\ \eta\zeta, \zeta\xi, \xi\eta),$$

the power-product $$\xi^\lambda \eta^\mu \zeta^\nu,$$

where $$\lambda+\mu+\nu = l+2p,$$

and $$\lambda \leqslant l+p, \quad \mu \leqslant l+p, \quad \nu \leqslant l+p.$$

No two distinct standard power-products give the same $\xi^\lambda \eta^\mu \zeta^\nu$. For

(i) if $\lambda < p$, $\mu < p$, the standard power-product is

$$x_2^l y_0^{p-\lambda} y_1^{p-\mu} y_2^{l+p-\nu};$$

(ii) if $\lambda < p$, $\mu > p$, the standard power-product is

$$x_1^{\mu-p} x_2^{l+p-\mu} y_0^{p-\lambda} y_2^{\lambda};$$

(iii) if $\lambda > p$, $\mu < p$, the standard power-product is

$$x_0^{\lambda-p} x_2^{l+p-\lambda} y_1^{p-\mu} y_2^{\mu};$$

(iv) if $\lambda > p$, $\mu > p$, the standard power-product is

$$x_0^{\lambda-p} x_1^{\mu-p} x_2^{\nu} y_2^{p}.$$

If $\lambda = p$, $\mu \gtrless p$; or $\mu = p$, $\lambda \gtrless p$ the above results still hold, and if $\lambda = \mu = p$ it follows that $\nu = l$, and the one standard power-product is $x_2^l y_2^p$.

Returning to the polynomial $f(x_0, x_1, x_2; y_0, y_1, y_2)$ of degree l in the x and p in the y, we can write

$$f \equiv A(x_0 y_0 - x_1 y_1) + B(x_1 y_1 - x_2 y_2) + F,$$

where $F(x_0, x_1, x_2; y_0, y_1, y_2)$ is a sum of standard power-products. If $f(\xi, \eta, \zeta; \eta\zeta, \zeta\xi, \xi\eta) = 0$, then $F(\xi, \eta, \zeta; \eta\zeta, \zeta\xi, \xi\eta) = 0$. But unless $F(x_0, x_1, x_2; y_0, y_1, y_2)$ is identically zero, it consists of one or more standard power-products, and we have proved that these give rise to the same number of distinct power-products $\xi^\lambda \eta^\mu \zeta^\nu$. Since the coefficients of the standard power-products and the $\xi^\lambda \eta^\mu \zeta^\nu$ to which they give rise are the same, it follows from the equation

$$F(\xi, \eta, \zeta; \eta\zeta, \zeta\xi, \xi\eta) = 0$$

that $$F(x_0, x_1, x_2; y_0, y_1, y_2) \equiv 0.$$

Hence the equation $f(x_0, x_1, x_2; y_0, y_1, y_2) = 0$ is satisfied by all points of the correspondence whenever $f(\xi, \eta, \zeta; \eta\zeta, \zeta\xi, \xi\eta) = 0$. The correspondence

$$x_0 y_0 = x_1 y_1 = x_2 y_2$$

is therefore irreducible.

5. A special correspondence. In practical applications it is usually easier to define an irreducible correspondence by finding a generic point than to show that a given set of equations defines an irreducible correspondence. An exception, which is of some importance in applications, is given by

THEOREM I. *Let the equations of a correspondence be*

$$f_i(x) = 0 \quad (i = 1, \dots, r), \tag{1}$$

$$\sum_{j=0}^{n} a_{ij}(x)\, y_j = 0 \quad (i = 1, \dots, s). \tag{2}$$

Then if (i) *the variety M in S_m given by* (1) *is irreducible, and* (ii) *the matrix $(a_{ij}(x'))$ is of rank $n-b$ $(b \geqslant 0)$ for all points x' on M, the correspondence is irreducible.*

Let ξ be a generic point of M, which we may suppose normalised so that $\xi_0 = 1$. If v_{ij} $(i = 1, \dots, n-b;\ j = 1, \dots, s)$ are independent indeterminates over $K(\xi)$, M is still irreducible over $K(v)$ [X, § 5, Th. VI], and ξ is still a generic point of M. Let

$$a'_{ij}(x) = \sum_{k=1}^{s} v_{ik}\, a_{kj}(x).$$

Then, clearly, the $(n-b) \times (n+1)$ matrix $(a'_{ij}(x'))$ is of rank $n-b$ for all proper algebraic specialisations x' of ξ.

We now adjoin to $K(v)$ the $b(n+1)$ independent indeterminates u_{ij} $(i = 1, \dots, b;\ j = 0, \dots, n)$. Then the matrix of the set of linear equations

$$\sum_{j=0}^{n} a'_{ij}(\xi)\, y_j = 0 \quad (i = 1, \dots, n-b), \tag{3}$$

$$\sum_{j=0}^{n} u_{ij}\, y_j = 0 \quad (i = 1, \dots, b) \tag{4}$$

is of rank n. Let $(-1)^k D_k(\xi, u, v)$ be the determinant formed from this matrix by omitting the $(k+1)$th column. Not all $D_k(\xi, u, v)$ are zero, and the equations have the solution $(\eta_0, \dots, \eta_n)$, where $\eta_i = D_i(\xi, u, v)$. We prove that (ξ, η) is a generic point-pair of the correspondence over K.

The equations of the correspondence are satisfied by (ξ, η), since ξ satisfies the equations (1) and η, which satisfies the equations (3), also satisfies the equivalent equations

$$\sum_{j=0}^{n} a_{ij}(\xi)\, y_j = 0 \quad (i = 1, \dots, s).$$

We now prove that if $F(\xi, \eta) = 0$ is any homogeneous relation which holds for (ξ, η), then $F(x', y') = 0$, where (x', y') is any solution of the equations (1) and (2). It is sufficient to consider algebraic solutions [cf. p. 5].

Let x' be any algebraic point of M, y' any solution of the equations

$$\sum_{j=0}^{n} a_{ij}(x')\, y_j = 0 \quad (i = 1, \ldots, s).$$

Then y' is also a solution of the equations

$$\sum_{j=0}^{n} a'_{ij}(x')\, y_j = 0 \quad (i = 1, \ldots, n-b).$$

Now let

$$\sum_{j=0}^{n} w_{ij} y_j = 0 \quad (i = 1, \ldots, b)$$

be the equations of b generic primes through the point y'. Since $(a'_{ij}(x'))$ is of rank $n-b$, the matrix

$$\begin{pmatrix} a'_{ij}(x') \\ w_{ij} \end{pmatrix}$$

is of rank n, and therefore not all the determinants

$$D_k(x', w, v) \quad (k = 0, \ldots, n)$$

can be zero. We therefore have the relations

$$\rho y'_k = D_k(x', w, v) \quad (k = 0, \ldots, n). \tag{5}$$

If $F(x, y)$ is any polynomial homogeneous in $(x_0, \ldots, x_m)$ and $(y_0, \ldots, y_n)$ such that $F(\xi, \eta) = 0$, then

$$F(\xi, D(\xi, u, v)) = 0.$$

Since ξ is a generic point of M, and the u_{ij}, v_{ij} are independent indeterminates over $K(\xi)$, it follows that

$$F(x', D(x', u, v)) = 0.$$

Replacing the indeterminates u_{ij} by the specialisations w_{ij},

$$F(x', D(x', w, v)) = 0,$$

so that, from (5),

$$F(x', y') = 0.$$

Hence (ξ, η) is a generic point-pair of the correspondence. The correspondence given by the equations (1) and (2) is therefore irreducible.

We apply this theorem and the Principle of Counting Constants to obtain a theorem already proved [X, § 8, Th. I]:

The intersection of an irreducible variety V of dimension $a > 0$ with a generic prime $\sum_{i=0}^{n} u_i x_i = 0$ is a variety irreducible over the field $K(u)$ and of dimension $a-1$.

We set up an algebraic correspondence between the points x of V and the primes y which pass through x. The equations of this correspondence C are the equations of V, say,

$$f_i(x) = 0 \quad (i = 1, \ldots, r),$$

together with the single equation which expresses the fact that the point x lies in the prime y. This is

$$\sum_{j=0}^{n} y_j x_j = 0.$$

By the theorem above, C is an irreducible correspondence, and a generic point-pair (ξ, η) of C is obtained by taking a generic prime η through a generic point ξ of V. The Principle of Counting Constants gives us, with the usual notation,

$$a + (n-1) = c + d. \tag{6}$$

The object- and image-varieties of C are the variety V, and some subvariety N of the totality of primes in S_n. We show that N coincides with this totality of primes, by proving that $c = n$. In fact, a generic prime η through a generic point ξ of V does not contain an arbitrary fixed point x' of V. η is therefore a prime which meets V but does not contain it. Hence the generic prime of the system of primes meeting V does not contain V. This means that the dimension of $M(\eta) \leqslant a-1$. That is, $d \leqslant a-1$. It follows from (6) that $c \geqslant n$. Since $c \leqslant n$, it follows that $c = n$. Hence also $d = a-1$.

We deduce that $M(\eta)$, which is the intersection with V of a generic prime of S_n, is irreducible over the field $K(\eta)$ and of dimension $a-1$.

In exactly the same way we can prove the more general

THEOREM II. *The intersection of an irreducible a-dimensional variety V $(a > 0)$ with a generic primal of order g is an irreducible variety of dimension $a-1$ relative to the field $K(u)$, where the $u_{i_0 \ldots i_n}$ are the indeterminate coefficients in the equation of the primal.*

The equation of a generic primal of order g is

$$\Sigma u_{i_0 \ldots i_n} x_0^{i_0} \ldots x_n^{i_n} = 0 \quad (i_0 + \ldots + i_n = g),$$

and we consider the correspondence between the x-space and the u-space defined by the equations of V and this equation. The proof then proceeds as above.

If we apply this theorem a times we have

THEOREM III. *The intersection of an irreducible k-dimensional variety V with k generic primals is a finite number of conjugate points.*

In particular we obtain a weaker form of X, § 7, Th. III:

THEOREM IV. *A generic linear subspace S_{n-a} of S_n meets an irreducible a-dimensional variety V in a finite number of points, conjugate over the field $K(u_{ij})$, where $\sum_{j=0}^{n} u_{ij} x_j = 0$ $(i = 1, \ldots, a)$ are the equations of the S_{n-a}.*

To find the stronger form of this theorem we must show that each point of the intersection is a generic point of V. This follows easily from two theorems:

(i) any point of an irreducible zero-dimensional variety is generic for that variety [X, § 5, Th. V],

(ii) with the notation used above, ξ is both a generic point of M over K and a generic point of $M(\eta)$ over $K(\eta)$ [§ 4].

We have already proved [X, § 5, Lemma II] that a generic linear subspace S_m of S_n has no points in common with an a-dimensional variety V if $a + m < n$. We use this theorem to prove

THEOREM V. *If in a correspondence C there corresponds to a generic point of the irreducible object-variety M a b-dimensional variety of image-points, there corresponds to any special point of M a variety of dimension at least b.*

If a variety is both reducible and impure, its dimension is defined to be the greatest dimension of any of its components. Let the image-variety of M lie in S_n. If to the equations of the correspondence we add b generic linear equations in the coordinates of the projective space S_n, we obtain a new correspondence in which there is assigned at least one image-point to a generic point of M, the object-variety of the original correspondence. A generic point of M therefore lies on the object-variety of the new correspondence, and therefore all points of M do so. Hence every point of M has an image-point in the new correspondence, which means that in the original correspondence the subvariety $N(x')$, where x' is any point of M, has at least one point in common with a generic S_{n-b} of the S_n in which $N(x')$ lies. The dimension of $N(x')$ is therefore at least b.

The result contained in Theorem V will be strengthened in § 6, Th. I.

6. Systems of algebraic varieties and related correspondences. In X, § 8, we introduced the notion of a system of algebraic varieties. Let $F(z, u_0, \ldots, u_d)$ be a form which is homogeneous of degree g in the indeterminates $(u_{i0}, \ldots, u_{in})$ $(i = 0, \ldots, d)$, with indeterminate coefficients $z_0, \ldots, z_D$. Then $F(z', u_0, \ldots, u_d)$ is the Cayley form of an algebraic variety of dimension d and order g (which is possibly reducible) if and only if z' satisfies the set of homogeneous algebraic equations

$$T_\omega(z) = 0 \quad (\omega = 1, 2, \ldots).$$

These equations determine a variety I in S_D, and if I_1 is any subvariety of I, the varieties whose Cayley forms are $F(z', u_0, \ldots, u_d)$, where z' is on I_1, form, by definition, an algebraic system of varieties.

Let the equations of the variety I_1 in S_D be

$$f_i(z) = 0 \quad (i = 1, 2, \ldots),$$

and let $S^0, \ldots, S^d$ be $d+1$ skew-symmetric matrices whose elements s^i_{jk} $(k > j)$ are independent indeterminates. Consider the algebraic correspondence between S_n and S_D defined by the equations

$$F(z, S^0x, \ldots, S^dx) = 0,$$

$$f_i(z) = 0 \quad (i = 1, 2, \ldots),$$

where in the first equation we equate to zero the various power-products of s^i_{jk} in $F(z, S^0x, \ldots, S^dx)$. The points x', z' in S_n and S_D correspond if and only if

$$F(z', S^0x', \ldots, S^dx') = 0,$$

$$f_i(z') = 0 \quad (i = 1, 2, \ldots),$$

that is, if and only if x' lies on the variety in S_n of dimension d and order g corresponding to the point z' of I_1. Thus an algebraic system of varieties of dimension d and order g defines an algebraic correspondence between the image-variety I_1 in S_D and S_n, in which the points of S_n which correspond to a point z' of I_1 are the points of the variety of dimension d and order g in S_n defined by z'. The object-variety in S_n contains all the points of S_n which correspond to points of I_1. Hence it contains all the varieties of the algebraic system, and through every point of the object-variety there passes at least one variety of the system. This object-variety is sometimes called the *carrier-variety* of the algebraic system.

We call I_1 the *Cayley image* of the algebraic system. We note that I_1 is uniquely defined (i.e. not merely defined to within birational transformation). If the system is irreducible, a generic point of the Cayley image of an algebraic system is called a *parameter* of the system. It is important to observe that there is a one-to-one correspondence without exception between the varieties of the system and specialisations of the parameter.

We now consider the relation between general correspondences and the special correspondences defined by algebraic systems of varieties. These special correspondences have the property that to every point z' of the image-variety there corresponds a variety of dimension d. But we have seen that there exist correspondences, even irreducible correspondences, in which the variety corresponding to a special point of the image-variety has a different dimension from that corresponding to a generic point [p. 106]. We propose, however, to associate with any given irreducible correspondence an irreducible system of varieties which will help us to obtain some general theorems on correspondences.

Let C be an irreducible correspondence given by the equations

$$f_i(x, y) = 0 \quad (i = 1, \ldots, r),$$

and let M in S_m and N in S_n be, respectively, its object- and image-varieties. The equations of the subvariety $N(\xi)$ of N which corresponds to a generic point ξ of M are

$$f_i(\xi, y) = 0 \quad (i = 1, \ldots, r).$$

We suppose that $N(\xi)$ is of dimension d and order g. We map $N(\xi)$ by means of its Cayley form $F(\zeta; u_0, \ldots, u_d)$ on an image-point ζ of the image-variety I in S_D which maps all varieties in S_n of order g and dimension d. The point ζ is the Cayley image of the variety $N(\xi)$. It is a generic point of some subvariety I_1 of I, and $N(\xi)$ is therefore a generic variety of some irreducible algebraic system of varieties.

Let the equations of I_1 be

$$T_\omega(z) = 0 \quad (\omega = 1, 2, \ldots),$$

this set of equations including the equations of I. We denote that variety of order g and dimension d in S_n which corresponds to a point z' of I_1 by the symbol $V(z')$. Then, by definition,

$$N(\xi) = V(\zeta).$$

If y' is any point of $N(x')$, the point-pair (x', y') is a pair of the correspondence C, and is therefore a proper specialisation of (ξ, η). This specialisation may be extended to a proper specialisation (x', y', z') of (ξ, η, ζ) [XI, § 1, Th. I]. Certain homogeneous algebraic relations which are preserved under this proper specialisation are of interest:

(i) $T_\omega(\zeta) = 0$ $(\omega = 1, 2, \ldots)$ becomes $T_\omega(z') = 0$ $(\omega = 1, 2, \ldots)$, and therefore z' is a point of I_1 and the Cayley image of some variety $V(z')$ of order g and dimension d.

(ii) η lies on $V(\zeta)$, so that $F(\zeta; S^0\eta, \ldots, S^d\eta) = 0$. After specialisation we have

$$F(z'; S^0y', \ldots, S^dy') = 0,$$

and therefore y' lies on $V(z')$.

(iii) $V(\zeta)$ is identical with $N(\xi)$. If

$$F(\zeta; u_0, \ldots, u_d) = A(u_1, \ldots, u_d) \prod_{\rho=1}^{g} \left(\sum_{i=0}^{n} u_{0i} \eta_i^{(\rho)} \right),$$

then $$f_i(\xi, \eta^{(\rho)}) = 0 \quad (i = 1, \ldots, r; \rho = 1, \ldots, g).$$

Eliminating the $\eta^{(\rho)}$ from these equations in the manner described in the proof of X, § 8, Th. II, we obtain a set of equations

$$H_\chi(\zeta, \xi) = 0$$

which are homogeneous in ζ and in ξ. Hence, after specialisation,

$$H_\chi(z', x') = 0.$$

Since z' satisfies the equations $T_\omega(z') = 0$, the Cayley form $F(z'; u_0, \ldots, u_d)$ defines a variety $V(z')$. This lies on the variety $N(x')$, for, because of the equations $H_\chi(z', x') = 0$, we have the factorisation

$$F(z'; u_0, \ldots, u_d) = B(u_1, \ldots, u_d) \prod_{\rho=1}^{g} \left(\sum_{i=0}^{n} u_{0i} \tau_i^{(\rho)} \right),$$

where we have

$$f_i(x', \tau^{(\rho)}) = 0 \quad (i = 1, \dots, r; \rho = 1, \dots, g),$$

and these equations show that the generic point $\tau^{(\rho)}$ of any component of $V(z')$ lies on $N(x')$.

This proves that there is a pure d-dimensional variety $V(z')$ which contains any given point y' of $N(x')$ and lies in $N(x')$. It follows that every point y' of $N(x')$ lies in an irreducible component of $N(x')$ of dimension not less than d. We have therefore proved

THEOREM I. *If in an irreducible correspondence between M and N there corresponds to a generic point ξ of M a d-dimensional variety of points $N(\xi)$, there corresponds to any special point x' of M a variety $N(x')$ no irreducible component of which has dimension less than d.*

We return now to the irreducible system of algebraic varieties which we wish to associate with the correspondence C. If we consider the irreducible correspondence C' defined by the generic point-pair (ξ, ζ), where ξ is a generic point of M and ζ is the Cayley image of $N(\xi)$, this correspondence has M as its object-variety, and its image-variety is an irreducible subvariety I_1 of I. The variety $N(\xi)$ is therefore a generic variety of an irreducible system of algebraic varieties $\mathfrak{S}_1$. It is this system which we associate with the given correspondence C.

A generic member of $\mathfrak{S}_1$ is $V(\zeta) = N(\xi)$. To any specialised point x' of M there corresponds, in the correspondence C', one or more points z'. To every such point z' in I_1 there corresponds a variety $V(z')$ of order g and dimension d. We can prove that *the aggregate of these varieties $V(z')$ is precisely the variety $N(x')$*. The argument given above shows that if y' is any point of $N(x')$, then y' lies on some variety $V(z')$ which corresponds to x', and conversely all points of $V(z')$ lie on $N(x')$. We have therefore proved

THEOREM II. *If in an irreducible correspondence between M and N there corresponds to a generic point ξ of M a d-dimensional variety $N(\xi)$ on N whose Cayley image is ζ, the generic point-pair (ξ, ζ) defines an irreducible correspondence between M and an irreducible variety I_1 on I. To every specialised point x' of M there corresponds one or more points z', which are the Cayley images of varieties $V(z')$ whose sum is precisely $N(x')$.*

With the help of Theorem I we can prove

THEOREM III. *The intersection of an irreducible variety V of dimension a with a primal $f = 0$ is either V itself or an unmixed variety of dimension $a-1$.*

We set up an algebraic correspondence between the points x of V and the primals of a fixed order g,

$$\Sigma u_{i_0 \ldots i_n} x_0^{i_0} \ldots x_n^{i_n} = 0 \quad (i_0 + \ldots + i_n = g),$$

which pass through x, and prove that $M(u)$, where u is a generic point of the linear space S_N on which the primals are mapped, is of dimension $a-1$. This is the content of § 5, Th. II. If u' is any rational point of S_N we learn from Theorem I that all components of $M(u')$ have dimension $\geqslant a-1$. But $M(u')$ is the intersection of V with the special primal

$$\Sigma u'_{i_0 \ldots i_n} x_0^{i_0} \ldots x_n^{i_n} = 0.$$

If any component of the intersection has dimension a, $M(u')$ coincides with V, and the primal contains V. If no component has dimension a, all components have dimension $a-1$, which proves the theorem for primals of any order g.

If we apply this theorem several times we obtain

THEOREM IV. *The intersection of an irreducible variety V of dimension a with k primals $f_1 = 0, \ldots, f_k = 0$ has no irreducible component of dimension $< a-k$.*

Exactly the same argument enables us to prove that if V is an irreducible variety in r-way space $S_{n_1 \ldots n_r}$, of dimension a, and

$$\phi_i(x, y, \ldots, t) \quad (i = 1, \ldots, k \leqslant a)$$

are k forms homogeneous in $x, y, \ldots, t$, the section of V by

$$\phi_i(x, y, \ldots, t) = 0 \quad (i = 1, \ldots, k)$$

has no component of dimension less than $a-k$.

We use this last remark to prove the *Principle of Plücker-Clebsch*:

THEOREM V. *Let $(\tau_0, \tau_1, \ldots, \tau_m)$ be a generic point of a variety U of dimension k. If*

$$f_\nu(\tau_0, \ldots, \tau_m; y_1, \ldots, y_n) = 0 \quad (\nu = 1, \ldots, n) \tag{1}$$

is a set of n equations in the n unknowns $y_1, \ldots, y_n$ which have an isolated solution $y'_1, \ldots, y'_n$ for some proper specialisation $\tau \to t'$, then the equations are soluble in an extension of the field of the coefficients when τ is not specialised.

By an *isolated* solution we mean one which does not lie on a variety of solutions of dimension greater than zero. Let

$$\phi_i(t_0, \ldots, t_m) = 0 \quad (i = 1, 2, \ldots) \tag{2}$$

be the equations of U, and let

$$g_i(t_0, \ldots, t_m; y_0, \ldots, y_n) = 0 \quad (i = 1, \ldots, n) \tag{3}$$

be the homogeneous equations obtained from (1) by replacing τ_i by t_i and by introducing y_0.

Equations (2) and (3) define a correspondence between the S_m containing U and the space S_n. By hypothesis these equations have a solution for $t' = (1, t'_1, \ldots, t'_m)$ which is isolated. Hence some irreducible component of this correspondence has an isolated solution for $t = t'$. Let C be such an irreducible component, q its dimension. In the 2-way space S_{mn} the equations (2) define a variety of dimension $k+n$, and (2) and (3) together define the intersection of this variety with n primals in S_{mn}. By the remark above, no component of this intersection is of dimension less than $(k+n)-n$. Hence, in particular, C is of dimension k at least; that is, $q \geqslant k$.

The object-variety M of the correspondence C lies on U. If it is of dimension a $(a \leqslant k)$ we know from the Principle of Counting Constants that to a generic point of M there corresponds a variety of dimension $q-a$. By Theorem I, when this generic point is specialised to t' no component of the corresponding variety has dimension less than $q-a$. But for this specialisation it has, by hypothesis, a component of dimension zero. Hence $q-a \leqslant 0$, and since $q \geqslant k \geqslant a$, it follows that $a = k = q$. The object-variety M therefore coincides with U.

We deduce that to a generic point of U there corresponds at least one point in C (and therefore in the correspondence defined by (2) and (3)). To prove our theorem, which refers to solutions of the non-homogeneous equations

$$f_\nu(\tau_0, \ldots, \tau_m; y_1, \ldots, y_n) = 0 \quad (\nu = 1, \ldots, n),$$

it is sufficient to show that not all the solutions of (3), with t replaced by τ, lie in $y_0 = 0$. If they did, the generic point (τ, η) of the component C of the correspondence would have $\eta_0 = 0$, and hence any proper specialisation of it would satisfy $y_0 = 0$. But (t', y') is a proper specialisation, where $y' = (1, y'_1, \ldots, y'_n)$, and this does not satisfy $y_0 = 0$. Hence $\eta_0 \neq 0$, and the theorem is proved.

For convenience in exposition we have used homogeneous coordinates $\tau_0, \ldots, \tau_m$. The conclusion remains valid, however, if we replace τ_0 by 1, provided that there exists an isolated solution of the specialised equations in which $(1, t'_1, \ldots, t'_m)$ is a proper specialisation of $(1, \tau_1, \ldots, \tau_m)$. In applications this non-homogeneous form is often more convenient.

We can prove in the same way that *if the set of equations* (1) *consists of* $n-r$ *equations, and possesses for* $\tau = t'$ *an isolated* r-*dimensional variety of solutions, then the set of equations* (1) *has at least* ∞^r *solutions for a generic* τ.

The Principle of Plücker-Clebsch has been used in the study of canonical forms, in which the main problem is that of the compatibility of n equations in n unknowns. For example, if we have a ternary cubic form $\Sigma a_{ijk} x_i x_j x_k$, we may ask whether it is possible to find three linear forms

$$X_i = b_{i0} x_0 + b_{i1} x_1 + b_{i2} x_2 \quad (i = 0, 1, 2)$$

and a constant λ so that

$$\Sigma a_{ijk} x_i x_j x_k \equiv X_0^3 + X_1^3 + X_2^3 + 6\lambda X_0 X_1 X_2.$$

If we equate coefficients of power-products in this identity in x_0, x_1, x_2 we obtain the equations

$$a_{ijk} = \phi_{ijk}(b, \lambda) \tag{4}$$

for the ten unknowns b_{ij}, λ. The theorem proved above says that if we can find a special value of the a_{ijk} for which there is an isolated solution b'_{ij}, λ', then the reduction of a ternary cubic form with generic coefficients to the canonical form shown is possible.

Let us take special values of the a_{ijk} so that

$$\Sigma a_{ijk} x_i x_j x_k = x_0^3 + x_1^3 + x_2^3.$$

We consider the equations obtained by equating coefficients on both sides of the identity

$$x_0^3 + x_1^3 + x_2^3 \equiv \sum_0^2 (b_{i0} x_0 + b_{i1} x_1 + b_{i2} x_2)^3 + 6\lambda \prod_{i=0}^{2} (b_{i0} x_0 + b_{i1} x_1 + b_{i2} x_2). \tag{5}$$

One solution of these equations is clearly

$$b'_{ij} = \delta_{ij}, \quad \lambda' = 0.$$

We prove that this is an isolated solution, and therefore that the given form is a canonical form.

We consider any solution b_{ij} of these equations. If $\det(b_{ij}) = 0$ there is a linear relation between the vectors (b_{i0}, b_{i1}, b_{i2}) $(i = 0, 1, 2)$, and the right-hand side of (5) can be expressed as a binary cubic form in, say, the indeterminates

$$u \equiv b_{10}x_0 + b_{11}x_1 + b_{12}x_2, \quad v \equiv b_{20}x_0 + b_{21}x_1 + b_{22}x_2.$$

Such a form is reducible in a suitable extension of the ground field, whereas the left-hand side of (5) is not reducible. Hence $\det(b_{ij}) \neq 0$.

Differentiating (5) with respect to x_0, we have the identity

$$\begin{aligned} 3x_0^2 \equiv 3\sum_0^2 b_{i0}(b_{i0}x_0 + b_{i1}x_1 + b_{i2}x_2)^2 \\ + 6\lambda b_{00}(b_{10}x_0 + b_{11}x_1 + b_{12}x_2)(b_{20}x_0 + b_{21}x_1 + b_{22}x_2) \\ + 6\lambda b_{10}(b_{00}x_0 + b_{01}x_1 + b_{02}x_2)(b_{20}x_0 + b_{21}x_1 + b_{22}x_2) \\ + 6\lambda b_{20}(b_{00}x_0 + b_{01}x_1 + b_{02}x_2)(b_{10}x_0 + b_{11}x_1 + b_{12}x_2). \end{aligned}$$

The left-hand side of this identity is a perfect square, and therefore the matrix

$$\begin{pmatrix} b_{00} & \lambda b_{20} & \lambda b_{10} \\ \lambda b_{20} & b_{10} & \lambda b_{00} \\ \lambda b_{10} & \lambda b_{00} & b_{20} \end{pmatrix}$$

is of rank one. Similar results are obtained by differentiating with respect to x_1 and x_2, and we see that the matrix

$$\begin{pmatrix} b_{0i} & \lambda b_{2i} & \lambda b_{1i} \\ \lambda b_{2i} & b_{1i} & \lambda b_{0i} \\ \lambda b_{1i} & \lambda b_{0i} & b_{2i} \end{pmatrix}$$

is of rank one, for $i = 0, 1, 2$.

If $\lambda = 0$, this implies that two of b_{0i}, b_{1i}, b_{2i} are zero for each value of i. Since $\det(b_{ij}) \neq 0$ it follows easily that in this case X_0, X_1, X_2 is a derangement of $\omega_0 x_0, \omega_1 x_1, \omega_2 x_2$, where $\omega_i^3 = 1$. Hence there is only a finite number of solutions with $\lambda = 0$.

If $\lambda \neq 0$ we have the equations

$$\begin{aligned} b_{0i}b_{1i} &= \lambda^2(b_{2i})^2, & b_{2i}^2 &= \lambda b_{0i}b_{1i}, \\ b_{1i}b_{2i} &= \lambda^2(b_{0i})^2, & b_{0i}^2 &= \lambda b_{1i}b_{2i}, \\ b_{0i}b_{2i} &= \lambda^2(b_{1i})^2, & b_{1i}^2 &= \lambda b_{0i}b_{2i}. \end{aligned}$$

If $b_{0i} = 0$, then $b_{2i} = b_{1i} = 0$. This is impossible, since $\det(b_{ij}) \neq 0$.

Hence it follows that $\lambda^3 = 1$. Therefore the solution $b'_{ij} = \delta_{ij}$, $\lambda' = 0$ is an isolated solution. It follows that $\Sigma a_{ijk} x_i x_j x_k$, where the a_{ijk} are indeterminates, can always be reduced to the canonical form $X_0^3 + X_1^3 + X_2^3 + 6\lambda X_0 X_1 X_2$.

7. Normal problems. In most problems which are of significance in geometry the data involve certain elements which can be varied, and what is sought is a solution of the problem when the variable elements are chosen as generally as possible; from the solution of the general problem deductions regarding the solutions of particular cases are then made. For instance, let us consider the elementary problem of finding the lines which meet four given lines l_1, l_2, l_3, l_4 in S_3. It is usually said that the number of solutions is two, but there are special configurations in which there is only one solution, and others in which there is an infinite number. The justification in this case for saying that there are two solutions in the general case is found in the fact that the aggregate of all curves in S_3 which consist of four lines is an algebraic system Ω, and the aggregate of those curves in S_3 which consist of four lines for which the number of transversals is different from two forms an algebraic system which is a proper subsvstem of Ω.

In this section we develop a general theory of geometrical problems with variable data, and examine the conditions under which such a problem may be said to have a solution 'in the general case'; we then go on to determine the conditions in which we can, from a knowledge of the solution in the general case, make assertions regarding particular cases. As a guide to the exact formulation of our theory we write down the equations to be solved in the problem of finding the transversals of four lines.

Let $\alpha_i, \alpha'_i, \beta_i, \beta'_i, \gamma_i, \gamma'_i, \delta_i, \delta'_i$ $(i = 0, 1, 2, 3)$ be 8×4 independent indeterminates over the ground field K. If

$$a_{ij} = \alpha_i \alpha'_j - \alpha_j \alpha'_i, \quad b_{ij} = \beta_i \beta'_j - \beta_j \beta'_i,$$
$$c_{ij} = \gamma_i \gamma'_j - \gamma_j \gamma'_i, \quad d_{ij} = \delta_i \delta'_j - \delta_j \delta'_i,$$

then $(\dots, a_{ij}, \dots), \quad (\dots, b_{ij}, \dots), \quad (\dots, c_{ij}, \dots), \quad (\dots, d_{ij}, \dots)$

are the coordinates of four independent generic lines in S_3. The lines which meet them are given by the solutions of the equations

$$\left.\begin{aligned} \Sigma a_{ij} p_{kl} = 0, \quad \Sigma b_{ij} p_{kl} = 0, \quad \Sigma c_{ij} p_{kl} = 0, \quad \Sigma d_{ij} p_{kl} = 0, \\ p_{01} p_{23} + p_{02} p_{31} + p_{03} p_{12} = 0. \end{aligned}\right\} \qquad (1)$$

If l_1, l_2, l_3, l_4 are any four lines in S_3 their coordinates are proper specialisations $(\ldots, a'_{ij}, \ldots)$ of $(\ldots, a_{ij}, \ldots)$, etc., and the problem of finding the transversals of l_1, l_2, l_3, l_4 is that of solving the specialised equations

$$\left.\begin{aligned} \Sigma a'_{ij} p_{kl} = 0, \quad \ldots, \quad \Sigma d'_{ij} p_{kl} = 0, \\ p_{01}p_{23} + p_{02}p_{31} + p_{03}p_{12} = 0. \end{aligned}\right\} \tag{2}$$

By saying that the problem has two solutions in the general case we mean that the equations (1) have two solutions. We cannot assert that (2) has two solutions, but, as we shall see later, we can use the fact that (1) has two solutions to make certain significant statements about the solutions of (2).

The problem considered is typical of the kind of problem with which we are concerned. These problems depend, for their solution, on the solution of a set of homogeneous equations

$$f_i(\xi, y) = 0 \quad (i = 1, 2, \ldots, r) \tag{3}$$

which are homogeneous in the unknowns $(y_0, \ldots, y_n)$, the coefficients in any $f_i(\xi, y)$ being homogeneous of the same degree in the coordinates $(\xi_0, \ldots, \xi_m)$ of the generic point of a variety M in S_m representing the variable elements of the data. In this way we see that the problem is a *correspondence* problem, and its solution depends on the properties of the correspondence between S_m and S_n given by the equations

$$\left.\begin{aligned} f_i(x, y) = 0 \quad (i = 1, 2, \ldots, r), \\ g_i(x) = 0 \quad (i = 1, 2, \ldots, s), \end{aligned}\right\} \tag{4}$$

where the second set of equations are the equations of M. It is essential for our theory that the equations of M, whose generic point is ξ, should include the resultant system of the set of equations $f_i(x, y) = 0$ with respect to y, and hence that the equations (3) have at least one solution. We note that while the correspondence may be reducible, the object-variety M must be irreducible, by definition. We do not exclude the possibility that the variety in S_n over $K(\xi)$ formed by the solutions of (3) may be reducible, or of dimension greater than zero. But it is found to be convenient to restrict ourselves to the case in which the variety of solutions of (3) is *unmixed*, of dimension b, say, where we assume $b \geqslant 0$, since the equations (3) have at least one solution.

A geometrical problem is called a *normal problem* if the variable elements in its data can be represented as the points of an irreducible variety M in S_m, whose generic point is ξ, and if its solution can be reduced to that of a set of equations of the form (3), the solutions for generic ξ being unmixed, of dimension $b \geqslant 0$.

This definition of a normal problem is sometimes generalised to include problems in which a solution is represented, not by a point in a space S_n, but by a finite set of points, one in S_{n_1}, one in S_{n_2}, ..., one in S_{n_r}. Then instead of the equations (3) we have equations

$$f_i(\xi, y^{(1)}, \ldots, y^{(r)}) = 0 \quad (i = 1, 2, \ldots), \tag{3)'$$

where $f_i(\xi, y^{(1)}, \ldots, y^{(r)})$ is homogeneous in $(y_0^{(i)}, \ldots, y_{n_i}^{(i)})$ $(i = 1, \ldots, r)$. This, however, merely means that a solution is represented by a point in r-way space $S_{n_1 \ldots n_r}$. If this, in turn, is represented by its Segre variety, we return to our original definition. There is thus no gain in generality in considering problems given by equations of type (3)′, and we shall confine ourselves to the case $r = 1$.

We begin with the case $b = 0$. The equations (3) then have a finite number of solutions, say $\eta^{(1)}, \ldots, \eta^{(g)}$. Let x' be a proper specialisation of ξ. Then there exist points $'y^{(1)}, \ldots, 'y^{(g)}$ in S_n such that $(x', 'y^{(1)}, \ldots, 'y^{(g)})$ is a proper specialisation of $(\xi, \eta^{(1)}, \ldots, \eta^{(g)})$ [§ 1, Th. I]. By the properties of proper specialisations, $'y^{(1)}, \ldots, 'y^{(g)}$ are solutions of

$$f_i(x', y) = 0 \quad (i = 1, 2, \ldots, r). \tag{5}$$

Certain questions present themselves at once: (i) for the specialisation $\xi \rightarrow x'$, are the points $'y^{(1)}, \ldots, 'y^{(g)}$ determined uniquely, at least when regarded as an unordered set? (ii) if they are uniquely determined as a set, is every solution of (5) contained in this set? That these properties need not hold can be seen in the following examples.

(1) We consider the cubic curve C in S_2 given by the equation

$$x_0 x_1 x_2 = x_0^3 + x_1^3.$$

Let us investigate the normal problem: to find where the join of $(0, 0, 1)$ to a point of C meets $x_2 = 0$. If the join to (x_0, x_1, x_2) meets $x_2 = 0$ in $(y_0, y_1, 0)$, we have

$$x_0 y_1 - x_1 y_0 = 0.$$

Since (x_0, x_1, x_2) lies on C and is not, in general, at $(0, 0, 1)$, we must have

$$\frac{x_0}{y_0^2 y_1} = \frac{x_1}{y_0 y_1^2} = \frac{x_2}{y_0^3 + y_1^3},$$

and these equations determine (y_0, y_1) for a generic point on C. A generic point of C is $\xi = (\lambda^2\mu, \lambda\mu^2, \lambda^3+\mu^3)$, where λ, μ are indeterminates over K. Corresponding to this form for a generic point the solution of the normal problem is $(\lambda, \mu, 0)$. For the specialisation $\xi \to (0, 0, 1)$, $(\lambda, \mu, 0) \to (1, 0, 0)$ or $(0, 1, 0)$, and hence there are *two* proper specialisations $(y_0', y_1', 0)$ of a solution of the normal problem.

(2) We consider the normal problem given by the equations

$$\xi_1 y_0 - \xi_0 y_1 = 0,$$
$$y_0 y_2 = 0,$$
$$y_2(y_1 - y_2) = 0,$$

ξ_0, ξ_1 being independent indeterminates. These equations have a unique solution $\eta = (\xi_0, \xi_1, 0)$. When $\xi = (\xi_0, \xi_1) \to (0, 1)$, $\eta \to (0, 1, 0)$, and $y' = (0, 1, 0)$ is determined uniquely. But the equations

$$y_0 = 0,$$
$$y_0 y_2 = 0,$$
$$y_2(y_1 - y_2) = 0$$

have solutions $(0, 1, 0)$ and $(0, 1, 1)$, the second of which is not a proper specialisation of η.

While it is true that in some cases results of geometrical interest can be obtained when the answers to our questions (i) and (ii) are in the negative, it is convenient, in those elementary parts of the theory of normal problems which we shall consider, to confine ourselves to the cases in which the specialisations $'y^{(1)}, \ldots, 'y^{(g)}$ of $\eta^{(1)}, \ldots, \eta^{(g)}$ are uniquely defined as a set. We therefore consider conditions for this to be the case. In fact, it is not easy to state, in any convenient form, conditions which are both necessary and sufficient, but some sufficient conditions are easily obtainable. These relate to the nature of the point x' on M. The best result we can give, which is not too complicated for general application, is the following:

THEOREM I. *If x' is a specialisation of ξ such that (a) the equations $f_i(x', y) = 0$ $(i = 1, 2, \ldots, r)$ have a finite number of solutions, and (b) the quotient ring of x' on M is integrally closed in its quotient field, then the specialisations $'y^{(1)}, \ldots, 'y^{(g)}$ of $\eta^{(1)}, \ldots, \eta^{(g)}$ corresponding to the specialisation $\xi \to x'$ are uniquely determined as a set.*

The concept 'quotient ring of a point' will be discussed in Vol. III. In the meantime it is sufficient to say that condition (*b*) means that if $\phi(x_0, \ldots, x_m)$ and $\psi(x_0, \ldots, x_m)$ are two homogeneous polynomials of the same degree, the second, at least, not vanishing on M, such that the form

$$a_0(x_0, \ldots, x_m)\,\phi^r + a_1(x_0, \ldots, x_m)\,\phi^{r-1}\psi + \ldots + a_r(x_0, \ldots, x_m)\,\psi^r$$

vanishes on M, while $a_0(x_0', \ldots, x_m') \neq 0$, there exist polynomials $b(x_0, \ldots, x_m)$ and $c(x_0, \ldots, x_m)$ such that

$$c(x_0, \ldots, x_m)\,\phi - b(x_0, \ldots, x_m)\,\psi$$

vanishes on M, while $c(x_0', \ldots, x_m') \neq 0$.

The equations

$$f_i(x', y) = 0 \quad (i = 1, 2, \ldots, r)$$

have, by hypothesis, only a finite number of solutions, and we may therefore suppose that none lies in $y_0 = 0$. Since each $\eta^{(\rho)}$ has a specialisation $'y^{(\rho)}$, it follows that $\eta_0^{(\rho)} \neq 0$, and we normalise so that $\eta_0^{(\rho)} = 1$. We can find a form $a_{00}(x_0, \ldots, x_m)$ such that

$$a_{00}(\xi)\prod_{\rho=1}^{g}\left(\sum_{i=0}^{n} u_i \eta_i^{(\rho)}\right) = a_{00}(\xi)\,u_0^g + a_{01}(\xi)\,u_0^{g-1}u_1 + a_{02}(\xi)\,u_0^{g-1}u_2 + \ldots, \qquad (6)$$

where the $a_{00}(x_0, \ldots, x_m)$, $a_{01}(x_0, \ldots, x_m)$, $a_{02}(x_0, \ldots, x_m)$, ... are forms in $K[x_0, \ldots, x_m]$ and are all of the same degree. For the coefficients of the various power-products of the u_i in $\prod_{\rho=1}^{g}\left(\sum_{i=0}^{n} u_i \eta_i^{(\rho)}\right)$ are symmetric functions of the coordinates $\eta_i^{(\rho)}$, and are therefore in $K(\xi)$, and can be written in the form $a_k(\xi)/a_0(\xi)$, where $a_i(\xi)$ is in $K[\xi]$. We choose homogeneous polynomials

$$a_{00}(x_0, \ldots, x_m), \quad a_{01}(x_0, \ldots, x_m), a_{02}(x_0, \ldots, x_m), \quad \ldots$$

of the same degree such that

$$a_{00}(\xi) = a_0(\xi), \quad a_{01}(\xi) = a_1(\xi), \quad a_{02}(\xi) = a_2(\xi), \quad \ldots,$$

and obtain (6).

We now consider the equations

$$z_{00}\prod_{\rho=1}^{g}\left(\sum_{i=0}^{n} u_i y_i^{(\rho)}\right) = \left(\prod_{\rho=1}^{g} y_0^{(\rho)}\right)[z_{00}u_0^g + z_{01}u_0^{g-1}u_1 + z_{02}u_0^{g-1}u_2 + \ldots], \quad (7)$$

$$z_{ij}\,a_{kl}(x) = z_{kl}\,a_{ij}(x). \qquad (8)$$

These equations are satisfied by $x = \xi$, $y^{(\rho)} = \eta^{(\rho)}$, $z_{ij} = a_{ij}(\xi)$, and are homogeneous in the unknowns x, $y^{(\rho)}$ and z respectively. Corresponding to the specialisation $\xi \to x'$, $\eta^{(\rho)} \to \bar{y}^{(\rho)}$ ($\rho = 1, \ldots, g$) (where we do not assume that $\bar{y}^{(1)}, \ldots, \bar{y}^{(g)}$ are $'y^{(1)}, \ldots, 'y^{(g)}$ in some order) there is at least one proper specialisation $z \to z'$, and (7) becomes

$$z'_{00} \prod_{\rho=1}^{g} \left(\sum_{i=0}^{n} u_i \bar{y}_i^{(\rho)} \right) = \left(\prod_{\rho=1}^{g} \bar{y}_0^{(\rho)} \right) [z'_{00} u_0^g + z'_{01} u_0^{g-1} u_1 + \ldots]. \tag{9}$$

The $\bar{y}^{(\rho)}$ are among the solutions of the equations $f_i(x', y) = 0$ ($i = 1, \ldots, r$), and therefore $\bar{y}_0^{(\rho)} \neq 0$. If $z'_{00} = 0$ it follows from (9) that

$$z'_{00} u_0^g + z'_{01} u_0^{g-1} u_1 + \ldots = 0.$$

Since the u_i are indeterminates we deduce that all $z'_{ij} = 0$, which contradicts the hypothesis that $z \to z'$ is a proper specialisation. Hence $z'_{00} \neq 0$.

Now let
$$\tau = (\tau_{00}, \tau_{ij}) = (a_{00}(\xi), a_{ij}(\xi)).$$

Then
$$z_{ij} \tau_{00} = z_{00} \tau_{ij}.$$

Consider any specialisation (t'_{00}, t'_{ij}) of τ corresponding to the specialisation $(\xi, \eta^{(\rho)}, z) \to (x', \bar{y}^{(\rho)}, z')$. Then

$$z'_{ij} t'_{00} = z'_{00} t'_{ij},$$

and since $z'_{00} \neq 0$, and t'_{00}, t'_{ij} are not both zero, it follows that $t'_{00} \neq 0$.

Since (τ_{00}, τ_{ij}) does not specialise to $(0, t'_{ij})$ there must exist a form $f(x; y_0, y_1)$, homogeneous in $(x_0, \ldots, x_m)$ and in (y_0, y_1) such that

$$f(\xi; \tau_{00}, \tau_{ij}) = 0,$$
$$f(x'; 0, t'_{ij}) \neq 0.$$

Hence $f(x; y_0, y_1) = b_0(x) y_0^\lambda + b_1(x) y_0^{\lambda-1} y_1 + \ldots + b_\lambda(x) y_1^\lambda$, where $b_\lambda(x') \neq 0$. By the property (b) of x' it follows that we can find forms $c_{00}(x)$, $c_{ij}(x)$ such that $c_{00}(x') \neq 0$, and

$$a_{00}(x) c_{ij}(x) - a_{ij}(x) c_{00}(x)$$

vanishes on M.

On M we therefore have

$$a_{00}(x) : a_{01}(x) : \ldots = c_{00}(x) : c_{01}(x) : \ldots,$$

and therefore

$$a_{00}(\xi) : a_{01}(\xi) : \ldots = c_{00}(\xi) : c_{01}(\xi) : \ldots.$$

Hence, from (6),

$$c_{00}(\xi) \prod_{\rho=1}^{g} \left(\sum_{i=0}^{n} u_i \eta_i^{(\rho)} \right) = c_{00}(\xi) u_0^g + c_{01}(\xi) u_0^{g-1} u_1 + \ldots.$$

Since $c_{00}(x') \neq 0$, it follows that any specialisation of $(\eta^{(1)}, \ldots, \eta^{(g)})$ corresponding to the specialisation $\xi \to x'$ is determined by the factors of

$$c_{00}(x')\, u_0^g + c_{01}(x')\, u_0^{g-1} u_1 + \ldots,$$

and hence that the specialisations of the points $\eta^{(1)}, \ldots, \eta^{(g)}$ are uniquely determined as a set.

It will be proved in Vol. III that the quotient ring of any simple point of an irreducible variety is integrally closed in its quotient field. Hence we have as a corollary to Theorem I:

Corollary. If x' is a simple point of M, and there is only a finite number of points in S_n corresponding to x', then the specialisations of $\eta^{(1)}, \ldots, \eta^{(g)}$ are uniquely determined as a set.

By an application of some of the ideas used above we may obtain some useful information concerning the specialisations $'y^{(1)}, \ldots, 'y^{(g)}$ of $\eta^{(1)}, \ldots, \eta^{(g)}$, even when these are not unique. We again suppose, as we may, that $\eta_0^{(\rho)} \neq 0$ $(\rho = 1, \ldots, g)$ and normalise $\eta^{(\rho)}$ so that $\eta_0^{(\rho)} = 1$. Then, as above, we can find a form $a(x_0, \ldots, x_m)$ such that

$$a(\xi) \prod_{\rho=1}^{g} \left(\sum_{i=0}^{n} u_i \eta_i^{(\rho)} \right) = F(\zeta, u)$$

is a form in $u_0, \ldots, u_n$ with coefficients which are homogeneous polynomials in $\xi_0, \ldots, \xi_m$. $F(\zeta, u)$ is the Cayley form of the zero-dimensional variety determined by $\eta^{(1)}, \ldots, \eta^{(g)}$. The coefficients of the various power-products of the u_i determine the Cayley image ζ of this zero-dimensional variety.

Now consider the point $(\xi, \zeta, \eta^{(1)}, \ldots, \eta^{(g)})$. Corresponding to any specialisation $\xi \to x'$, $\eta^{(\rho)} \to {}'y^{(\rho)}$ $(\rho = 1, \ldots, g)$ there corresponds a specialisation $\zeta \to z'$. Conversely, to the specialisation $(\xi, \zeta) \to (x', z')$ there corresponds a specialisation of $(\xi, \eta^{(1)}, \ldots, \eta^{(g)})$ to $(x', {}'y^{(1)}, \ldots, {}'y^{(g)})$. Hence the specialisations of $\eta^{(1)}, \ldots, \eta^{(g)}$ corresponding to the specialisation $\xi \to x'$ are obtained by considering the specialisations $F(z', u)$ of the form $F(\zeta, u)$ which correspond to $\xi \to x'$. In particular, the specialisations of $\eta^{(1)}, \ldots, \eta^{(g)}$ are uniquely determined as a set if and only if there is a unique specialisation of ζ corresponding to the specialisation $\xi \to x'$. From these considerations and § 6, Th. II, we deduce

THEOREM II. *If C is an irreducible correspondence such that (a) the quotient ring of every point of the object-variety is integrally closed in its quotient field; (b) the variety $N(x')$ corresponding to every point*

x' of M is zero-dimensional, then the varieties $N(x')$ form an irreducible algebraic system.

If the specialisations $'y^{(1)}, \ldots, 'y^{(g)}$ of $\eta^{(1)}, \ldots, \eta^{(g)}$ are uniquely determined as a set when $\xi \to x'$, the points $'y^{(\rho)}$ are not necessarily all distinct. Suppose that only s of them are distinct, and that the number of specialised points with coordinates equal to those of $y'^{(\rho)}$ is r_ρ. Then

$$\sum_{\rho=1}^{s} r_\rho = g. \tag{10}$$

The number r_ρ attached to $'y^{(\rho)}$ is uniquely determined. It is called the *multiplicity* of $'y^{(\rho)}$ as a solution of the equations

$$f_i(x', y) = 0 \quad (i = 1, \ldots, r).$$

We note that the multiplicity of $'y^{(\rho)}$ is only defined when the set $'y^{(1)}, \ldots, 'y^{(g)}$ is uniquely determined. Equation (10) is sometimes described in the following terms:

Principle of Conservation of the Number. The sum of the multiplicities of the solutions of the equations

$$f_i(\xi, y) = 0 \quad (i = 1, \ldots, r)$$

for all specialisations $\xi \to x'$ is the same, provided that the multiplicities are defined.

We have not, of course, excluded the possibility of a solution of the equations

$$f_i(x', y) = 0 \quad (i = 1, \ldots, r)$$

not appearing among the specialised solutions $'y^{(1)}, \ldots, 'y^{(g)}$. Indeed, we had an example earlier in this section:

$$\xi_1 y_0 - \xi_0 y_1 = y_0 y_2 = y_2(y_1 - y_2) = 0.$$

The multiplicities of the solutions corresponding to the specialisation $(\xi_0, \xi_1) \to (0, 1)$ are uniquely defined, but the equations which arise from the specialisation have a solution which is not a proper specialisation of the solution $(\xi_0, \xi_1, 0)$ of the generic equations, namely, $(0, 1, 1)$.

According to our definition, we cannot attach a multiplicity to such a solution, but we shall say that such a solution has *multiplicity zero*, and speak of those solutions for which a multiplicity is defined as having *positive multiplicity*. A sufficient condition for all solutions to have positive multiplicity is given by

THEOREM III. *If the multiplicities of solutions of a normal problem be defined, a sufficient condition for every solution to have positive multiplicity is that the correspondence defined by the problem be irreducible.*

This follows at once from the fact that if the correspondence defined by the equations

$$f_i(x, y) = 0 \quad (i = 1, \ldots, r)$$

be irreducible, any solution (x', y') of the equations is a proper specialisation of a generic point-pair (ξ, η) of the correspondence.

So far, we have only dealt with normal problems for which the number of solutions in the generic case is finite. The removal of this restriction is a simple matter. Let

$$f_i(\xi, y) = 0 \quad (i = 1, \ldots, r) \tag{11}$$

be the equations of a normal problem for which the solutions, for generic ξ, form a variety of dimension b (unmixed, by definition of a normal problem). Let u_{ij} $(i = 1, \ldots, b; j = 0, \ldots, n)$ be $b(n+1)$ independent indeterminates over $K(\xi)$. We add to equations (11) the equations

$$\sum_{j=0}^{n} u_{ij} y_j = 0 \quad (i = 1, \ldots, b).$$

This reduces the problem to the case $b = 0$. If $\eta^{(1)}, \ldots, \eta^{(g)}$ are the solutions of the new problem (as usual, we may assume that $\eta_0^{(\rho)} = 1$),

$$F(\zeta, u_0, \ldots, u_b) = A(u_1, \ldots, u_b) \prod_{\rho=1}^{g} \left(\sum_{i=0}^{n} u_{0i} \eta_i^{(\rho)} \right)$$

is the Cayley form of the variety of solutions of (11). We can read off from this the properties of (11). We leave to the reader the proof of

THEOREM IV. *If the solution of* (11) *for the specialisation* $\xi \to x'$ *is purely b-dimensional, and the quotient ring of the point x' of the object-variety is integrally closed in its quotient field, then the multiplicities of the components $V_1, \ldots, V_k$ of the solutions of the equations*

$$f_i(x', y) = 0 \quad (i = 1, \ldots, r)$$

are uniquely determined. If the order of V_i is g_i and its multiplicity is r_i,

$$\Sigma r_i g_i = g,$$

where g is the order of the variety of solutions of (11).

If the correspondence defined by the normal problem be irreducible, $r_i > 0$ $(i = 1, \ldots, k)$.

We can apply this theorem to the problem of the intersection of an irreducible algebraic variety V of dimension d $(d \geq 1)$ and order g in S_n and a prime Π. We saw in § 5 that this problem can be expressed as a normal problem, the associated correspondence C being irreducible, its object-variety M being an n-space. To a generic point of M there corresponds an irreducible variety in S_n of order g [X, § 8, Th. I]. Since the conditions of Theorem IV are fulfilled, we conclude that the intersection of V by any prime not containing V is an unmixed variety of total order g.

This result could be extended to the section of V by a k-space $(k \geq n-d)$, but we do not do this here, since the result is contained in the general theory of intersections studied in the next chapter.

8. Multiplicative varieties. The association of the concept of multiplicity with a variety forms the basis of a multiplicative theory of varieties which plays an important role in numerous aspects of algebraic geometry. Let V_d be any unmixed variety of dimension d in S_n,

$$V_d = V_d^{(1)} \dotplus \ldots \dotplus V_d^{(h)}, \tag{1}$$

where $V_d^{(i)}$ is an irreducible variety of dimension d and order g_i. Let $F_i(u_0, \ldots, u_d)$ be the Cayley form of $V_d^{(i)}$; it is irreducible over K and of degree g_i in $u_{j0}, \ldots, u_{jn}$, for $j = 0, \ldots, d$. The form

$$F(u_0, \ldots, u_d) = \prod_{i=1}^{h} [F_i(u_0, \ldots, u_d)]^{a_i}, \tag{2}$$

where $a_1, \ldots, a_h$ are positive integers, satisfies the conditions for a Cayley form [X, § 8] of an algebraic variety which, regarded as a set of points, coincides with V_d. We now consider a new entity, consisting of the variety V_d associated with the form $F(u_0, \ldots, u_d)$, for a given choice of the exponents $a_1, \ldots, a_h$. If V_d is irreducible and the exponent $a = 1$, we denote this entity by $\mathbf{V}_d$, using Clarendon type, and continue to speak of $\mathbf{V}_d$ as an irreducible variety (in the multiplicative sense). More generally, if V_d is given by (1), and the associated form is given by (2), we write

$$\mathbf{V}_d = a_1 \mathbf{V}_d^{(1)} + \ldots + a_h \mathbf{V}_d^{(h)},$$

and say that V_d is *a variety in the multiplicative sense*. The order in which the terms of the sum are written is immaterial. The convention

$$1V_d = V_d$$

enables us to include the definition of V_d, where V_d is irreducible, in this definition. It is also convenient to extend our definition to include the case in which some of the exponents a_i in (2) are zero; we do this by the convention that if, for instance, a_h is zero,

$$\begin{aligned} V_d &= a_1 V_d^{(1)} + \dots + a_h V_d^{(h)} \\ &= a_1 V_d^{(1)} + \dots + a_{h-1} V_d^{(h-1)}. \end{aligned}$$

We now define aV_d, where a is any non-negative integer, and V_d is any multiplicative variety

$$V_d = a_1 V_d^{(1)} + \dots + a_h V_d^{(h)},$$

by the equation

$$aV_d = aa_1 V_d^{(1)} + \dots + aa_h V_d^{(h)}.$$

Thus the point-set variety associated with aV_d is that associated with V_d, while the Cayley form of aV_d is the ath power of that of V_d. We define the sum of two varieties of dimensions d in the multiplicative sense by the equations

$$aV_d + bV_d = (a+b)\,V_d,$$

$$\begin{aligned} (a_1 V_d^{(1)} + \dots + a_h V_d^{(h)}) &+ (b_1 U_d^{(1)} + \dots + b_k U_d^{(k)}) \\ &= a_1 V_d^{(1)} + \dots + a_h V_d^{(h)} + b_1 U_d^{(1)} + \dots + b_k U_d^{(k)}. \end{aligned}$$

In fact, the addition of varieties of dimension d, in the multiplicative sense, is given by the product of their Cayley forms. With this law of addition, the addition of varieties in the multiplicative sense is clearly associative and commutative.

At a later stage, we shall find it convenient to extend our concept of a variety in the multiplicative sense by allowing the coefficients $a_1, a_2, \dots$ to take negative integral values. The operation of addition is then defined as above, so that if the zero variety is that in which all the coefficients a_i are zero, the set of d-dimensional varieties in S_n, in the multiplicative sense, forms an additive semi-group.

If

$$V_d = a_1 V_d^{(1)} + \dots + a_h V_d^{(h)},$$

where $V_d^{(1)}, \dots, V_d^{(h)}$ are irreducible, we speak of $V_d^{(1)}, \dots, V_d^{(h)}$ as the *irreducible components* of V_d, and of $a_1, \dots, a_h$ as the *multiplicities* of these components. If $V_d^{(i)}$ is of order g_i, V_d is said to be of order

$\Sigma a_i g_i$; in some works it is customary to speak of this as the order of V_d 'when the components are counted with their proper multiplicities'.

The multiplicative theory of varieties is the proper medium in which to interpret the concept of a normal problem. In any normal problem the generic case has a solution

$$V_d(\xi) = V_d^{(1)}(\xi) \dotplus \ldots \dotplus V_d^{(h)}(\xi),$$

where $V_d^{(1)}, \ldots, V_d^{(h)}$ are distinct irreducible varieties of dimension d. What we are really doing is to define from our generic solution the multiplicative variety

$$\mathbf{V}_d(\xi) = \mathbf{V}_d^{(1)}(\xi) + \ldots + \mathbf{V}_d^{(h)}(\xi),$$

in which each component is *simple*, that is, of multiplicity one. If $\xi \to x'$ is a specialisation of ξ for which it is possible to assign multiplicities to the solution of the derived problem, let $\bar{V}_d^{(1)} \ldots \bar{V}_d^{(k)}$ be the components of this solution, a_i the multiplicity assigned to $\bar{V}_d^{(i)}$. Then if $\mathbf{V}_d = a_1 \mathbf{V}_d^{(1)} + \ldots + a_k \mathbf{V}_d^{(k)}$, $\mathbf{V}_d$ is called the proper specialisation of $\mathbf{V}_d(\xi)$ corresponding to the specialisation $\xi \to x'$, and it is clear from our definition of multiplicity that this is equivalent to defining the proper specialisation of $\mathbf{V}_d(\xi)$ to be the (unique) multiplicative variety defined by the proper specialisation of the Cayley form of $\mathbf{V}_d(\xi)$. This concept can, of course, easily be extended to define the proper specialisation of any variety

$$\mathbf{V}_d(\xi) = a_1 \mathbf{V}_d^{(1)}(\xi) + \ldots + a_h \mathbf{V}_d^{(h)}(\xi),$$

corresponding to a specialisation of ξ, as the variety whose Cayley form is the proper specialisation of the Cayley form of $V_d(\xi)$.

Thus it is clear that in a normal problem we are simply choosing the generic multiplicative variety $\mathbf{V}_d(\xi)$ to be that obtained by assigning unit multiplicity to each component of the point-set solution of the generic case of the normal problem. For any specialisation $\xi \to x'$, the multiplicities of the specialised problem are then defined as the multiplicities of the components of the corresponding specialisation of $V_d(\xi)$, and are properly defined if and only if this is unique.

9. A criterion for unit multiplicity. We now know some conditions under which the solutions of a normal problem have a positive multiplicity. As we saw in § 7, these conditions cover some important cases. But in applications of the Principle of Conservation of the Number, in particular to problems in enumerative geometry, we must also be able to calculate the exact multiplicity of a solution. The theorem we are about to prove tells us that under certain conditions the multiplicity of a solution does not exceed unity. If these conditions are satisfied, and if we also know that the solution has positive multiplicity, it follows that the solution has unit multiplicity exactly. If this is the case for every solution of a selected special case of a normal problem it then follows, by the Principle of Conservation of the Number, that the number of distinct solutions in the generic case is exactly equal to the number of solutions in the special case selected. This special case is often easier to study than the generic case.

THEOREM I. *If a normal problem is given by the equations*

$$f_i(\xi, y) = 0 \quad (i = 1, \ldots, r), \tag{1}$$

and has the property that for generic ξ it has a finite number of solutions, and if, under the proper specialisation $\xi \to x'$, two distinct solutions η', η'' specialise to the same solution y', the specialised primals

$$f_i(x', y) = 0 \quad (i = 1, \ldots, r) \tag{2}$$

have a common tangent line at the point $y = y'$.

We prove that the line joining η' to η'' specialises to a tangent as $\xi \to x'$.

We may assume that if $y' = (y'_0, \ldots, y'_n)$, then $y'_0 \neq 0$. It follows that $\eta'_0 \neq 0$, and $\eta''_0 \neq 0$, and we may put $y'_0 = \eta'_0 = \eta''_0 = 1$. We now write

$$\tau_k = \eta''_k - \eta'_k.$$

Then $\tau_0 = 0$. The proper specialisation $(\xi, \eta', \eta'') \to (x', y', y')$ can be extended to a specialisation $(\xi, \eta', \eta'', \tau) \to (x', y', y', t')$, where $t'_0 = 0$. We have the equations

$$f_i(\xi, \eta') = 0,$$

$$f_i(\xi, \eta'') = f_i(\xi, \eta' + \tau) = 0.$$

This last equation can be expanded in powers of $\tau_1, \ldots, \tau_n$ and becomes

$$f_i(\xi, \eta') + \sum_1^n \tau_k \frac{\partial f_i(\xi, \eta')}{\partial \eta'_k} + \text{higher powers} = 0,$$

that is

$$\sum_1^n \tau_k \frac{\partial f_i(\xi, \eta')}{\partial \eta'_k} + \text{higher powers} = 0. \tag{3}$$

The degree of each term of (3) in the τ_k is at least one. By replacing τ_k by $\eta''_k - \eta'_k$ we can make the left-hand side of (3) homogeneous of degree one in the τ_k. If we then also make the left-hand side of (3) homogeneous in the η'_i and the η''_j respectively by introducing η'_0 and η''_0, we have an equation which continues to hold after the specialisation $(\xi, \eta', \eta'', \tau) \to (x', y', y', t')$. Since $\eta''_k - \eta'_k$ (or $\eta''_k \eta'_0 - \eta'_k \eta''_0$) vanishes under the specialisation, because both η'' and η' specialise to y', only the first term of (3) remains, and this is now

$$\sum_1^n t'_k \frac{\partial f_i(x', y')}{\partial y'_k} = 0.$$

This equation shows that the tangent primes at y' to the specialised primals $f_i(x', y) = 0$ have in common the point

$$t' = (0, t'_1, \ldots, t'_n).$$

Since these tangent primes all pass through the point

$$y' = (1, y'_1, \ldots, y'_n),$$

it follows that the line $y't'$ lies in all the tangent primes. This proves the theorem.

Corollary. If (x', y') is a proper specialisation of (ξ, η), and the matrix $\left(\frac{\partial f_i(x', y')}{\partial y'_j}\right)$ is of rank n, y' counts simply in the solutions of the equations

$$f_i(x', y) = 0 \quad (i = 1, \ldots, r).$$

This follows from the fact that the linear equations

$$\sum_{j=0}^n z_j \frac{\partial f_i(x', y')}{\partial y'_j} = 0$$

define primes with exactly one common point of intersection, the point y'.

We can extend this corollary to the case in which the solutions of the normal problem are unmixed varieties of dimension b. If we add to the equations

$$f_i(\xi, y) = 0 \quad (i = 1, \ldots, r)$$

the equations of b generic primes

$$\sum_{j=0}^n u_{ij} y_j = 0 \quad (i = 1, \ldots, b),$$

we obtain a set of points $\eta^{(1)}, \ldots, \eta^{(g)}$ which are generic points for the various components of $N(\xi)$. If $N(x')$ has a component which has multiplicity greater than unity, the generic S_{n-b} defined above meets this component in a generic point y' which is a proper specialisation of more than one $\eta^{(i)}$. Hence, by the corollary, the matrix

$$\begin{pmatrix} \dfrac{\partial f_i(x', y')}{\partial y'_j} \\ \\ u_{ij} \end{pmatrix}$$

is of rank less than n. Since the u_{ij} are indeterminates, we deduce that the matrix

$$\left(\frac{\partial f_i(x', y')}{\partial y'_j}\right)$$

is of rank less than $n-b$. Hence we have

THEOREM II. *If in the normal problem given by the equations*

$$f_i(x, y) = 0 \quad (i = 1, \ldots, r)$$

the variety $N(\xi)$ is unmixed of dimension b, sufficient conditions for the component of $N(x')$ with generic point y' to have unit multiplicity are (i) *the component has a unique positive multiplicity, and* (ii) *the matrix $\left(\dfrac{\partial f_i(x', y')}{\partial y'_j}\right)$ is of rank $n-b$.*

10. Simple points. The proof of Theorem II of § 5 shows that the problem of finding the intersection of an irreducible variety V of d dimensions in S_n with a prime is a normal problem for which the associated correspondence is irreducible (in short, an irreducible normal problem). Repeating the argument, we see that the problem of finding the intersection of the irreducible variety V with an $(n-d)$-space is an irreducible normal problem. The object-variety M represents the system of all $(n-d)$-spaces in S_n. Since every point of this is simple [cf. Chapter XIV], it follows that whenever a space S'_{n-d} meets V in a finite number of points, a unique positive multiplicity can be attached to each of the intersections, so that the sum of these multiplicities is g, the order of V.

It often happens that we have to consider the intersection of V with a space S_{n-d} which is a generic $(n-d)$-space through an assigned point x' of S_n, and it is convenient to establish here some results for this case. In particular, we shall consider the case in which x' is a simple point of V, and we shall obtain a new characterisation of a simple point of an irreducible variety.

If x' is any point of S_n, we adjoin its coordinates to the ground field K. Over $K(x')$, V may be reducible, but each of its components is of dimension d [X, § 11, Th. I]. We need only consider one of these irreducible components V'. We then make a transformation of coordinates so that x' becomes the point $(1, 0, \ldots, 0)$, and consider the intersections of V' with $(n-d)$-spaces through this point. A prime through x' has an equation of the form

$$u_1 y_1 + \ldots + u_n y_n = 0. \tag{1}$$

Let ξ be a generic point of V', and let η be a generic solution of

$$u_1 \xi_1 + \ldots + u_n \xi_n = 0.$$

The correspondence given by the equations of V' and (1) is not necessarily irreducible if x' lies on V, but the proof of Theorem I, § 5 shows that if x'' is any point of V' other than x', and u'' is any prime through x' and x'', then (x'', u'') is a proper specialisation of (ξ, η). Continuing this argument, we find that if ζ is a generic $(n-d)$-space through x' at ξ, and if x'' is any point of V' different from x', and z'' is any $(n-d)$-space through x' and x'', (x'', z'') is a proper specialisation of (ξ, ζ). Applying the arguments of § 4 to the correspondence whose generic point-pair is (ξ, ζ), we see that if a generic $(n-d)$-space through x' meets V' in any point different from x', the number of such points is finite, and each is a generic point of V' over $K(x')$. Since a generic point of V' over $K(x')$ is a generic point of V over K, we obtain

THEOREM I. *If x' is any point of S_n, which may or may not be on V, a generic S_{n-d} through x' meets V elsewhere in a finite number of points at most, each of which is a generic point of V.*

We have now to consider the multiplicities of the intersection of V with a generic $(n-d)$-space through x', as determined by specialisation from the normal problem of the intersection of V with $(n-d)$-spaces of S_n.

We only consider the case in which x' is a simple point of V. Let

$$f_i(x) = 0 \quad (i = 1, \ldots, r)$$

be a basis for the equations of V. Then, if

$$\pi = (\ldots, \pi_{i_0 \ldots i_{n-d}}, \ldots)$$

is a generic $(n-d)$-space of S_n, the normal problem is given by the equations

$$\sum_{j=0}^{n-d+1} (-1)^j \pi_{i_0 \ldots i_{j-1} i_{j+1} \ldots i_{n-d+1}} x_{i_j} = 0,$$

$$f_i(x) = 0 \quad (i = 1, \ldots, r).$$

If we specialise π to be a generic $(n-d)$-space π' through x', the point x' is a solution of the resulting system of equations, and has multiplicity one if the equations

$$\sum_{j=0}^{n-d+1} (-1)^j \pi'_{i_0\ldots i_{j-1}i_{j+1}\ldots i_{n-d+1}} x_{i_j} = 0, \tag{2}$$

$$\sum_j x_j \frac{\partial f_i}{\partial x'_j} = 0 \quad (i = 1, \ldots, r) \tag{3}$$

have x' as an isolated solution. Since x' is a simple point of V, equations (3) determine the tangent d-space to V at x'. Equations (2) are the equations of a generic S_{n-d} through x', which meets this tangent space only in x'. Hence x' is an isolated solution of (2) and (3). Thus we have

THEOREM II. *If x' is a simple point of V, a generic $(n-d)$-space through x' has a simple intersection with V at x'.*

We next consider those points which are different from x' and are common to V and a generic $(n-d)$-space through the simple point x'. We prove

THEOREM III. *If x' is a simple point of V, each point common to V and a generic $(n-d)$-space through x' has multiplicity one.*

The case of x' itself has been covered by Theorem II. If this is the only intersection, nothing further is required. Assume that there is another intersection ξ. By Theorem I, ξ is generic on V. If ξ counts multiply among the points of intersection, a generic S_{n-d} through x' and ξ meets the tangent S_d at ξ in a line, at least [§9, Th. I]. If x'' is any simple point of V, distinct from x', and we consider *any* S_{n-d} through the line $x'x''$, it follows that this specialised S_{n-d} meets the tangent S_d at x'' in a line, at least. In fact, the geometrical property we are considering can be expressed by the vanishing of expressions homogeneous in the coordinates of x', ξ, S_{n-d} and S_d, and is therefore preserved by proper specialisation of any of the coordinates considered. But if x' does not lie in the tangent S_d at a given point x'', we can always find an S_{n-d} through $x'x''$ which meets S_d at x'' and in no other point. It follows that if a generic S_{n-d} through x' meets V again in a point which counts multiply among the further intersections, the tangent space at every simple point of V contains x'.

In particular, the tangent space at every generic point of V contains x'. We prove that this can only happen when V is a linear space S_d, in which case our hypothesis, that a generic S_{n-d} through a point x' of V meets V again, breaks down.

Let the forms which vanish on V have the basis

$$f_\nu(x) \quad (\nu = 1, \ldots, r).$$

By hypothesis x' lies in the tangent space to V at a generic point ξ.

Hence $$\sum_{j=0}^{n} x'_j \frac{\partial f_\nu}{\partial \xi_j} = 0 \quad (\nu = 1, \ldots, r).$$

The form $$\Sigma x'_j \frac{\partial f_\nu}{\partial x_j} \quad (\nu = 1, \ldots, r)$$

therefore vanishes at ξ. It therefore vanishes on V, and can be added to the equations of V. In particular the argument given can be repeated, and we find that

$$\Sigma x'_i x'_j \frac{\partial^2 f_\nu}{\partial x_i \partial x_j} \quad (\nu = 1, \ldots, r)$$

vanishes on V. If $f_\nu(x)$ is of order g_ν, an elementary identity shows that if $k = g_\nu - 1$,

$$\sum_{i_1, \ldots, i_k} x'_{i_1} x'_{i_2} \ldots x'_{i_k} \frac{\partial^k f_\nu}{\partial x_{i_1} \ldots \partial x_{i_k}} = \rho \sum_i x_i \frac{\partial f_\nu}{\partial x'_i},$$

where $\rho = k!$ Therefore, by a continuation of the above argument, V satisfies the equations

$$\Sigma x_i \frac{\partial f_\nu}{\partial x'_i} = 0 \quad (\nu = 1, \ldots, r).$$

But since x' is a simple point on V, these are the equations of the tangent S_d to V at x', and therefore V lies in this tangent space. Since V is of dimension d, we must have $V = S_d$.

Finally we prove the converse of Theorem II:

THEOREM IV. *If x' is any point of V such that a generic S_{n-d} through x' meets V in a set of points amongst which x' counts simply, then V has a tangent S_d at x', which is therefore a simple point.*

Let the order of V be g. We first prove the theorem when V is a primal; that is when $d = n-1$, and V is given by a single irreducible equation of degree g,

$$f(x) \equiv f(x_0, x_1, \ldots, x_n) = 0.$$

We assume that a generic line through x' meets V in a set of points amongst which x' counts simply. Let ξ, η be two generic points of S_n. The line joining them meets V in the g points $\xi+\lambda_i\eta$, where $\lambda_1, \ldots, \lambda_g$ are the roots of the equation

$$f(\xi+\lambda\eta) = 0.$$

To obtain the multiplicities of a specialised intersection we must obtain the Cayley form of the set of points $\xi+\lambda_i\eta$, and then specialise ξ to x'. The Cayley form of this zero-dimensional set is

$$F(u) = A\prod_{i=1}^{g}\left[\sum_{j=0}^{n} u_j(\xi_j+\lambda_i\eta_j)\right] = A\prod_{i=1}^{g}\left[\sum_{j=0}^{n} u_j\xi_j+\lambda_i\sum_{j=0}^{n} u_j\eta_j\right].$$

Now
$$f(\xi+\lambda\eta) = B\prod_{i=1}^{g}(\lambda-\lambda_i),$$

and if in this identity we write

$$\lambda = -\left(\sum_{j=0}^{n} u_j\xi_j\right)\Big/\left(\sum_{j=0}^{n} u_j\eta_j\right),$$

and note that $F(u)$ is a polynomial in $u_0, \ldots, u_n$, we can take

$$F(u) = f\left(\xi\sum_{j=0}^{n} u_j\eta_j-\eta\sum_{j=0}^{n} u_j\xi_j\right).$$

It follows immediately from this that if x', y' are any two distinct points of S_n, the multiplicity of $x'+\mu y'$ as an intersection of the line $x'y'$ with V is the multiplicity of μ as a root of

$$f(x'+\lambda y') = 0.$$

Hence if η is a generic point of S_n, x' counts simply amongst the intersections of the line $x'\eta$ with V if and only if

$$\sum_{i=0}^{n}\eta_i\frac{\partial f(x')}{\partial x_i'} \neq 0.$$

The η_i being indeterminates, this condition implies that at least one derivative $\dfrac{\partial f(x')}{\partial x_i'} \neq 0$. Hence the tangent prime

$$\sum_{i=0}^{n} x_i\frac{\partial f(x')}{\partial x_i'} = 0$$

to V at x' exists, and this is what we wished to prove.

Now let V be an irreducible variety of dimension d, and let Σ_{n-d} be a generic $(n-d)$-dimensional linear space through x'.

By hypothesis Σ_{n-d} meets V in g distinct points, of which x' is one. A generic $(n-d-1)$-dimensional linear space in Σ_{n-d} through x' does not contain any of these g points except x'. Let Σ_{n-d-1} be such a space, and let Σ_{n-d-2} be a generic $(n-d-2)$-dimensional linear space in Σ_{n-d-1}. Then Σ_{n-d-2} is generic in S_n.

We construct the locus C of spaces S_{n-d-1} through Σ_{n-d-2} which meet V. By X, § 7, p. 46, this locus C is a primal of order g, irreducible since V is irreducible. We show that a generic line through x' meets C simply at x'.

In fact, a line l meets C in a number of points not greater than the number g of points in which the S_{n-d} joining Σ_{n-d-2} to l meets V. If l is a generic line through x' which lies in Σ_{n-d}, l meets C in g distinct points, of which x' is one. For in this case the S_{n-d} which joins Σ_{n-d-2} to l is a generic S_{n-d} through x', and by hypothesis this meets V in g distinct points. The joins of Σ_{n-d-2} to these points meet l in g distinct points, of which x' is one.

Hence there is a tangent prime at x' to C. This prime contains Σ_{n-d-1}, which lies on C, and its intersection with Σ_{n-d} is precisely Σ_{n-d-1}. If we now vary the Σ_{n-d-1} through x', keeping Σ_{n-d} fixed, the intersection of all the spaces Σ_{n-d-1} is just the point x'. Hence the tangent primes at x' to the various primals C through V have only the point x' in common with Σ_{n-d}. But the common intersection of these tangent primes is necessarily a linear space. Hence the dimension of this common linear space cannot exceed d. It follows from X, § 14, Th. II, that the linear space common to the tangent primes at x' to all primals through V for which x' is a simple point has a dimension not greater than d. But this dimension is not less than d [X, § 14, Th. I]. Hence the dimension of this common linear space is exactly d, so that there is a tangent S_d to V at x'. The point x' is therefore a simple point.

We therefore have

THEOREM V. *A simple point x' of an irreducible variety V of dimension d can be defined in two equivalent ways:* (i) *x' is simple if there exists a tangent S_d to V at x';* (ii) *x' is simple if a generic S_{n-d} through x' meets V simply at x'.*

CHAPTER XII

INTERSECTION THEORY

1. Introduction. In Chapter X we discussed some properties of the intersection of algebraic varieties considered as sets of points. Here we develop a theory of intersections of algebraic varieties in a multiplicative sense [XI, § 8] which will lead, eventually, to a quantitative theory of systems of (multiplicative) varieties lying on an irreducible variety V. Our results will be confined, almost entirely, to the case in which V has no singular points. It is possible to carry the theory further than we do, and for such developments we refer the reader to Professor Weil's book (see Bibliography), and to the papers of Professor Zariski which are listed in the Bibliography at the end of volume III. Nevertheless, the results which we obtain are sufficient to provide a sound foundation for those developments of classical algebraic geometry which are based on intersection properties.

Our objective is to obtain a natural definition of the intersection of two varieties V_a and V_b. We begin with a definition which appears to be arbitrary, depending on the particular choice of a normal problem. We shall show, however, that if certain conditions are satisfied our definition does not, in fact, depend in any essential way on the normal problem selected, and has an intrinsic meaning.

We shall first define the intersection of two varieties V_a and V_b in a projective space S_n, and show that the associative, commutative and distributive laws hold. We then pass on to the consideration of the intersections of varieties lying on an irreducible V_n, and prove a general theorem on the intersection of varieties which vary in irreducible systems. From this theorem we shall be able to deduce that the intersection of the two varieties is independent of the normal problem chosen.

Until this point is reached our definition of the intersection of two varieties V_a and V_b (of respective dimensions a and b) on a variety V_n of dimension n will be confined to the case in which the corresponding point-set varieties have in common an unmixed variety of dimension $a+b-n$, if $a+b \geqslant n$. If $a+b<n$, we assume that there is no intersection. But with the introduction of systems of varieties, we can formally remove these restrictions. The interest

then shifts from the actual intersection of two varieties V_a and V_b to the unique 'intersection system' which is defined by the systems of which V_a and V_b are members, and a calculus of systems is developed. The chapter concludes with a study of the systems of varieties which exist on certain special varieties. A general property of varieties is illustrated. This general property has, as yet, only been proved in special cases.

Before proceeding to develop the theory of intersections of multiplicative varieties systematically, we consider [§§ 2, 3] certain constructions relating to irreducible varieties in the point-set sense which will prove to be useful tools in our general theory.

2. The degeneration of an irreducible variety in S_n. Let V_d be an irreducible variety in the point-set sense in S_n of dimension d and order g, and let

$$f_i(x) = 0 \quad (i = 1, \ldots, r) \tag{1}$$

be a basis for the equations of V_d. If ξ is a generic point of V_d the matrix

$$\left(\frac{\partial f_i(\xi)}{\partial \xi_j}\right)$$

is of rank $n-d$, and the equations

$$\sum_{j=0}^{n} x_j \frac{\partial f_i(\xi)}{\partial \xi_j} = 0 \quad (i = 1, \ldots, r) \tag{2}$$

are the equations of the tangent space to V_d at ξ [X, § 14]. Let S_N be a linear space of dimension $N = n(n+2)$, in which the homogeneous coordinates are t_{ij} $(i,j = 0, \ldots, n)$. Let τ_{ij} $(i,j = 0, \ldots, n)$ be $(n+1)^2$ indeterminates over $K(\xi)$, and consider the correspondence between S_N and S_n defined by the generic point $(\tau, \tau^{-1}\xi)$, where $\tau = (\tau_{ij})$. This irreducible correspondence is of dimension $n(n+2)+d$, since the τ_{ij} are indeterminate over $K(\xi)$. The object-variety is S_N. Hence to a generic point $\tau = (\tau_{00}, \ldots, \tau_{nn})$ of S_N there corresponds in S_n an irreducible variety of d dimensions over $K(\tau)$, say V_d^τ.

Among the equations of V_d^τ are the equations

$$f_i(\tau x) = 0 \quad (i = 1, \ldots, r), \tag{3}$$

for, by the properties of a generic point ξ of V_d, we have

$$f_i(\tau\eta) = f_i(\tau\tau^{-1}\xi) = f_i(\xi) = 0 \quad (i = 1, \ldots, r),$$

where $\eta = \tau^{-1}\xi$ is a generic point of V_d^τ over $K(\tau)$ [XI, § 4]. Conversely, let $\bar{x}$ be any solution of (3). Then $\tau\bar{x}$ is a point of V_d, and is therefore a proper specialisation of ξ. Hence $\bar{x}$ is a proper specialisation of $\tau^{-1}\xi$ and lies on V_d^τ. The equations (3) therefore define V_d^τ.

The tangent space to V_d^τ at η is given by the equations

$$\sum_{j=0}^{n} x_j \frac{\partial f_i(\tau\eta)}{\partial \eta_j} = 0 \quad (i = 1, \ldots, r),$$

that is, by the equations

$$\sum_{j=0}^{n} \sum_{k=0}^{n} x_j \tau_{kj} \frac{\partial f_i(\xi)}{\partial \xi_k} = 0 \quad (i = 1, \ldots, r).$$

Hence the tangent space to V_d^τ at η is the transform of the tangent space to V_d at ξ, given by the equations (2), under the collineation $\bar{x} \to x$, where

$$\bar{x}_i = \sum_{j=0}^{n} \tau_{ij} x_j \quad (i = 0, \ldots, n). \tag{4}$$

It is, of course, evident that V_d^τ is the transform of V_d under the collineation (4). It follows that V_d^τ is of order g.

Returning to the correspondence with generic point $(\tau, \tau^{-1}\xi)$, we have seen that to a generic point τ of S_N there corresponds an irreducible variety of dimension d and order g over $K(\tau)$. But we cannot assume that these varieties V_d^τ describe an irreducible algebraic system as τ describes S_N. On the other hand, by XI, § 6, Th. II, we can define an irreducible algebraic system of varieties whose generic member is V_d^τ. We let ζ be the Cayley image of V_d^τ [XI, § 6], and consider the irreducible correspondence whose generic point-pair is (τ, ζ). The image-variety of this correspondence gives us an irreducible algebraic system of varieties of dimension d and order g. A generic variety of this system is V_d^τ. We call this system *the system obtained by projective transformation from* V_d.

To every specialised point t' of S_N there corresponds one or more points z' which are the Cayley images of varieties of dimension d and order g, the carrier-variety of which is the variety whose equations are

$$f_i(t'x) = 0 \quad (i = 1, \ldots, r). \tag{5}$$

If these equations define an irreducible variety of dimension d over $K(t')$, it follows from XI, § 6, Th. II that this variety is the unique specialisation of V_d^τ corresponding to $\tau \to t'$. It is therefore a member of the system obtained from V_d by projective transformation.

In particular, when t' is the unit matrix, we obtain the original variety V_d itself, and have therefore proved that the system obtained from V_d by projective transformation contains V_d.

If the Cayley form of V_d is $F(u_0, \ldots, u_d)$, in which $u_0, \ldots, u_d$ are regarded as column vectors, the Cayley form of V_d^τ is

$$F(\tilde{\tau} u_0, \ldots, \tilde{\tau} u_d),$$

where $\tilde{\tau} = (\tau')^{-1}$. To prove this result we note that for specialisations $\bar{u}_0, \ldots, \bar{u}_d$, $F(\bar{u}_0, \ldots, \bar{u}_d) = 0$ is the condition that the S_{n-d-1} given by the equations

$$(\bar{u}_i)' x = 0 \quad (i = 0, \ldots, d)$$

should meet the variety V_d given by the equations

$$f_i(x) = 0 \quad (i = 1, \ldots, r).$$

Hence, if $\bar{v}_i = \tau' \bar{u}_i$, $F(\bar{u}_0, \ldots, \bar{u}_d) = 0$ is also the condition that the S_{n-d-1} given by the equations

$$(\bar{v}_i)' x = (\bar{u}_i)' \tau x = 0 \quad (i = 0, \ldots, d)$$

should meet the variety V_d^τ given by the equations

$$f_i(\tau x) = 0 \quad (i = 1, \ldots, r).$$

Since $\bar{v}_i = \tau' \bar{u}_i$, $F(\tilde{\tau} \bar{v}_0, \ldots, \tilde{\tau} \bar{v}_d) = 0$ is the condition that the S_{n-d-1} given by

$$(\bar{v}_i)' x = 0 \quad (i = 0, \ldots, d)$$

should meet the variety V_d^τ. Replacing the v_i by u_i, the result follows.

Having established these preliminary results, we are now concerned with a specialisation t_0 of τ in which t_0 is a singular matrix, defined as follows.

Let $P_0, \ldots, P_n$ be $n+1$ independent generic points of S_n, let S_{d-1}^0 be the space spanned by $P_0, \ldots, P_{d-1}$, and S_{n-d}^0 that spanned by $P_d, \ldots, P_n$. Each of these spaces is generic in S_n. If the coordinates of P_i $(i = 0, \ldots, n)$ are

$$P_i = [\alpha_{0i}, \alpha_{1i}, \ldots, \alpha_{ni}],$$

we write down the matrix in which the coordinates of P_i are column vectors:

$$\alpha = (\alpha_{ij}).$$

Now let σ be the diagonal matrix with zero in the first d places on the principal diagonal, and 1 in the remaining $n-d+1$ places.

Finally, let β be a generic $(n+1)\times(n+1)$ matrix. Then we define the singular matrix t_0 to be

$$t_0 = \beta\alpha\sigma\alpha^{-1}.$$

Let V' be the variety whose equations are

$$f_i(t_0 x) = 0 \quad (i = 1, \ldots, r). \tag{6}$$

Any point which satisfies the equations

$$t_0 x = 0$$

lies on V'. These are the points which satisfy the equations

$$\sigma\alpha^{-1}x = 0,$$

and these equations define S^0_{d-1}. If x' is any point satisfying (6) but not in S^0_{d-1}, the point $\alpha\sigma\alpha^{-1}x'$ lies on V^β_d. It is easily seen that $\alpha\sigma\alpha^{-1}x'$ is the projection of x' from S^0_{d-1} on to S^0_{n-d}.

Conversely, if the projection $\alpha\sigma\alpha^{-1}x'$ of x' from S^0_{d-1} on to S^0_{n-d} lies on V^β_d,

$$f_i(\beta\alpha\sigma\alpha^{-1}x') = 0 \quad (i = 1, \ldots, r),$$

and x' lies on V'.

Hence the points of V' consist of the points of S^0_{d-1} and those points which project from S^0_{d-1} into the intersections of S^0_{n-d} and V^β_d. Since S^0_{n-d} is a generic $(n-d)$-space over $K(\beta)$, there are g conjugate points common to S^0_{n-d} and V^β_d, each of which is simple for V^β_d. It follows that V' consists of g conjugate d-spaces over $K(\beta)$ which each contain S^0_{d-1}. The variety V' is therefore irreducible and of dimension d over the field $K(t_0)$. By what we have said above, $V' = V^{t_0}_d$ is therefore a member of the system obtained from V_d by projective transformation.

At a generic point η of $V^{t_0}_d$ the tangent space is given by the equations

$$\sum_{j=0}^{n} x_j \frac{\partial f_i(t_0\eta)}{\partial \eta_j} = 0 \quad (i = 1, \ldots, r). \tag{7}$$

Since η is generic it does not lie in S^0_{d-1}. It must therefore project from S^0_{d-1} into one of the g simple points, Q_i say, in which S^0_{n-d} meets V^β_d. Since the tangent space is of dimension d, equations (7) represent that component of $V^{t_0}_d$ which contains η and meets S^0_{n-d} in Q_i, that is, the join of S^0_{d-1} to Q_i.

The g points $Q_1, \ldots, Q_g$ in which S^0_{n-d} meets V^β_d are the transform, under the collineation $\bar{x} \to x$ given by $x = \beta^{-1}\bar{x}$ of the points of intersection of some S_{n-d} which is generic over K and V_d. Since β is generic, each point Q_i is a generic point of S_n over K. The d-space $S_d^{(i)}$ joining S_{d-1}^0 to Q_i is the join of a generic S_{d-1}^0 to a generic point of S_n. Hence each $S_d^{(i)}$ must be generic over K.

The system obtained by projective transformation from V_d therefore has the following properties:

(i) it contains V_d;

(ii) it contains a variety consisting of g d-spaces which all contain a generic $(d-1)$-space; each of the g d-spaces is a generic d-space of S_n over K.

If we are given an irreducible variety V_d defined over K, and can construct an irreducible algebraic system of varieties which contains

(i) V_d;

(ii) a variety (possibly defined over some extension of K) which is reducible to a sum $U_d \dotplus W_d$,

we say that V_d can be *degenerated* into $U_d \dotplus W_d$. We state the results of this section in

THEOREM I. *Any irreducible variety V_d of dimension d and order g can be degenerated into a sum*

$$S_d^{(1)} \dotplus \dots \dotplus S_d^{(g)},$$

where $S_d^{(1)}, \dots, S_d^{(g)}$ are g distinct d-spaces through a generic S_{d-1}, each of the S_d's being generic in S_n over K.

Since V_d and $\sum_{i=1}^{g} S_d^{(i)}$ are each of order g, we deduce immediately that if $\mathbf{V}_d$ and $\mathbf{S}_d^{(i)}$ are regarded as *irreducible* multiplicative varieties, there is an irreducible system of multiplicative varieties containing $\mathbf{V}_d$ and $\mathbf{S}_d^{(1)} + \dots + \mathbf{S}_d^{(g)}$.

3. The product and cross-join of two irreducible varieties in S_n. Let V_a and V_b be two irreducible varieties in S_n defined over K. We consider certain loci associated with them which are of considerable use in intersection problems.

Let
$$f_i(x) = 0 \quad (i = 1, \dots, r)$$
be a basis for the equations of V_a, and let
$$g_i(x) = 0 \quad (i = 1, \dots, s)$$
be a basis for the equations of V_b. We also assume that ξ, η are normalised generic points of V_a, V_b respectively.

We begin by defining the *direct product* of two varieties. Let S'_n be a second n-dimensional projective space in which the coordinates are $(y_0, \ldots, y_n)$, and let V'_b be the variety in S'_n whose equations are

$$g_i(y) = 0 \quad (i = 1, \ldots, s).$$

In the two-way projective space S_{nn} in which the coordinates are (x, y) consider the variety whose equations are

$$\left.\begin{aligned} f_i(x) &= 0 \quad (i = 1, \ldots, r), \\ g_i(y) &= 0 \quad (i = 1, \ldots, s). \end{aligned}\right\} \tag{1}$$

We call this variety the *direct product* of V_a and V'_b, and denote it by $V_a \times V'_b$. Evidently S_{nn} may be regarded as the direct product of S_n and S'_n.

Over the field $K(\xi)$ it is possible that V'_b is reducible. Each component is then of dimension b [X, § 11, Th. I], and we assume that

$$V'_b = V_b^{(1)} \dotplus \ldots \dotplus V_b^{(k)} = \sum_{i=1}^{k} V_b^{(i)},$$

where Σ denotes summation in the point-set sense. Let $\eta^{(1)}$ be a normalised generic point of $V_b^{(1)}$ over $K(\xi)$. Then [X, § 11, p. 74] $\eta^{(1)}$ is a generic point of V'_b over K. In S_{nn} the point-pair $(\xi, \eta^{(1)})$ is a generic point over K of a variety $M^{(1)}$, say. Since $K(\xi)$ is of dimension a over K, and $K(\xi, \eta^{(1)})$ is of dimension b over $K(\xi)$, $K(\xi, \eta^{(1)})$ is of dimension $a+b$ over K; that is, $M^{(1)}$ is of dimension $a+b$ over K.

The variety $M^{(1)}$ defines an irreducible correspondence between S_n and S'_n. The object-variety in S_n has generic point ξ over K, and is therefore V_a, and the image-variety in S'_n has generic point $\eta^{(1)}$ over K, and is therefore V'_b. To the generic point ξ of V_a there corresponds the b-dimensional variety $V_b^{(1)}$, which is irreducible over $K(\xi)$.

In the correspondence $\sum_i M^{(i)}$ between S_n and S'_n there corresponds to the generic point ξ of V_a the variety $\sum_i V_b^{(i)} = V'_b$. In the correspondence between S_n and S'_n defined by the equations (1) there also corresponds to the generic point ξ of V_a the variety $V'_b = \sum_i V_b^{(i)}$. The two correspondences are therefore identical; that is

$$V_a \times V'_b = \sum_i M^{(i)}. \tag{2}$$

Hence the direct product of two varieties irreducible over K may be reducible over K.

Since $\sum_i M^{(i)}$ is defined by equations (1), it must be symmetrically related to V_a and V'_b. We could therefore have defined it by choosing a generic point η of V'_b, and by considering the components $V_a^{(i)}$ of V_a over $K(\eta)$. If there are k' such components, we conclude, as above, that the variety given by (1) has k' components over K. Hence $k' = k$. An immediate consequence of this is that if either V_a or V_b is absolutely irreducible, $k = 1$, and hence the direct product is irreducible.

Let K^* be any algebraic extension of K over which V_a and V'_b are completely reducible. Then if, over K^*, we have

$$V_a = \sum_{i=1}^{s} v_a^{(i)}, \quad V'_b = \sum_{j=1}^{t} v_b^{(j)},$$

it is clear that

$$V_a \times V'_b = \sum_{i,j} v_a^{(i)} \times v_b^{(j)}.$$

Since $\sum_i M^{(i)} = V_a \times V'_b$, it follows that

$$M^{(i)} = \Sigma^{(i)} v_a^{(l)} \times v_b^{(m)},$$

where in the k summations $\Sigma^{(i)}$ $(i = 1, \ldots, k)$ each $v_a^{(l)} \times v_m^{(m)}$ $(l = 1, \ldots, s;$ $m = 1, \ldots, t)$ occurs exactly once. The variety $M^{(i)}$ defines a correspondence between the x-space and the y-space in which the object- and image-varieties are V_a and V_b, and $v_a^{(l)} \times v_b^{(m)}$ defines a correspondence in which the object- and image-varieties are $v_a^{(l)}$ and $v_b^{(m)}$. Hence it follows that in each summation $\Sigma^{(i)}$, l takes all values from 1 to s, and m all values from 1 to t. If $x^{(l)}$ is a generic point over K^* of $v_a^{(l)}$ and $y^{(m)}$ a generic point of $v_b^{(m)}$, we see that to $x^{(l)}$ there corresponds in $M^{(i)}$ the points of all varieties $v_b^{(m)}$ such that $v_a^{(l)} \times v_b^{(m)}$ is a term of $\Sigma^{(i)}$. Since $K(x^{(p)})$ and $K(x^{(q)})$ are equivalent extensions of K, it follows that the varieties in the x-space which correspond to $x^{(p)}$ and $x^{(q)}$ in $M^{(i)}$ have the same order; hence it follows that the number of terms $v_a^{(l)} \times v_b^{(m)}$ in $\Sigma^{(i)}$ for which $l = p$ is the same as the number of terms for which $l = q$. If this number is σ_i, $M^{(i)}$ has $s \times \sigma_i$ components. Similarly, we show that it has $t \times \tau_i$ components, where τ_i is the number of terms in $\Sigma^{(i)}$ in which a given $v_b^{(m)}$ appears. The numbers σ_i, τ_i may vary with i.

The following example will illustrate the way in which $V_a \times V'_b$ may be reducible.

Let K be a field which is not algebraically closed, and let

$$f(x,y) = a_0x^3 + a_1x^2y + a_2xy^2 + a_3y^3$$
$$= a_3(y-\alpha_1 x)(y-\alpha_2 x)(y-\alpha_3 x),$$

be any cubic which is irreducible over K, α_1, α_2, α_3 being in some extension of K. Let V_a and V'_b be the varieties in one-space given by

$$f(x_0, x_1) = 0, \quad f(y_0, y_1) = 0,$$

respectively. $(1, \alpha_1)$ is a generic point of V_a. We consider V'_b defined over $K(\alpha_1)$. It is clearly reducible, since it contains the point $(1, \alpha_1)$ which has coordinates in $K(\alpha_1)$. The remaining component is reducible or not according as

$$a_3(y_1-\alpha_2 y_0)(y_1-\alpha_3 y_0) = a_3 y_1^2 + (a_2 + a_3\alpha_1) y_0 y_1 - \frac{a_0}{\alpha_1} y_0^2$$

is or is not reducible over $K(\alpha_1)$. If this equation is irreducible, we have $k = 2$ and the two components of $V_a \times V'_b$ over K consist of the sets of conjugate points.

$$(1, \alpha_1;\ 1, \alpha_1), \quad (1, \alpha_2;\ 1, \alpha_2), \quad (1, \alpha_3;\ 1, \alpha_3)$$

and

$$(1, \alpha_2;\ 1, \alpha_3), \quad (1, \alpha_3;\ 1, \alpha_2), \quad (1, \alpha_3;\ 1, \alpha_1),$$
$$(1, \alpha_1;\ 1, \alpha_3), \quad (1, \alpha_1;\ 1, \alpha_2), \quad (1, \alpha_2;\ 1, \alpha_1).$$

The first set satisfies the equations

$$f(x_0, x_1) = 0, \quad f(y_0, y_1) = 0, \quad x_0y_1 - x_1y_0 = 0,$$

and the second set satisfies the equations

$$f(x_0, x_1) = 0, \quad f(y_0, y_1) = 0,$$
$$a_3x_1y_1(x_0y_1 + x_1y_0) + a_2x_0x_1y_0y_1 - a_0x_0^2y_0^2 = 0.$$

If $V_a \times V'_b$ has three components, the roots $\alpha_1, \alpha_2, \alpha_3$ of $f(x, y)$ have the property that there exist polynomials $g(x)$, $h(x)$ of degree 2 in $K[x]$ such that

$$(\alpha_1, \alpha_2, \alpha_3) = (\alpha_1, g(\alpha_1), h(\alpha_1)) = (h(\alpha_2), \alpha_2, g(\alpha_2)) = (g(\alpha_3), h(\alpha_3), \alpha_3).$$

Then it is easily seen that each component of $V_a \times V'_b$ satisfies the equations $f(x_0, x_1) = 0, f(y_0, y_1) = 0$ and *one* of the equations

$$x_0y_1 - x_1y_0 = 0, \quad y_1x_0^2 = y_0x_0^2 g\left(\frac{x_1}{x_0}\right), \quad y_1x_0^2 = y_0x_0^2 h\left(\frac{x_1}{x_0}\right).$$

Let S_N $(N = n(n+2))$ be a projective space in which the coordinates are $(t_{00}, \ldots, t_{nn})$. As in § 2 we shall represent a point of S_N by the $(n+1)\times(n+1)$ matrix $t = (t_{ij})$ $(i,j = 0, \ldots, n)$. We consider the variety $U^{(1)}$ in the three-way space S_{nnN} whose equations over K consist of the equations of $M^{(1)}$ and the set of equations

$$y_j \sum_{l=0}^{n} t_{kl}x_l - y_k \sum_{l=0}^{n} t_{jl}x_l = 0 \quad (j,k = 0, \ldots, n). \tag{3}$$

For *any* point (x', y') on $M^{(1)}$ the corresponding set of equations (3) for t is of rank n. Hence, by the reasoning of XI, § 5, Th. I, $U^{(1)}$ is irreducible, and if $\tau^{(1)}$ is a generic solution of the equations

$$\eta_j^{(1)} \sum_{l=0}^{n} t_{kl}\xi_l - \eta_k^{(1)} \sum_{l=0}^{n} t_{jl}\xi_l = 0 \quad (j,k = 0, \ldots, n),$$

where ξ is a normalised generic point of V_a, $\eta^{(1)}$ a normalised generic point of $V_b^{(1)}$, the point $(\xi, \eta^{(1)}, \tau^{(1)})$ is a generic point of $U^{(1)}$. The dimension of $U^{(1)}$ is

$$a+b+n(n+2)-n = a+b+n^2+n.$$

The projective transformation

$$y = \tau^{(1)}x$$

is the generic transformation between S_n and S'_n over K which maps ξ on $\eta^{(1)}$. Hence it follows that

(*a*) $\tau^{(1)}$ is non-singular, and

(*b*) the inverse transformation

$$x = (\tau^{(1)})^{-1} y$$

maps the tangent space to V'_b at $\eta^{(1)}$ on a b-space through ξ which is generic over $K(\xi)$ in the system of b-spaces through ξ.

Reasoning similar to the above applies to the varieties $M^{(2)}, \ldots, M^{(k)}$, and enables us to define varieties $U^{(2)}, \ldots, U^{(k)}$.

The variety $U^{(i)}$ defines a correspondence between the spaces S_{nN} and S'_n. Let $T^{(i)}$ be the object-variety of this correspondence in S_{nN}. It has $(\xi, \tau^{(i)})$ as a generic point. The equations

$$y_j \sum_{l=0}^{n} \tau_{kl}^{(i)}\xi_l - y_k \sum_{l=0}^{n} \tau_{jl}^{(i)}\xi_l = 0 \quad (j,k = 0, \ldots, n)$$

have a unique solution $y = \eta^{(i)}$, since $\tau^{(i)}$ is non-singular. Hence, to a generic point of $T^{(i)}$ there corresponds in $U^{(i)}$ a variety of dimension zero. By the Principle of Counting Constants [§ 7, p. 104], $T^{(i)}$ is therefore also of dimension $a+b+n(n+1)$.

The equations of $\sum_i U^{(i)}$ in S_{nnN} are clearly

$$\left.\begin{aligned} f_i(x) &= 0 \quad (i = 1,\ldots,r), \\ g_i(y) &= 0 \quad (i = 1,\ldots,s), \\ y_j\sum_{l=0}^{n} t_{kl}x_l - y_k\sum_{l=0}^{n} t_{jl}x_l &= 0 \quad (j,k = 0,\ldots,n). \end{aligned}\right\} \tag{4}$$

The equations of the object-variety $\sum_i T^{(i)}$ in S_{nN} are obtained by eliminating $(y_0,\ldots,y_n)$ from these equations. We show that the equations

$$\left.\begin{aligned} f_i(x) &= 0 \quad (i = 1,\ldots,r), \\ g_i(tx) &= 0 \quad (i = 1,\ldots,s), \end{aligned}\right\} \tag{5}$$

are a resultant system for the set of equations (4) with respect to y. Clearly, if (x',y',t') satisfies (4), (x',t') satisfies (5). Conversely, let (x',t') be a solution of (5). If

$$\sum_{j=0}^{n} t'_{ij}x'_j = 0 \quad (i = 0,\ldots,n),$$

and y' is any point of V'_b, then (x',y',t') satisfies the equations (4). If $\sum_{j=0}^{n} t'_{ij}x'_j$ is not zero for all values of i, put

$$y'_i = \sum_{j=0}^{n} t'_{ij}x'_j \quad (i = 0,\ldots,n).$$

Then again (x',y',t') satisfies the equations (4). Hence the equations (5) define the object-variety $\sum_i T^{(i)}$.

The variety $T^{(i)}$ in S_{nN} defines an irreducible correspondence between S_N and S_n. If c_i denotes the dimension of the object-variety of this correspondence in S_N, d_i the dimension of the variety in S_n which corresponds to a generic point of the object-variety, the Principle of Counting Constants gives us the equation

$$c_i + d_i = \dim(T^{(i)}) = a+b+n^2+n.$$

If $a+b<n$, $c_i \leqslant c_i+d_i < n^2+2n$, and the object-variety in S_N does not fill the space. This is the case for each value of i, and it follows that if τ is a generic point of S_N, that is, a generic $(n+1)\times(n+1)$ matrix over K, the equations

$$\left.\begin{aligned} f_i(x) &= 0 \quad (i = 1,\ldots,r),\\ g_i(\tau x) &= 0 \quad (i = 1,\ldots,s),\end{aligned}\right\}$$

have no solution in x. Hence the varieties V_a and V_b^τ, both regarded as varieties in S_n, have no points in common. We now prove that if $a+b \geqslant n$, then $c_i = n^2+2n$, $d_i = a+b-n$, so that $V_a \wedge V_b^\tau$ is purely $(a+b-n)$-dimensional.

Let us consider first the case in which $k = 1$. Suppose that $c = c_1 = n^2+2n-\delta$ $(\delta \geqslant 0)$. Then $d = d_1 = a+b-n+\delta$. The object-variety of the correspondence between S_N and S_n defined by T is of dimension c, and to a generic point τ' of it there corresponds a variety of dimension d. This variety is given by the equations

$$\left.\begin{aligned} f_i(x) &= 0 \quad (i = 1,\ldots,r),\\ g_i(\tau' x) &= 0 \quad (i = 1,\ldots,s).\end{aligned}\right\}$$

If t' is any point of S_N, the equations

$$\left.\begin{aligned} f_i(x) &= 0 \quad (i = 1,\ldots,r),\\ g_i(t' x) &= 0 \quad (i = 1,\ldots,s)\end{aligned}\right\} \tag{6}$$

have no solution if t' is not on the object-variety, and define an algebraic variety no component of which is of dimension less than d if t' is on the object-variety [XI, § 6, Th. I]. It follows that if we can find a point t' of S_N for which the equations (6) define a variety of dimension $a+b-n$, then $\delta = 0$, and hence $c = n^2+2n$, $d = a+b-n$. Take t' to be the matrix t_0 of § 2. Then the equations

$$g_i(t_0 x) = 0 \quad (i = 1,\ldots,s)$$

define a variety consisting of h b-spaces, each of which is generic over K, where h is the order of V_b [§2, p. 145]. Each of these b-spaces meets V_a in a variety of dimension $a+b-n$; hence the equations (6), with t' replaced by t_0, define a variety of dimension $a+b-n$.

If $k>1$, we consider the effect of extending the ground field K to K^*, over which

$$V_a = \sum_i v_a^{(i)}, \quad V_b = \sum_i v_b^{(i)},$$

where $v_a^{(i)}$, $v_b^{(i)}$ are absolutely irreducible [cf. p. 147]. We have seen that

$$M^{(i)} = \Sigma^{(i)} v_a^{(l)} \times v_b^{(m)}.$$

Just as the variety $U = \Sigma U^{(i)}$ is constructed from $V_a \times V_b$ (in the case $k = 1$), so we construct a variety $u^{(ij)}$ from $v_a^{(i)} \times v_b^{(j)}$, and it is immediately obvious that

$$U^{(i)} = \Sigma^{(i)} u^{(lm)}.$$

Again, as we pass from U to T, so can we pass from $u^{(ij)}$ to a variety $T^{(ij)}$, and we see that

$$T^{(i)} = \Sigma^{(i)} T^{(lm)}.$$

Now if

$$f_l^{(i)}(x) = 0 \quad (l = 1, 2, \ldots)$$

are the equations of $v_a^{(i)}$, and

$$g_m^{(i)}(x) = 0 \quad (m = 1, 2, \ldots)$$

are the equations of $v_b^{(i)}$, $T^{(ij)}$ has the equations

$$f_l^{(i)}(x) = 0 \quad (l = 1, 2, \ldots),$$
$$g_m^{(j)}(tx) = 0 \quad (m = 1, 2, \ldots).$$

Applying the result already proved for the case $k = 1$, we see that if τ is a generic $(n+1) \times (n+1)$ matrix the equations

$$f_l^{(i)}(x) = 0 \quad (l = 1, 2, \ldots),$$
$$g_m^{(j)}(\tau x) = 0 \quad (m = 1, 2, \ldots),$$

define a variety of dimension $a + b - n$ over $K(\tau)$. Since

$$T^{(i)} = \Sigma^{(i)} T^{(lm)},$$

we conclude that the object-variety of the correspondence between S_N and S_n defined by $T^{(i)}$ is of dimension $n^2 + 2n$. Hence $c_i = n^2 + 2n$. The result is thus established.

It is clear that if, over $K(\xi)$, the variety V'_b in S'_n is reducible,

$$V'_b = \Sigma V_b^{(i)},$$

then V_b is reducible over $K(\xi)$, and we write

$$V_b = \Sigma V_b^{(i)}.$$

ξ is a generic point of V_a, and we assume that $\eta^{(i)}$ is a generic point of $V_b^{(i)}$. Let $\tau^{(i)}$ be a generic solution of the equations

$$\eta_j^{(i)} \sum_{l=0}^{n} t_{kl} \xi_l - \eta_k^{(i)} \sum_{l=0}^{n} t_{jl} \xi_l = 0 \quad (j, k = 0, \ldots, n).$$

Then, as we have seen, the collineation

$$\bar{x} = (\tau^{(i)})^{-1} x$$

transforms the tangent space at $\eta^{(i)}$ to V_b into a b-space through ξ, generic with respect to $K(\xi)$, which is the tangent space to $V_b^{\tau^{(i)}}$ at $(\tau^{(i)})^{-1} \eta^{(i)} = \xi$. This generic b-space meets the tangent space to V_a at ξ in a linear space of dimension $a+b-n$. We have proved that $\tau^{(i)}$ is a generic point of S_N over K, and $(\xi, \tau^{(i)})$ is a generic point of $T^{(i)}$. The point ξ is a generic point over $K(\tau^{(i)})$ of the variety which corresponds to $\tau^{(i)}$ in $T^{(i)}$, and equations (5) show that this variety is a component of $V_a \wedge V_b^{\tau^{(i)}}$. It follows that $V_a \wedge V_b^{\tau}$ consists of k components each of dimension $a+b-n$ such that at a generic point of any component there exist tangent spaces to V_a and to V_b^{τ}, and these meet in an $(a+b-n)$-space.

If two irreducible varieties U_c, U_d intersect in an irreducible variety of dimension $c+d-n$ which is such that at a generic point of it tangent spaces to U_c and to U_d exist, and these tangent spaces meet in a $(c+d-n)$-space, we say that U_c and U_d *intersect simply* along this variety, or have the variety as a *simple intersection*. We sum up the results proved above in

THEOREM I. *If V_a, V_b are two irreducible varieties, and τ is a generic $(n+1)\times(n+1)$ matrix, V_a and V_b^{τ} do not intersect if $a+b<n$, and intersect simply along a finite number of varieties if $a+b \geqslant n$.*

Before leaving this section we introduce some concepts which are closely related to the ideas developed above. Using the notation introduced earlier, let ξ be a generic point of V_a, $\eta^{(i)}$ a generic point of $V_b^{(i)}$. Let $\zeta^{(i)} = \lambda\xi + \mu\eta^{(i)}$, where λ, μ are indeterminates over the field $K(\xi, \eta^{(i)})$.

We consider the variety in three-way space of which $(\xi, \eta^{(i)}, \zeta^{(i)})$ is a generic point, and call it $J^{(i)}$. The dimension of $J^{(i)}$ is $a+b+1$. Let $J = \sum_i J^{(i)}$. If z' is a proper specialisation of $\zeta^{(i)}$, we can find x', y', such that (x', y', z') is a proper specialisation of $(\xi, \eta^{(i)}, \zeta^{(i)})$. Since x', y' are proper specialisations of ξ, $\eta^{(i)}$, they lie on V_a and V_b respectively. Hence any specialisation of $\zeta^{(i)}$ has the property that through it there passes a line meeting V_a and V_b. Conversely, let x' be a point of V_a, y' a point of V_b, where $x' \neq y'$, and let z' be a point of the line $x'y'$. We show that (x', y', z') is a point of J. The point y' lies on one of the components $V_b^{(i)}$ of V_b, say $V_b^{(1)}$. Then (x', y') is a point of $M^{(1)}$ and is therefore a proper specialisation of $(\xi, \eta^{(1)})$. Let $z' = lx' + my'$.

It can be verified at once that (x', y', z') is a proper specialisation of $(\xi, \eta^{(i)}, \zeta^{(i)})$. Hence any point on a line joining two points x', y' of V_a, V_b, respectively, where $x' \neq y'$, lies on J. It is more difficult to determine the specialisations which correspond to specialisations (x', y') of $(\xi, \eta^{(1)})$ where $x' = y'$, but we do not have to consider these.

A number of important loci are connected with the variety J in three-way space.

(1) Let $W^{(i)}$ be the variety in S_n which has $\zeta^{(i)} = \lambda\xi + \mu\eta^{(i)}$ as generic point, and write $W = \sum_i W^{(i)}$. Then W is called the *cross-join* of V_a and V_b.

Lemma I. If the tangent space at ξ to V_a meets the tangent space at $\eta^{(i)}$ to $V_b^{(i)}$ in a space of dimension $a+b-m$ $(\max(a, b) \leqslant m \leqslant a+b+1)$, *then $W^{(i)}$ is of dimension m.*

Since $\zeta^{(i)} = \lambda\xi + \mu\eta^{(i)}$ is a generic point of $W^{(i)}$, it is a simple point of $W^{(i)}$. We consider the line joining $\zeta^{(i)}$ to a point $\zeta' = \lambda'\xi' + \mu'\eta'$ of $W^{(i)}$, where ξ', η' are new generic points of V_a, $V_b^{(i)}$, and any specialisation of this line as $\zeta' \to \zeta^{(i)}$. It is easily seen that the line specialises to the line joining $\zeta^{(i)}$ to a point

$$\rho\xi + \sigma\eta^{(i)} + \lambda\xi^* + \mu\eta^*,$$

where ξ^* is in the tangent space to V_a at ξ, and η^* is in the tangent space to $V_b^{(i)}$ at $\eta^{(i)}$, and ρ, σ are scalars. Hence any tangent line to $W^{(i)}$ at $\zeta^{(i)}$ lies in the join of the tangent space at ξ to V_a and the tangent space at $\eta^{(i)}$ to $V_b^{(i)}$. This join is an m-space. Hence the dimension of the tangent space at $\zeta^{(i)}$ to $W^{(i)}$, which is equal to the dimension of $W^{(i)}$, cannot exceed m.

On the other hand, since $\xi\xi^*$ can be any line through ξ in the tangent space to V_a, and $\eta^{(i)}\eta^*$ can be any line through $\eta^{(i)}$ in the tangent space to $V_b^{(i)}$, any line through $\zeta^{(i)}$ in the join of these tangent spaces is tangent to $W^{(i)}$. Hence the dimension of $W^{(i)}$ is at least m. The dimension of $W^{(i)}$ is therefore m.

(2) We consider the correspondence $A^{(i)}$ with generic point $(\xi, \zeta^{(i)})$, and let $A = \sum_i A^{(i)}$. We call A *the correspondence determined by joining V_a to V_b*. Since $K(\xi)$ is of dimension a over K, and $\zeta^{(i)}$ is of dimension $b+1$ over $K(\xi)$, $A^{(i)}$ is of dimension $a+b+1$ over K. The correspondence $A^{(i)}$ is an irreducible algebraic variety in the two-way space S_{nn} in which the coordinates are $(x_0, \ldots, x_n; z_0, \ldots, z_n)$. The object-variety of the correspondence $A^{(i)}$ is V_a, and the image-

variety $W^{(i)}$. Now consider the correspondence between V_a and $(u_0, \ldots, u_n)$ defined by the equations of $A^{(i)}$ and the equation

$$\sum_{i=0}^{n} u_i z_i = 0.$$

The proof [XI, § 5, Th. II] that a generic prime section of an irreducible d-dimensional variety is irreducible of dimension $d-1$ (unless $d = 0$) can be applied here to prove that the correspondence $A_1^{(i)}$ derived from $A^{(i)}$ by requiring z to lie in a generic prime is irreducible of dimension $a+b$, provided that $m > 0$. This follows from the fact that the dimension of the image-variety $W^{(i)}$ in the z-space is m. If $m = 0$, the correspondence $A_1^{(i)}$ is vacuous. A space in S_{nn} given by a set of independent equations

$$\sum_{j=0}^{n} u_{ij} z_j = 0 \quad (i = 1, \ldots, n-d)$$

will be called a z-space of S_{nn}. It has dimension $n+d$. If the coefficients u_{ij} are independent indeterminates over K, it is a generic z-space of S_{nn} of dimension $n+d$. By repeating the argument given above, we see that a section of A by a generic z-space of dimension $n+\rho$ is vacuous if $\rho < n-m$, and if $\rho \geqslant n-m$ it is an unmixed variety A_ρ of dimension $a+b+1-n+\rho$. The object-variety of the correspondence defined by $A_\rho \subseteq V_a$, and the image-variety is of dimension $m-n+\rho$. If x' is any point of V_a, z' any point of the image-variety such that $x'z'$ is a line in S_n meeting V_b in a point different from x', then (x', z') is on A_ρ. Thus we have

Lemma II. Let S_ρ be the generic space

$$\sum_{j=0}^{n} u_{ij} x_j = 0 \quad (i = 1, \ldots, n-\rho)$$

in S_n, and let A_ρ be the section of A by the z-space

$$\sum_{j=0}^{n} u_{ij} z_j = 0 \quad (i = 1, \ldots, n-\rho)$$

in S_{nn}. If $\rho \geqslant n-m$, A_ρ defines a correspondence in S_n of which the object-variety is contained in V_a and the image-variety is of dimension $m-n+\rho$ in S_ρ. If x' is any point of V_a, z' any point of S_ρ such that $x'z'$ meets V_b in a point distinct from x', then x' and z' correspond in A_ρ.

In the definitions of J, W, A, a preferential position has been given to V_a. Let ξ be a normalised generic point of V_a, η a normalised generic point of V_b. If, over $K(\xi)$, V_b splits into k components $\Sigma V_b^{(i)}$, then, over $K(\eta)$, V_a splits into k components $\Sigma V_a^{(i)}$. If $\eta^{(i)}$ is a generic point of $V_b^{(i)}$ over $K(\xi)$, $K(\eta^{(i)})$ and $K(\eta)$ are equivalent extensions of K, and hence there exists a point ξ' such that $K(\xi, \eta^{(i)})$ and $K(\xi', \eta)$ are equivalent extensions of K. (ξ', η) is therefore a generic point of $M^{(i)}$, and ξ' is therefore a generic point of the component $V_a^{(i)}$ of V_a over $K(\eta)$. If $\zeta^{(i)} = \lambda\xi + \mu\eta^{(i)}$, $\zeta' = \lambda\xi' + \mu\eta$, it follows that (ξ', η, ζ') is a generic point of $J^{(i)}$, and ζ' is a generic point of $W^{(i)}$. From this it easily follows that J and W are symmetrically related to V_a and V_b. On the other hand, A is not symmetically related to V_a and V_b, but we can define the correspondence B determined by joining V_b to V_a whose properties can be read off by interchanging the roles of V_a and V_b in A.

4. The intersection of two irreducible varieties in S_n. In this section we consider the intersection of two irreducible varieties in S_n, V_a, of dimension a and order g, and V_b, of dimension b and order h. To begin with we consider these varieties and their intersections in a point-set sense, but later we shall use the point-set results to define the intersection of the multiplicative varieties $\mathbf{V}_a$ and $\mathbf{V}_b$.

Let V_a have the equations

$$f_i(x) = 0 \quad (i = 1, \ldots, r), \tag{1}$$

and let V_b have the equations

$$g_i(x) = 0 \quad (i = 1, \ldots, s). \tag{2}$$

Let S_N be a space of dimension $N = n^2 + 2n$ in which the point with coordinates $(t_{00}, \ldots, t_{nn})$ represents the matrix $t = (t_{ij})$. The equations

$$\left.\begin{aligned} f_i(x) &= 0 \quad (i = 1, \ldots, r), \\ g_i(tx) &= 0 \quad (i = 1, \ldots, s), \end{aligned}\right\} \tag{3}$$

define a correspondence T between S_N and S_n consisting of k irreducible components $T^{(i)}$ $(i = 1, \ldots, k)$ [§ 3]. If $a + b < n$, the object-variety of T in S_N has dimension less than N. Hence if t is a generic point of S_N, equations (3) have no solution for x, that is, $V_a \wedge V_b^t$ is empty. If t' is any point of S_N, the equations (3) with $t = t'$ have a solution if and only if t' lies on the object-variety of T. Thus we say that *in general* two irreducible varieties V_a and V_b do not intersect if $a+b<n$.

Now suppose that $a+b \geqslant n$. Then the object-variety of each $T^{(i)}$ $(i = 1, \ldots, k)$ is S_N, and hence equations (3) have a solution for any proper specialisation of t. If t is generic, the solution is an unmixed variety of dimension $a+b-n$ over $K(t)$ [§ 3, Th. I]. For any specialisation of t, the specialised equations (3) have a solution, and it follows from XI, § 6, Th. I, that no component of this solution is of dimension less than $a+b-n$. Hence we have

THEOREM I. *If V_a and V_b are irreducible varieties of dimensions a and b respectively in S_n, $V_a \wedge V_b$ is 'in general' empty if $a+b<n$. If $a+b \geqslant n$, the intersection has no component of dimension less than $a+b-n$.*

In this section we are concerned with the case in which $a+b \geqslant n$, and every component $V_a \wedge V_b$ is of dimension $a+b-n$. Let τ be a generic point of S_N and let $\Gamma^{(i)}$ be the irreducible variety of dimension $a+b-n$ over $K(\tau)$ which corresponds to τ in $T^{(i)}$. Then [§ 3]

$$\Gamma^{(1)} \dot{+} \Gamma^{(2)} \dot{+} \ldots \dot{+} \Gamma^{(k)} = V_a \wedge V_b^{\tau},$$

and [§ 3, Th. I] V_a and V_b^{τ} meet simply along each $\Gamma^{(i)}$. From XI, § 7, Th. IV, we conclude that if t' is any specialisation of τ such that the variety defined by equations (3), with t replaced by t', is pure $(a+b-n)$-dimensional, then the multiplicities of the components of the specialisation of $\Gamma^{(i)}$ in the normal problem given by (3) are uniquely determined and are all positive. If, further, V_a and $V_b^{t'}$ intersect simply along each of the components of $V_a \wedge V_b^{t'}$, it follows from XI, § 9, Th. II, that each of these components counts once in this normal problem.

We now consider the case in which $t' = t_0$, t_0 being the singular matrix defined in § 2. $V_b^{t_0}$ consists of h b-spaces $S_b^{(i)}$ $(i = 1, \ldots, h)$ through a $(b-1)$-space S_{b-1}^0, where S_{b-1}^0 and the $S_b^{(i)}$ are generic spaces of S_n over K. We note, however, that the h spaces $S_b^{(i)}$ are conjugate over $K(t_0)$, and when this field is extended so that $V_b^{t_0}$ splits up, V_a may become reducible. Let K^* be an algebraic extension of K over which V_a and $V_b^{t_0}$ are completely reducible:

$$V_a = \sum_1^l v_a^{(i)}, \quad V_b^{t_0} = \sum_1^h S_b^{(i)}.$$

It is easily verified that $S_b^{(i)}$ is a generic b-space of S_n over K^*; hence it meets $v_a^{(j)}$ simply, in an irreducible variety of dimension $a+b-n$ and order g_j, where g_j is the order of $v_a^{(j)}$. We now show that $v_a^{(i)}{}_\wedge S_b^{(j)} = v_a^{(i')}{}_\wedge S_b^{(j')}$ if and only if $i = i', j = j'$. If $j = j'$, the equation $v_a^{(i)}{}_\wedge S_b^{(j)} = v_a^{(i')}{}_\wedge S_b^{(j)}$ implies that a generic b-space of S_n meets $v_a^{(i)}$ and $v_a^{(i')}$ in the same variety. Writing down the Cayley forms of $v_a^{(i)}{}_\wedge S_b^{(j)}$ and $v_a^{(i')}{}_\wedge S_b^{(j)}$ from those of $v_a^{(i)}$ and $v_a^{(i')}$ [X, § 8, Th. I, Cor.], we see that this implies that $v_a^{(i)} = v_a^{(i')}$, that is, $i = i'$. If $j \neq j'$, the intersection of $v_a^{(i)}{}_\wedge S_b^{(j)}$ and $v_a^{(i')}{}_\wedge S_b^{(j')}$ is contained in $V_{a\wedge} S_{b-1}^0$, and since S_{b-1}^0 is generic over K, this is of dimension $a+b-n-1$. Since $v_a^{(i)}{}_\wedge S_b^{(j)}$ and $v_a^{(i')}{}_\wedge S_b^{(j')}$ are of dimension $a+b-n$, we conclude that $v_a^{(i)}{}_\wedge S_b^{(j)} \neq v_a^{(i')}{}_\wedge S_b^{(j')}$.

It follows that V_a and $V_b^{t_0}$ intersect simply along the hl distinct components $v_a^{(i)}{}_\wedge S_b^{(j)}$ $(i = 1, \ldots, l;\ j = 1, \ldots, h)$. Hence each of these components has multiplicity one in our normal problem. The sum of the orders of these components is $h\Sigma g_i = gh$. It follows that V_a and V_b^τ intersect simply along the components $\Gamma^{(i)}$ of $V_{a\wedge} V_b^\tau$, and that the sum of the orders of these components is gh.

We are now in a position to define intersections of multiplicative varieties. Let $\boldsymbol{V}_a$ be the variety obtained by attaching unit multiplicity to V_a, that is, by associating with it its irreducible Cayley form, which is of order g, and similarly let $\boldsymbol{V}_b$ be the variety obtained by attaching unit multiplicity to V_b. $\boldsymbol{V}_b^\tau$ is defined similarly. We define $\boldsymbol{V}_a \cdot \boldsymbol{V}_b^\tau$ to be the variety obtained by attaching unit multiplicity to each of the components of $V_{a\wedge} V_b^\tau$. The order of $\boldsymbol{V}_a \cdot \boldsymbol{V}_b^\tau$ is then gh. As we have already seen, for any specialisation t' of τ such that $V_{a\wedge} V_b^{t'}$ is pure $(a+b-n)$-dimensional, the specialisation of $\boldsymbol{V}_a \cdot \boldsymbol{V}_b^\tau$ is uniquely determined, and is obtained by attaching a positive multiplicity to each of the components of $V_{a\wedge} V_b^{t'}$. Moreover, if V_a and $V_b^{t'}$ intersect simply along each of the components of $V_{a\wedge} V_b^{t'}$, each of these multiplicities is one. Assuming, as we have done, that $V_{a\wedge} V_b$ is purely $(a+b-n)$-dimensional, we define $\boldsymbol{V}_a \cdot \boldsymbol{V}_b$ as the unique specialisation of $\boldsymbol{V}_a \cdot \boldsymbol{V}_b^\tau$ as $\tau \to I$. We note that the order of $\boldsymbol{V}_a \cdot \boldsymbol{V}_b$ is equal to that of $\boldsymbol{V}_a \cdot \boldsymbol{V}_b^\tau$, which is gh. Our reasoning implies that if t' is any specialisation of t such that $V_{a\wedge} V_b^{t'}$ is $(a+b-n)$-dimensional, then $\boldsymbol{V}_a \cdot \boldsymbol{V}_b^\tau \to \boldsymbol{V}_a \cdot \boldsymbol{V}_b^{t'}$ as $t \to t'$.

We can now state

THEOREM II (*Bezout's Theorem*). *If* $\boldsymbol{V}_a$, $\boldsymbol{V}_b$ *are irreducible varieties of orders* g *and* h, *and if* $V_{a\wedge} V_b$ *is pure* $(a+b-n)$*-dimensional, then* $\boldsymbol{V}_a \cdot \boldsymbol{V}_b$ *is a multiplicative variety of order* gh. *Each component of*

$V_a \wedge V_b$ has positive multiplicity in $V_a . V_b$. If V_a and V_b intersect simply along each component of $V_a \wedge V_b$, then $V_a . V_b$ is the sum of these components, each counted with multiplicity one.

Let τ, σ be two $(n+1)\times(n+1)$ matrices whose elements are independent indeterminates over K. Then each component of $V_a^\sigma \wedge V_b^\tau$ has multiplicity one in $V_a^\sigma . V_b^\tau$, since V_a^σ and V_b^τ intersect simply. Further, if we consider the normal problem given by the equations

$$\left.\begin{aligned} f_i(\sigma x) &= 0 \quad (i = 1, \ldots, r), \\ g_i(\tau x) &= 0 \quad (i = 1, \ldots, s), \end{aligned}\right\}$$

the object-variety of the associated correspondence is S_{NN} (if $a+b \geqslant n$). It follows immediately, as above, that if $V_a^{s'} \wedge V_b^{t'}$ is pure $(a+b-n)$-dimensional, the specialisation of $V_a^\sigma . V_b^\tau$ is uniquely determined as $(\sigma, \tau) \to (s', t')$. If we specialise $\sigma \to I$, we have

$$V_a^\sigma . V_b^\tau \to V_a . V_b^\tau,$$

since V_a and V_b^τ intersect simply along each component of $V_a \wedge V_b^\tau$, and the sum of the orders of these components is gh. We now specialise $\tau \to I$, and we see that $V_a^\sigma . V_b^\tau \to V_a . V_b$. Also, since V_a^σ and V_b^τ intersect simply along each component of $V_a^\sigma \wedge V_b^\tau$, we have

$$V_a^\sigma . V_b^\tau = V_b^\tau . V_a^\sigma.$$

We conclude that $V_a . V_b = V_b . V_a$. Hence we have *the commutative law of intersection.*

THEOREM III. *The intersection $V_a . V_b$ is equal to $V_b . V_a$.*

While it would have been more symmetrical to define $V_a . V_b$ from $V_a^\sigma . V_b^\tau$, the definition based on $V_a . V_b^\tau$ is more convenient from the notational point of view in what follows, when we allow one of the varieties V_a, V_b to vary in an irreducible system.

5. Intersection theory in S_n. We now extend our definition of intersections to reducible unmixed varieties. Let

$$V_a = \sum_{i=1}^{k} \alpha_i V_a^{(i)},$$

and

$$V_b = \sum_{i=1}^{l} \beta_i V_b^{(i)}$$

be two varieties, in the multiplicative sense, of dimensions a and b and orders g and h respectively, where $V_a^{(i)}$, $V_b^{(j)}$ are irreducible (that is, $V_a^{(i)}$, $V_b^{(j)}$ are irreducible, as are the associated Cayley forms). We assume that $V_a \wedge V_b$ is of dimension $a+b-n$; and hence that $V_a^{(i)} \wedge V_b^{(j)}$ is of dimension $a+b-n$. Let g_i be the order of $V_a^{(i)}$, h_j the order of $V_b^{(j)}$. Then

$$g = \sum_{i=1}^{k} \alpha_i g_i, \quad h = \sum_{i=1}^{l} \beta_i h_i.$$

We define $\mathbf{V}_a . \mathbf{V}_b$ by the formula

$$\mathbf{V}_a . \mathbf{V}_b = \sum_{i=1}^{k} \sum_{j=1}^{l} \alpha_i \beta_j \mathbf{V}_a^{(i)} . \mathbf{V}_b^{(j)},$$

where $\mathbf{V}_a^{(i)} . \mathbf{V}_b^{(j)}$ is defined as in § 4. It is clear that with this definition of intersection $\mathbf{V}_a . \mathbf{V}_b = \mathbf{V}_b . \mathbf{V}_a$, since we have proved that

$$\mathbf{V}_a^{(i)} . \mathbf{V}_b^{(j)} = \mathbf{V}_b^{(j)} . \mathbf{V}_a^{(i)},$$

and it is also evident that

$$\mathbf{V}_a . (\mathbf{V}_b' + \mathbf{V}_b'') = \mathbf{V}_a . \mathbf{V}_b' + \mathbf{V}_a . \mathbf{V}_b''.$$

Hence the commutative and distributive laws hold. Further, the order of $\mathbf{V}_a . \mathbf{V}_b$ is $\Sigma \alpha_i \beta_j g_i h_j = gh$.

Again, if τ is a $(n+1) \times (n+1)$ matrix of indeterminates over K, and the equations of V_b are

$$g_i(x) = 0 \quad (i = 1, \ldots, s),$$

the equations $\quad g_i(\tau x) = 0 \quad (i = 1, \ldots, s)$

define the variety $\quad V_b^\tau = \sum_i V_b^{(i)\tau},$

and we can define $\quad \mathbf{V}_b^\tau = \sum_{i=1}^{l} \beta_i \mathbf{V}_b^{(i)\tau}.$

It then follows that as $\tau \to I$,

$$\mathbf{V}_a . \mathbf{V}_b^\tau \to \mathbf{V}_a . \mathbf{V}_b.$$

We now prove the invariance of our definition of intersection over an extension of the ground field. We need only consider the case in which V_a, V_b are irreducible over K. We begin by supposing that K^* is an extension of K over which V_a, V_b each split into a sum of absolutely irreducible varieties. Over K^* we define $\mathbf{V}_a$ to be the variety whose Cayley form is the same as that of V_a, regarded as a variety over K. Since the factors of the Cayley form are all distinct, $\mathbf{V}_a$ over K^* is the sum of the components of V_a each taken with unit multiplicity. That is

$$\mathbf{V}_a = \mathbf{V}_a^{(1)} + \ldots + \mathbf{V}_a^{(k)},$$

where the $V_a^{(i)}$ are all distinct and irreducible. Similarly, let

$$V_b = V_b^{(1)} + \ldots + V_b^{(l)},$$

where the $V_b^{(j)}$ are all distinct and irreducible. The variety $V_a . V_b$, defined over K^*, is the proper specialisation of $\sum_i \sum_j V_a^{(i)} . V_b^{(j)\tau}$ corresponding to the specialisation $\tau \to I$. The variety $V_a . V_b^\tau$, defined as an intersection over K, becomes a variety over K^* which specialises to the variety obtained from $V_a . V_b$, defined over K, when K is extended to K^*. This follows from the fact that the specialisation of the Cayley form corresponding to the specialisation $\tau \to I$ is the same whether the ground field is extended before or after specialisation. Hence, in order to prove our result, we have only to show that $V_a . V_b^\tau$, defined over K, becomes $\sum_i \sum_j V_a^{(i)} . V_b^{(j)\tau}$ when K is extended to K^*.

It is clear that these varieties, regarded as point sets, coincide. Furthermore, since V_a and V_b^τ intersect simply along all the components of intersection, when regarded as varieties over K, their various components intersect simply when V_a, V_b^τ are regarded as varieties over K^*. Therefore $V_a . V_b^\tau$, defined over K, becomes the sum of the various components of $V_a \wedge V_b^\tau$, each counted once, when K is extended to K^*. Hence to prove that $V_a . V_b^\tau$, defined over K, is equal to $\sum_i \sum_j V_a^{(i)} . V_b^{(j)\tau}$ when K is extended to K^*, it is sufficient to show that if

$$V_a^{(i)} \wedge V_b^{(j)\tau} = V_a^{(i')} \wedge V_b^{(j')\tau},$$

then $i = i'$, $j = j'$. For this will imply that the kl varieties $V_a^{(i)} \wedge V_b^{(j)\tau}$ are the distinct irreducible components of $V_a \wedge V_b^\tau$.

If

$$V_a^{(i)} \wedge V_b^{(j)\tau} = V_a^{(i')} \wedge V_b^{(j')\tau},$$

then

$$V_a^{(i)} \wedge V_b^{(j)\tau} \subseteq V_a^{(i')}.$$

If $i \neq i'$, we may assume that $V_a^{(i)} \not\subseteq V_a^{(i')}$, so that there exists a point P on $V_a^{(i)}$ which is not on $V_a^{(i')}$. We can specialise $\tau \to t$ so that $V_b^{(j)t}$ contains P. Then

$$V_a^{(i)} \wedge V_b^{(j)t} \subseteq V_a^{(i')},$$

that is, P is on $V_a^{(i')}$, contrary to hypothesis. Hence $i = i'$. Similarly, we have

$$V_a^{(i)} \wedge V_b^{(j)\tau} \subseteq V_b^{(j')\tau},$$

and therefore

$$V_a^{(i)\sigma} \wedge V_b^{(j)} \subseteq V_b^{(j')},$$

where $\sigma = \tau^{-1}$. We then repeat the argument to show that $j = j'$.

We have therefore proved that when K^* is an extension of K over which V_a, V_b become the sum of absolutely irreducible varieties, $V_a.V_b$, defined over K becomes, when K is extended to K^*, the intersection $V_a.V_b$ defined over K^*.

Now let $\bar{K}$ be any extension of K, K^* an extension of $\bar{K}$ over which V_a, V_b are expressible as sums of absolutely irreducible varieties. Let $V_a.V_b$, defined over K, become the variety V_{a+b-n} when K is extended to $\bar{K}$. By what is proved above, $V_a.V_b$, defined over $\bar{K}$, becomes $V_a.V_b$, defined over K^*, when $\bar{K}$ is extended to K^*. On the other hand, V_{a+b-n}, when K is extended to K^*, becomes $V_a.V_b$, defined over K^*. Hence, since V_{a+b-n} and $V_a.V_b$, defined over K^*, are varieties over $\bar{K}$ which coincide when $\bar{K}$ is extended to K^*, they must coincide as varieties over $\bar{K}$. We have therefore proved

THEOREM I. *If $V_{a+b-n} = V_a.V_b$, defined over K, becomes V^*_{a+b-n} when K is extended to K^*, then $V^*_{a+b-n} = V_a.V_b$, defined over K^*.*

This result is of considerable use in proving theorems on intersections. It enables us, in many cases, to confine ourselves to the study of the intersection of absolutely irreducible varieties. This is illustrated by our next result, which establishes the associative property of intersections. Let V_a, V_b, V_c be three varieties of dimension a, b, c respectively. $V_a \wedge V_b$ has no component of dimension less than $a+b-n$, and therefore no component of $V_a \wedge V_b \wedge V_c$ has dimension less than $a+b+c-2n$. If every component of $V_a \wedge V_b \wedge V_c$ is of dimension $a+b+c-2n$ (which implies the inequality $a+b+c \geqslant 2n$, and consequently the inequalities $a+b \geqslant n$, $b+c \geqslant n$, $c+a \geqslant n$), no component of $V_a \wedge V_b$ is of dimension greater than $a+b-n$. Similar results hold for $V_b \wedge V_c$ and $V_c \wedge V_a$.

Hence, if the components of $V_a \wedge V_b \wedge V_c$ are all of dimension $a+b+c-2n$, we can define $(V_a.V_b).V_c$ and $V_a.(V_b.V_c)$. We now prove that

$$(V_a.V_b).V_c = V_a.(V_b.V_c).$$

This formula remains invariant under extensions of the ground field, by Theorem I, so that we need only consider the case in which the components of V_a, V_b, V_c are absolutely irreducible. If

$$V_a = \Sigma \alpha_i V_a^{(i)}, \quad V_b = \Sigma \beta_i V_b^{(i)}, \quad V_c = \Sigma \gamma_i V_c^{(i)},$$

$$(V_a.V_b).V_c = \sum_i \sum_j \sum_k \alpha_i \beta_j \gamma_k (V_a^{(i)}.V_b^{(j)}).V_c^{(k)},$$

and

$$V_a.(V_b.V_c) = \sum_i \sum_j \sum_k \alpha_i \beta_j \gamma_k V_a^{(i)}.(V_b^{(j)}.V_c^{(k)}).$$

Hence it is sufficient to prove the result when V_a, V_b, V_c are absolutely irreducible varieties. Let the respective orders of these varieties be g, h, e.

We first consider the case in which V_c is a generic c-space, with equations

$$\sum_{j=0}^{n} v_{ij} x_j = 0 \quad (i = 1, \ldots, n-c),$$

where the v_{ij} are indeterminate over K. Let the Cayley form of $\boldsymbol{V}_a . \boldsymbol{V}_b^{\tau}$ be $F(\tau, u_0, \ldots, u_{a+b-n})$. This form is irreducible over $K(\tau)$, and of order gh in the indeterminates

$$(u_{i0}, \ldots, u_{in}) \quad (i = 0, \ldots, a+b-n).$$

If λ, μ are matrices of independent indeterminates, the Cayley form of $\boldsymbol{V}_a^{\lambda} . \boldsymbol{V}_b^{\mu}$ is $F(\mu\lambda^{-1}, \tilde{\lambda} u_0, \ldots, \tilde{\lambda} u_{a+b-n})$, by § 2, p. 143, and by X, § 8, Th. I, Cor.,

$$F(\mu\lambda^{-1}, \tilde{\lambda} u_0, \ldots, \tilde{\lambda} u_{a+b+c-2n}, \tilde{\lambda} v_1, \ldots, \tilde{\lambda} v_{n-c})$$

is the Cayley form of a variety $W^{\lambda,\mu}$, which consists of the components of $V_a^{\lambda} \wedge V_b^{\mu} \wedge V_c$ each counted once. Let us first specialise $\mu \to \lambda$. Then $\boldsymbol{W}^{\lambda,\mu} \to \boldsymbol{W}^{\lambda,\lambda}$, where $W^{\lambda,\lambda}$ is given by the Cayley form

$$F(1, \tilde{\lambda} u_0, \ldots, \tilde{\lambda} u_{a+b+c-2n}, \tilde{\lambda} v_1, \ldots, \tilde{\lambda} v_{n-c}).$$

Hence
$$\boldsymbol{W}^{\lambda,\lambda} = (\boldsymbol{V}_a . \boldsymbol{V}_b)^{\lambda} . \boldsymbol{V}_c.$$

Now specialise $\lambda \to I$. By definition (and using the commutative law)

$$\boldsymbol{W}^{\lambda,\lambda} = (\boldsymbol{V}_a . \boldsymbol{V}_b)^{\lambda} . \boldsymbol{V}_c \to (\boldsymbol{V}_a . \boldsymbol{V}_b) . \boldsymbol{V}_c.$$

If instead we consider the specialisation of $\boldsymbol{W}^{\lambda,\mu}$ corresponding to $\mu \to I$, we see that $\boldsymbol{W}^{\lambda,I}$ is a variety which, as a point-set, coincides with $V_a^{\lambda} \wedge V_b \wedge V_c$, and every component of this point-intersection counts with positive multiplicity in $\boldsymbol{W}^{\lambda,I}$. Since λ is a matrix of indeterminates, V_a^{λ} and V_b have simple intersections, the sum of whose orders is gh, and V_c meets each of these components simply, since V_c is a generic c-space. The sum of the orders of the components of $V_a^{\lambda} \wedge V_b \wedge V_c$ is therefore gh. Hence each component counts once in $\boldsymbol{W}^{\lambda,I}$.

Since V_c is a generic c-space, each component of $V_b \wedge V_c$ is simple, and is met simply by V_a^{λ}. Hence

$$\boldsymbol{W}^{\lambda,I} = \boldsymbol{V}_a^{\lambda} . (\boldsymbol{V}_b . \boldsymbol{V}_c).$$

If we now specialise $\lambda \to I$,

$$\boldsymbol{W}^{\lambda,I} = \boldsymbol{V}_a^{\lambda} . (\boldsymbol{V}_b . \boldsymbol{V}_c) \to \boldsymbol{V}_a . (\boldsymbol{V}_b . \boldsymbol{V}_c).$$

Thus it will follow that $V_a \cdot (V_b \cdot V_c) = (V_a \cdot V_b) \cdot V_c$ if we can show that the specialisation of $W^{\lambda,\mu}$ corresponding to $(\lambda,\mu) \rightarrow (I,I)$ is uniquely determined. But $W^{\lambda,\mu}$ arises in a correspondence which is irreducible and has S_{NN} as object-variety. Since all points of S_{NN} are simple we can once more apply the criteria of XI, § 7, Th. IV, the equations of $W^{\lambda,\mu}$ when $(\lambda,\mu) = (I,I)$ determining an unmixed variety of dimension $a+b+c-2n$. Hence the associative law is proved in the case when V_c is a generic c-space.

Now let V_a, V_b, V_c be any absolutely irreducible varieties. λ, μ, ν are independent $(n+1)\times(n+1)$ matrices of indeterminates, and we suppose that t_0 is the specialisation of ν such that $V_c^{t_0}$ consists of e c-spaces each of which is generic over K. Let $W^{\lambda,\mu,\nu}$ be the variety obtained by assigning unit multiplicity to each component of $V_a^\lambda \wedge V_b^\mu \wedge V_c^\nu$. We specialise $\nu \rightarrow t_0$. Since the components of $V_a^\lambda \wedge V_b^\mu$ are simple, and each c-space of $V_c^{t_0}$ is generic, W^{λ,μ,t_0} is the variety which consists of each component of $V_a^\lambda \wedge V_b^\mu \wedge V_c^{t_0}$ counted once. Hence, if $V_c^{t_0} = \sum_{i=0}^{e} S_c^{(i)}$,

$$W^{\lambda,\mu,t_0} = \sum_{i=1}^{e} V_a^\lambda \cdot V_b^\mu \cdot S_c^{(i)} = \sum_{i=1}^{e} (V_a^\lambda \cdot V_b^\mu) \cdot S_c^{(i)}.$$

By what has been proved above, if we now specialise $(\lambda,\mu) \rightarrow (I,I)$,

$$V_a^\lambda \cdot V_b^\mu \cdot S_c^{(i)} \rightarrow (V_a \cdot V_b) \cdot S_c^{(i)}.$$

Hence
$$W^{\lambda,\mu,t_0} \rightarrow (V_a \cdot V_b) \cdot V_c^{t_0}.$$

Let $V_a \cdot V_b = \sum_i \alpha_i V_{a+b-n}^{(i)}$. When $(\lambda,\mu) \rightarrow (I,I)$, $W^{\lambda,\mu,\nu}$ clearly specialises to a variety which, as a point-set, coincides with $\sum_i V_{a+b-n}^{(i)} \wedge V_c^\nu$. Since $V_{a+b-n}^{(i)} \wedge V_c^\nu$ is irreducible, we must have

$$W^{I,I,\nu} = \sum_i \beta_i V_{a+b-n}^{(i)} \cdot V_c^\nu.$$

Now specialise $\nu \rightarrow t_0$. Then

$$W^{I,I,\nu} \rightarrow \Sigma \beta_i V_{a+b-n}^{(i)} \cdot V_c^{t_0}.$$

As above, we can show that the specialisation of $W^{\lambda,\mu,\nu}$ is uniquely determined when $(\lambda,\mu,\nu) \rightarrow (I,I,t_0)$. It follows that $\alpha_i = \beta_i$, and therefore

$$W^{I,I,\nu} = \Sigma \alpha_i V_{a+b-n}^{(i)} \cdot V_c^\nu = (V_a \cdot V_b) \cdot V_c^\nu.$$

The unique specialisation of $W^{\lambda,\mu,\nu}$ corresponding to

$$(\lambda,\mu,\nu) = (I,I,I)$$

is therefore $(V_a \cdot V_b) \cdot V_c$.

By the commutative law,

$$V_a.(V_b.V_c) = (V_b.V_c).V_a = (V_c.V_b).V_a.$$

An exactly similar argument to that just given shows that

$$W^{\lambda,\mu,\nu} \to (V_c.V_b).V_a$$

as $(\lambda, \mu, \nu) \to (I, I, I)$. Hence

$$(V_a.V_b).V_c = (V_c.V_b).V_a = V_a.(V_b.V_c).$$

This proves the associative law for three absolutely irreducible varieties, and therefore for any three varieties V_a, V_b, V_c.

We sum up the results of this section in

THEOREM II. *The definition of the intersection of varieties in S_n satisfies the associative, commutative and distributive laws.*

6. The intersection of irreducible varieties on a V_n in S_r. As a preliminary to the extension of the theory of § 5 to varieties on V_n, a variety of n dimensions lying in a projective space S_r, we consider some properties of the point-set intersection of two irreducible varieties V_a, V_b lying on V_n.

Let V_n be an irreducible variety lying in S_r, V_a an irreducible variety on V_n. We shall speak of a space S_k in S_r which is *independent of a point* P (or a set of points $P_1, P_2, \ldots$). By this we mean that S_k is generic in S_r over the field obtained by adjoining the coordinates of P (or $P_1, P_2, \ldots$) to K.

Let ξ be a generic point of V_a, and let S_{r-n-1} be an $(r-n-1)$-space of S_r which is independent of ξ. We obtain the *cone joining* S_{r-n-1} to V_a. This is a particular case of the cross-join of two varieties discussed in § 3. Since S_{r-n-1} is absolutely irreducible, the cross-join of S_{r-n-1} and V_a is irreducible. Let η be a generic point of S_{r-n-1}, and let $\zeta = \lambda\xi + \mu\eta$, where λ, μ are new indeterminates. Consider the variety W in S_n whose generic point is ζ. It is of dimension $a+(r-n-1)+1 = a+r-n$. This follows from Lemma I of § 4, by observing that since S_{r-n-1} is independent of the generic point ξ of V_a, the tangent space to V_a at ξ does not meet S_{r-n-1}, since $r-n-1+a<r$. Let x' be any point of V_a, y' any point of S_{r-n-1}, and λ', μ' any constants not both zero. We note that S_{r-n-1} does not meet V_n, and therefore does not meet V_a, so that $x' \neq y'$. Clearly $z' = \lambda' x' + \mu' y'$ is a proper specialisation of ζ, and therefore lies on W. It follows that if x' is any point of V_a, the $(r-n)$-space

joining x' to S_{r-n-1} lies on W. Conversely, since S_{r-n-1} and V_a do not meet, the points of W are the points on the join of any point of S_{r-n-1} to any point of V_a.

Lemma I. The intersection $W \wedge V_n$ is pure a-dimensional, and if not all the points of V_a are singular points of V_n, then W and V_n intersect simply along V_a.

Since W is of dimension $a+r-n$, and V_n of dimension n, no component of $W \wedge V_n$ is of dimension less than

$$a+r-n+n-r = a,$$

by § 4, Th. I. If any component of $W \wedge V_n$ is of dimension greater than a, a generic S_{r-a-1} of S_r, independent of S_{r-n-1}, must meet $W \wedge V_n$. Now $U = V_n \wedge S_{r-a-1}$ is an irreducible variety of dimension $n-a-1$, and U does not meet V_a, since S_{r-a-1} does not meet V_a. The cross-join of U and V_a is a variety of dimension n at most, which is not met by S_{r-n-1}, which is generic and independent of S_{r-a-1}. Hence there is no point of U which is on the join of a point of S_{r-n-1} and a point of V_a. It follows that S_{r-a-1} does not meet $W \wedge V_n$, and therefore every component of $W \wedge V_n$ is of dimension a.

Since not all points of V_a are singular points of V_n, a generic point ξ of V_a is both simple on V_a and on V_n. The tangent space at ξ to W is the join of the tangent space at ξ to V_a with S_{r-n-1} [§ 3, p. 154]. Since S_{r-n-1} is independent of ξ, this join is a generic $(a+r-n)$-space through the tangent space at ξ to V_a, and therefore meets the tangent space to V_n at ξ in an a-space, which is the tangent space to V_a at ξ. Hence the intersection of W and V_n along V_a is simple.

We may now write

$$W \wedge V_n = V_a \dotplus \sum_i {}^*V_a^{(i)}.$$

Lemma II. If P is a point of V_a which is simple for V_n and is independent of S_{r-n-1}, P is not on any ${}^*V_a^{(i)}$.

There is nothing to prove if $W \wedge V_n = V_a$. Suppose that P lies on ${}^*V_a^{(i)}$, and let Q^* be a generic point of this variety. Since ${}^*V_a^{(i)} \neq V_a$, Q^* is not on V_a. Q^* is, however, on W. Hence the $(r-n)$-space S^*_{r-n} which joins Q^* to S_{r-n-1} must meet V_a. Let P^* be a point of the intersection. Specialise Q^* to P. Then S^*_{r-n} specialises to the $(r-n)$-space S_{r-n} which joins P to S_{r-n-1}. Since S_{r-n-1} is independent of P, S_{r-n} is a generic $(r-n)$-space through P. Since $r-n<r-a$, this space does not meet V_a outside P. But S^*_{r-n} meets V_a in P^*, and therefore P^* specialises to P. Hence P must count twice as an

intersection of S_{r-n} and V_n, and this contradicts the fact that P is simple for V_a and S_{r-n} is a generic $(r-n)$-space through P. It follows that P cannot lie on $*V_a^{(i)}$.

Now let V_b be any irreducible variety of dimension b on V_n. We may suppose that S_{r-n-1} is also independent of a generic point of V_b, besides being independent of a generic point of V_a. We have the relation

$$W_\wedge V_n = V_a \dotplus \sum_i {}^*V_a^{(i)},$$

from which we deduce the relation

$$W_\wedge V_{n\wedge} V_b = V_{a\wedge} V_b \dotplus \sum_i {}^*V_a^{(i)}{}_\wedge V_b,$$

which, since $V_b \subseteq V_n$, becomes

$$W_\wedge V_b = V_{a\wedge} V_b \dotplus \sum_i {}^*V_a^{(i)}{}_\wedge V_b.$$

By § 4, Th. I, no component of $W_\wedge V_b$ has dimension less than

$$(a+r-n)+b-r = a+b-n.$$

It follows that any irreducible component of $U = V_{a\wedge} V_b$ which is of dimension less than $a+b-n$ is necessarily embedded in a component of some ${}^*V_a^{(i)}{}_\wedge V_b$. We show that if we assume that there is a component of $V_{a\wedge} V_b$ of dimension less than $a+b-n$ which is simple for V_n, we are led to a contradiction.

If there is a simple point P on a component of $U = V_{a\wedge} V_b$, it may be assumed to be independent of S_{r-n-1}, and it follows from Lemma II that P is not on any ${}^*V_a^{(i)}$. Hence P does not lie on $\sum_i {}^*V_a^{(i)}{}_\wedge V_b$. The component of U under consideration cannot, therefore, be embedded. Hence we have proved

THEOREM I. *If V_a and V_b are two irreducible varieties on V_n of dimensions a and b respectively, any component of $V_{a\wedge} V_b$ which does not consist entirely of singular points of V_n is of dimension $a+b-n$, at least. On the other hand, if V_b is the section of V_n by a sufficiently general $(r-n+b)$-space S_{r-n+b} of S_r, $V_{a\wedge} V_b = V_{a\wedge} S_{r-n+b}$ is of dimension $a+b-n$ precisely.*

We show by an example that it is possible to have a component of $V_{a\wedge} V_b$ (necessarily consisting of singular points of V_n) which is of dimension less than $a+b-n$. Consider in S_4 the irreducible variety V_3 of dimension 3 given by the single equation

$$x_0 x_1 = x_2 x_3.$$

$x_0 = x_2 = 0$ is an irreducible V_2 on this variety, and so is $x_1 = x_3 = 0$. These two varieties meet only in the point $(0,0,0,0,1)$, whilst $a+b-n = 1$. The point is easily shown to be singular on V_3.

We saw in § 4 that if V_a, V_b are irreducible varieties of dimensions a, b in S_n, and $a+b \geqslant n$, there are, necessarily, points common to V_a and V_b. This need not happen if V_a and V_b lie on a V_n.

Consider, for example, the surface in S_3 given by the equation

$$x_0 x_1 = x_2 x_3.$$

The lines $x_0 = x_2 = 0$ and $x_1 = x_3 = 0$ both lie on it, but do not meet, whilst $a+b = n$.

We assume, for the remainder of this section, that V_n has no singular points, and that $a+b \geqslant n$.

We now examine the intersection of V_b with the varieties $*V_a^{(i)}$ in which W meets V_n outside V_a. It is important to note that we do not assume that every component of $V_a \wedge V_b$ is of dimension $a+b-n$, though from Theorem I we know that no component has dimension less than $a+b-n$. The varieties $*V_a^{(i)}$ were defined by the relation

$$W \wedge V_n = V_a \dot{+} \sum_i *V_a^{(i)}.$$

We first consider the intersections, if any, which do not lie in $V_a \wedge V_b$. Let $B = \sum_i B^{(i)}$ be the correspondence obtained by joining V_b to V_a [§ 3, p. 154], and define $C = \sum_i C^{(i)}$ to be the section of B by S_{r-n-1}, regarded as a z-space. Suppose that the join of the tangent spaces at independent generic points of V_a and V_b is an m-space $(m \leqslant a+b+1)$. B is of dimension $a+b+1$, and by § 3, Lemma II, C is vacuous if $m \leqslant n$ and is of dimension $a+b-n$ if $m > n$.

We consider first the case in which $m \leqslant n$. Since C is vacuous, there is no line which meets S_{r-n-1} and also meets V_a and V_b in distinct points. Hence there is no line on W which meets V_b in a point which is not also on V_a. It follows that $*V_a^{(i)} \wedge V_b \subseteq V_a \wedge V_b$.

Now consider the case $m > n$. Then C is not vacuous. If y' is any point on $*V_a^{(i)} \wedge V_b$ which is not on V_a, y' is on W, and is therefore on a line which meets S_{r-n-1}, in z', say, and V_a, in x', say By hypothesis y' is not on V_a, so that $x' \neq y'$. Hence (y', z') correspond in C, and therefore y' is on the object-variety of C, which has dimension $a+b-n$ at most, this being the dimension of C itself. It follows that any component of $*V_a^{(i)} \wedge V_b$ which is not embedded in $V_a \wedge V_b$ is of dimension $a+b-n$, at most. Since V_n is assumed to have no singular points, no isolated component of $*V_a^{(i)} \wedge V_b$ can be of dimension less than $a+b-n$ [Th. I]. We conclude that any component of $*V_a^{(i)}{}_\wedge V_b$ which is not embedded in $V_a{}_\wedge V_b$ has dimension $a+b-n$.

We can go further and show that if C is not vacuous, and V_b is not contained in V_a, W and V_b intersect simply as varieties in S_r, along any intersection not contained in $V_a{}_\wedge V_b$.

If y' is any point of $W_\wedge V_b$ not on V_a, it is a point on $*V_a^{(i)}{}_\wedge V_b$ for some value of i, and hence, as above, it is on the object-variety of the correspondence C. Conversely, it is seen that if y' is a point of the object-variety of C and is not on V_a, it is a point of $W_\wedge V_b$ not on V_a. Hence the components of $W_\wedge V_b$ not embedded in $V_a{}_\wedge V_b$ are the components of the object-variety not so embedded. These components are the object-varieties, which are not embedded, of the components $C^{(i)}$ of C. Let $\eta^{(i)}$ be a generic point of the object-variety of $C^{(i)}$ with respect to the ground field K^* obtained by adjoining the coefficients of the equations of S_{r-n-1} to K. Then, over K, $\eta^{(i)}$ is a generic point of the object-variety of the correspondence $B^{(i)}$, which is V_b. Thus $\eta^{(i)}$ is a generic point of V_b over K.

Let $\zeta^{(i)}$ be a generic point over $K^*(\eta^{(i)})$ of the variety which corresponds to $\eta^{(i)}$ in $C^{(i)}$. Then $\eta^{(i)}\zeta^{(i)}$ meets V_a in a point $\xi^{(i)}$, which is a generic point of V_a over K.

Let α be the tangent space to V_a at $\xi^{(i)}$, and let β be the tangent space to V_b at $\eta^{(i)}$. The tangent space γ to W at $\xi^{(i)}$ is the join of α to S_{r-n-1}. Suppose that β and γ intersect in a k-space. Then

$$k \geqslant b+a+r-n-r = a+b-n.$$

We show that $k = a+b-n$. We can specialise S_{r-n-1} so that it becomes the join of $\zeta^{(i)}$ to a generic S'_{r-n-2} of S_r (independent of $\xi^{(i)}$ and $\eta^{(i)}$). The $(a+r-n)$-space π joining $\zeta^{(i)}$, α, S'_{r-n-2} will meet β in a space of dimension k at least (by the properties of proper specialisation). Let S_m be the join of α and β. Then π meets S_m in a space Π of dimension $a+m-n$, which is a generic space of the system of $(a+m-n)$-spaces through α and $\zeta^{(i)}$, when $m>n$. Since the join of α and $\zeta^{(i)}$ meets β in a space of dimension $a+b+1-m$, it follows, when $m>n$, that Π meets β in a space of dimension $a+b-n$. Hence π meets β in a space of dimension $a+b-n$. Therefore we have $k \leqslant a+b-n$. We know that $k \geqslant a+b-n$. It follows that $k = a+b-n$.

We deduce that W and V_b intersect simply along the components $*V_a^{(i)}{}_\wedge V_b$ which are not embedded in $V_a{}_\wedge V_b$. Taking first the case in which $V_b = V_n$, we have

Lemma III. The intersections of W and V_n are all simple.

We next consider the case in which V_b is a proper subvariety of V_n. Suppose that $\eta^{(i)}$ lies on the component $*V_a^{(j)}$ of $W_\wedge V_n$. The tangent space to $*V_a^{(j)}$ at $\eta^{(i)}$ is the intersection of the tangent space τ at $\eta^{(i)}$ to V_n (which contains the tangent space β to V_b at $\eta^{(i)}$) with the tangent space γ to W at $\eta^{(i)}$. But β and γ meet in a space of dimension $a+b-n$. It follows that τ, β, γ intersect in a space of dimension $a+b-n$; that is, β meets the tangent space to $*V_a^{(j)}$ at $\eta^{(i)}$ in a space of dimension $a+b-n$.

If U_c and U_d are two varieties on V_n having a point P in common which is simple on each variety, the tangent spaces at P to U_c and to U_d each lie in the tangent space to V_n at P, and therefore meet in a space of dimension $c+d-n$, at least. When they meet in a space of dimension $c+d-n$, we say that U_c and U_d *meet simply* along the component of their intersection which contains P. We then have

Lemma IV. Any component of $*V_a^{(i)}{}_\wedge V_b$ which is not embedded in $V_a{}_\wedge V_b$ is a simple intersection of $*V_a^{(i)}$ and V_b.

There is a point P on any component of $V_a{}_\wedge V_b$ which is independent of S_{r-n-1}. By Lemma II this point is not on $*V_a^{(i)}$, and therefore not on $*V_a^{(i)}{}_\wedge V_b$. There follows

Lemma V. If $V_a{}_\wedge V_b$ is of dimension l, any component of $*V_a^{(i)}{}_\wedge V_b$ embedded in $V_a{}_\wedge V_b$ is of dimension less than l.

Finally, we consider the case in which $V_a{}_\wedge V_b$ is unmixed of dimension $a+b-n$. By Theorem I, any component of $*V_a^{(i)}{}_\wedge V_b$ is of dimension $a+b-n$, at least. It follows from Lemma V that no component of $*V_a^{(i)}{}_\wedge V_b$ is embedded in $V_a{}_\wedge V_b$. Using Lemma IV we have

Lemma VI. If $V_a{}_\wedge V_b$ is unmixed of dimension $a+b-n$, then the components of $*V_a^{(i)}{}_\wedge V_b$ are all simple and of dimension $a+b-n$. None is embedded in $V_a{}_\wedge V_b$.

7. Intersection theory on a non-singular V_n. It is clear from the preceding section that serious difficulties arise in the development of a theory of intersections on a variety V_n if this variety has singular points. We shall therefore confine our attention to *non-singular* varieties V_n, that is, to varieties V_n without singular points, leaving the reader who wishes to go further to consult the works mentioned on p. 140.

We proceed now to develop a theory of intersections of multiplicative varieties on a non-singular variety V_n. We assume, then, that V_n is an irreducible variety free of singular points. Let V_a, V_b be two irreducible varieties lying on V_n which are such that every component of $V_a \wedge V_b$ is of dimension $a+b-n$. We define $\mathbf{V}_n$, $\mathbf{V}_a$, $\mathbf{V}_b$ as usual, by associating with them their irreducible Cayley forms. Taking independent generic spaces S_{r-n-1}, S'_{r-n-1} in S_r, as in § 6, we construct the irreducible varieties W_{a+r-n}, W_{b+r-n} which join S_{r-n-1} to V_a, S'_{r-n-1} to V_b. The corresponding irreducible varieties in the multiplicative sense are denoted by $\mathbf{W}_{a+r-n}$, $\mathbf{W}_{b+r-n}$.

We denote the intersection of $\mathbf{V}_a$ and $\mathbf{V}_b$ on V_n by

$$(\mathbf{V}_a \cdot \mathbf{V}_b)_{V_n},$$

and define this symbol to mean the variety whose components are the components of $V_a \wedge V_b$, each taken with the multiplicity with which it appears in

$$\mathbf{W}_{a+r-n} \cdot \mathbf{V}_n \cdot \mathbf{W}_{b+r-n}.$$

Using the commutative law for intersections in S_r [§ 4], we see at once that

$$(\mathbf{V}_a \cdot \mathbf{V}_b)_{V_n} = (\mathbf{V}_b \cdot \mathbf{V}_a)_{V_n}.$$

From Lemmas I, III of § 6,

$$\mathbf{W}_{a+r-n} \cdot \mathbf{V}_n = \mathbf{V}_a + \sum_i {}^*\mathbf{V}_a^{(i)}.$$

Hence
$$\mathbf{W}_{a+r-n} \cdot \mathbf{V}_n \cdot \mathbf{W}_{b+r-n} = \mathbf{V}_a \cdot \mathbf{W}_{b+r-n} + \sum_i {}^*\mathbf{V}_a^{(i)} \cdot \mathbf{W}_{b+r-n}. \tag{1}$$

It follows from the proof of Lemma V, § 6, that no component of ${}^*V_a^{(i)} \wedge W_{b+r-n}$ coincides with any component of $V_a \wedge V_b$. Hence we deduce from (1) that *we can define* $(\mathbf{V}_a \cdot \mathbf{V}_b)_{V_n}$ *to be the sum of the components of* $V_a \wedge V_b$*, each taken with the multiplicity with which it appears in* $\mathbf{V}_a \cdot \mathbf{W}_{b+r-n}$*, or in* $\mathbf{V}_b \cdot \mathbf{W}_{a+r-n}$.

We now extend our definition to cover unmixed varieties $\mathbf{V}_a$, $\mathbf{V}_b$. If

$$\mathbf{V}_a = \Sigma \alpha_i \mathbf{V}_a^{(i)},$$

$$\mathbf{V}_b = \Sigma \beta_j \mathbf{V}_b^{(j)},$$

and $V_a \wedge V_b$ is pure $(a+b-n)$-dimensional, we define

$$(\mathbf{V}_a \cdot \mathbf{V}_b)_{V_n} = \Sigma\Sigma \alpha_i \beta_j (\mathbf{V}_a^{(i)} \cdot \mathbf{V}_b^{(j)})_{V_n}.$$

From this definition it follows immediately that the distributive law holds.

Alternatively, we may define $W^{(i)}_{a+r-n}$ to be the irreducible variety joining a generic S_{r-n-1} to $V^{(i)}_a$, $W^{(j)}_{b+r-n}$ to be the irreducible variety joining an independent generic S'_{r-n-1} to $V^{(j)}_b$, and $\boldsymbol{W}^{(i)}_{a+r-n}$, $\boldsymbol{W}^{(j)}_{b+r-n}$ to be the corresponding irreducible varieties in the multiplicative sense. We then define

$$\boldsymbol{W}_{a+r-n} = \sum_i \alpha_i \boldsymbol{W}^{(i)}_{a+r-n},$$

$$\boldsymbol{W}_{b+r-n} = \sum_j \beta_j \boldsymbol{W}^{(j)}_{b+r-n},$$

and see at once that $(\boldsymbol{V}_a . \boldsymbol{V}_b)_{V_n}$ is equal to the sum of those components (with the same multiplicity) of $\boldsymbol{W}_{a+r-n} . \boldsymbol{V}_n . \boldsymbol{W}_{b+r-n}$ which are components of $V_a \wedge V_b$.

The next question which arises is whether our definition of intersections on V_n is invariant, in the sense of § 5, Th. I, under an extension of the ground field. This is clearly the case when V_n is absolutely irreducible, since any decomposition of $\boldsymbol{V}_a$ or $\boldsymbol{V}_b$ determines a similar decomposition of $\boldsymbol{W}_{a+r-n}$, $\boldsymbol{W}_{b+r-n}$. But if V_n is not absolutely irreducible, it may become reducible. In such a case we require a theory of intersections on a reducible variety.

However, our assumption that V_n has no singular points implies that no two components of V_n have a point in common. For if $V^{(i)}_n$, $V^{(j)}_n$ have a point P in common, a generic S_{r-n} through P meets $V^{(i)}_n$ and $V^{(j)}_n$ at P, and therefore meets V_n at least twice at P. Hence P cannot be a simple point on V_n.

It can now be shown that if $\boldsymbol{V}_n = \sum_i \boldsymbol{V}^{(i)}_n$, and

$$\boldsymbol{V}_a = \sum_i \boldsymbol{V}^{(i)}_a \quad (V^{(i)}_a \text{ on } V^{(i)}_n),$$

$$\boldsymbol{V}_b = \sum_j \boldsymbol{V}^{(j)}_b \quad (V^{(j)}_b \text{ on } V^{(j)}_n),$$

then

$$(\boldsymbol{V}_a . \boldsymbol{V}_b)_{V_n} = \sum_i (\boldsymbol{V}^{(i)}_a . \boldsymbol{V}^{(i)}_b)_{V^{(i)}_n}.$$

Finally, we prove the associative law for intersections on an irreducible V_n. Let V_a, V_b, V_c be three varieties on V_n whose intersection $V_a \wedge V_b \wedge V_c$ is pure $(a+b+c-2n)$-dimensional (or vacuous), and let W_{a+r-n}, W_{b+r-n}, W_{c+r-n} be the related cones, defined in each case by joining the variety to a generic $(r-n-1)$-space. Now consider the variety $\boldsymbol{U}_{a+b+c-2n}$ which is obtained by attaching to the components of $V_a \wedge V_b \wedge V_c$ the multiplicities which they have in

$$\boldsymbol{W}_{a+r-n} . \boldsymbol{W}_{b+r-n} . \boldsymbol{W}_{c+r-n} . \boldsymbol{V}_n.$$

Since $$W_{c+r-n} \cdot V_n = V_c + \sum_i {}^*V_c^{(i)},$$

$$W_{a+r-n} \cdot W_{b+r-n} \cdot W_{c+r-n} \cdot V_n = W_{a+r-n} \cdot W_{b+r-n} \cdot V_c + \sum_i W_{a+r-n} \cdot W_{b+r-n} \cdot {}^*V_c^{(i)}.$$

Using Lemma V, § 6, as above, no component of

$$W_{a+r-n} \cdot W_{b+r-n} \cdot {}^*V_c^{(i)}$$

coincides with any component of $V_a \wedge V_b \wedge V_c$. Hence $U_{a+b+c-2n}$ is equal to the variety obtained by attaching to the components of $V_a \wedge V_b \wedge V_c$ their multiplicities in $W_{a+r-n} \cdot W_{b+r-n} \cdot V_c$. Again,

$$W_{b+r-n} \cdot V_c = (V_b \cdot V_c)_{V_n} + \sum_i R^{(i)},$$

where no variety $R^{(i)}$ appears among the components of $V_b \wedge V_c$. Hence $U_{a+b+c-2n}$ is obtained by attaching to the components of $V_a \wedge V_b \wedge V_c$ their multiplicities in $W_{a+r-n} \cdot (V_b \cdot V_c)_{V_n}$, and we may write

$$U_{a+b+c-2n} = (V_a \cdot (V_b \cdot V_c)_{V_n})_{V_n}.$$

Similarly, $$U_{a+b+c-2n} = ((V_a \cdot V_b)_{V_n} \cdot V_c)_{V_n}.$$

Hence the associative law holds. We have proved

THEOREM I. *Provided the intersections are defined, the associative, commutative and distributive laws hold for intersections on a non-singular variety V_n.*

8. Intersections of systems of varieties. In the preceding sections we have examined the intersections of two individual varieties V_a and V_b in S_n or on an irreducible V_n, and have been concerned only with the case in which every component of $V_a \wedge V_b$ is of dimension $a+b-n$. The next stage in our argument is to consider the intersection $V_a \cdot V_b$ as V_a, say, varies in an irreducible algebraic system, subject still to the condition that every component of $V_a \wedge V_b$ is of dimension $a+b-n$. When intersections are taken on a variety V_n, it will be assumed that V_n is irreducible and has no singular points.

Let $\mathfrak{S}$ denote an irreducible system of varieties of dimension a on V_n, and let V_a^* denote a generic member of $\mathfrak{S}$. Let V_b be a variety of dimension b on V_n, and let V_a' be a specialisation of V_a^* such that every component of $V_a' \wedge V_b$ is of dimension $a+b-n$. No component of $V_a^* \wedge V_b$ is of dimension less than $a+b-n$ [§ 6, Th. 1]. On the other hand, if there exists a component of dimension greater than $a+b-n$, a generic space S $(\rho = r+n-a-b-1)$ meets $V_a^* \wedge V_b$. Hence the resultant system of the equations of V_a^*, V_b, S_ρ vanishes. This remains true when we specialise V_a^* to V_a', so that a generic ρ-space meets $V_a' \wedge V_b$. This implies that $V_a' \wedge V_b$ has a component of dimension greater than $a+b-n$, and we have a contradiction. Hence every component of $V_a^* \wedge V_b$ is of dimension $a+b-n$. The theorem which we seek to prove is as follows:

THEOREM I. *Let* $\mathbf{V}_a^*$ *be a generic member of an irreducible system* $\mathfrak{S}$ *of varieties of dimension a on* V_n, *and let* $\mathbf{V}_b$ *be a variety of dimension b on* V_n. *If there exists a specialisation* $\mathbf{V}_a'$ *of* $\mathbf{V}_a^*$ *such that every component of* $V_a' \wedge V_b$ *is of dimension* $a+b-n$, *then as* $\mathbf{V}_a^* \to \mathbf{V}_a'$,

$$(\mathbf{V}_a^* . \mathbf{V}_b)_{V_n} \to (\mathbf{V}_a' . \mathbf{V}_b)_{V_n}.$$

We first reduce the proof of this result to the case in which V_n is an n-space S_n. Let S_{r-n-1} be a generic $(r-n-1)$-space in S_r, and let W_{r-n+a}^*, W_{r-n+a}' be the cross-joins of V_a^*, V_a' to S_{r-n-1}. Each irreducible component of $V_a^*(V_a')$ determines an irreducible component of $W_{r-n+a}^*(W_{r-n+a}')$. We assign to each component of $W_{r-n+a}^*(W_{r-n+a}')$ the multiplicity of the corresponding component of V_a^* in $\mathbf{V}_a^*$ (V_a' in $\mathbf{V}_a'$), and so obtain a multiplicative variety $\mathbf{W}_{r-n+a}^*$ $(\mathbf{W}_{r-n+a}')$. We show that when $\mathbf{V}_a^* \to \mathbf{V}_a'$, $\mathbf{W}_{r-n+a}^* \to \mathbf{W}_{r-n+a}'$. Let $F_i^*(u_0, \ldots, u_a)$ $(i = 1, \ldots, \sigma)$ be the Cayley form of any irreducible component $V_a^{(i)}$ of V_a^*. Then if $V_a^{(i)}$ is of order g_i,

$$F_i^*(u_0, \ldots, u_a) = G_i(\ldots, |\, u_{0i_0} \ldots u_{ai_a} \,|, \ldots),$$

a form in the determinants $|\, u_{0i_0} \ldots u_{ai_a} \,|$ of order g_i. G_i has the property that if $(\ldots |\, u_{0i_0}' \ldots u_{ai_a}' \,|, \ldots)$ are the dual coordinates of any S_{r-a-1}', a necessary and sufficient condition that this space meets $V_a^{(i)}$ is

$$G_i(\ldots, |\, u_{0i_0}' \ldots u_{ai_a}' \,|, \ldots) = 0.$$

Now let S_{n-a-1}' be any $(n-a-1)$-space, $(\ldots, |\, u_{0i_0}' \ldots u_{\rho i_\rho}' \,|, \ldots)$ its dual coordinates, where $\rho = r+a-n$. A necessary and sufficient condition that S_{n-a-1}' meets the component $W_{r-n+a}^{(i)}$ of W_{r-n+a}^* which corresponds to $V_a^{(i)}$ is that the join of S_{n-a-1}' to S_{r-n-1} meets $V_a^{(i)}$. If $(\ldots p_{i_0 \ldots i_{r-n-1}}, \ldots)$ are the Grassmann coordinates of S_{r-n-1}, the dual coordinates of this join are [VII, § 5, p. 307]

$$(\ldots, \sum_{j_0 \ldots j_{r-n-1}} p_{j_0 \ldots j_{r-n-1}} |\, u_{0j_0}' \ldots u_{r-n-1\, j_{r-n-1}}' u_{r-n i_0}' \ldots u_{\rho i_a}' \,|, \ldots).$$

S'_{n-a-1} therefore meets $W^{(i)}_{r-n+a}$ if and only if

$$G_i(\ldots, \sum_j p_{j_0\ldots j_{r-n-1}} \,|\, u'_{0j_0} \ldots u'_{\rho i_a}\,|, \ldots) = 0.$$

Since $\quad \Phi^*_i(u_0, \ldots, u_\rho) = G_i(\ldots, \sum_j p_{j_0\ldots j_{r-n-1}} \,|\, u_{0j_0} \ldots u_{\rho i_a}\,|, \ldots),$

is clearly irreducible because F^*_i is irreducible, Φ^*_i is the Cayley form of the component $W^{(i)}_{r-n+a}$ of W^*_{r-n+a}.

If the Cayley form of V^*_a is $\prod_{i=1}^{\sigma} [F^*_i(u_0, \ldots, u_a)]^{\rho_i}$ then, by the definition of W^*_{r-n+a}, its Cayley form is $\prod_{i=1}^{\sigma} [\Phi^*_i(u_0, \ldots, u_\rho)]^{\rho_i}$. Similarly, we obtain the Cayley form of W'_{r-n+a} from that of V'_a, and it is clear from the Cayley forms that as $V^*_a \to V'_a$, $W^*_{r-n+a} \to W'_{r-n+a}$.

Now $(V^*_a . V_b)_{V_n}$ and $(V'_a . V_b)_{V_n}$ are the components of $W^*_{a+r-n} \wedge V_b$ and $W'_{a+r-n} \wedge V_b$ contained in $V^*_a \wedge V_b$ and $V'_a \wedge V_b$ taken with their multiplicities in $W^*_{a+r-n} . V_b$ and $W'_{a+r-n} . V_b$ respectively. Hence in order to prove that as $V^*_a \to V'_a$ then $(V^*_a . V_b)_{V_n} \to (V'_a . V_b)_{V_n}$, it is sufficient to prove that as $W^*_{r-n+a} \to W'_{r-n+a}$, then

$$W^*_{r-n+a} . V_b \to W'_{r-n+a} . V_b.$$

This, however, is just Theorem I, with V_n replaced by S_r. Hence it is sufficient to prove

THEOREM II. *Let V^*_a be a generic member of an irreducible algebraic system $\mathfrak{S}$ of varieties of dimension a in S_n and let V_b be a variety of dimension b in S_n. Then if V'_a is a specialisation of V^*_a such that every component of $V'_a \wedge V_b$ is of dimension $a+b-n$, then as $V^*_a \to V_a$, $V^*_a . V_b \to V'_a . V_b$.*

The proof of this theorem is similar to the proof of the associative law for intersections in S_n [§ 5]. We first prove it in the case in which V_b is a generic b-space in S_n. As in the proof of the associative law, we see that if

$$\sum_j \nu_{ij} x_j = 0 \quad (i = 1, \ldots, n-b)$$

are the equations of V_b, and if $F(u_0, \ldots, u_a)$ is the Cayley form of V^*_a, the Cayley form of $V^*_a . V_b$ is $F(u_0, \ldots, u_{a+b-n}, \nu_1, \ldots, \nu_{n-b})$. The Cayley form of $V'_a . V_b$ is constructed similarly. From the properties of the Cayley form it follows that as $V^*_a \to V'_a$, $V^*_a . V_b \to V'_a . V_b$. This result extends at once to the case in which V_b is a set of conjugate b-spaces through a generic S_{b-1}.

Now consider a general variety V_b. For the reasons given in § 7 it is sufficient to consider the case in which V_b is an absolutely irreducible variety counted with multiplicity one. Let

$$V'_a = \Sigma \rho_i V_a^{(i)},$$

where $V_a^{(i)}$ is irreducible and has multiplicity one. We define V_b^τ as in § 2. Then as $\tau \to I$,

$$V_a^* . V_b^\tau \to V_a^* . V_b, \quad V'_a . V_b^\tau \to V'_a . V_b.$$

We first show that as $V_a^* \to V'_a$, $V_a^* . V_b^\tau \to V'_a . V_b^\tau$. It is clear that as $V_a^* \to V'_a$, the specialisation of $V_a^* . V_b^\tau$ coincides with $V'_a . V_b^\tau$ as a point-set. Since the components of $V_a^{(i)} \wedge V_b^\tau$ are conjugate over $K(\tau)$, it is also clear that each of these components has the same multiplicity in the specialisation of $V_a^* . V_b^\tau$. (This follows at once from a consideration of the Cayley form.) Hence, as $V_a^* \to V'_a$,

$$V_a^* . V_b^\tau \to \Sigma \sigma_i V_a^{(i)} . V_b^\tau.$$

Now specialise $\tau \to t_0$, where t_0 is the singular matrix defined in § 2.

Then $$\Sigma \sigma_i V_a^{(i)} . V_b^\tau \to \Sigma \sigma_i V_a^{(i)} . V_b^{t_0}.$$

Thus, by specialising $V_a^* \to V'_a$ and then $\tau \to t_0$, we obtain

$$V_a^* . V_b^\tau \to \Sigma \sigma_i V_a^{(i)} . V_b^{t_0}.$$

On the other hand, if we specialise $\tau \to t_0$,

$$V_a^* . V_b^\tau \to V_a^* . V_b^{t_0},$$

and, by the special case considered above, when we specialise $V_a^* \to V'_a$,

$$V_a^* . V_b^{t_0} \to V'_a . V_b^{t_0}.$$

Hence if we specialise $\tau \to t_0$ and then $V_a^* \to V'_a$,

$$V_a^* . V_b^\tau \to \Sigma \rho_i V_a^{(i)} . V_b^{t_0}.$$

Hence, if we can prove that the specialisation of $V_a^* . V_b^\tau$ corresponding to the specialisation $(V_a^*, \tau) \to (V'_a, t_0)$ is independent of the order of specialisation, we obtain $\rho_i = \sigma_i$ for each value of i.

Let us, for the moment, assume that this is true. Then since as $V_a^* \to V'_a$ we have

$$V_a^* . V_b^\tau \to \Sigma \rho_i V_a^{(i)} . V_b^\tau,$$

it follows that when we specialise V_a^* to V'_a, and then $\tau \to I$,

$$V_a^* . V_b^\tau \to \Sigma \rho_i V_a^{(i)} . V_b = V'_a . V_b.$$

On the other hand, if we specialise $\tau \to I$, then $\boldsymbol{V}_a^* \to \boldsymbol{V}_a'$, $\boldsymbol{V}_a^* . \boldsymbol{V}_b^\tau$ specialises to the required specialisation of $\boldsymbol{V}_a^* . \boldsymbol{V}_b$. If it is true that the specialisation of $\boldsymbol{V}_a^* . \boldsymbol{V}_b^\tau$ as $(\boldsymbol{V}_a^*, \tau) \to (\boldsymbol{V}_a', I)$ does not depend on the order in which the specialisation is carried out, we deduce that $\boldsymbol{V}_a^* . \boldsymbol{V}_b$ specialises to $\boldsymbol{V}_a' . \boldsymbol{V}_b$ as $\boldsymbol{V}_a^* \to \boldsymbol{V}_a'$.

Thus Theorem II, and hence Theorem I, will be established if we can prove that the specialisation of $\boldsymbol{V}_a^* . \boldsymbol{V}_b^\tau$ corresponding to the specialisation of $(\boldsymbol{V}_a^*, \tau)$ to $(\boldsymbol{V}_a', t_0)$ and $(\boldsymbol{V}_a', I)$ are determined irrespective of the order in which $\boldsymbol{V}_a^*$ and τ are specialised. We prove this most simply by using the arithmetic methods to be developed in Chapter XVI, Vol. III. Let U be the Cayley image of the system $\mathfrak{S}$, and let W be a normal variety derived from U [XVI]. Let P be the point of U which represents V_a^*. It is a generic point of U, and to it there corresponds a unique point P^* of W, which is a generic point of this variety. We may assume that the coordinates of P^* are in the function field of U. To each point of W there corresponds exactly one point of U, and to each point of U there corresponds at least one and only a finite number of points of W. Let P' be the point of U which represents V_a' and let Q' be any one of the points of W which corresponds to P'. We can then establish an algebraic correspondence between the space containing W and S_n in which the object-variety is W and the variety corresponding to any point R of W is the unique member of $\mathfrak{S}$ which corresponds to the point of U corresponding to R. We shall write the equations of this in the form

$$f_i(R, x) = 0 \quad (i = 1, 2, \ldots).$$

Now let S_N be a space of dimension $n^2 + 2n$, in which the coordinates are $(t_{00}, \ldots, t_{nn})$. In it, we represent the matrix τ by the point $(\tau_{00}, \ldots, \tau_{nn})$. Let Π be the direct product of W and S_N. This is irreducible, since W is irreducible and S_N is absolutely irreducible. We establish a correspondence between Π and S_n in which to (P^*, τ) there corresponds the variety $\boldsymbol{V}_a^* . \boldsymbol{V}_b^\tau$, and to any point of Π there corresponds all the proper specialisations of this corresponding to the specialisation of (P^*, τ).

To the generic point (P^*, τ) of Π there corresponds the variety of dimension $a + b - n$ having the equations

$$f_i(P^*, x) = 0 \quad (i = 1, 2, \ldots),$$

$$g_i(\tau x) = 0 \quad (i = 1, 2, \ldots),$$

where $$g_i(x) = 0 \quad (i = 1, 2, \ldots)$$

are the equations of V_b. Let t_1 be any specialisation of τ such that $V'_a \wedge V_b^{t_1}$ is of dimension $a+b-n$. Then the equations

$$f_i(Q',x) = 0 \quad (i = 1,2,\ldots),$$

$$g_i(t_1,x) = 0 \quad (i = 1,2,\ldots),$$

define a variety of dimension $a+b-n$. Any specialisation of $V_a^*.V_b^\tau$ corresponding to the specialisation (Q',t_1) of (P^*,τ) satisfies the equations, and hence the total variety corresponding to (Q',t_1) in the correspondence is of dimension $a+b-n$. Since the quotient ring of (Q',t_1) on Π is integrally closed† it follows that the specialisation of $V_a^*.V_b^\tau$ corresponding to the specialisation $(P^*,\tau)\to(Q',t_1)$ is uniquely determined, and in particular it does not matter whether we first specialise $P^*\to Q'$ and then $\tau\to t_1$, or specialise $\tau\to t_1$ and then $P^*\to Q'$. It follows that the specialisation of $V_a^*.V_b^\tau$ corresponding to

$$(V_a^*,\tau)\to(V_a^*,t_1)\to(V'_a,t_1)$$

is the same as the specialisation corresponding to

$$(V_a^*,\tau)\to(V'_a,\tau)\to(V'_a,t_1).$$

Taking $t_1 = t_0$ and $t_1 = I$, in turn, we obtain the results necessary to complete the proof of Theorem II.

Theorem I is the key theorem to an intersection theory of systems

† Choose a non-homogeneous coordinate system in the space containing W so that Q' is at a finite distance, and let $(\zeta_1,\ldots,\zeta_m)$ be the coordinates of a generic point of W. Similarly, choose the coordinate system in S_N so that t_1 is at a finite distance, and let $(\eta_1,\ldots,\eta_N)$ be the coordinates of a generic point of S_N. We may suppose that $\eta_1,\ldots,\eta_N$ are independent over $K(\zeta)$. By a fundamental property of normal varieties $K[\zeta_1,\ldots,\zeta_m]$ is integrally closed in its quotient field. If we can prove that $K[\zeta_1,\ldots,\zeta_m,\eta_1,\ldots,\eta_N]$ is integrally closed in its quotient field, it will follow that the quotient ring of (Q',t_1) is integrally closed.

If α is any element of the quotient field of $K[\zeta_1,\ldots,\zeta_m,\eta_1,\ldots,\eta_N]$ which is integrally dependent on this ring, α is *a fortiori* integrally dependent on $K(\zeta)[\eta_1,\ldots,\eta_N]$. Since $\eta_1,\ldots,\eta_N$ are algebraically independent over $K(\zeta)$ it follows that α is in $K(\zeta)[\eta_1,\ldots,\eta_N]$, that is

$$\alpha = A(\zeta,\eta)$$

where $A(\zeta,\eta)$ is a polynomial in $\eta_1,\ldots,\eta_N$ with coefficients in $K(\zeta)$. Let $(y'_1,\ldots,y'_N)$ be any specialisation of $(\eta_1,\ldots,\eta_N)$ in K. Then since α is integrally dependent on $K[\zeta_1,\ldots,\zeta_m,\eta_1,\ldots,\eta_N]$, $A(\zeta,y')$ is integrally dependent on $K[\zeta_1,\ldots,\zeta_m,y'_1,\ldots,y'_N] = K[\zeta_1,\ldots,\zeta_m]$. Since $K[\zeta_1,\ldots,\zeta_m]$ is integrally closed in its quotient field, $A(\zeta,y')$ lies in $K[\zeta_1,\ldots,\zeta_m]$. If there are r power products $y_1^{\lambda_1},\ldots,y_N^{\lambda_N}$ in $A(\zeta,y)$ we can choose r specialisations of η so that the determinant of the specialised power products is non-singular. Solving the equations we see that each coefficient of $A(\zeta,y)$ is in $K[\zeta_1,\ldots,\zeta_m]$. Hence it is in $K[\zeta_1,\ldots\zeta_m,\eta_1,\ldots,\eta_N]$ and the result follows.

of varieties, as we shall see in the next section. It also enables us to remove the apparent arbitrariness in our definition of the intersection of two varieties in S_n (see § 1). The definition given in § 5 of the intersection $\boldsymbol{V}_a . \boldsymbol{V}_b$ of two irreducible varieties V_a, V_b, where $V_a \wedge V_b$ is purely $(a+b-n)$-dimensional, seems to depend on the choice of a particular normal problem. We considered the intersection $\boldsymbol{V}_a . \boldsymbol{V}_b^\tau$, defined to consist of the components of $V_a \wedge V_b^\tau$, the multiplicity one being attached to each component, and defined $\boldsymbol{V}_a . \boldsymbol{V}_b$ to be the specialisation of $\boldsymbol{V}_a . \boldsymbol{V}_b^\tau$ which corresponds to the specialisation $\tau \to I$.

Theorem I now tells us that if $\boldsymbol{V}_b(\zeta)$ is a generic member of any irreducible system of varieties containing $\boldsymbol{V}_b$, where

$$\boldsymbol{V}_b(\zeta) \to \boldsymbol{V}_b(z_0) = \boldsymbol{V}_b$$

as $\zeta \to z_0$, and if $V_a \wedge V_b(\zeta)$ is purely $(a+b-n)$-dimensional, and all components are simple intersections, then the variety obtained by attaching unit multiplicity to each component of $V_a \wedge V_b(\zeta)$ specialises to $\boldsymbol{V}_a . \boldsymbol{V}_b$ when $\zeta \to z_0$.

Thus $\boldsymbol{V}_a . \boldsymbol{V}_b$ may be defined by varying $\boldsymbol{V}_b$ in any irreducible algebraic system $\boldsymbol{V}_b(\zeta)$, subject only to the condition that V_a and a generic member $V_b(\zeta)$ of the system intersect simply. We made use in our definition of a particular system $\boldsymbol{V}_b(\zeta)$, that was obtained from $\boldsymbol{V}_b$ by projective transformation. This was a matter of convenience, and we now see that the definition does not depend on this particular system. But we also found that the system obtained from $\boldsymbol{V}_b$ by projective transformation is an example of a system $\boldsymbol{V}_b(\zeta)$ whose generic member intersects V_a simply.

Similar reasoning to that just given shows that the intersection $(\boldsymbol{V}_a . \boldsymbol{V}_b)_{V_n}$ is an intrinsic concept and does not depend on the particular method we have used to define it *provided that we can vary V_b (or, in view of the commutative law, V_a) in an irreducible system whose generic member meets V_a (or V_b) simply.* When V_n is not a projective space S_n it is not apparent that V_b or V_a can be so varied, and, indeed, examples can be given in which this possibility is excluded. This difficulty will be overcome in the following sections.

Finally, we note that when the intersection $\boldsymbol{V}_a . \boldsymbol{V}_b$ can be defined by varying V_a in an irreducible system whose generic member meets V_b simply, the argument used in the proof of Theorem II of § 4 enables us to show that each component of $V_a \wedge V_b$ has positive multiplicity in $\boldsymbol{V}_a . \boldsymbol{V}_b$.

9. Equivalence on an algebraic variety. We consider varieties which lie on a given non-singular algebraic variety V_n. Since all intersections considered will be intersections on V_n, we shall omit references to V_n in what follows. V_n is defined over K, but the varieties $V_a, \ldots$ which we consider on V_n may be defined over extensions of K. We should like to say that two varieties V_a, V'_a are *equivalent* if they belong to the same irreducible algebraic system of varieties on V_n. But although this definition has the properties of being reflexive and symmetric, the property of transitivity may be lacking. If V_a and V'_a belong to an irreducible system, and V'_a, V''_a belong to an irreducible system, it does not follow that V_a and V''_a belong to an irreducible system.

To obtain a true equivalence relation we must find a definition which implies that the relation is reflexive, symmetric and transitive. This is now given.

V_a and V'_a are said to be *equivalent* if we can find a finite sequence of varieties on V_n,

$$V_a^{(1)} = V_a, \quad V_a^{(2)}, \ldots, V_a^{(k-1)}, \quad V_a^{(k)} = V'_a,$$

such that $V_a^{(i)}$ and $V_a^{(i+1)}$ $(i = 1, \ldots, k-1)$ belong to some irreducible system of varieties on V_n.

This definition clearly has the three required properties. All the varieties $V_a^{(i)}$ must have the same order, and if we denote by $\mathfrak{S}^{(i)}$ the Cayley image of the system to which $V_a^{(i)}$, $V_a^{(i+1)}$ belong, the varieties $\mathfrak{S}^{(1)}, \ldots, \mathfrak{S}^{(k-1)}$ form a *connected* set; that is, they cannot be divided into two distinct non-vacuous sets which have no points in common. A system of varieties on V_n which can be represented by a connected set of varieties (on the Cayley image of all varieties of the order and dimension of those considered) is said to be a *continuous* system. Our definition of equivalence may now be reworded:

V_a and V'_a are equivalent if they belong to the same continuous system.

If V_a and V'_a are equivalent, we write

$$V_a \equiv V'_a.$$

If $V_a^{(1)} \equiv V_a^{(2)}$ and $V_a^{(3)} \equiv V_a^{(4)}$, we prove that

$$\text{(i)} \qquad V_a^{(1)} + V_a^{(3)} \equiv V_a^{(2)} + V_a^{(4)}.$$

This will follow from the equivalences

$$V_a^{(1)} + V_a^{(3)} \equiv V_a^{(2)} + V_a^{(3)} \equiv V_a^{(2)} + V_a^{(4)}$$

if we can prove that the equivalence

$$V_a \equiv V'_a$$

implies the equivalence

$$V_a + V''_a \equiv V'_a + V''_a.$$

To prove this last result we show that if V_a and V'_a are contained in the same irreducible algebraic system of varieties, then $V_a + V''_a$ and $V'_a + V''_a$ are contained in the same irreducible algebraic system of varieties. On account of the transitivity of the equivalence relation it is allowable to assume that V_a is a generic member of the system, and V'_a any specialisation of V_a.

Let ζ and ζ' be the Cayley images of V_a and V'_a respectively. Then ζ' is a proper specialisation of ζ. If the Cayley images of $V_a + V''_a$ and $V'_a + V''_a$ are Z, Z' respectively, then

$$Z_i = \sum_j a_{ij} \zeta_j,$$

$$Z'_i = \sum_j a_{ij} \zeta'_j,$$

where the a_{ij} are coefficients of the Cayley form of V''_a. For if the Cayley form of V_a is $F(u_0, \ldots, u_a)$, and the Cayley form of V''_a is $G(u_0, \ldots, u_a)$, the Cayley form of $V_a + V''_a$ is $F(u_0, \ldots, u_a)\, G(u_0, \ldots, u_a)$.

We now wish to show that Z' is a proper specialisation of Z. This is evident, for if

$$H(\ldots, Z_i, \ldots) = H(\ldots, \sum_j a_{ij} \zeta_j, \ldots) = 0$$

is a homogeneous relation satisfied by Z, then, ζ' being a proper specialisation of ζ,

$$H(\ldots, Z'_i, \ldots) = H(\ldots, \sum_j a_{ij} \zeta'_j, \ldots) = 0.$$

This proves (i). In particular we deduce that if $V_a^{(1)} \equiv V_a^{(2)}$, then $2V_a^{(1)} \equiv 2V_a^{(2)}$, and by simple induction we have

(ii) $$pV_a^{(1)} \equiv pV_a^{(2)}$$

for any positive integer p.

Again, if $V_a^{(1)} \equiv V_a^{(2)}$, and $V_a^{(3)} \equiv V_a^{(4)}$, then

(iii) $$pV_a^{(1)} + qV_a^{(3)} \equiv pV_a^{(2)} + qV_a^{(4)},$$

for any positive integers p, q.

It follows that if

$$V_a = \sum_{i=1}^{k} a_i V_a^{(i)},$$

and

$$V_a^{(i)} \equiv {}'V_a^{(i)},$$

then if we define the variety V'_a by the equation

$$V'_a = \sum_{i=1}^{k} a_i {}'V_a^{(i)},$$

we have the relation

$$V_a \equiv V'_a.$$

Hence, in dealing with continuous systems of reducible varieties, we may replace any component by an equivalent variety.

If $V_a \equiv V'_a$ and $V_b \equiv V'_b$, Theorem I of the last section gives us the equivalence

$$V_a . V_b \equiv V'_a . V_b,$$

if we assume that $V_a^{(i)} {}_\wedge V_b$ $(i = 1, \ldots, k)$ are unmixed of dimension $a+b-n$ [see p. 189]. (The suffix V_n is omitted since all equivalences and intersections are taken to be on V_n.)

With a similar assumption we also have

$$V'_a . V_b \equiv V'_a . V'_b,$$

and therefore

$$V_a . V_b \equiv V'_a . V'_b.$$

The equivalence which we have defined above is only one type of equivalence which can be defined for algebraic varieties on V_n, and it is sometimes called *equivalence in the narrow sense*. A wider type of equivalence can now be introduced. We have seen that if

$$V_a \equiv V'_a,$$

and V''_a is any other variety on V_n of dimension a, then

$$V_a + V''_a \equiv V'_a + V''_a.$$

The converse of this result, that

$$V_a + V''_a \equiv V'_a + V''_a$$

implies

$$V_a \equiv V'_a,$$

may not be true. We therefore extend our definition of equivalence, and write

$$V_a \equiv V'_a,$$

to imply that there exists a variety V''_a such that

$$V_a + V''_a \equiv V'_a + V''_a.$$

The relation $\equiv$ is clearly reflexive and symmetric. To prove that it is transitive, we suppose that

$$V_a \equiv V'_a, \quad V'_a \equiv V''_a.$$

Then there exist varieties $V_a^{(3)}$, $V_a^{(4)}$ such that

$$V_a + V_a^{(3)} \equiv V'_a + V_a^{(3)},$$

and
$$V'_a + V_a^{(4)} \equiv V''_a + V_a^{(4)}.$$

Then, by (i),
$$\begin{aligned} V_a + V_a^{(3)} + V_a^{(4)} &\equiv V'_a + V_a^{(3)} + V_a^{(4)} \\ &\equiv V'_a + V_a^{(4)} + V_a^{(3)} \\ &\equiv V''_a + V_a^{(4)} + V_a^{(3)} \\ &\equiv V''_a + V_a^{(3)} + V_a^{(4)}. \end{aligned}$$

Hence
$$V_a \equiv V''_a.$$

Thus $\equiv$ is a true equivalence relation. We call this *equivalence in the wide sense.*

Again if
$$V_a^{(1)} \equiv V_a^{(2)}, \quad V_a^{(3)} \equiv V_a^{(4)},$$

there exist varieties $V_a^{(5)}$, $V_a^{(6)}$ such that

$$V_a^{(1)} + V_a^{(5)} \equiv V_a^{(2)} + V_a^{(5)},$$

$$V_a^{(3)} + V_a^{(6)} \equiv V_a^{(4)} + V_a^{(6)}.$$

Hence
$$V_a^{(1)} + V_a^{(3)} + V_a^{(5)} + V_a^{(6)} \equiv V_a^{(2)} + V_a^{(4)} + V_a^{(5)} + V_a^{(6)},$$

and therefore
$$V_a^{(1)} + V_a^{(3)} \equiv V_a^{(2)} + V_a^{(4)},$$

that is, equivalences in the wide sense may be added. From this we conclude that if

$$V_a = \sum_1^k \rho_i V_a^{(i)},$$

and
$$V_a^{(1)} \equiv {}'V_a^{(i)} \quad (i = 1, \ldots, k),$$

then
$$V_a \equiv \sum_1^k \rho_i {}'V_a^{(i)}.$$

Finally, if
$$V_a \equiv V'_a,$$

and V''_a is such that
$$V_a + V''_a \equiv V'_a + V''_a,$$

then
$$V_a . V_b + V''_a . V_b \equiv V'_a . V_b + V''_a . V_b$$

and hence
$$V_a . V_b \equiv V'_a . V_b,$$

with an assumption similar to that on p. 182.

10. Virtual varieties. The equivalence relations, both in the narrow sense and in the wide sense, defined in § 9 are true equivalence relations. Hence they serve to divide the a-dimensional varieties (in the multiplicative sense) on V_n into mutually exclusive sets, the varieties in any set being all those equivalent (in the sense selected) to any one of them. Let us, for the moment, concentrate on equivalence in the wide sense. We denote the class containing V_a by $[V_a]$.

Since the equivalences

$$U_a \equiv U'_a, \quad V_a \equiv V'_a$$

imply the equivalence $U_a + V_a \equiv U'_a + V'_a$,

it follows that the classes $[U_a]$ and $[V_a]$ determine, by the addition law for multiplicative varieties, a unique class $[U_a + V_a]$, which we denote by $[U_a] + [V_a]$. Since

$$U_a + V_a \equiv V_a + U_a,$$

and
$$(U_a + V_a) + W_a \equiv U_a + (V_a + W_a),$$

we have
$$[U_a] + [V_a] = [V_a] + [U_a]$$

and
$$([U_a] + [V_a]) + [W_a] = [U_a] + ([V_a] + [W_a]),$$

that is, the addition law for classes is both commutative and associative. But under this law the classes of equivalent varieties do not form a group, since in general we cannot solve the equation

$$[U_a] + x = [V_a].$$

Similar difficulties would arise if we defined our classes by equivalence in the narrow sense.

In order to obtain the group property, we have to increase the number of classes by introducing a generalisation of the notion of a variety. Let $V_a^{(i)}$ $(i = 1, \ldots, k)$ be any finite set of irreducible varieties of dimension a and multiplicity one and let ρ_i $(i = 1, \ldots, k)$ be any set of integers, positive, negative, or zero. Formally, we define the sum $\sum_1^k \rho_i V_a^{(i)}$ to be a *virtual algebraic variety* of dimension a:

$$V_a = \sum_1^k \rho_i V_a^{(i)}.$$

In this sum, the order of the terms is irrelevant, and we shall agree to define the sum and difference of V_a and the virtual variety

$$V'_a = \sum_1^k \sigma_i V_a^{(i)}$$

as

$$V_a \pm V'_a = \sum_1^k (\rho_i \pm \sigma_i) V_a^{(i)}.$$

It is clear that the varieties previously considered are included among the virtual varieties, and are given by the special cases in which none of the coefficients ρ_i is negative. Any such variety

$$V_a = \sum_1^k \rho_i V_a^{(i)}$$

in which $\rho_i \geqslant 0$ $(i = 1, \ldots, k)$ is called an *effective variety*. In the special case in which each ρ_i is zero the variety is a *zero variety*. Two varieties which differ by a zero variety are assumed to be equal.

If $V_a = \sum_1^k \rho_i V_a^{(i)}$ is any virtual variety, we define an effective variety

$$V'_a = \Sigma' \rho_i V_a^{(i)},$$

where the summation is over the values of i for which $\rho_i > 0$, and a second effective variety

$$V''_a = \Sigma''(-\rho_i) V_a^{(i)},$$

summed over the values of i for which $\rho_i < 0$. Either V'_a or V''_a may be zero. We can then write

$$V_a = V'_a - V''_a,$$

so that any virtual variety can be written as the difference of two effective varieties.

If V_a^*, V_a^{**} are any other effective varieties such that

$$V_a = V_a^* - V_a^{**},$$

then clearly

$$V'_a + V_a^{**} = V''_a + V_a^*.$$

If

$$V_a^* = \sum_1^{k'} \sigma_i V_a^{(i)}, \quad V_a^{**} = \sum_1^{k'} \tau_i V_a^{(i)} \quad (k' \geqslant k),$$

where $\sigma_i \geqslant 0$, $\tau_i \geqslant 0$, it follows, from the fact that the coefficient of $V_a^{(i)}$ is zero either in V'_a or in V''_a, that

$$V_a^* = V'_a + \overline{V}_a, \quad V_a^{**} = V''_a + \overline{V}_a,$$

where $\overline{V}_a$ is an effective variety.

Two virtual varieties U_a, V_a are said to be *equivalent in the virtual sense*, $U_a \sim V_a$ if and only if

$$U'_a + V''_a \equiv U''_a + V'_a,$$

where $U_a = U'_a - U''_a$, and $V_a = V'_a - V''_a$, and U'_a, U''_a, V'_a, V''_a are effective varieties. Since this condition is equivalent to

$$U'_a + \bar{U}_a + V''_a + \bar{V}_a \equiv U''_a + \bar{U}_a + V'_a + \bar{V}_a,$$

where $\bar{U}_a$, $\bar{V}_a$ are any effective varieties, the condition for $U_a \sim V_a$ does not depend on the particular choice of the varieties U'_a, U''_a used to express U_a as the difference of effective varieties, or on the corresponding choice for V_a.

The condition $U_a \sim V_a$ is clearly reflexive and symmetric. To prove that it is transitive, we suppose that

$$U_a = U'_a - U''_a \sim V_a = V'_a - V''_a,$$

and

$$V_a \sim W_a = W'_a - W''_a,$$

where U'_a, U''_a, V'_a, V''_a, W'_a, W''_a are all effective. Then

$$U'_a + V''_a \equiv U''_a + V'_a,$$

$$V'_a + W''_a \equiv V''_a + W'_a.$$

We add these equivalences and obtain

$$U'_a + V''_a + V'_a + W''_a \equiv U''_a + V'_a + V''_a + W'_a,$$

and hence

$$U'_a + W''_a \equiv U''_a + W'_a.$$

Hence $U_a \sim W_a$, and the relation $\sim$ is a true equivalence relation.

Again, a similar argument to that given in § 9 shows that if

$$U_a \sim U'_a, \quad V_a \sim V'_a,$$

then

$$U_a \pm V_a \sim U'_a \pm V'_a.$$

Thus equivalences in the virtual sense can be added and subtracted.

Let us consider the particular case of two effective varieties U_a, V_a. To write them as the difference of two effective varieties we merely take $U'_a = U_a$, $U''_a = 0$, and $V'_a = V_a$, $V''_a = 0$. Hence a necessary and sufficient condition for $U_a \sim V_a$ is $U_a \equiv V_a$. Thus equivalence in the virtual sense, when applied to effective varieties, is exactly the same as equivalence in the wide sense, as defined in § 9.

Just as in the case of equivalence for effective varieties (see the beginning of this section), we can use equivalence in the virtual sense to divide up the virtual varieties on V_n into mutually exclusive sets, those in any one set being the virtual varieties equivalent to any one of them. The set which includes $\boldsymbol{V}_a$ is denoted by $\{\boldsymbol{V}_a\}$ and is called the *system of equivalence* defined by $\boldsymbol{V}_a$. If the system of equivalence $\{\boldsymbol{V}_a\}$ contains an effective variety $\boldsymbol{V}'_a$, then from the fact that equivalence in the wide sense for effective varieties is the same as equivalence in the virtual sense it follows that

$$[\boldsymbol{V}'_a] \subseteq \{\boldsymbol{V}_a\} = \{\boldsymbol{V}'_a\}.$$

There are, however, systems $\{\boldsymbol{V}_a\}$ which contain no effective varieties; for instance if $\boldsymbol{V}_a$ is effective and not zero, $\{-\boldsymbol{V}_a\}$ contains no effective varieties.

Just as above, addition can be defined for systems of equivalence,

$$\{\boldsymbol{U}_a\} + \{\boldsymbol{V}_a\} = \{\boldsymbol{U}_a + \boldsymbol{V}_a\},$$

and this addition is associative and commutative. Now consider the equation

$$\{\boldsymbol{U}_a\} + \boldsymbol{x} = \{\boldsymbol{V}_a\}, \tag{1}$$

where $\{\boldsymbol{U}_a\}$ and $\{\boldsymbol{V}_a\}$ are any two systems of equivalence. Let $\boldsymbol{U}_a$ be any virtual variety of $\{\boldsymbol{U}_a\}$, $\boldsymbol{V}_a$ any virtual variety of $\{\boldsymbol{V}_a\}$. Then $\boldsymbol{V}_a - \boldsymbol{U}_a$ is a virtual variety, and clearly

$$\{\boldsymbol{U}_a\} + \{\boldsymbol{V}_a - \boldsymbol{U}_a\} = \{\boldsymbol{V}_a\}.$$

Hence the equations have a solution. So we have

THEOREM I. *The systems of equivalence on V_n of virtual varieties of dimension a form a commutative group under addition.*

By a property of groups [I, § 1] the solution of the equation (1) is unique. Hence the virtual varieties $\boldsymbol{W}_a$ which satisfy the equivalence

$$\boldsymbol{U}_a + \boldsymbol{W}_a \sim \boldsymbol{V}_a$$

form a single system of equivalence.

In the additive group of systems of equivalence, there is a unique zero-element. We denote it by $\boldsymbol{O}$. If $\boldsymbol{V}_a$ is any zero variety, we have

$$\boldsymbol{U}_a + \boldsymbol{V}_a \sim \boldsymbol{U}_a,$$

and hence $\boldsymbol{V}_a \in \boldsymbol{O}$. Therefore $\boldsymbol{O} = \{\boldsymbol{V}_a\}$. Let $\boldsymbol{W}_a$ be any virtual variety in $\boldsymbol{O}$. We write

$$\boldsymbol{W}_a = \boldsymbol{W}'_a - \boldsymbol{W}''_a,$$

where W'_a and W''_a are effective. If V_a is the zero variety, we have

$$W_a \sim V_a,$$

and hence

$$W'_a \sim W''_a + V_a = W''_a.$$

Since W'_a and W''_a are effective, we conclude that

$$W'_a \equiv W''_a.$$

Conversely, if $W'_a \equiv W''_a$, $W_a = W'_a - W''_a \sim V_a$. Hence a necessary and sufficient condition that a virtual variety W_a be equivalent to zero in the virtual sense is that it be the difference of two effective varieties which are equivalent in the wide sense of § 9. Moreover, since

$$W'_a \equiv W''_a$$

if and only if there exists a variety U_a such that

$$W'_a + U_a \equiv W''_a + U_a,$$

and $W_a = (W'_a + U_a) - (W''_a + U_a)$, it follows that if $\{W_a\} = O$ then W_a can be expressed as the difference of varieties equivalent in the narrow sense of § 9. Since equivalence in the narrow sense implies equivalence in the wide sense, the converse is obvious.

We now consider the intersection of two virtual varieties V_a, V_b. Let

$$V_a = V'_a - V''_a, \quad V_b = V'_b - V''_b,$$

where V'_a, V''_a, V'_b, V''_b are effective varieties. Suppose that the intersections $V'_a \wedge V'_b$, $V'_a \wedge V''_b$, $V''_a \wedge V'_b$, $V''_a \wedge V''_b$ are all of dimension $a+b-n$. We define $V_a . V_b$ by the equations

$$V_a . V_b = V'_a . V'_b + V''_a . V''_b - V'_a . V''_b - V''_a . V'_b.$$

It is trivial to show that the virtual variety $V_a . V_b$ defined in this way from V_a and V_b does not depend on the representation of V_a and V_b as differences of effective varieties. Also, it follows immediately from the corresponding properties for effective varieties that the associative, commutative and distributive laws hold:

$$V_a . (V_b . V_c) = (V_a . V_b) . V_c,$$

$$V_a . V_b = V_b . V_a,$$

$$V_a . (V_b + V'_b) = V_a . V_b + V_a . V'_b,$$

always assuming that the intersections can be defined. The proofs of these statements may be left to the reader.

Again, let us suppose that

$$V_a = V'_a - V''_a \sim U_a = U'_a - U''_a,$$

where U'_a, U''_a are effective varieties. Then, with the assumption made on p. 182, we have

$$V'_a + U''_a \equiv V''_a + U'_a,$$

and hence $$V'_a . V'_b + U''_a . V'_b \equiv V''_a . V'_b + U'_a . V'_b,$$

and $$V'_a . V''_b + U''_a . V''_b \equiv V''_a . V''_b + U'_a . V''_b.$$

Therefore, by addition,

$$V'_a . V'_b + V''_a . V''_b + U'_a . V''_b + U''_a . V'_b$$
$$\equiv U'_a . V'_b + U'' . V''_b + V'_a . V''_b + V''_a . V'_b,$$

and this is the condition for

$$V_a . V_b \sim U_a . V_b.$$

From this it follows that, if $V_b \sim U_b$,

$$V_a . V_b \sim U_a . V_b \sim U_a . U_b.$$

Thus the systems of equivalence $\{V_a\}$ and $\{V_b\}$ define uniquely a system of equivalence $\{V_a . V_b\}$ of virtual varieties of dimension $a+b-n$, which we may denote by

$$\{V_a\} . \{V_b\}.$$

In order that we may define $\{V_a\} . \{V_b\}$ for *any* two systems of virtual varieties, without any restrictions, we must first prove that $\{V_a\}$ and $\{V_b\}$ contain varieties V_a and V_b such that

$$\dim V_a \wedge V_b = a+b-n,$$

and secondly we must justify the assumption made on p. 182.

The proof of the first result is given below [Theorem II]. In this proof, V_b is any variety of $\{V_b\}$, and starting from any variety V_a of $\{V_a\}$ we construct a variety $\overline{V}_a$, equivalent to V_a, whose components meet the components of V_b simply. In constructing $\overline{V}_a$, we use a series of projective transformations in S_r, and if these are all generic and independent over K, $\overline{V}_a$ is a generic member of an irreducible system of virtual varieties which contains V_a.

To justify the assumption of p. 182, it is only necessary to show that if $V_a \equiv V'_a$, and $V_a \wedge V_b$ and $V'_a \wedge V_b$ are of dimension $a+b-n$, then $V_a . V_b \sim V'_a . V_b$. By hypothesis, there exists a sequence of varieties

$$V_a^{(1)} = V_a,\ V_a^{(2)},\ \ldots,\ V_a^{(k-1)},\ V_a^{(k)} = V'_a$$

such that $V_a^{(i)}$ and $V_a^{(i+1)}$ belong to an irreducible system. Let $U^{(i)}$ be a generic member of this system. As in Theorem II, we construct $\overline{V}_a^{(i)}$, $\overline{U}^{(i)}$. Then $\overline{U}^{(i)}$ is a generic member over K of an irreducible system of virtual varieties containing $\overline{V}_a^{(i)}$ and $\overline{V}_a^{(i+1)}$. Then if

$$W_a^{(1)} = V_a,\ W_a^{(i+1)} = \overline{V}_a^{(i)} \quad (i = 1, \ldots, k),\ W_a^{(k+2)} = V'_a,$$

$W_a^{(1)}, \ldots, W_a^{(k+2)}$ is a sequence of varieties connecting V_a and V'_a such that $W_a^{(i)} \wedge V_b$ is of dimension $a+b-n$, and $W_a^{(i)}$, $W_a^{(i+1)}$ belong to an irreducible system of virtual varieties. Since the proof of Theorem II of § 8 is valid for systems of virtual varieties, we conclude that $V_a . V_b \sim V'_a . V_b$.

Thus, even when $V_a . V_b$ is not defined, $\{V_a\} . \{V_b\}$ is a uniquely defined system of equivalence of dimension $a+b-n$. Further we note that, in the case $\dim V_a \wedge V_b = a+b-n$, there exist varieties V'_a, V'_b belonging to the same irreducible systems as V_a, V_b such that V_a and V_b intersect simply. As

$$V'_a \to V_a,\ V'_b \to V_b,\ V'_a . V'_b \to V_a . V_b.$$

Hence, as in the case of intersections in S_n, the intersection of V_a, V_b is defined intrinsically on V_n.

It only remains to prove

THEOREM II. *Let V_b be an irreducible point-set variety on V_n, of dimension b, and let V_a be any virtual variety on V_n of dimension a, where $a+b \geqslant n$. Then we can find on V_n a variety $U_a \sim V_a$, such that if*

$$U_a = \sum_i \sigma_i U_a^{(i)},$$

then $U_a^{(i)} \wedge V_b$ is of dimension $a+b-n$, and the intersections of $U_a^{(i)}$ and V_b are simple.

Let $V_a = \Sigma \rho_i V_a^{(i)}$, where $V_a^{(i)}$ is effective and irreducible. If each $V_a^{(i)}$ meets V_b simply, there is nothing to prove. Suppose that $V_a^{(i)}$ and V_b do not meet simply. Then either $\dim [V_a^{(i)} \wedge V_b] > a+b-n$, or else $V_a^{(i)} . V_b$ has multiple components. We consider first the case in which $\dim [V_a^{(i)} \wedge V_b] = l > a+b-n$. Let S_{r-n-1} be a generic $(r-n-1)$-space of S_r, independent of a generic point of $V_a^{(i)}$, and let $W_{r-n+a}^{(i)}$ be the cross-join of S_{r-n-1} and $V_a^{(i)}$ [§ 6]. Then

$$W_{r-n+a}^{(i)} . V_n = V_a^{(i)} + \sum_j{}^* V_a^{(i,j)},$$

where $V_a^{(i)}$, $*V_a^{(i,j)}$ are all distinct and irreducible. Then by § 6, Lemma IV, if $*V_a^{(i,j)} {}_\wedge V_b$ is not embedded in $V_a^{(i)} {}_\wedge V_b$, $*V_a^{(i,j)}$ and V_b intersect simply. If $*V_a^{(i,j)} {}_\wedge V_b$ is embedded in $V_a^{(i)} {}_\wedge V_b$ it follows from Lemma V of the same section that $\dim [*V_a^{(i,j)} {}_\wedge V_b] < l$. We also know that $\dim [*V_a^{(i,j)} {}_\wedge V_b] \geq a+b-n$.

Let $W_{r-n+a}^{(i),\tau}$ be the variety obtained from $W_{r-n+a}^{(i)}$ by a generic projective transformation. Then [§ 3, Th. I] $\bar{V}_a^{(i)} = W_{r-n+a}^{(i),\tau} {}_\wedge V_n$ and V_b meet simply. Hence

$$\bar{\boldsymbol{V}}_a^{(i)} \equiv \boldsymbol{V}_a^{(i)} + \sum_j *\boldsymbol{V}_a^{(i,j)},$$

and therefore

$$\boldsymbol{V}_a^{(i)} \sim \bar{\boldsymbol{V}}_a^{(i)} - \sum_j *\boldsymbol{V}_a^{(i,j)}.$$

We therefore replace $\boldsymbol{V}_a = \Sigma \rho_i \boldsymbol{V}_a^{(i)}$ by the equivalent variety

$$\bar{\boldsymbol{V}}_a = \sum_{j \neq i} \rho_j \boldsymbol{V}_a^{(j)} + \rho_i \bar{\boldsymbol{V}}_a^{(i)} - \rho_i \sum_j *\boldsymbol{V}_a^{(i,j)}.$$

Proceeding in this way, it is clear that after a finite number of stages we can replace $\boldsymbol{V}_a$ by an equivalent variety $\boldsymbol{V}_a^* = \Sigma \rho_i^* \boldsymbol{V}_a^{*(i)}$, such that $V_a^{*(i)} {}_\wedge V_b$ is of dimension $a+b-n$ exactly.

Now consider the case in which $V_a^{(i)} {}_\wedge V_b$ is of dimension $a+b-n$ exactly, but where $V_a^{(i)}$ and V_b do not intersect simply. We construct $W_{r-n+a}^{(i)}$ as above. In this case, it follows from Lemma V and VI of § 6, that if

$$\boldsymbol{W}_{r-n+a}^{(i)} \cdot \boldsymbol{V}_n = \boldsymbol{V}_a^{(i)} + \sum_j \boldsymbol{V}_a^{(i,j)},$$

no component of $V_a^{(i,j)} {}_\wedge V_b$ is embedded in $V_a^{(i)} {}_\wedge V_b$, and hence eac $V_a^{(i,j)}$ meets V_b simply. If $\bar{V}_a^{(i)} = W_{r-n+a}^{(i),\tau} {}_\wedge V_n$, $\bar{V}_a^{(i)}$ and V_b meet simply, and hence

$$\boldsymbol{V}_a^{(i)} \sim \bar{\boldsymbol{V}}_a^{(i)} - \sum_j \boldsymbol{V}_a^{(i,j)},$$

where $\bar{V}_a^{(i)}$ and $V_a^{(i,j)}$ meet V_b simply. Treating each of the $\boldsymbol{V}_a^{(i)}$ in this way we eventually obtain an equivalence $\boldsymbol{V}_a \sim \boldsymbol{U}_a$, where

$$\boldsymbol{U}_a = \sum_i \sigma_i \boldsymbol{U}_a^{(i)},$$

where $U_a^{(i)}$ is irreducible and meets V_b simply. This completes the proof of Theorem II.

This theorem, as we have already seen, enables us to give a complete theory of intersections of systems of equivalence on a non-singular variety V_n of dimension n. One further remark on the subject of systems of equivalence on V_n may be made. Let R be the set of all $(n+1)$-tuples $[\{\boldsymbol{V}_0\}, \{\boldsymbol{V}_1\}, \ldots, \{\boldsymbol{V}_n\}]$, where $\{\boldsymbol{V}_i\}$ is any system of equivalence of virtual varieties of dimension i, possibly the zero system $\boldsymbol{O}_i$. We can define addition and multiplication in R as follows:

$$[\{\boldsymbol{V}_0\},\{\boldsymbol{V}_1\},\ldots,\{\boldsymbol{V}_n\}]+[\{\boldsymbol{U}_0\},\{\boldsymbol{U}_1\},\ldots,\{\boldsymbol{U}_n\}] = [\{\boldsymbol{V}_0+\boldsymbol{U}_0\},\{\boldsymbol{V}_1+\boldsymbol{U}_1\},\ldots,\{\boldsymbol{V}_n+\boldsymbol{U}_n\}],$$

$$[\{\boldsymbol{V}_0\},\{\boldsymbol{V}_1\},\ldots,\{\boldsymbol{V}_n\}][\{\boldsymbol{U}_0\},\{\boldsymbol{U}_1\},\ldots,\{\boldsymbol{U}_n\}] = [\{\boldsymbol{W}_0\},\{\boldsymbol{W}_1\},\ldots,\{\boldsymbol{W}_n\}]$$

where

$$\{\boldsymbol{W}_i\} = \{\boldsymbol{V}_i\}.\{\boldsymbol{U}_n\}+\{\boldsymbol{V}_{i+1}\}.\{\boldsymbol{U}_{n-1}\}+\ldots+\{\boldsymbol{V}_n\}.\{\boldsymbol{U}_i\}.$$

From the properties of systems of equivalence, it follows immediately that with these two laws of composition the set R is a ring. We note also that if

$$\alpha = [\{\boldsymbol{V}_0\},\{\boldsymbol{V}_1\},\ldots,\{\boldsymbol{V}_n\}]$$

is an element of R, so is

$$\rho\alpha = [\{\rho\boldsymbol{V}_0\},\{\rho\boldsymbol{V}_1\},\ldots,\{\rho\boldsymbol{V}_n\}],$$

where ρ is any integer. Thus R is a linear set over the ring of integers. An important set of properties of a variety V_n is bound up with the properties of the ring R for V_n. An example of such a property is given in the next section.

The property discussed in § 11 is equivalent to the theorem that R is *a finite linear set over the ring of integers.*

Since two effective varieties belonging to the same irreducible system on V_n have the same order, it follows from our definition that if two virtual varieties $\boldsymbol{V}_a$ and $\boldsymbol{V}'_a$ belong to the same system of equivalence on V_n they must have the same order, the order of

$$\boldsymbol{V}_a = \sum_i \rho_i \boldsymbol{V}_a^{(i)},$$

where $\boldsymbol{V}_a^{(i)}$ is effective and irreducible and of order g_i, being defined as $\sum_i \rho_i g_i$. If $\boldsymbol{V}_a$ and $\boldsymbol{V}_b$ are any two virtual varieties on V_n such that $a+b=n$, the intersection $\{\boldsymbol{V}_a\}.\{\boldsymbol{V}_b\}$ is a system of equivalence of varieties of zero dimension. All varieties of the system have the same order. Now it is clear that any two points on V_n are equivalent, since all the points of V_n form an irreducible system to which they both belong. Hence $\{\boldsymbol{V}_a\}.\{\boldsymbol{V}_b\}$ contains varieties $g\boldsymbol{P}$, where $\boldsymbol{P}$ is any point of V_n counted with multiplicity one, and g is the order of the varieties of $\{\boldsymbol{V}_a\}.\{\boldsymbol{V}_b\}$. g is called *the intersection-number* of $\boldsymbol{V}_a$ and $\boldsymbol{V}_b$ and we write

$$(\boldsymbol{V}_a.\boldsymbol{V}_b) = g.$$

It is uniquely defined by $\boldsymbol{V}_a$ and $\boldsymbol{V}_b$, whatever the point-set intersection $V_a \wedge V_b$; but we note that if V_a and V_b are effective, and intersect simply in a finite number of points, $(\boldsymbol{V}_a . \boldsymbol{V}_b)$ is the number of points common to V_a and V_b. If $\boldsymbol{V}_a \sim \boldsymbol{V}'_a$, $\boldsymbol{V}_b \sim \boldsymbol{V}'_b$, then $(\boldsymbol{V}_a . \boldsymbol{V}_b) = (\boldsymbol{V}'_a . \boldsymbol{V}'_b)$.

If $n = 2m$ and $\boldsymbol{V}_m$ is any m-dimensional variety on V_n, we can define $(\boldsymbol{V}_m . \boldsymbol{V}_m)$. This is called the *grade* of $\boldsymbol{V}_m$. It is, by definition, an integer, but examples can be given to show that it may be negative, even when $\boldsymbol{V}_m$ is effective. Similarly, if $a+b+c = 2n$, we can define $(\boldsymbol{V}_a . \boldsymbol{V}_b . \boldsymbol{V}_c)$, and so on.

11. Theory of the base. The results of § 2, on the degeneration of an algebraic variety in S_n, when expressed in the terminology of § 9 are as follows:

Any variety $\boldsymbol{V}_a$ in S_n of dimension a and order g satisfies the equivalence relation

$$\boldsymbol{V}_a \equiv \boldsymbol{S}_a^{(1)} + \ldots + \boldsymbol{S}_a^{(g)},$$

where $\boldsymbol{S}_a^{(1)}, \ldots, \boldsymbol{S}_a^{(g)}$ are linear spaces of dimension a with a common S_{a-1}. Each $S_a^{(i)}$ is a generic space of S_n over the ground field. Since, if $\boldsymbol{S}_a$ is any a-space in S_n, we also have the equivalence

$$\boldsymbol{S}_a^{(i)} \equiv \boldsymbol{S}_a \quad (i = 1, \ldots, g),$$

we obtain the relation

$$\boldsymbol{V}_a \equiv g\boldsymbol{S}_a. \tag{1}$$

We notice that in reducing $\boldsymbol{V}_a$ to $g\boldsymbol{S}_a$ we have considered varieties over an extension of the ground field. In this particular example we do not need to extend the ground field, provided that it is algebraically closed. We merely choose S_{a-1} to be any $(a-1)$-space over K, and S_{n-a} to be any $(n-a)$-space over K which does not meet S_{a-1}, and meets V_a in g distinct points. Then the methods of § 2 enable us to reduce V_a to g a-spaces through S_{a-1}. Irreducible systems of a-spaces which contain each of the g a-spaces and any assigned S_a are easily constructed, and so we obtain once more the equivalence

$$\boldsymbol{V}_a \equiv g\boldsymbol{S}_a.$$

Nevertheless, it is convenient to be able to extend the ground field when we wish. Better still, we shall assume that the ground field is replaced by the 'universal field' Σ^* discussed in X, § 1.

The results of § 2, as stated in (1) above, are usually described in the following way:

THEOREM I. *The system of a-spaces in S_n forms a base for varieties of dimension a in S_n.*

The problem of generalising this result is one which has exercised the minds of many geometers. The general result sought is:

If V_n is an absolutely irreducible variety of dimension n, defined over a field K, there exists, for each value of d ($0 \leqslant d \leqslant n$) a finite number of varieties $\mathbf{V}_d^{(1)}, \ldots, \mathbf{V}_d^{(k)}$ on V_n, defined over the universal field Σ^*, such that, over Σ^*, any variety $\mathbf{V}_d$ on V_n satisfies an equivalence

$$a\mathbf{V}_d \sim a_1 \mathbf{V}_d^{(1)} + \ldots + a_k \mathbf{V}_d^{(k)},$$

where $a, a_1, \ldots, a_k$ are integers, and $a \neq 0$.

The varieties $\mathbf{V}_d^{(1)}, \ldots, \mathbf{V}_d^{(k)}$ are said to form a *base* for the varieties of dimension d. If the equivalence

$$b_1 \mathbf{V}_d^{(1)} + \ldots + b_k \mathbf{V}_d^{(k)} \sim \mathbf{0}$$

implies that $b_1 = \ldots = b_k = 0$, the base is said to be *minimal*.

Most attempts to prove this theorem have dealt with the case in which the ground field is the field of complex numbers. Then transcendental and topological methods are available. Using these methods, the existence of a finite base has been proved in the case $d = n-1$. The cases $d = 0$, $d = n$ are trivial. Beyond this little progress has been made, except in special cases. In Chapters XIII and XIV the theorem of the base will be proved for quadrics and for Grassmann varieties. We conclude this chapter by proving the theorem for a Segre variety [XI, § 2] defined over a ground field K.

We shall work in a k-way space $S_{n_1 \ldots n_k}$, but when we consider intersections and equivalences we shall be thinking of the corresponding result on the Segre variety. The correspondence set up in XI, § 2 makes it evident that we may do this. We prove

THEOREM II. *If $\mathbf{V}$ is an effective variety of dimension d in a k-way space $S_{n_1 \ldots n_k}$, then*

$$\mathbf{V} \equiv \Sigma \alpha_{r_1 \ldots r_k} \mathbf{S}_{r_1 \ldots r_k},$$

where the summation is over the finite set of values of $r_1, \ldots, r_k$ which are such that $\sum_{i=1}^{k} r_i = d$, and the $\mathbf{S}_{r_1 \ldots r_k}$ are arbitrarily chosen in $S_{n_1 \ldots n_k}$.

It should be noted that the result stated in Theorem II is stronger than the general theorem stated above. While it is interesting to prove the stronger result in the present case, the important properties of the base follow from the general form, which is more flexible.

We suppose that $(x, y, \ldots, t)$ is an allowable coordinate system in $S_{n_1 \ldots n_k}$. Consider the variety defined by the equations

$$\left.\begin{array}{ll} \sum\limits_{j=0}^{n_1} a_{ij} x_j = 0 & (i = 1, \ldots, n_1 - r_1), \\ \sum\limits_{j=0}^{n_2} b_{ij} y_j = 0 & (i = 1, \ldots, n_2 - r_2), \\ \cdot \quad \cdot \quad \cdot & \cdot \quad \cdot \\ \sum\limits_{j=0}^{n_k} d_{ij} t_j = 0 & (i = 1, \ldots, n_k - r_k), \end{array}\right\} \tag{2}$$

where $(a_{ij}), (b_{ij}), \ldots, (d_{ij})$ are matrices of rank $n_1 - r_1, n_2 - r_2, \ldots, n_k - r_k$. This variety is easily seen to be irreducible, and is, in fact, an $S_{r_1 \ldots r_k}$. It is also evident that for given $r_1, \ldots, r_k$, all $S_{r_1 \ldots r_k}$ in $S_{n_1 \ldots n_k}$ are equivalent in the narrow sense, and hence also in the virtual sense.

If r_1 is zero, $S_{r_1 \ldots r_k}$ is also a $(k-1)$-way space, $S_{r_2 \ldots r_k}$.

Consider another k-way $S_{s_1 \ldots s_k}$, in $S_{n_1 \ldots n_k}$, given by the equations

$$\left.\begin{array}{ll} \sum\limits_{j=0}^{n_1} \alpha_{ij} x_j = 0 & (i = 1, \ldots, n_1 - s_1), \\ \sum\limits_{j=0}^{n_2} \beta_{ij} x_j = 0 & (i = 1, \ldots, n_2 - s_2), \\ \cdot \quad \cdot \quad \cdot & \cdot \quad \cdot \\ \sum\limits_{j=0}^{n_k} \delta_{ij} x_j = 0 & (i = 1, \ldots, n_k - s_k). \end{array}\right\} \tag{3}$$

Their intersection is given by the equations (2) and (3) taken simultaneously. Let $n_1 - t_1, \ldots, n_k - t_k$ be the ranks of the matrices $\begin{pmatrix} a \\ \alpha \end{pmatrix}, \ldots, \begin{pmatrix} d \\ \delta \end{pmatrix}$. Then $t_i \geqslant \max[-1, r_i + s_i - n_i]$. If any $t_i = -1$, the equations (2) and (3) have no solution; then $S_{r_1 \ldots r_k} \wedge S_{s_1 \ldots s_k}$ is vacuous. If every $t_i > -1$, the solutions of (2) and (3) define a k-way space $S_{t_1 \ldots t_k}$. The dimension of this space is

$$\Sigma t_i \geqslant \Sigma r_i + \Sigma s_i - \Sigma n_i,$$

and hence

$$\dim S_{t_1 \ldots t_k} = \dim S_{r_1 \ldots r_k} + \dim S_{s_1 \ldots s_k} - \dim S_{n_1 \ldots n_k},$$

if and only if $\quad t_i = r_i + s_i - n_i \geqslant 0 \quad (i = 1, \ldots, k).$

In this case $\quad S_{r_1 \ldots r_k} \wedge S_{s_1 \ldots s_k} = S_{r_1 + s_1 - n_1 \ldots r_k + s_k - n_k}.$

If the coefficients in (2) and (3) are taken as independent indeterminates over K, we have

$$t_i = \max[-1, r_i + s_i - n_i] \quad (i = 1, \ldots, k);$$

hence there exist k-way spaces $S'_{r_1 \ldots r_k}$ and $S'_{s_1 \ldots s_k}$ whose intersection is either zero or of dimension equal to

$$\dim S_{r_1 \ldots r_k} + \dim S_{s_1 \ldots s_k} - \dim S_{n_1 \ldots n_k}.$$

Now all k-way spaces $S_{r_1 \ldots r_k}$ contained in $S_{n_1 \ldots n_k}$ are equivalent in the virtual sense. It follows that if, for any value of i, we have $r_i + s_i < n_i$,

$$\{\mathbf{S}_{r_1 \ldots r_k}\} . \{\mathbf{S}_{s_1 \ldots s_k}\} = \mathbf{0},$$

and if

$$r_i + s_i \geqslant n_i \quad (i = 1, \ldots, k)$$

we have

$$\{\mathbf{S}_{r_1 \ldots r_k}\} . \{\mathbf{S}_{s_1 \ldots s_k}\} = \{\rho \mathbf{S}_{r_1+s_1-n_1 \ldots r_k+s_k-n_k}\},$$

where ρ is an integer. Now choose in $\{\mathbf{S}_{r_1 \ldots r_k}\}$, $\{\mathbf{S}_{s_1 \ldots s_k}\}$ two k-way spaces $S'_{r_1 \ldots r_k}$ and $S'_{s_1 \ldots s_k}$ such that

$$S'_{r_1 \ldots r_k} \wedge S'_{s_1 \ldots s_k} = S'_{r_1+s_1-n_1 \ldots r_k+s_k-n_k}.$$

Consider the images of these varieties on the Segre variety. It is easy to determine the tangent spaces to the images of $S'_{r_1 \ldots r_k}$ and $S'_{s_1 \ldots s_k}$ at any point common to them, and to show that these images meet simply. Hence $\rho = 1$. Thus, if $r_i + s_i \geqslant n_i$ $(i = 1, \ldots, k)$,

$$\{\mathbf{S}_{r_1 \ldots r_k}\} . \{\mathbf{S}_{s_1 \ldots s_k}\} = \{\mathbf{S}_{r_1+s_1-n_1 \ldots r_k+s_k-n_k}\}.$$

Let V be an irreducible variety of dimension d in $S_{n_1 \ldots n_k}$ given by the equations

$$f_i(x, y, \ldots, t) = 0 \quad (i = 1, \ldots, r). \tag{4}$$

Theorem II will be proved by induction on k and on n_1. We know that it is true in ordinary or 1-way space [(1), p. 193]. Hence we may assume that it is true for $(k-1)$-way spaces. Since this implies the truth of the theorem for the space $S_{0 n_2 \ldots n_k}$, we may assume that the theorem holds in $S_{n_1 - 1 \ldots n_k}$.

If $d = 0$, V is a set of conjugate points, and we have $\boldsymbol{V} \sim k\boldsymbol{P}$, which is the required result. So we may assume $d > 0$. If $d > 0$, consider the correspondence defined by (4) between S_{n_i} and $S_{n_1 \ldots n_{i-1} n_{i+1} \ldots n_k}$. If the object-variety is of dimension zero for each value of i, the correspondence is of dimension zero, and we have a contradiction. Without loss of generality, we shall assume that the object-variety in S_{n_1} is of dimension greater than zero.

Let α, β, ..., δ now represent square matrices of orders n_1+1, n_2+1, ..., n_k+1 respectively, whose elements are independent indeterminates over K, and consider the variety V^* whose equations are

$$f_i(\alpha x, \beta y, \ldots, \delta t) = 0 \quad (i = 1, \ldots, r). \tag{5}$$

Generalising from the case $k = 1$ we can prove that V^* is irreducible of dimension d over the field $K(\alpha, \beta, \ldots, \delta)$, and is a generic member of an irreducible system of varieties which contains V.

Assuming these evident generalisations of the case $k = 1$, we now define the (n_1+1, n_1+1) matrix

$$E = \begin{pmatrix} \epsilon & & & \\ & 1 & & \\ & & \ddots & \\ & & & 1 \end{pmatrix},$$

where ϵ is a new indeterminate, and consider the variety V_ϵ^* whose equations are

$$f_i(\alpha Ex, \beta y, \ldots, \delta t) = 0 \quad (i = 1, \ldots, r),$$

or, equivalently,

$$f_i^*(Ex, y, \ldots, t) = 0 \quad (i = 1, \ldots, r), \tag{6}$$

where the coefficients of f_i^* are in $K^* = K(\alpha, \ldots, \delta)$. The section of V_ϵ^* by $x_0 = 0$ is the transform by $(x, \ldots, t) \to (x', \ldots, t')$, where

$$x' = \alpha^{-1}x, \ldots, \quad t' = \delta^{-1}t,$$

of the section of V by

$$\alpha_{00}x_0 + \ldots + \alpha_{0n_1}x_{n_1} = 0.$$

Since this is a generic x-prime, the section of V by it is irreducible and of dimension $d-1$, unless the intersection is vacuous. This can only happen when the object-variety of the correspondence between S_{n_1} and $S_{n_2 \ldots n_k}$ defined by V is zero-dimensional. This is contrary to hypothesis. Here the section of V_ϵ^* by $x_0 = 0$ is an irreducible variety of dimension $d-1$.

Now specialise $\epsilon \to 0$. Any corresponding specialisation of V_ϵ^* satisfies the equations

$$f_i^*(E_0 x, y, \ldots, t) = 0 \quad (i = 1, \ldots, r), \tag{7}$$

where

$$E_0 = \begin{pmatrix} 0 & & & \\ & 1 & & \\ & & \ddots & \\ & & & 1 \end{pmatrix},$$

that is, the equations obtained from (6) by putting $x_0 = 0$. Let equations (7) define a variety V'. Consider the correspondence, over $K(\alpha, \ldots, \delta)$, between the line $(1, \epsilon)$ and $S_{n_1 \ldots n_k}$. Since every point of the line is simple, it follows from XI, § 7, Th. IV, that if V' is of dimension d, the specialisation V_0^* of V_ϵ^* corresponding to $\epsilon \to 0$ is uniquely determined. We therefore consider the variety V' defined by (7). It is clear that no component of V' is of dimension less than d. Let $(x'_0, \ldots, x'_{n_1}; y'_0, \ldots, y'_{n_2}; \ldots; t'_0, \ldots, t'_{n_k})$ be a point of V' for which $x'_1, \ldots, x'_{n_1}$ are not all zero. Then, since this point satisfies (7), $(0, x'_1, \ldots, x'_{n_1}; \ldots; t'_0, \ldots, t'_{n_k})$ satisfies (6), that is, lies on V_ϵ^* (any ϵ). Conversely, if $(0, x'_1, \ldots, x'_{n_1}, \ldots; t'_0, \ldots, t'_{n_k})$ lies on V_ϵ^*, $(x'_0, \ldots, x'_{n_1}; \ldots; t'_0, \ldots, t'_{n_k})$, where x'_0 is arbitrary, lies on V'. Let $U^{(1)}$ be the section of V_ϵ^* by $x_0 = 0$. It is an irreducible variety of dimension $d-1$ contained in an $S_{n_1-1\, n_2 \ldots n_k}$ lying in $S_{n_1 \ldots n_k}$.

Let $(\xi_0, \ldots, \xi_{n_1}; \ldots; \tau_0, \ldots, \tau_{n_k})$ be a generic point of any component U' of V' which does not lie in

$$x_1 = \ldots = x_{n_1} = 0.$$

We may suppose that it is normalised with respect to some x_i $(i \geqslant 1)$, and some $y_j, \ldots, t_l$. Then $(0, \xi_1, \ldots, \xi_{n_1}; \ldots; \tau_0, \ldots, \tau_{n_k})$ lies on $U^{(1)}$. Hence $K^*(\xi_1, \ldots, \xi_{n_1}, \ldots, \tau_0, \ldots, \tau_{n_k})$ is of dimension $d-1$ at most over K^*. Therefore $K^*(\xi_0, \ldots, \xi_{n_1}; \ldots; \tau_0, \ldots, \tau_{n_k})$ is of dimension d at most over K^*. Hence the component U' of V' is not of dimension greater than d. But it cannot be of dimension less than d, and hence its dimension is d exactly. This implies that

$$K^*(\xi_1, \ldots, \xi_{n_1}; \ldots; \tau_0, \ldots, \tau_{n_k}),$$

is of dimension $d-1$ over K^*, and hence $(0, \xi_1, \ldots, \xi_{n_1}; \ldots; \tau_0, \ldots, \tau_{n_k})$ is a generic point of $U^{(1)}$. Also ξ_0 is indeterminate over

$$K(\xi_1, \ldots, \xi_{n_1}; \ldots; \tau_0, \ldots, \tau_{n_k}).$$

Thus U' can be obtained as the variety having the generic point $(\xi_0, \ldots, \xi_{n_1}; \ldots; \tau_0, \ldots, \tau_{n_k})$, where $(0, \xi_1, \ldots, \xi_{n_1}; \ldots; \tau_0, \ldots, \tau_{n_k})$ is a generic point of $U^{(1)}$, and ξ_0 is a new indeterminate. It follows that V' has only one component U' not in $x_1 = \ldots = x_{n_1} = 0$, and this is of dimension d. It further follows that if

$$g_i(x_1, \ldots, x_{n_1}; \ldots; t_0, \ldots, t_{n_k}) = 0 \quad (i = 1, \ldots, s), \tag{8}$$
$$x_0 = 0,$$

is a basis for the equations of $U^{(1)}$, the equations (8) are a basis for the equations of U'. From this we see that if $U^{(1)}$ varies in an irreducible system of varieties in $x_0 = 0$, U' varies in an irreducible system of varieties in $S_{n_1 \dots n_k}$; and from this it follows that if $U^{(1)}$ and $U^{(2)}$ are varieties in $x_0 = 0$ such that $U^{(1)} \equiv U^{(2)}$ in this $S_{n_1 - 1\, n_2 \dots n_k}$, then $U' \equiv U''$ in $S_{n_1 \dots n_k}$, where U'' is the variety obtained from $U^{(2)}$ in the same way that U' is obtained from $U^{(1)}$.

We have now to consider the points of V' which lie in

$$x_1 = \dots = x_{n_1} = 0.$$

The point $(1, 0, \dots, 0;\ \eta_0, \dots, \eta_{n_2};\ \dots, \tau_0, \dots, \tau_{n_k})$ lies on V' if and only if $(\eta_0, \dots, \eta_{n_2}; \dots; \tau_0, \dots, \tau_{n_k})$ satisfies all the equations of V_ϵ^* which do not involve x; that is, if it lies on the image-variety in $S_{n_2 \dots n_k}$ of the correspondence between S_{n_1} and $S_{n_2 \dots n_k}$. This image-variety is irreducible, and of dimension $c \leqslant d$. It is of dimension d if and only if to a generic point of the image-variety in $S_{n_2 \dots n_k}$ there corresponds a zero-dimensional variety in S_{n_1}. It follows that the points of V' for which $x_1 = \dots = x_{n_1} = 0$ form an irreducible variety W' of dimension c in a space $S_{0 n_2 \dots n_k}$ lying in $S_{n_1 \dots n_k}$. Equivalence (in any sense) for W' as a variety in $S_{n_2 \dots n_k} = S_{0 n_2 \dots n_k}$ implies equivalence in $S_{n_1 \dots n_k}$. It follows that

$$V' = U' + W'.$$

Since no component of V' can be of dimension less than d [XI, § 6, Th. I], $W' \subset U'$ unless $c = d$. V' is of dimension d; hence the specialisation V_0^* of V_ϵ^* as $\epsilon \to 0$ is uniquely determined. Moreover, since $V_0^* \subseteq V'$, we must have

$$\boldsymbol{V}_0^* = \lambda \boldsymbol{U}' + \mu \boldsymbol{W}',$$

where λ and μ are integers (and $\mu = 0$ if W' is of dimension less than d). Since

$$\boldsymbol{V} \equiv \boldsymbol{V}_1^*$$

and

$$\boldsymbol{V}_1^* \equiv \boldsymbol{V}_0^*,$$

we have

$$\boldsymbol{V} \equiv \lambda \boldsymbol{U}' + \mu \boldsymbol{W}'.$$

We now apply our induction hypothesis. The variety $U^{(1)}$ is a variety of dimension d in $S_{n_1 - 1\, n_2 \dots n_k}$. Hence, by hypothesis

$$\boldsymbol{U}^{(1)} \equiv \Sigma a_{i_1 \dots i_k} \boldsymbol{S}^{(1)}_{i_1 \dots i_k},$$

where the $a_{i_1 \dots i_k}$ are integers, and $\boldsymbol{S}^{(1)}_{i_1 \dots i_k}$ is a k-way space in $S_{n_1 - 1\, n_2 \dots n_k}$, and $\Sigma i_j = d - 1$.

As $\boldsymbol{U}^{(1)}$ varies in a continuous system and becomes $\Sigma a_{i_1 \dots i_k} \boldsymbol{S}^{(1)}_{i_1 \dots i_k}$, U' varies in a continuous system and becomes a variety which lies in the point-set variety $\Sigma S_{i_1+1\,i_2\ldots i_k}$. Since both varieties are of dimension d, we must have

$$U' \equiv \Sigma a'_{i_1\ldots i_k} \mathbf{S}_{i_1+1\,i_2\ldots i_k},$$

where the $a'_{i_1\ldots i_k}$ are integers. It is easy to show that $a'_{i_1\ldots i_k} = a_{i_1\ldots i_k}$, but this is not necessary for the proof of our theorem.

Since $\boldsymbol{W}'$ is a variety of dimension d in $S_{n_2\ldots n_k}$,

$$\boldsymbol{W}' \equiv \Sigma b_{i_2\ldots i_k} \mathbf{S}_{i_2\ldots i_k}.$$

As $S_{n_2\ldots n_k} = S_{0\,n_2\ldots n_k} \subseteq S_{n_1\ldots n_k}$, we have the corresponding equivalence in $S_{n_1\ldots n_k}$,

$$\boldsymbol{W}' \equiv \Sigma b'_{i_2\ldots i_k} \mathbf{S}_{0\,i_2\ldots i_k}.$$

Taking the two results together, we immediately have the equivalence

$$\boldsymbol{V} \equiv \Sigma \alpha_{r_1\ldots r_k} \mathbf{S}_{r_1\ldots r_k}, \tag{9}$$

summed over all values of $r_1, \ldots, r_k$ such that $\Sigma r_i = d$.

The result obviously extends to the equivalence $\equiv$ or to $\sim$ for virtual varieties.

It is easy to show that the coefficients $\alpha_{r_1\ldots r_k}$ in (9) are uniquely determined by $\boldsymbol{V}$. Indeed, if we also have

$$\boldsymbol{V} \equiv \Sigma \beta_{r_1\ldots r_k} \mathbf{S}_{r_1\ldots r_k},$$

we have
$$\Sigma \alpha_{r_1\ldots r_k} \mathbf{S}_{r_1\ldots r_k} \equiv \Sigma \beta_{r_1\ldots r_k} \mathbf{S}_{r_1\ldots r_k}. \tag{10}$$

Let $S_{s_1\ldots s_k}$ be a generic space, where $\Sigma s_i = \Sigma n_i - d$. Then the intersection of $S_{r_1\ldots r_k}$, $S_{s_1\ldots s_k}$ is vàcuous unless $r_i + s_i = n_i$ $(i = 1, \ldots, k)$, and is a single point if these equalities hold. Moreover, we have seen that any two complementary spaces $S_{r_1\ldots r_k}$ and $S_{s_1\ldots s_k}$ which meet in a point intersect simply. Hence, taking the intersection of each side of (10) by $\mathbf{S}_{s_1\ldots s_k}$, we obtain

$$\alpha_{n_1-s_1\ldots n_k-s_k} = \beta_{n_1-s_1\ldots n_k-s_k}.$$

As this holds for all admissible values of $s_1, \ldots, s_k$ we deduce that

$$\alpha_{r_1\ldots r_k} = \beta_{r_1\ldots r_k}$$

for all values of $r_1, \ldots, r_k$. Hence the form (9) is unique.

Corollary. The varieties $\mathbf{S}_{r_1\ldots r_k}$ defined in Theorem II form a minimal base for the varieties of dimension d in $S_{n_1\ldots n_k}$.

BOOK IV

QUADRICS AND GRASSMANN VARIETIES

CHAPTER XIII

QUADRICS

In the preceding three chapters we have considered a general theory of algebraic varieties in space of n dimensions. The methods developed in these chapters are applied in this and the following chapter to certain particular varieties which are of importance in the classical theory of algebraic varieties. The present chapter is devoted to the geometry of quadrics.

The theory of quadrics is so large that any attempt to cover it all would make this chapter disproportionately long. We have therefore exercised a considerable degree of selection in our choice of the subjects with which we deal. Since our object is primarily to illustrate how the methods developed in earlier chapters can be used to investigate properties of quadrics, we confine ourselves to those aspects of the subject which are of most interest to algebraic geometers, and we shall seek to show how these questions can be treated by the methods previously described. We shall be mainly concerned with showing how such questions as the properties of linear spaces on a quadric, stereographic projection, transformation of a quadric into itself, etc., can be treated by algebraic means. But in order to make the chapter self-contained, we shall have to begin with some elementary properties of quadrics which will be familiar to most readers; these, however, will be treated as briefly as possible. Any apparent unevenness of emphasis is to be explained on these grounds.

We shall assume throughout this chapter that the *ground field K is algebraically closed*. This will exclude the consideration of certain problems, particularly those dealing with questions of reality; but as we have explained, we do not attempt to cover all problems relating to quadrics. The word 'point', without any qualifying adjective, will mean an algebraic point, and hence, since the ground field is algebraically closed, we may, and shall, assume that its coordinates are in K. The only other points which we shall consider will be generic points, either of the quadric or of a subvariety of it.

1. Definitions and elementary properties.

A quadric is sometimes defined as a d-dimensional variety of order two. This, however, is too general; a variety in S_n, consisting of two lines which do not meet, is a variety of dimension one and order two, but we shall find it convenient to restrict our definition so that two lines make up a quadric of dimension one only when they meet. Before giving a formal definition of a quadric, we prove the following theorem.

THEOREM I. *Any irreducible variety V of dimension d and order two in S_n is contained in a space of $d+1$ dimensions.*

Since the ground field K is algebraically closed, V must be absolutely irreducible [X, § 11, Th. IV, Cor.]. Also, we must assume $d \geqslant 1$, since any irreducible zero-dimensional variety over an algebraically closed ground field is a point, and is therefore of order one. Not all points of V are singular points, and the singular points of V lie on a subvariety U of V of dimension e $(e < d)$ [X, § 14]. A generic S_{n-d} does not meet U. Hence we can find an $(n-d)$-space in S_n, defined over K, which does not meet U. This meets V in a variety of dimension zero at least, every point of which is a simple point of V. Hence we can find an algebraic point on V which is simple. Since K is algebraically closed, we may choose the coordinate system in S_n so that this point is $(1, 0, \ldots, 0)$. Let $\xi = (\xi_0, \ldots, \xi_n)$ be a generic point of V. Since V contains a point not in $x_0 = 0$, $\xi_0 \neq 0$, and we may normalise the coordinates of ξ so that $\xi_0 = 1$. Let η be the point of $x_0 = 0$ whose coordinates are $(0, \xi_1, \ldots, \xi_n)$, and consider the correspondence, whose generic point is (ξ, η), between S_n and the prime $x_0 = 0$. The object-variety in S_n is the variety with generic point ξ, that is V, which is of dimension d, and to the generic point ξ of V there corresponds the unique point η of $x_0 = 0$; hence the correspondence is of dimension d. Let V' be the image-variety, d' its dimension. If $y' = (0, y'_1, \ldots, y'_n)$ is any point of V', it follows from the equations of the correspondence that the corresponding points of V lie on the line

$$\frac{x_1}{y'_1} = \frac{x_2}{y'_2} = \ldots = \frac{x_n}{y'_n}.$$

It follows that $d' < d$ if and only if any line through $(1, 0, \ldots, 0)$ which meets V' lies on V. Now since $(1, 0, \ldots, 0)$ is simple on V and V is of order two, a generic S_{n-d} through this point meets V elsewhere in only one point ξ', which is a generic point of V [XI, § 10, Th. I]. Thus the line joining $(1, 0, \ldots, 0)$ to ξ' meets V' and does not lie on V. Hence $d' = d$, and to a generic point of V' there corresponds a zero-dimensional variety on V, each of whose points is a generic point of V.

V' is a d-dimensional variety in $x_0 = 0$. To find its order, we consider a generic S_{n-d-1} in $x_0 = 0$, and determine the number of points in which it meets V'. Each of these intersections is a generic point of V' [X, § 7, Th. III], and the point of V corresponding to any of them is a generic point of V. Now the join of S_{n-d-1} to $(1, 0, \ldots, 0)$ is a generic S_{n-d} through this point. It therefore meets V in $(1, 0, \ldots, 0)$ and in one other point ξ'. Since $(1, 0, \ldots, 0)$ is not a generic point of V, it does not correspond to an intersection of V' and S_{n-d-1}. Hence it follows that ξ' is the unique point of V which corresponds to an intersection of S_{n-d-1} and V', and therefore $(0, \xi'_1, \ldots, \xi'_n)$ is the only point of V' which lies in S_{n-d-1}. Hence V' is of order one, and it is therefore a d-space [X, § 7, p. 48].

Let the equations of V' be

$$\sum_{j=1}^{n} a_{ij} x_j = 0 \quad (i = 1, \ldots, n-d-1),$$
$$x_0 = 0.$$

If $(y'_0, \ldots, y'_n)$ be any point of V other than $(1, 0, \ldots, 0)$, $(0, y'_1, \ldots, y'_n)$ is on V', and hence

$$\sum_{j=1}^{n} a_{ij} y'_j = 0 \quad (i = 1, \ldots, n-d-1).$$

Hence every point of V other than $(1, 0, \ldots, 0)$ lies in

$$\sum_{j=1}^{n} a_{ij} x_j = 0 \quad (i = 1, \ldots, n-d-1). \tag{1}$$

Since $(1, 0, \ldots, 0)$ also satisfies these equations, every point of V satisfies (1), and hence V is contained in the $(d+1)$-space defined by these equations. It cannot be contained in any subspace S_d of this, for since V is of dimension d this would imply that $V = S_d$; but S_d is of order one. Theorem I is therefore proved.

In view of this theorem, it follows that if we are given any irreducible variety V in S_n of dimension d and order two, we can choose the coordinate system in S_n so that V lies in the $(d+1)$-space given by the equations

$$x_{d+2} = x_{d+3} = \dots = x_n = 0,$$

and V_n is a primal of this space given by an irreducible equation

$$F(x_0, \dots, x_{d+1}) = 0,$$

of order two. Any form $f(x_0, \dots, x_n)$ can be written as

$$f(x_0, \dots, x_n) = \sum_{j=d+2}^{n} g_j(x_0, \dots, x_n)\, x_j + g(x_0, \dots, x_{d+1}),$$

and if $f(x_0, \dots, x_n)$ vanishes on V,

$$g(x_0, \dots, x_{d+1}) = 0$$

on V. Hence we must have

$$g(x_0, \dots, x_{d+1}) = b(x_0, \dots, x_{d+1})\, F(x_0, \dots, x_{d+1}),$$

and it follows that the equations

$$F(x_0, \dots, x_{d+1}) = 0,$$

$$x_{d+2} = \dots = x_n = 0,$$

form a basis for the equations of V. Thus V is the intersection of a $(d+1)$-space and an irreducible primal of order two not containing that space.

We now define a quadric of dimension d in S_n as *the intersection of a $(d+1)$-space and a primal of order two not containing it, whether the primal is irreducible or not.* This intersection is always of order two, but may reduce to two d-spaces, or to a single d-space counted with multiplicity two. An immediate corollary is

THEOREM II. *The section of a quadric by an r-space not contained in it is a quadric.*

We shall usually use the symbol Q_d to denote a quadric of d dimensions.

We now obtain canonical forms for the equations of a quadric Q_d. By definition, Q_d lies in a $(d+1)$-space, and we can choose the coordinate system so that this is the space

$$x_{d+2} = \dots = x_n = 0,$$

and the quadric is defined by these equations and one other equation

$$F(x_0, \dots, x_{d+1}) = \sum_{i=0}^{d+1} \sum_{j=0}^{d+1} b_{ij} x_i x_j \quad (b_{ij} = b_{ji}),$$

which is uniquely determined save for a multiplier in K. Let $b = (b_{ij})$ be of rank $d+2-\rho$. Then [IX, § 2, Th. II] we can find a non-singular $(d+2)\times(d+2)$ matrix $p = (p_{ij})$ such that

$$p'bp = \begin{pmatrix} 1 & 0 & 0 & . & 0 \\ 0 & 1 & 0 & . & 0 \\ . & . & . & . & . \\ 0 & . & . & 0 & . \\ 0 & . & . & . & 0 \end{pmatrix},$$

in which there are $d+2-\rho$ non-zero elements on the principal diagonal. Now make the transformation of coordinates

$$\left.\begin{aligned} x_i &= \sum_{j=0}^{d+1} p_{ij} y_j \quad (i = 0, \ldots, d+1), \\ x_i &= y_i \qquad\qquad (i = d+2, \ldots, n), \end{aligned}\right\}$$

and we see that the equations of the quadric can be written as

$$\left.\begin{aligned} f_1(y) &\equiv y_0^2 + \ldots + y_{d+1-\rho}^2 = 0, \\ f_2(y) &\equiv y_{d+2} = 0, \\ &\vdots \\ f_{n-d}(y) &\equiv y_n = 0. \end{aligned}\right\} \qquad (2)$$

The matrix $\left(\dfrac{\partial f_i}{\partial y_j}\right)$ has the form

$$\begin{pmatrix} 2y_0 & 2y_1 & . & 2y_{d+1-\rho} & 0 & . & . & 0 & 0 & . & 0 \\ 0 & 0 & . & 0 & 0 & . & . & 1 & 0 & . & 0 \\ 0 & 0 & . & 0 & 0 & . & . & 0 & 1 & . & 0 \\ . & . & . & . & . & . & . & . & . & . & . \\ 0 & 0 & . & 0 & 0 & . & . & 0 & 0 & . & 1 \end{pmatrix}.$$

This is of rank less than $n-d$ at points of the quadric if and only if

$$y_0 = \ldots = y_{d+1-\rho} = 0, \quad y_{d+2} = \ldots = y_n = 0.$$

Hence the singular points of the quadric are the points of a linear space $S_{\rho-1}$. The number ρ is therefore a geometrical character of the quadric. If $\rho = 0$, the quadric is said to be *non-singular* (since all its points are simple), and if $\rho > 0$ the quadric is said to be *of type* ρ

(or to be a *quadric cone of type* ρ). The equations of a quadric of dimension d and type ρ can always be written in the form (2) by a suitable choice of coordinate system. From this we have

THEOREM III. *Two quadrics of the same dimension and type are projectively equivalent.*

The geometry of a quadric of dimension d can be completely determined when its geometry as a primal in the S_{d+1} of S_n which contains it is known. We shall therefore confine our attention to the geometry of quadric primals.

If a quadric Q_d is reducible, it must consist of two d-spaces, or of a d-space counted twice. If P is a point common to the two d-spaces in the former case, or any point of the d-space in the latter case, it is clear that a generic S_{n-d} through P does not meet the quadric Q_d except in P. Hence P is a singular point of Q_d. Now Q_d lies in a $(d+1)$-space; hence if it consists of two d-spaces, these meet in a $(d-1)$-space; therefore $\rho \geqslant d$; and similarly if Q_d is a d-space counted twice. It follows that if $\rho < d$, Q_d is irreducible. If $\rho = d$ its equation can be taken in the form

$$y_0^2 + y_1^2 = 0, \quad y_{d+2} = \dots = y_n = 0,$$

and it clearly consists of the d-spaces

$$y_0 \pm i y_1 = 0, \quad y_{d+2} = \dots = y_n = 0.$$

If $\rho = d+1$, the equation of Q_d can be written in the form

$$y_0^2 = 0, \quad y_{d+2} = \dots = y_n = 0,$$

and this is clearly a d-space (counted twice).

2. Quadric primals in S_n. By the results of § 1, a quadric Q_{n-1} in S_n is given by a single quadratic relation

$$\sum_{i,j=0}^{n} b_{ij} x_i x_j = 0 \quad (b_{ij} = b_{ji}).$$

It is often convenient to write this in matrix notation, writing x for the $(n+1) \times 1$ matrix whose elements are $x_0, \dots, x_n$, and $b = (b_{ij})$. Then the equations can be written in the form

$$x'bx = 0. \tag{1}$$

We shall assume that $n > 1$, since the case $n = 1$ is of no particular interest.

To find where the line joining the two distinct points y and z meets Q_{n-1}, we have to find the values of the ratio $\lambda:\mu$ such that $\lambda y+\mu z$ satisfies (1). We obtain the equation

$$\lambda^2 y'by+\lambda\mu(y'bz+z'by)+\mu^2 z'bz = 0,$$

or, since $$y'bz = \Sigma b_{ij}y_i z_j = z'by,$$

we have the equation

$$\lambda^2 y'by+2\lambda\mu y'bz+\mu^2 z'bz = 0.$$

Let us suppose, first, that z is a generic point of S_n, and hence that it is not on Q_{n-1}. Then $z'bz$ is not zero. The line yz is a generic line through y. If y is on Q_{n-1}, $y'by = 0$, and this point is simple if and only if a generic line through it meets Q_{n-1} once at y and once at another point. Hence y is simple if and only if

$$\Sigma b_{ij}y_i z_j \neq 0$$

for generic z; in other words, the singular points of Q_{n-1} are those which satisfy the equation

$$bx = \sum_j b_{ij}x_j = 0 \quad (i = 0,\ldots,n). \tag{2}$$

Conversely, if y satisfies these equations,

$$y'by = 0,$$

and $$z'by = y'bz = 0,$$

and y is a point of Q_{n-1} and is singular on it. If b is of rank $n+1-\rho$, the equations (2) define an $S_{\rho-1}$, and Q_{n-1} is of type ρ [§ 1]. The $S_{\rho-1}$ defined by (2) is called the *vertex* of Q_{n-1}.

Two points y and z of S_n are said to be *conjugate* with regard to Q_{n-1} if

$$ybz' = zby' = 0.$$

This relation is clearly symmetric. It is also projectively invariant; for if we make the transformation of coordinates

$$x = pX,$$

the new coordinates Y, Z of y, z, are, respectively, $p^{-1}y$ and $p^{-1}z$, and the equation of Q_{n-1} is

$$X'p'bpX = 0.$$

But $$Y'p'bpZ = y'\tilde{p}p'bpp^{-1}z = y'bz = 0.$$

It is clear that y is conjugate to every point of S_n if and only if $y'b = 0$, that is, if and only if y satisfies the equation (2); that is, if and only if y lies in the vertex of Q_{n-1}.

If y is not in the vertex, the locus of points conjugate to y is the prime

$$y'bx = 0;$$

this is called the *polar prime* of y, and, as above, its definition is clearly independent of the coordinate system chosen. The polar prime of y always passes through the vertex of Q_{n-1}, since each point of the vertex is conjugate to y. Moreover, writing down the equation of the tangent prime, we see that if y is on Q_{n-1} but not in the vertex, the polar prime of y is the tangent prime at y. If y is on Q_{n-1} but not in the vertex, and z is any point common to its tangent prime and Q_{n-1}, that is, if z is any point of Q_{n-1} conjugate to y, we have

$$y'by = 0, \quad y'bz = 0, \quad z'bz = 0,$$

and hence the line yz lies on Q_{n-1}. In particular, the line joining y to any point of the vertex lies on Q_{n-1}, and hence the ρ-space joining any simple point of Q_{n-1} to the vertex lies on Q_{n-1}. Now let $S_{n-\rho}$ be an $(n-\rho)$-space not meeting the vertex, and y any point of Q_{n-1} in $S_{n-\rho}$. The tangent space to Q_{n-1} at y passes through the vertex; if it contained $S_{n-\rho}$, $S_{n-\rho}$ would meet the vertex, since an $(n-\rho)$-space and a $(\rho-1)$-space in an $(n-1)$-space meet. Hence the section has a well-defined tangent space at y, and therefore the section of $S_{n-\rho}$ and Q_{n-1} has y as a simple point. If $\rho = n$, the vertex is an $(n-1)$-space, and hence Q_{n-1} is the vertex counted twice; if $\rho \leqslant n-1$, $S_{n-\rho}$ and Q_{n-1} meet in a $Q_{n-\rho-1}$, every point of which is simple. If y is any point of Q_{n-1} not in the vertex, the ρ-space joining the vertex to y lies on Q_{n-1}, and meets $S_{n-\rho}$. The space joining the vertex to y therefore meets $S_{n-\rho}$ in a point of $Q_{n-\rho-1}$. Hence Q_{n-1} is the join of the vertex to the points of a non-singular quadric $Q_{n-\rho-1}$.

Finally, if y and z are two points of S_n, neither of which lies on Q_{n-1}, which are conjugate with respect to Q_{n-1}, the points $\lambda_1 y + \mu_1 z$ and $\lambda_2 y + \mu_2 z$ in which their join meets Q_{n-1} are given by the roots of

$$\lambda^2 y'by + \mu^2 z'bz = 0,$$

and hence satisfy the relation

$$\lambda_1 \mu_2 + \lambda_2 \mu_1 = 0.$$

They are harmonically conjugate with respect to y and z. Conversely, if y and z are not on Q_{n-1} and their join meets Q_{n-1} in points harmonically conjugate with respect to y and z, y and z are conjugate with respect to Q_{n-1}.

From these geometrical interpretations of the relation of conjugacy, we deduce easily that if S_r is any linear space in S_n containing two points y and z which are conjugate with regard to Q_{n-1}, then y and z are conjugate with respect to the section of Q_{n-1} by S_r, and conversely.

If A_r is the point $(\delta_{r0}, \ldots, \delta_{rn})$ $(r = 0, \ldots, n)$, (δ_{ij} is the Kronecker delta), the condition for conjugacy of A_r and A_s is

$$b_{rs} = 0.$$

It follows that if we choose a coordinate system in S_n, as in § 1, so that the equation of Q_{n-1} is

$$y_0^2 + \ldots + y_{n-\rho}^2 = 0,$$

the vertices of the new simplex of reference are conjugate in pairs with respect to Q_{n-1}. Any simplex whose vertices are conjugate in pairs with regard to Q_{n-1} is called a *self-polar simplex* of Q_{n-1}. If $D_0, \ldots, D_n$ is a self-polar simplex and D_r is not in the vertex of Q_{n-1}, the polar prime of D_r goes through $D_0, \ldots, D_{r-1}, D_{r+1}, \ldots, D_n$ (since D_r and D_s are conjugate); hence it must be the face of the simplex which contains these n points.

We now consider how to construct a self-polar simplex of Q_{n-1}. Let D_0 be any point of S_n not on the quadric, and choose the coordinate system $(y_0, \ldots, y_n)$ such that D_0 is $(1, 0, \ldots, 0)$ and its polar prime (which does not pass through D_0, since D_0 is not on the quadric) is $y_0 = 0$. Let the equation of Q_{n-1} in the new coordinate system be

$$y'cy \equiv \Sigma c_{ij} y_i y_j = 0.$$

Since $(1, 0, \ldots, 0)$ is conjugate to each of the other vertices of the simplex of reference (which lie in its polar prime $y_0 = 0$),

$$c_{0i} = 0 \quad (i = 1, \ldots, n),$$

and

$$c = \begin{pmatrix} c_{00} & 0 & . & 0 \\ 0 & c_{11} & . & c_{1n} \\ . & . & . & . \\ 0 & c_{n1} & . & c_{nn} \end{pmatrix},$$

while $c_{00} \neq 0$, since D_0 is not on Q_{n-1}. Since c is of rank $n+1-\rho$,

$$\begin{pmatrix} c_{11} & \cdot & c_{1n} \\ \cdot & \cdot & \cdot \\ c_{n1} & \cdot & c_{nn} \end{pmatrix}$$

is of rank $n-\rho$. Hence the section of Q_{n-1} by the polar prime of D_0 whose equations are

$$y_0 = 0, \quad \sum_{i,j=1}^{n} c_{ij} y_i y_j = 0$$

is a quadric of type ρ, having the same vertex as Q_{n-1}. If $\rho = n$, $c_{ij} = 0$ $(i,j = 1, \ldots, n)$, and the equation of Q_{n-1} is

$$y_0^2 = 0.$$

If $\rho \leqslant n-1$, we choose any point D_1 of $y_0 = 0$ not on Q_{n-1}. The polar prime of D_1 passes through D_0, since D_0 and D_1 are conjugate, and does not pass through D_1, since D_1 is not on Q_{n-1}. If we choose the coordinate system so that D_0 is $(1, 0, \ldots, 0)$ and D_1 is $(0, 1, 0, \ldots, 0)$, while the polar primes of D_0 and D_1 are $y_0 = 0$ and $y_1 = 0$, we find that the equation of Q_{n-1} is

$$c_{00} y_0^2 + c_{11} y_1^2 + \sum_{i,j=2}^{n} c_{ij} y_i y_j = 0,$$

where (c_{ij}) $(i,j = 2, \ldots, n)$ is of rank $n-1-\rho$, and c_{00} and c_{11} are not zero. The intersection of the polar prime of D_0 and D_1 meets Q_{n-1} in the quadric

$$y_0 = 0, \quad y_1 = 0, \quad \sum_{i,j=2}^{n} c_{ij} y_i y_j = 0,$$

which is again a quadric of type ρ, whose vertex is the vertex of Q_{n-1}. Proceeding in this way we see that we can find $n+1-\rho$ points $D_0, \ldots, D_{n-\rho}$, none of which is on Q_{n-1}, such that D_0 is any point of S_n (not on Q_{n-1}), D_1 any point conjugate to D_0 (but not on Q_{n-1}), $\ldots$, $D_{n-\rho}$ is any point (not on Q_{n-1}) conjugate to $D_0, \ldots, D_{n-\rho-1}$. If $D_{n-\rho+1}, \ldots, D_n$ are any ρ independent points in the vertex of Q_{n-1}, $D_0, \ldots, D_n$ is a self-polar simplex of Q_{n-1}, and any self-polar simplex of Q_{n-1} is obtained in this way. If $D_0, \ldots, D_n$ are the vertices of the simplex of reference of the coordinate system $(z_0, \ldots, z_n)$, the equation of Q_{n-1} is of the form

$$c_0 z_0^2 + \ldots + c_{n-\rho} z_{n-\rho}^2 = 0 \quad (c_i \neq 0),$$

and if we write

$$y_i = c_i^{\frac{1}{2}} z_i \quad (i = 0, \ldots, n-\rho),$$

$$y_i = z_i \qquad (i = n-\rho+1, \ldots, n),$$

we obtain the equation $\quad y_0^2 + \ldots + y_{n-\rho}^2 = 0.$

So the reduction of the equation of Q_{n-1} to the canonical form is essentially equivalent to finding a self-polar simplex of Q_{n-1}.

Let Q_{n-1} be a quadric of type ρ $(\rho < n)$. Then there are on it points which are simple for Q_{n-1}. Let A be such a point, and choose the coordinate system so that A is $(1, 0, \ldots, 0)$. If the equation of Q_{n-1} is

$$\sum_{i=0}^{n} \sum_{j=0}^{n} a_{ij} x_i x_j = 0,$$

we have $a_{00} = 0$. The tangent prime at A is

$$\sum_{j=1}^{n} a_{0j} x_j = 0,$$

and not all the coefficients of this are zero, since A is simple. Since this prime goes through A, we may choose the coordinate system so that the tangent prime at A is $x_n = 0$. Then $a_{0j} = 0$ $(j < n)$, $a_{0n} \neq 0$. The matrix (a_{ij}) $(i, j = 0, \ldots, n)$ is of rank $n+1-\rho$, and hence, since $a_{0i} = 0$ $(i < n)$, $a_{0n} \neq 0$, the matrix (a_{ij}) $(i, j = 1, \ldots, n-1)$ is of rank $n-1-\rho$. The section of Q_{n-1} by $x_n = 0$ is the quadric Q_{n-2} given by the equation

$$x_n = 0, \quad \sum_{i=1}^{n-1} \sum_{j=1}^{n-1} a_{ij} x_i x_j = 0,$$

and is therefore a quadric of type $\rho + 1$. Since any point of the vertex of Q_{n-1} is in $x_n = 0$ and is conjugate to every point of S_{n-1} with respect to Q_{n-1}, it is conjugate to any point of $x_n = 0$ with respect to Q_{n-2}. Hence the vertex of Q_{n-1} is contained in the vertex of Q_{n-2}. Also, since $x_n = 0$ is the polar prime of A, and contains A, A is conjugate to any point of $x_n = 0$ with respect to Q_{n-1}, and hence with respect to Q_{n-2}. Therefore A is in the vertex of Q_{n-2}. Since Q_{n-2} has a vertex of dimension ρ, its vertex is the join of A to the vertex of Q_{n-1}.

We use these results to give another form to the equation of a quadric Q_{n-1} of type ρ $(\rho < n-1)$. Let A and B be any two simple points of Q_{n-1} which are not conjugate with regard to Q_{n-1}. The section of Q_{n-1} by the tangent prime at A is a quadric Q_{n-2} of dimension $n-2$ and of type $\rho+1$, whose vertex is the join of A to the vertex of Q_{n-1}. The tangent prime to Q_{n-1} at B does not contain A, hence it meets Q_{n-2} in a quadric Q_{n-3} of dimension $n-3$ and type ρ. Since the tangent primes at A and B both contain the vertex of Q_{n-1}, it follows that the vertex of Q_{n-3} is the same as that of Q_{n-1}. Let $A_1, \ldots, A_{n-1}$ be a simplex in the space S_{n-2} containing Q_{n-3} which is self-polar with regard to Q_{n-3}. Since A and B are not conjugate AB, cannot meet S_{n-2}. For if AB meets S_{n-2} in P, lying in the polar primes of A and B, P is conjugate to A and B. Hence the polar prime of P passes through A and B and hence through the line AB, and therefore through P. P therefore lies on the quadric. Since A and B are not conjugate, their join does not lie on Q_{n-1}, and therefore meets Q_{n-1} only in A and B. But P is to be conjugate to A and B, and A and B are not conjugate. Hence P cannot lie at A or B. Therefore AB does not meet S_{n-1} and hence $A, A_1, \ldots, A_{n-1}, B$ is a simplex. Take this as the simplex of reference. If the equation of Q_{n-1} is now

$$\Sigma b_{ij} x_i x_j = 0 \quad (b_{ij} = b_{ji}),$$

the polar of $A = (1, 0, \ldots, 0)$ is to be $x_n = 0$. Hence

$$b_{00}x_0 + \ldots + b_{0n}x_n = \lambda x_n \quad (\lambda \neq 0),$$

and therefore $b_{00} = \ldots = b_{0n-1} = 0$, $b_{0n} \neq 0$. Similarly, since the polar of $B = (0, \ldots, 0, 1)$ is to be $x_0 = 0$, we have $b_{n1} = \ldots = b_{nn} = 0$. Also since A_i and A_j are conjugate,

$$b_{ij} = 0 \quad (i, j = 1, \ldots, n-1;\ i \neq j).$$

Remembering that Q_{n-3} is of type ρ, we see that the equation of Q_{n-1} must be of the form

$$x_0 x_n = \sum_{j=1}^{n-1-\rho} b_j x_j^2 \quad (b_j \neq 0).$$

Putting

$$y_0 = x_0,$$

$$y_j = \sqrt{b_j}\, x_j \quad (j = 1, \ldots, n-1),$$

$$y_n = x_n,$$

the equation of Q_{n-1} becomes

$$y_0 y_n = \sum_{1}^{n-1-\rho} y_i^2,$$

a form which we shall often find useful.

Finally, we consider the concept dual to a quadric. Let Q_{n-1} be a quadric in S_n of type ρ, and suppose that the coordinate system is chosen so that its equation is

$$x_0^2 + \ldots + x_{n-\rho}^2 = 0.$$

The tangent prime at any point ξ of this (not in the vertex) is of the form

$$\xi_0 x_0 + \ldots + \xi_{n-\rho} x_{n-\rho} = 0,$$

where $\xi_0, \ldots, \xi_{n-\rho}$ are not all zero and

$$\xi_0^2 + \ldots + \xi_{n-\rho}^2 = 0.$$

Conversely, if $\xi_0, \ldots, \xi_{n-\rho}$ are not all zero, and

$$\xi_0^2 + \ldots + \xi_{n-\rho}^2 = 0,$$

then

$$\xi_0 x_0 + \ldots + \xi_{n-\rho} x_{n-\rho} = 0$$

is a tangent prime to Q_{n-1}. Hence the tangent primes to Q_{n-1} are the primes whose coordinates satisfy the equations

$$u_0^2 + \ldots + u_{n-\rho}^2 = 0,$$

$$u_{n-\rho+1} = \ldots = u_n = 0.$$

Hence if we represent the primes of S_n by the points of a space Σ_n, in the usual way, the locus corresponding to the tangent primes of Q_{n-1} is represented by a non-singular quadric of dimension $n-\rho-1$. The space containing the quadric is the dual of the vertex of Q_{n-1}.

More generally, if Q_d is a quadric of type ρ in S_n, and we choose the coordinates so that its equations are

$$x_0^2 + \ldots + x_{d+1-\rho}^2 = 0,$$

$$x_{d+2} = \ldots = x_n = 0,$$

the prime

$$\xi_0 x_0 + \ldots + \xi_n x_n = 0$$

contains a tangent d-space to Q_d if either $\xi_0 = \ldots = \xi_{d+1} = 0$, that is, if the prime contains the $(d+1)$-space containing Q_d, or else if

$$\xi_0 x_0 + \ldots + \xi_{d+1} x_{d+1} = 0,$$

$$x_{d+2} = \ldots = x_n = 0,$$

is a tangent d-space to Q_d. For this we require

$$\xi_0^2 + \ldots + \xi_{d+1-\rho}^2 = 0,$$

$$\xi_{d+2-\rho} = \ldots = \xi_{d+1} = 0.$$

Hence the locus of points in Σ_n representing primes of S_n which contain the tangent spaces of Q_d is a quadric of dimension $n-\rho-1$ and type $n-d-1$. Conversely, we can verify at once that the primes of S_n corresponding to the points of Σ_n which lie on a quadric of dimension $n-\rho-1$ and type $n-d-1$ are the primes which contain d-spaces tangent to a quadric Q_d of type ρ.

It is customary to omit all reference to Σ_n, and simply to consider the primes in S_n whose coordinates satisfy equations of the form

$$\sum_{i,j=0}^{n} \alpha_{ij} u_i u_j = 0,$$

$$\sum_{j=0}^{n} \beta_{ij} u_j = 0 \quad (i = 1, \ldots, n-e-1),$$

where we assume that the solutions form a system of primes of dimension e. Such a system of primes is called a *quadric-envelope* (in contrast with a *quadric-locus*), and the results obtained by considering Σ_n amount to saying that the primes of a quadric-envelope of dimension $n-\rho-1$ and type $n-d-1$ are just the primes which contain the tangent spaces to a quadric-locus of dimension d and type ρ. This quadric-locus is usually called the *envelope* of the quadric-envelope.

The most important case arises when $d = n-1$, $\rho = 0$. Then the quadric-envelope formed by the tangent spaces to the non-singular quadric Q_{n-1} is of dimension $n-1$, and is non-singular. Let the equation of the quadric-locus Q_{n-1} be

$$x'bx \equiv \sum_{i,j=0}^{n} b_{ij} x_i x_j = 0, \tag{1}$$

where b is non-singular. If $(u'_0, \ldots, u'_n)$ is a tangent prime to this, there exists a point x' of Q_{n-1} such that

$$\sum_{j=0}^{n} b_{ij} x'_j = \lambda u'_i \quad (i = 0, \ldots, n)$$

for some non-zero λ, and since x' lies on Q_{n-1},

$$\lambda \sum_{0}^{n} u'_i x'_i = \sum_{i,j=0}^{n} b_{ij} x'_i x'_j = 0.$$

Hence we have the $n+2$ equations

$$\sum_{j=0}^{n} b_{ij} x'_j - \lambda u'_i = 0 \quad (i = 0, \ldots, n),$$

$$\sum_{j=0}^{n} u'_j x'_j \qquad = 0.$$

Eliminating $x'_0, \ldots, x'_n$, λ we see that u' satisfies the equation

$$\begin{vmatrix} b_{00} & . & b_{0n} & u_0 \\ b_{10} & . & b_{1n} & u_1 \\ . & . & . & . \\ b_{n0} & . & b_{nn} & u_n \\ u_0 & . & u_n & 0 \end{vmatrix} = 0,$$

that is,
$$\sum_{i,j=0}^{n} \beta_{ij} u_i u_j = 0, \tag{3}$$

where β_{ij} is the cofactor of b_{ij} in $\det b$. Since b is non-singular, it follows [II, § 8, Th. VIII] that β is non-singular. The equation (3) therefore represents a non-singular quadric-envelope, which contains the system of tangent primes to Q_{n-1}. As this system is an irreducible system of dimension $n-1$, it must therefore coincide with the system of primes given by (3). Hence (3) is the equation of the quadric-envelope defined by (1).

Conversely, let
$$\Sigma \beta_{ij} u_i u_j = 0$$

be any non-singular quadric-envelope of dimension $n-1$, and let Q_{n-1} be the quadric

$$x'bx \equiv \begin{vmatrix} \beta_{00} & . & \beta_{0n} & x_0 \\ . & . & . & . \\ \beta_{n0} & . & \beta_{nn} & x_n \\ x_0 & . & x_n & 0 \end{vmatrix} = 0. \tag{4}$$

Form the equation of the quadric-envelope of this:

$$\sum_{i,j=0}^{n} \gamma_{ij} u_i u_j = 0.$$

By an elementary property of determinants [II, § 8, Th. VIII],

$$\gamma_{ij} = \Delta^{n-1} \beta_{ij},$$

where Δ ($\Delta \neq 0$) is the determinant $|\beta_{ij}|$. Hence the quadric-locus which is the envelope of the given quadric envelope is given by (4).

We can construct a polar theory for quadric-envelopes corresponding to that already given for quadric-loci. Confining ourselves to non-singular quadrics we define the *pole* of the prime v with regard to (3) as the point

$$v'\beta u = 0.$$

If this is the point y, we can write

$$y = \beta v,$$
$$\Delta v = by.$$

Hence v is the polar of y in the corresponding quadric-locus. Two primes are conjugate if the pole of each lies in the other. It follows that two primes are conjugate with regard to a quadric-envelope if and only if their poles are conjugate with respect to the corresponding quadric-locus.

The polar theory of a non-singular quadric primal is indeed the same as the polar theory of the corresponding quadric envelope. In the next section we give a brief account of this theory.

3. Polar theory of quadrics. Let Q_{n-1} be a non-singular quadric in S_n, with equation

$$x'bx = \Sigma b_{ij} x_i x_j = 0, \tag{1}$$

and let

$$u'\beta u = \Sigma \beta_{ij} u_i u_j = 0 \tag{2}$$

be the equation of the quadric-envelope formed by the tangent primes to it. We may suppose that β_{ij} is the cofactor of b_{ij} in $\det b$.

Let S_d be any d-space of S_n, and let $y_0, \ldots, y_d$ be $d+1$ independent points of it. Any point y of S_d can be written in the form

$$y = \lambda_0 y_0 + \ldots + \lambda_d y_d.$$

The polar prime of y with respect to Q_{n-1} is

$$\lambda_0 y_0' bx + \ldots + \lambda_d y_d' bx = 0.$$

It follows that the polar primes of all points of S_d intersect in the linear space whose equations are

$$y_0' bx = 0, \quad \ldots, \quad y_d' bx = 0.$$

Since b is non-singular, and

$$\begin{pmatrix} y_0' \\ \cdot \\ y_d' \end{pmatrix} = \begin{pmatrix} y_{00} & \cdot & y_{0n} \\ \cdot & \cdot & \cdot \\ y_{d0} & \cdot & y_{dn} \end{pmatrix}$$

is of rank $d+1$, this space is of dimension $n-d-1$. We denote it by S_{n-d-1} and call it the *polar space* of S_d with respect to Q_{n-1}. Every point of S_d is conjugate to every point of S_{n-d-1}. S_{n-d-1} is called the polar space of S_d. The polar space of S_{n-d-1} is a space of dimension $n-(n-d-1)-1 = d$, and consists of all the points which are conjugate to every point of S_{n-d-1}; hence it must be S_d. Thus the relation between S_d and S_{n-d-1} is symmetrical. The polar space of S_{n-d-1} with regard to the quadric envelope (2) is defined, dually, as the locus of poles of all primes through S_{n-d-1}. But if the pole of the prime p with regard to (2) is P, the polar of P with regard to (1) is p. Hence we deduce that the locus of poles of primes through S_{n-d-1} is S_d, and similarly the locus of poles of primes through S_d is S_{n-d-1}. Thus if S_d and S_{n-d-1} are in the polar relation with regard to (1), they are also in the polar relation with regard to (2), and conversely.

Suppose that S_d and S_{n-d-1} meet in an S_k. Then their join is an S_{n-k-1}. If y is any point of S_k, we see by regarding it as a point of S_d that its polar prime contains S_{n-d-1}, and by regarding it as a point of S_{n-d-1} we see that its polar prime contains S_d. Hence the polar prime of y contains S_{n-k-1}, and since y is any point of S_k, it follows that the polar space of S_k with respect to (1) (and hence with respect to (2)) is S_{n-k-1}. Since S_k is contained in S_{n-k-1}, it follows that the polar space of every point y of S_k contains y; hence every point of S_k is on Q_{n-1}. Moreover, the tangent space to Q_{n-1} at any point of S_k contains S_{n-k-1} and hence S_d. Thus S_d lies in the tangent space at every point of S_k (and contains S_k); S_d is said to *touch* Q_{n-1} *along* S_k. Conversely, if y is any point of S_d which lies on Q_{n-1} and is such that S_d lies in the tangent prime at y, y is the pole of a prime through S_d. It therefore lies in S_{n-d-1}, and therefore also in S_k. Thus we have

THEOREM I. *If a d-space S_d touches Q_{n-1}, it touches it at points of a linear space S_k. The polar space of S_d meets S_d in this S_k.*

In particular, if S_d lies on Q_{n-1} the polar prime of any point of it contains S_d. Hence S_d is contained in its polar space S_{n-d-1}, and therefore $2d \leqslant n-1$.

Now let S'_d be any other d-space of S_n, S'_{n-d-1} its polar space with regard to Q_{n-1}. Suppose that S_d and S'_{n-d-1} meet in an a-space S_a. If y is any point of S_a, the polar prime of y goes through S_{n-d-1} and S'_d. Hence S_{n-d-1} and S'_d lie in the polar space S_{n-a-1} of S_a, and therefore meet in a b-space $(b \geqslant a)$. But if S'_d and S_{n-d-1} meet in a b-space, similar reasoning shows that their polar spaces meet in a space of dimension b at least. Hence $a \geqslant b$, and we conclude that

$a = b$. If S_d and S'_{n-d-1} meet at all, we say that *S_d and S'_d are conjugate*; if they meet in an a-space we say that their *degree of conjugacy* is $a+1$. The result just proved shows that the degree of conjugacy of two spaces is a symmetrical property of these spaces. If $d = 0$, the conjugacy now defined simply implies that the polar of S'_0 goes through S_0, and hence S_0 and S'_0 are conjugate according to the definition of § 2. Conversely, if y and z are conjugate according to the definition of § 2, they are conjugate in our present sense. Again, a space S_d is *self-conjugate* to a degree $k+1$ if and only if it meets its polar space in a k-space, that is, if and only if it touches Q_{n-1} along a k-space.

The condition that two d-spaces with Grassmann coordinates $(\ldots, p_{i_0 \ldots i_d}, \ldots)$ and $(\ldots, q_{i_0 \ldots i_d}, \ldots)$ are conjugate to a degree $k+1$ at least with regard to Q_{n-1} can easily be written down. Let $y_0, \ldots, y_d$ be a set of independent points of the first space. Then its polar space is given by the equations

$$\sum_{j=0}^{n} \sum_{k=0}^{n} y_{ij} b_{jk} x_k = 0 \quad (i = 0, \ldots, d).$$

The dual coordinates $(\ldots, p^{j_0 \ldots j_d}, \ldots)$ of this S_{n-d-1} are given by the equations

$$p^{j_0 \ldots j_d} = \sum_{k_0 \ldots k_d} \begin{vmatrix} y_{0k_0} & \cdot & y_{0k_d} \\ \cdot & \cdot & \cdot \\ y_{dk_0} & \cdot & y_{dk_d} \end{vmatrix} \begin{vmatrix} b_{k_0 j_0} & \cdot & b_{k_0 j_d} \\ \cdot & \cdot & \cdot \\ b_{k_d j_0} & \cdot & b_{k_d j_d} \end{vmatrix}$$

$$= \sum_{k_0 \ldots k_d} \begin{vmatrix} b_{k_0 j_0} & \cdot & b_{k_0 j_d} \\ \cdot & \cdot & \cdot \\ b_{k_d j_0} & \cdot & b_{k_d j_d} \end{vmatrix} p_{k_0 \ldots k_d},$$

and the conditions that this $(n-d-1)$-space should meet $(\ldots, q_{i_0 \ldots i_d}, \ldots)$ in a k-space at least can be written down by the methods of VII, § 5.

In particular, the condition that they should be conjugate is seen to be

$$\sum_i \sum_j \begin{vmatrix} b_{i_0 j_0} & \cdot & b_{i_0 j_d} \\ \cdot & \cdot & \cdot \\ b_{i_d j_0} & \cdot & b_{i_d j_d} \end{vmatrix} p_{i_0 \ldots i_d} q_{j_0 \ldots j_d} = 0.$$

The polar theory of quadrics which we have outlined can be extended to singular quadrics, but since the dual of a quadric of type ρ and dimension $n-1$ is a non-singular quadric of dimension

$n-\rho-1$, the theory becomes less interesting. We shall merely indicate how it can be developed. If y is any point of S_n not in the vertex of Q_{n-1}, its polar prime η passes through the vertex [§ 2]. If y' is any other point, not in the vertex, but in the ρ-space joining y to the vertex, we see, on writing down the equation of the polar prime of y', that this polar prime is η, and that any point whose polar prime is η lies in this ρ-space. Now take the section of Q_{n-1} by an $S_{n-\rho}$ not meeting the vertex. This is a non-singular quadric $Q_{n-\rho-1}$ of dimension $n-\rho-1$. Two points y and z of S_n are conjugate with regard to Q_{n-1} if and only if their projections from the vertex on to $S_{n-\rho}$ are conjugate with respect to Q_{n-1} (that is, with respect to $Q_{n-\rho-1}$). We therefore define polar and conjugate spaces in $S_{n-\rho}$ with regard to $Q_{n-\rho-1}$, and call the joins of these to the vertex of Q_{n-1} polar and conjugate spaces with regard to Q_{n-1}. Thus the polar properties of Q_{n-1} can be obtained at once from the polar properties of a non-singular quadric. By duality, we can obtain a polar theory for singular quadric-envelopes.

We now return to the case in which Q_{n-1}, given by (1), is non-singular. Let $P_0, \ldots, P_n$ be a self-polar simplex with regard to Q_{n-1}, and suppose that the coordinates of P_r are $(p_{r0}, \ldots, p_{rn})$. Since $P_0, \ldots, P_n$ is a simplex,

$$p = \begin{pmatrix} p_{00} & \cdot & p_{0n} \\ \cdot & \cdot & \cdot \\ p_{n0} & \cdot & p_{nn} \end{pmatrix}$$

is a non-singular matrix. Consider the transformation of coordinates given by

$$x = p'X.$$

In the new system of coordinates, $P_0, \ldots, P_n$ is the simplex of reference. The new dual coordinates are given by

$$U = pu.$$

Since the simplex of reference is self-polar with regard to Q_{n-1}, the equation of the quadric-envelope defined by Q_{n-1} is of the form

$$\sum_0^n \lambda_i U_i^2 = 0;$$

that is, equation (2) can be written in the form

$$\sum_0^n \lambda_i (p_{i0} u_0 + \ldots + p_{in} u_n)^2 = 0.$$

Conversely, if we can find p_{ij} so that (2) can be written in this form, the forms $\sum_j p_{ij}u_j$ $(i = 0, \ldots, n)$ are linearly independent, otherwise (2) would be singular. The pole of v is

$$\sum_0^n \lambda_i(p_{i0}u_0 + \ldots + p_{in}u_n)(p_{i0}v_0 + \ldots + p_{in}v_n) = 0,$$

and if v is the prime containing

$$(p_{i0}, \ldots, p_{in}) \quad (i = 0, \ldots, r-1, r+1, \ldots, n),$$

this becomes $$p_{r0}u_0 + \ldots + p_{rn}u_n = 0,$$

that is, the pole is $(p_{r0}, \ldots, p_{rn})$. Thus the $n+1$ points $(p_{i0}, \ldots, p_{in})$ form a self-polar simplex for (2), and hence for (1).

This idea can be generalised. Let $P_i = (p_{i0}, \ldots, p_{in})$ $(i = 1, \ldots, r)$ be r points, and let

$$U_i \equiv p_{i0}u_0 + \ldots + p_{in}u_n = 0 \quad (i = 1, \ldots, r)$$

be their equations. Suppose that the equation (2) can be written in the form

$$\sum_1^r \lambda_i U_i^2 = 0.$$

Without loss of generality we may assume that the points have been chosen so that no λ_i is zero. Then

$$\beta = p'\Lambda p,$$

and since β is of rank $n+1$, p must be of rank $n+1$ at least, and $n+1$ of the points P_i must be linearly independent; in particular $r \geqslant n+1$.

Now let v be a prime containing $P_{i_1}, \ldots, P_{i_n}$, and let $i_{n+1}, \ldots, i_r$ be the remaining suffixes. The pole of v has equation

$$\sum_{s=1}^r \lambda_{i_s} U_{i_s}(p_{i_s 0}v_0 + \ldots + p_{i_s n}v_n) \equiv \sum_{s=n+1}^r \lambda_{i_s} U_{i_s}(p_{i_s 0}v_0 + \ldots + p_{i_s n}v_n) = 0,$$

and is therefore linearly dependent on $P_{i_{n+1}}, \ldots, P_{i_r}$.

The most interesting case of this arises when $n+1 \leqslant r \leqslant 2n$, and the points $P_1, \ldots, P_r$ are such that any $n+1$ of them are linearly independent (and therefore any s, where $s \leqslant n$, of them are linearly independent). In this case it follows that the pole of the prime defined by any n of them lies in the S_{r-n-1} joining the remaining points of the set. We shall say that a set of points $P_1, \ldots, P_r$

$(n+1 \leqslant r \leqslant 2n)$ is a *polar r-ad* of Q_{n-1} if (*a*) no subset of $n+1$ points lies in an S_{n-1}; (*b*) the pole of the prime containing any n of them lies in the space which spans the remaining $r-n$ of them. The existence of polar r-ads for Q_{n-1} is easily established. It can easily be shown that in S_n there exists a set of r points $R_1, \ldots, R_r$ with coordinates $(p_{i0}, \ldots, p_{in})$ $(i = 1, \ldots, r)$ such that (*a*) any $n+1$ of them are linearly independent; (*b*) the $r \times (n+1)$ matrix (p_{ij}) is such that $p'p$ is of rank $n+1$. Then

$$\sum_1^r (p_{i0}u_0 + \ldots + p_{in}u_n)^2 = 0$$

is a non-singular quadric envelope defining a non-singular quadric locus Q'_{n-1}. Since all non-singular quadric primals in S_n are projectively equivalent [§ 1, Th. III] we can find a non-singular collineation of S_n which carries Q'_{n-1} into Q_{n-1}. It is trivial to verify that this carries $R_1, \ldots, R_r$ into a polar r-ad $P_1, \ldots, P_r$ of Q_{n-1}. This establishes the existence of polar r-ads.

Now suppose that $P_1, \ldots, P_r$ is a polar r-ad for Q_{n-1}. We choose the coordinate system so that $P_1, \ldots, P_{n+1} = A_0, \ldots, A_n$ is the simplex of reference, and for convenience denote $P_{n+2}, \ldots, P_r$ by $B_1, \ldots, B_s$ $(s = r-n-1)$. In this coordinate system the equation of Q_{n-1}, regarded as an envelope, is

$$\sum_{i,j=0}^{n} \gamma_{ij} U_i U_j = 0.$$

The pole of the prime $A_0, \ldots, A_{i-1}, A_{i+1}, \ldots, A_n$ is given by the equation

$$\sum_{j=0}^{n} \gamma_{ij} U_j = 0,$$

and is therefore the point $(\gamma_{i0}, \ldots, \gamma_{in})$. This is to lie in the s-space joining the points $A_i, B_1, \ldots, B_s$. Hence if B_i has coordinates $(b_{i0}, \ldots, b_{in})$,

$$\gamma_{ij} = \lambda_i \delta_{ij} + \sum_{\rho=1}^{s} \lambda_{i\rho} b_{\rho j}, \tag{3}$$

where δ_{ij} is the Kronecker delta. Since $\gamma_{ij} = \gamma_{ji}$, we have

$$\sum_{\rho=1}^{s} \lambda_{i\rho} b_{\rho j} = \sum_{\rho=1}^{s} \lambda_{j\rho} b_{\rho i}. \tag{4}$$

From (3) we see that the equation of Q_{n-1} is of the form

$$0 = \lambda_0 U_0^2 + \ldots + \lambda_n U_n^2 + \sum_{i,j=0}^{n} \sum_{\rho=1}^{s} \lambda_{i\rho} b_{\rho j} U_i U_j$$

$$= \lambda_0 U_0^2 + \ldots + \lambda_n U_n^2 + \sum_{i=0}^{n} \sum_{\rho=1}^{s} \lambda_{i\rho} U_i U_{n+\rho}, \tag{5}$$

where $$U_{n+\rho} \equiv b_{\rho 0} U_0 + \ldots + b_{\rho n} U_n = 0 \tag{6}$$

is the equation of B_ρ.

Now let $i_0, \ldots, i_n$ be any derangement of $0, 1, \ldots, n$, and let h be any integer, where $1 \leqslant h \leqslant s$. Let V be the prime containing $A_{i_s}, \ldots, A_{i_n}, B_1, \ldots, B_{h-1}, B_{h+1}, \ldots, B_s$. Then

$$V_{i_s} = \ldots = V_{i_n} = 0,$$

$$V_{n+\rho} = 0 \quad (\rho \neq h),$$

where $V_{n+\rho}$ is obtained from $U_{n+\rho}$ by replacing U_j in $\Sigma b_{ij} U_j$ by V_j. Since this prime does not contain B_h (any $n+1$ points of the r-ad being independent), $V_{n+h} \neq 0$. The pole of V has equation

$$2\lambda_0 V_0 U_0 + \ldots + 2\lambda_n V_n U_n + \sum_{i=0}^{n} \sum_{\rho=1}^{s} \lambda_{i\rho} U_i V_{n+\rho} + \sum_{i=0}^{n} \sum_{\rho=1}^{s} \lambda_{i\rho} V_i U_{n+\rho} = 0.$$

Now $$\sum_{i=0}^{n} \sum_{\rho=1}^{s} \lambda_{i\rho} V_i U_{n+\rho} = \sum_{i,j=0}^{n} \sum_{\rho=1}^{s} \lambda_{i\rho} b_{\rho j} V_i U_j$$

$$= \sum_{i,j=0}^{n} \sum_{\rho=1}^{s} \lambda_{j\rho} b_{\rho i} V_i U_j \qquad \text{(from (4))}$$

$$= \sum_{i=0}^{n} \sum_{\rho=1}^{s} \lambda_{i\rho} U_i V_{n+\rho}.$$

Hence, since

$$V_{i_s} = \ldots = V_{i_n} = V_{n+1} = \ldots = V_{n+h-1} = V_{n+h+1} = \ldots = V_{r-1} = 0,$$

the pole of V is given by the equation

$$\lambda_{i_0} V_{i_0} U_{i_0} + \ldots + \lambda_{i_{s-1}} V_{i_{s-1}} U_{i_{s-1}} + V_{n+h} \sum_{i=0}^{n} \lambda_{ih} U_i = 0.$$

By the property of a polar r-ad, this lies in $A_{i_0} \ldots A_{i_{s-1}} B_h$. Since

$$\lambda_{i_0} V_{i_0} U_{i_0} + \ldots + \lambda_{i_{s-1}} U_{i_{s-1}} V_{i_{s-1}} = 0$$

lies in this space, so must

$$\sum_{i=0}^{n} \lambda_{ih} U_i = 0,$$

unless $\lambda_{ih} = 0$ $(i = 0, \ldots, n)$. Assuming for the moment that not all λ_{ih} are zero, $(\lambda_{0h}, \ldots, \lambda_{nh})$ lies in $A_{i_0} \ldots A_{i_{s-1}} B_h$, for all derangements $i_0, \ldots, i_n$ of $0, \ldots, n$. But the $\binom{n+1}{s}$ spaces $A_{i_0} \ldots A_{i_{s-1}} B_h$ have only B_h in common. Hence, there exists a constant λ_{n+h} such that

$$\lambda_{ih} = \lambda_{n+h} b_{hi} \quad (i = 0, \ldots, n).$$

This result includes the case in which each λ_{ih} is zero, by taking $\lambda_{n+h} = 0$. Hence in any case the equation (5) can be written in the form

$$\sum_{i=0}^{n} \lambda_i U_i^2 + \sum_{i=0}^{n} \sum_{\rho=1}^{s} \lambda_{n+\rho} (b_{\rho i} U_i) U_{n+\rho} = 0,$$

that is
$$\sum_{i=0}^{r-1} \lambda_i U_i^2 = 0, \qquad \text{from (6).}$$

Returning to the original coordinate system, the equation of Q_{n-1} can be written in the form

$$\sum_{i=1}^{r} \mu_i (p_{i0} u_0 + \ldots + p_{in} u_n)^2 = 0,$$

where $P_i = (p_{i0}, \ldots, p_{in})$. It will be observed that we have only used certain of the conditions of conjugacy of a polar r-ad. It follows that the other conditions are satisfied automatically.

We must also consider the possibility of an identical relation

$$\sum_{i=1}^{r} \lambda_i (p_{i0} u_0 + \ldots + p_{in} u_n)^2 \equiv 0, \tag{7}$$

where $P_i = (p_{i0}, \ldots, p_{in})$ $(i = 1, \ldots, r)$ are r points of S_n. If such a relation holds, it follows that $\lambda_1, \ldots, \lambda_r$ satisfy the $\frac{1}{2}(n+1)(n+2)$ equations

$$\sum_{i=1}^{r} \lambda_i p_{ij} p_{ik} = 0 \quad (j, k = 0, \ldots, n), \tag{8}$$

and conversely, if $\lambda_1, \ldots, \lambda_r$ is a solution of these equations,

$$\sum_{i=1}^{r} \lambda_i (p_{i0} u_0 + \ldots + p_{in} u_n)^2 \equiv 0.$$

A non-zero set of λ's can always be found if $r > \frac{1}{2}(n+1)(n+2)$. Suppose now that $r \leqslant \frac{1}{2}(n+1)(n+2)$. We assume, as we may, that each λ_i is different from zero. There always exists a quadric containing $r-1$ of the points P_i. Then if

$$\sum_{i,j=0}^{n} b_{ij} x_i x_j = 0$$

is a quadric through $P_1, \ldots, P_{h-1}, P_{h+1}, \ldots, P_r$,

$$\sum_{j,k=0}^{n} b_{jk}\, p_{ij} p_{ik} = 0 \quad (i = 1, \ldots, h-1, h+1, \ldots, r).$$

From equation (8) we see that

$$\sum_{i=1}^{r} \sum_{j,k=0}^{n} \lambda_i b_{jk} p_{ij} p_{ik} = 0,$$

and therefore

$$\sum_{j,k=0}^{n} b_{jk} p_{hj} p_{hk} = 0.$$

Hence the quadric passes through P_h. The identity (7) therefore implies that any quadric through all but one of the points $P_1, \ldots, P_r$ passes through the remaining one. Conversely, if $P_1, \ldots, P_r$ is a set of points such that any quadric through all but one of the points passes through the remaining one, a reversal of the argument shows that there exists a relation of the form (7).

Finally, we show that if there exists a relation of the form (7) with each λ_i different from zero, and not all points P_i lie in a prime, then $r \geqslant 2n+2$. Suppose indeed that $P_1, \ldots, P_{n+1}$ are linearly independent. Then

$$\sum_{i=1}^{n+1} \lambda_i (p_{i0} u_0 + \ldots + p_{in} u_n)^2 = 0$$

is a non-singular quadric. It is also given by the equation

$$\sum_{i=n+2}^{r} \lambda_i (p_{i0} u_0 + \ldots + p_{in} u_n)^2 = 0,$$

and if $r < 2n+2$, this quadric-envelope is necessarily singular, and so we have a contradiction. On the other hand, if $r = 2n+2$ we can find $P_1, \ldots, P_r$ for which there exists an identity of the form (7). Let Q_{n-1} be a non-singular quadric, $A_0, \ldots, A_n$ a self-polar simplex with regard to Q_{n-1}, where $A_i = (a_{i0}, \ldots, a_{in})$, and $B_0, \ldots, B_n$ is another self-polar simplex whose vertices are distinct from $A_0, \ldots, A_n$. We take the coordinates of B_i as $(b_{i0}, \ldots, b_{in})$. The equation of Q_{n-1} can be written in the form

$$\sum_{i=0}^{n} \lambda_i (a_{i0} u_0 + \ldots + a_{in} u_n)^2 = 0,$$

and in the form

$$\sum_{i=0}^{n} \mu_i (b_{i0} u_0 + \ldots + b_{in} u_n)^2 = 0.$$

Hence

$$\sum_{i=0}^{n} \lambda_i(a_{i0}u_0 + \ldots + a_{in}u_n)^2 \equiv \rho \sum_{i=0}^{n} \mu_i(b_{i0}u_0 + \ldots + b_{in}u_n)^2,$$

and therefore

$$\sum_{i=0}^{n} \lambda_i(a_{i0}u_0 + \ldots + a_{in}u_n)^2 + \sum_{i=0}^{n} -\rho\mu_i(b_{i0}u_0 + \ldots + b_{in}u_n)^2 \equiv 0,$$

which is an identity of the required form.

4. Linear spaces on a quadric, I. We have already seen [§ 2] that if y is any point of a quadric Q_{n-1}, and z is any point of Q_{n-1} which lies in the tangent prime to Q_{n-1} at y and is different from y, the line yz lies on Q_{n-1}. The tangent prime at y cuts Q_{n-1} in a quadric of dimension $n-2$, and hence, if $n > 2$, we can find a point z of the intersection different from y. Thus if $n > 2$, there are always lines of S_n contained in Q_{n-1}. On the other hand, we saw in § 3 that if S_k is a k-space contained in the non-singular quadric Q_{n-1}, the polar space of S_k is an S_{n-k-1} containing S_k, so that $2k \leqslant n-1$. In this and the following section we study the properties of the linear spaces which lie on a quadric. The present section is devoted to a study of the spaces of maximum dimension which lie on a non-singular quadric Q_{n-1} in S_n. We must consider separately the cases in which n is odd and in which n is even. We first consider the case n odd: $n = 2m+1$, where m is an integer.

In this case Q_{n-1} cannot contain a space of dimension greater than m. Let S_m be any m-space of S_n, and choose the coordinate system $(y_0, \ldots, y_n)$ in S_n so that S_m has the equations

$$y_0 = y_1 = \ldots = y_m = 0 \quad (m+1 = n-m).$$

If

$$f(y_0, \ldots, y_n) = 0$$

is the equation of any quadric through S_m, $f(0, \ldots, 0, y_{m+1}, \ldots, y_n)$ must be identically zero, and hence we must have

$$f(y_0, \ldots, y_n) = a_0(y)\, y_0 + \ldots + a_m(y)\, y_m,$$

where

$$a_i(y) = \sum_{j=0}^{n} a_{ij} y_j \quad (i = 0, \ldots, m),$$

and a_{ij} lies in K. Without loss of generality, we may assume

$a_{ij} = a_{ji}$ $(i, j = 0, \ldots, m)$. The quadric is non-singular if and only if the determinant

$$\begin{vmatrix} 2a_{00} & . & 2a_{0m} & a_{0\,m+1} & . & a_{0n} \\ . & . & . & . & . & . \\ 2a_{m0} & . & 2a_{mm} & a_{m\,m+1} & . & a_{mn} \\ a_{0\,m+1} & . & a_{m\,m+1} & 0 & . & 0 \\ . & . & . & . & . & . \\ a_{0n} & . & a_{mn} & 0 & . & 0 \end{vmatrix}$$

is not zero. This is true if and only if

$$\begin{vmatrix} a_{0\,m+1} & . & a_{0n} \\ . & . & . \\ a_{m\,m+1} & . & a_{mn} \end{vmatrix}$$

is not zero (since $n - m = m + 1$). Assuming that this determinant is not zero, we write

$$x_i = \sum_{j=0}^{n} a_{ij} y_j \quad (i = 0, \ldots, m).$$

It follows that $(x_0, \ldots, x_m, y_0, \ldots, y_m)$ can be chosen as a coordinate system in S_n, and referred to this system the quadric has the equation

$$x_0 y_0 + \ldots + x_m y_m = 0. \tag{1}$$

Since all non-singular quadrics in S_n are projectively equivalent, it follows that every non-singular quadric in S_n contains at least one m-space. Hence we have

THEOREM I. *If $n = 2m + 1$, every non-singular quadric in S_n contains at least one m-space. If S_m is an m-space on the non-singular quadric Q_{n-1}, the coordinate system in S_n can be chosen so that S_m has equations*

$$y_0 = \ldots = y_m = 0,$$

while Q_{n-1} has the equation (1).

We shall use the form (1) of the equation of Q_{n-1} to discuss the m-spaces which lie on it. Any m-space is given by a set of $(m+1)$ equations

$$\sum_{j=0}^{m} a_{ij} x_j + \sum_{j=0}^{m} b_{ij} y_j = 0 \quad (i = 0, \ldots, m), \tag{2}$$

where the $(m+1) \times (2m+2)$ matrix of coefficients is of rank $m+1$.

These equations can be solved to yield a set of equations

$$x_{i_r} = \sum_{s=0}^{k} \alpha_{i_r j_s} y_{j_s} + \sum_{s=k+1}^{m} \beta_{i_r i_s} x_{i_s} \quad (r = 0, \ldots, k), \tag{3}$$

$$y_{j_r} = \sum_{s=0}^{k} \gamma_{j_r j_s} y_{j_s} + \sum_{s=k+1}^{m} \delta_{j_r i_s} x_{i_s} \quad (r = k+1, \ldots, m),$$

where $(i_0, \ldots, i_m)$ and $(j_0, \ldots, j_m)$ are derangements of the integers $0, 1, \ldots, m$ (the $\delta_{j_r i_s}$ are *not* Kronecker deltas in this case). This is possible whenever the matrix of coefficients of $x_{i_0}, \ldots, x_{i_k}, y_{j_{k+1}}, \ldots, y_{j_m}$ in (2) is non-singular, and in general there are several ways of choosing $i_0, \ldots, i_k, j_{k+1}, \ldots, j_m$ to satisfy this requirement. We first show that if the m-space given by (2) lies on Q_{n-1}, then we can choose $(i_0, \ldots, i_m)$, $(j_0, \ldots, j_m)$ so that $i_r = j_r$ $(r = 0, \ldots, m)$.

Suppose that in (3) we have one of the numbers, $i_0, \ldots, i_k$ equal to one of the numbers $j_{k+1}, \ldots, j_m$, say $i_0 = j_{k+1}$. Then not all the numbers $i_{k+1}, \ldots, i_m$ are distinct from each of the numbers $j_0, \ldots, j_k$. We suppose, for instance, that $j_0 = i_{k+1}$. If $\alpha_{i_0 j_0} \neq 0$, we can solve the first of the equations (3) to express y_{j_0} in terms of $x_{i_0}, y_{j_1}, \ldots, y_{j_k}, x_{i_{k+1}}, \ldots, x_{i_m}$. Hence we can replace equations (3) by equations expressing $x_{i_1}, \ldots, x_{i_k}, y_{j_0}, y_{j_{k+1}}, \ldots, y_{j_m}$ in terms of $y_{j_1}, \ldots, y_{j_k}, x_{i_0}, x_{i_{k+1}}, \ldots, x_{i_m}$. In these new equations the number of integers common to the sets $i_1, \ldots, i_k$ and $j_0, j_{k+1}, \ldots, j_m$ is reduced by one. If $\alpha_{i_0 j_0} = 0$, but $\delta_{j_{k+1} i_{k+1}} \neq 0$, we can follow an analogous procedure to transfer $x_{i_{k+1}}$ to the left-hand side of (3) and $y_{j_{k+1}}$ to the right-hand side, so reducing the number of x's and y's on the left-hand side of (3) which have the same suffix. A similar argument holds if $\beta_{i_0 i_{k+1}}$ or $\gamma_{j_{k+1} j_0}$ is different from zero.

We proceed in this way to reduce the number of x's and y's on the left-hand side of (3) which have a common suffix. Eventually we obtain a set of equations which we write in an obvious matrix notation as

$$\begin{pmatrix} x_a \\ x_c \\ y_b \\ y_c \end{pmatrix} = \begin{pmatrix} \alpha_{11} & \alpha_{12} & \alpha_{13} & \alpha_{14} \\ \alpha_{21} & 0 & \alpha_{23} & 0 \\ \alpha_{31} & \alpha_{32} & \alpha_{33} & \alpha_{34} \\ \alpha_{41} & 0 & \alpha_{43} & 0 \end{pmatrix} \begin{pmatrix} y_a \\ y_d \\ x_b \\ x_d \end{pmatrix}, \tag{4}$$

where a, b, c, d denote distinct sets of integers which, taken together, form a derangement of $0, 1, \ldots, m$. Further reduction of the type described above is no longer possible. Suppose now that the m-space given by (4) lies on the quadric given by (1). This condition can be written in matrix form

$$y'_a x_a + y'_b x_b + y'_c x_c + y'_d x_d = 0.$$

When we use (4) to eliminate x_a, x_c, y_b, y_c from this equation, the result must be an identity in y_a, y_d, x_b, x_c. On substitution, the left-hand side of the equation (1) becomes

$$y'_a(\alpha_{11}y_a + \alpha_{12}y_d + \alpha_{13}x_b + \alpha_{14}x_d) + x'_b(\alpha_{31}y_a + \alpha_{32}y_d + \alpha_{33}x_b + \alpha_{34}x_d) + (y'_a\alpha'_{41} + x'_b\alpha'_{43})(\alpha_{21}y_a + \alpha_{23}x_b) + y'_d x_d.$$

However the elements of the matrices α_{ij} are chosen, this expression can only reduce to zero if there are no elements in the sum $y'_d x_d$. Hence the suffixes in the sets a and b must be a derangement of the set $0, 1, \ldots, m$. It follows also that the set c of suffixes is empty. Hence if (2) is any S_m on (1), its equation can be written in the form

$$\begin{pmatrix} x_{i_0} \\ \vdots \\ x_{i_r} \\ y_{i_{r+1}} \\ \vdots \\ y_{i_m} \end{pmatrix} = \begin{pmatrix} A & B \\ & \\ C & D \end{pmatrix} \begin{pmatrix} y_{i_0} \\ \vdots \\ y_{i_r} \\ x_{i_{r+1}} \\ \vdots \\ x_{i_m} \end{pmatrix}, \tag{5}$$

where $i_0, \ldots, i_m$ is a derangement of $0, \ldots, m$. When we substitute from these equations in (1), as above, we find at once that the equations are satisfied identically if and only if the matrix

$$\begin{pmatrix} A & B \\ C & D \end{pmatrix}$$

is skew-symmetric.

Let P be a skew-symmetric $(m+1)\times(m+1)$ matrix, whose elements above the principal diagonal are independent indeterminates. The m-space given by the equations

$$\begin{pmatrix} x_{i_0} \\ \vdots \\ x_{i_r} \\ y_{i_{r+1}} \\ \vdots \\ y_{i_m} \end{pmatrix} = P \begin{pmatrix} y_{i_0} \\ \vdots \\ y_{i_r} \\ x_{i_{r+1}} \\ \vdots \\ x_{i_m} \end{pmatrix} \tag{6}$$

lies on Q_{n-1}, and is a generic member of a system $\mathfrak{S}_{i_0\dots i_r;\, i_{r+1}\dots i_m}$ of m-spaces of which the m-space given by (5) is a member. Since the equations of (6) contain $\frac{1}{2}m(m+1)$ independent indeterminates, we readily deduce that the dimension of $\mathfrak{S}_{i_0\dots i_r;\, i_{r+1}\dots i_m}$ is $\frac{1}{2}m(m+1)$.

We thus obtain altogether 2^{m+1} systems $\mathfrak{S}_{i_0\dots i_r;\, i_{r+1}\dots i_m}$, such that any S_m on Q_{n-1} belongs to at least one of them. However, we show that the systems are not all distinct, and indeed they can be divided into two sets, all the systems in the same set being identical. Let (6) be a generic member of the system $\mathfrak{S}_{i_0\dots i_r;\, i_{r+1}\dots i_m}$, and consider another system $\mathfrak{S}_{j_0\dots j_s;\, j_{s+1}\dots j_m}$ such that $m-1$ of the coordinates $x_{i_0}, \dots, x_{i_r}, y_{i_{r+1}}, \dots, y_{i_m}$ are in the set $x_{j_0}, \dots, x_{j_s}, y_{j_{s+1}}, \dots, y_{j_m}$. We prove that the m-space (6) belongs to $\mathfrak{S}_{j_0\dots j_s;\, j_{s+1}\dots j_m}$.

Let us suppose, for instance, that x_i, x_j are the two coordinates in the set $x_{i_0}, \dots, x_{i_r}, y_{i_{r+1}}, \dots, y_{i_m}$ which are not in the set $x_{j_0}, \dots, x_{j_s}$, $y_{j_{s+1}}, \dots, y_{j_m}$; then we obtain this latter set by writing y_i, y_j for x_i, x_j in the former. Since the matrix P in (6) is skew and its elements above the principal diagonal are independent indeterminates, the elements p_{ij} and p_{ji} are not zero. We can therefore apply the method described earlier in this section, first to change x_i to the right-hand side of (6) and y_j to the left, and then x_j to the right and y_i to the left. The equations of (6) are then of the form

$$\begin{pmatrix} x_{j_0} \\ \vdots \\ x_{j_s} \\ y_{j_{s+1}} \\ \vdots \\ y_{j_m} \end{pmatrix} = R \begin{pmatrix} y_{j_0} \\ \vdots \\ y_{j_s} \\ x_{j_{s+1}} \\ \vdots \\ x_{j_m} \end{pmatrix},$$

and since (6) is on Q_{n-1}, R must be skew. A similar argument holds if the coordinates in $x_{i_0}, \dots, x_{i_r}, y_{i_{r+1}}, \dots, y_{i_m}$ not in the set $x_{j_0}, \dots, x_{j_s}$, $y_{j_{s+1}}, \dots, y_{j_m}$ are y_i, y_j or x_i, y_j. Hence in all cases (6) is a proper specialisation of the generic member of $\mathfrak{S}_{j_0\dots j_s;\, j_{s+1}\dots j_m}$. Since, however, (6) is a generic member of $\mathfrak{S}_{i_0\dots i_r;\, i_{r+1}\dots i_m}$, we have

$$\mathfrak{S}_{i_0\dots i_r;\, i_{r+1}\dots i_m} \subseteq \mathfrak{S}_{j_0\dots j_s;\, j_{s+1}\dots j_m}.$$

Similarly, we prove that

$$\mathfrak{S}_{j_0\dots j_s;\, j_{s+1}\dots j_m} \subseteq \mathfrak{S}_{i_0\dots i_r;\, i_{r+1}\dots i_m},$$

and hence we deduce that

$$\mathfrak{S}_{i_0 \ldots i_r;\, i_{r+1} \ldots i_m} = \mathfrak{S}_{j_0 \ldots j_s;\, j_{s+1} \ldots j_m}.$$

Applying this result repeatedly, we find that any system $\mathfrak{S}_{i_0 \ldots i_r;\, i_{r+1} \ldots i_m}$ is equal either to the system $\mathfrak{S}$ or to the system $\mathfrak{S}'$, where

$$\mathfrak{S} = \mathfrak{S}_{;01 \ldots m}, \quad \mathfrak{S}' = \mathfrak{S}_{m;0 \ldots m-1}.$$

We have not yet proved that $\mathfrak{S}$ and $\mathfrak{S}'$ are distinct systems, but this will be proved immediately. Assuming this result, we have

THEOREM II. *On a non-singular quadric Q_{2m} there exist two systems of m-spaces, each irreducible and of dimension $\frac{1}{2}m(m+1)$, and every m-space lying on Q_{2m} belongs to one or the other of them, but not to both.*

The fact that the systems $\mathfrak{S}$ and $\mathfrak{S}'$ are distinct, and in fact have no m-space in common, follows from a study of the intersection of two m-spaces lying on Q_{2m}. Let π be any m-space on Q_{2m}. Then, by Theorem I, we can choose the coordinate system in S_n so that π has the equation

$$y_0 = y_1 = \ldots = y_m = 0,$$

and Q_{2m} has the equation

$$x_0 y_0 + \ldots + x_m y_m = 0.$$

If the coordinates of π lie in the ground field K, the transformation of coordinates necessary to reduce the equations of S_m and of Q_{2m} is a transformation over K; but if the coordinates of π are transcendental over K Theorem I is still valid, provided we adjoin the coordinates of π to the ground field, since Theorem I does not depend on K being algebraically closed. In particular, π may be taken as a generic member of either system on Q_{2m}; we note, however, that over the extended ground field π is no longer generic, and that in the new coordinate system we have renamed the system to which π belongs so that it is the system $\mathfrak{S}$.

Now consider any other m-space on Q_{2m}. In the new coordinate system its equations can be written in the form

$$\begin{pmatrix} x_{i_0} \\ \vdots \\ x_{i_r} \\ y_{i_{r+1}} \\ \vdots \\ y_{i_m} \end{pmatrix} = \begin{pmatrix} A & B \\ & \\ C & D \end{pmatrix} \begin{pmatrix} y_{i_0} \\ \vdots \\ y_{i_r} \\ x_{i_{r+1}} \\ \vdots \\ x_{i_m} \end{pmatrix}, \qquad (5)$$

where $i_0, \ldots, i_m$ is a derangement of $0, \ldots, m$, and A, B, C, D are matrices of $r+1$, $r+1$, $m-r$, $m-r$ rows and $r+1$, $m-r$, $r+1$, $m-r$ columns, respectively, while the matrix

$$\begin{pmatrix} A & B \\ C & D \end{pmatrix} \tag{7}$$

is skew. If r is odd, (5) belongs to the system $\mathfrak{S}$, and if r is even it belongs to $\mathfrak{S}'$.

We consider the intersection of π with (5). It is given by

$$y_0 = \ldots = y_m = 0,$$

$$\begin{pmatrix} x_{i_0} \\ \vdots \\ x_{i_r} \end{pmatrix} = B \begin{pmatrix} x_{i_{r+1}} \\ \vdots \\ x_{i_m} \end{pmatrix},$$

$$D \begin{pmatrix} x_{i_{r+1}} \\ \vdots \\ x_{i_m} \end{pmatrix} = 0.$$

It is clear that if the rank of D is ρ, these equations define a space of dimension $m-r-\rho-1$. Now D is skew; hence its rank is even, say $\rho = 2\sigma$.

Case (*i*). Suppose that m is odd. If the m-space given by (5) belongs to the same system as π, r is odd, and hence $m-r-2\sigma-1$ is odd. Hence if the space given by (5) and π have any points in common they meet in a space of an odd number of dimensions. In the case considered, D is skew-symmetric of order $m-r$, which is even; hence there can exist spaces given by (5) for which $2\sigma = m-r$, and therefore there exist m-spaces belonging to the same system as π which do not meet it. In particular, if π is a generic member of the system (with respect to K), and the space given by (5) is an independent generic member of the same system, the matrix (7) may be taken as a skew-symmetric matrix of indeterminates over the field obtained by adjoining the coefficients of π to K. Then D is non-singular. Hence two independent generic members of the same system do not meet.

Now suppose that the m-space given by (5) belongs to the opposite system to π. Then r is even, and hence $m-r-2\sigma-1$ is even. Hence the space (5) and π meet in a space of an even number of dimensions, and in particular always have at least one point in common.

In particular, if the space (5) is a generic member of the system $\mathfrak{S}'$ independent of π, then the space (5) and π meet in a single point.

Case (ii), m even. Then $m-r-2\sigma-1$ is odd or even according as the m-space given by (5) belongs to the opposite or to the same system as π. By considerations similar to the above, we see that two independent generic m-spaces on Q_{2m} belonging to the same system have one point in common, and independent generic m-spaces of opposite systems do not meet.

From these properties it is clear that $\mathfrak{S}$ and $\mathfrak{S}'$ are indeed distinct systems, and have no m-spaces in common. Summing up, we have

THEOREM III. *If m is odd, two independent generic m-spaces of the same system on Q_{2m} do not meet, and any two m-spaces of the same system which meet do so in a space of an odd number of dimensions. Two independent generic m-spaces of opposite systems meet in a point, and any two m-spaces of opposite systems meet in a space of an even number of dimensions.*

If m is even, two independent generic m-spaces of the same system on Q_{2m} meet in a point, and any two m-spaces of the same system meet in a space of an even number of dimensions. Two independent generic m-spaces of opposite systems do not meet, and any two m-spaces of opposite systems meet in a space of an odd number of dimensions.

We now consider briefly how to find the m-spaces on a non-singular quadric Q_{2m} when the equation of Q_{2m} is given in the form

$$\sum_{i,j=0}^{2m+1} a_{ij} x_i x_j = 0.$$

An m-space S_m lies on Q_{2m} if and only if it is its own polar space with regard to Q_{2m} [§ 3, Th. I]. Now if $p = (\ldots, p_{i_0 \ldots i_m}, \ldots)$ is an m-space of S_{2m+1}, its polar space p' has dual coordinates $(\ldots, p^{i_0 \ldots i_m}, \ldots)$ where [p. 218]

$$p^{i_0 \ldots i_m} = \sum_k \begin{vmatrix} a_{i_0 k_0} & \cdot & a_{i_0 k_m} \\ \cdot & \cdot & \cdot \\ a_{i_m k_0} & \cdot & a_{i_m k_m} \end{vmatrix} p_{k_0 \ldots k_m}.$$

Hence p is its own polar space with regard to Q_{2m} if and only if there exists an element ρ of K such that

$$\rho p_{i_{m+1} \ldots i_{2m}} = \sum_k \begin{vmatrix} a_{i_0 k_0} & \cdot & a_{i_0 k_m} \\ \cdot & \cdot & \cdot \\ a_{i_m k_0} & & a_{i_m k_m} \end{vmatrix} p_{k_0 \ldots k_m},$$

where $i_0, \ldots, i_{2m}$ is an even derangement of $0, \ldots, 2m$. This holds for all even derangements. From this equation we deduce that

$$(-1)^{m+1}\rho p_{k_0\ldots k_m} = \sum_j \begin{vmatrix} a_{k_{m+1}j_0} & \cdot & a_{k_{m+1}j_m} \\ \cdot & \cdot & \cdot \\ a_{k_{2m}j_0} & \cdot & a_{k_{2m}j_m} \end{vmatrix} p_{j_0\ldots j_m},$$

and on substituting this expression for $p_{k_0\ldots k_m}$ in the first equation we obtain the equation

$$(-1)^{m+1}\rho^2 p_{i_{m+1}\ldots i_{2m}} = |a|\, p_{i_{m+1}\ldots i_{2m}},$$

for all choices of $i_{m+1}, \ldots, i_{2m}$, where $|a|$ is the determinant of the matrix a. Hence $\rho = \pm[(-1)^{m+1}|a|]^{\frac{1}{2}}$. Denoting the two values of ρ given here by ρ_1 and ρ_2, it follows that $p = (\ldots, p_{i_0\ldots i_m}, \ldots)$ is an m-space on Q_{2m} if and only if it satisfies one of the systems of equations

$$\rho_1 p_{i_{m+1}\ldots i_{2m}} = \sum_k \begin{vmatrix} a_{i_0k_0} & \cdot & a_{i_0k_m} \\ \cdot & \cdot & \cdot \\ a_{i_mk_0} & \cdot & a_{i_mk_m} \end{vmatrix} p_{k_0\ldots k_m},$$

and

$$\rho_2 p_{i_{m+1}\ldots i_{2m}} = \sum_k \begin{vmatrix} a_{i_0k_0} & \cdot & a_{i_0k_m} \\ \cdot & \cdot & \cdot \\ a_{i_mk_0} & \cdot & a_{i_mk_m} \end{vmatrix} p_{k_0\ldots k_m}.$$

No m-space can satisfy both sets of equations, since if p did, we could deduce that

$$(\rho_1 - \rho_2)\, p_{i_{m+1}\ldots i_{2m}} = 0,$$

that is, $p_{i_{m+1}\ldots i_{2m}}$ is zero for all choices of the suffixes. A transformation of coordinates which reduces the equations of the quadric to the form

$$y_0 y_1 + y_2 y_3 + \ldots + y_{2m} y_{2m+1} = 0$$

shows that the two sets of equations define the two systems of m-spaces on Q_{2m}.

It is possible to use these equations to obtain the properties of the two systems of m-spaces on Q_{2m}, but this method of investigation is more difficult. As an illustration, we show that two m-spaces of the same system meet if m is even and in general do not meet if m is odd. Let p, q belong to the first system. They meet if

$$\sum_i (p_{i_0\ldots i_m} q_{i_{m+1}\ldots i_{2m}} + q_{i_0\ldots i_m} p_{i_{m+1}\ldots i_{2m}}) = 0,$$

that is if

$$\rho_1^{-1}\sum_i\left[(-1)^{m+1}\sum_j\begin{vmatrix} a_{i_{m+1}j_0} & . & a_{i_{m+1}j_m} \\ . & . & . \\ a_{i_{2m}j_0} & . & a_{i_{2m}j_m}\end{vmatrix} p_{j_0\dots j_m}q_{i_{m+1}\dots i_{2m}} + \sum_j\begin{vmatrix} a_{i_0j_0} & . & a_{i_0j_m} \\ . & . & . \\ a_{i_mj_0} & . & a_{i_mj_m}\end{vmatrix} p_{j_0\dots j_m}q_{i_0\dots i_m}\right] = 0,$$

and this is clearly the case when m is even, but is not the case, in general, when m is odd.

An immediate corollary of Theorem III tells us that if two m-spaces on Q_{2m} meet in an $(m-1)$-space, they necessarily belong to opposite systems. Now consider any $(m-1)$-space S_{m-1} on Q_{2m}. The tangent prime at any point of S_{m-1} to Q_{2m} meets Q_{2m} in a quadric having the point as vertex and containing S_{m-1}. From this it follows that the polar S_{m+1} of S_{m-1} with respect to Q_{2m} meets Q_{2m} in a quadric having S_{m-1} as vertex. The section is a quadric of dimension m and type m; hence it consists of a pair of m-spaces, each lying on Q_{2m} and meeting in S_{m-1}. One of these must belong to each system of m-spaces on Q_{2m}.

This result enables us to study the spaces of maximum dimension lying on a non-singular quadric Q_{2m-1} in S_{2m}. It can be verified at once that the quadric in S_{2m}

$$x_0^2+x_1y_1+\dots+x_my_m = 0 \tag{8}$$

is non-singular. Since any two non-singular quadrics of the same dimension are projectively equivalent, we need only study the spaces of maximum dimension on the quadric (8). This, however, is the section of the quadric (1) by the prime

$$x_0 = y_0. \tag{9}$$

There are no spaces on (8) of dimension greater than $m-1$, since $\dim Q_{2m-1} = 2m-1$. On the other hand, if S_m is any m-space on (1), it meets the prime (9) in an $(m-1)$-space (since it cannot lie in it, the section of (1) by (9) not containing any m-spaces). Hence the quadric (8) contains $(m-1)$-spaces. Any $(m-1)$-space on the quadric (8) is also an $(m-1)$-space on the quadric (1), and through it there passes one m-space of each system of (1). The $(m-1)$-spaces of (8) are therefore in one-to-one correspondence, without exception, with the m-spaces of either system on (1). From this, we easily deduce by the theory of correspondences that the $(m-1)$-spaces of the quadric (8) form an irreducible system of dimension $\frac{1}{2}m(m+1)$, and that the $(m-1)$-space

$$y_i + \alpha_i x_0 = \tau_{i1}x_1 + \ldots + \tau_{im}x_m \quad (i = 1, \ldots, m),$$
$$x_0 = \alpha_1 x_1 + \ldots + \alpha_m x_m,$$

where $\tau_{ij} = -\tau_{ji}$, and the τ_{ij} $(i<j)$ and $\alpha_1, \ldots, \alpha_m$ are independent indeterminates, is a generic $(m-1)$-space of (8). Thus we have

THEOREM IV. *On a quadric Q_{2m-1}, the spaces of dimension $m-1$ form a single irreducible system of dimension $\frac{1}{2}m(m+1)$.*

We conclude this section with some remarks on the case $n = 3$. In this case Q_{n-1} is a quadric surface, and we assume that the reader is familiar with its properties. In the notation used in this section, the equation of Q_{n-1} is

$$x_0 y_0 + x_1 y_1 = 0,$$

and the equations of any line (generator) of it of the system $\mathfrak{S}$ are

$$ay_0 = bx_1, \quad ay_1 = -bx_0,$$

while the equations of any generator of the system $\mathfrak{S}'$ are

$$cx_1 = dx_0, \quad cy_0 = -dy_1.$$

If (x_0', x_1', y_0', y_1') is any point of Q_2, the equations

$$ay_0' = bx_1', \quad ay_1' = -bx_0'$$

determine the ratio $a:b$ uniquely, since $x_0'y_0' + x_1'y_1' = 0$, and not all of the coordinates x_0', x_1', y_0', y_1' are zero. Thus one generator of the system $\mathfrak{S}$ passes through the point, and, similarly, one generator of the system $\mathfrak{S}'$ passes through the point. Conversely, given $a:b$ and $c:d$, we obtain a unique point of the Q_2 common to the two generators, namely, $(ac, ad, bd, -bc)$. Representing the ratio $a:b$ as the point (a, b) of the line S_1, and the ratio $c:d$ as the point (c, d) of the line S_1', we see at once that Q_2 is in one-to-one correspondence, without exception, with the points of the two-way space S_{11}, and indeed is the Segre variety associated with this space (referred to a system of coordinates obtained from those used in Chapter XI, § 2, by an elementary transformation). The generators of the system $\mathfrak{S}$ represent the point-pairs of S_1, S_1' in which the point on S_1 is kept fixed and the point on S_1' varies, and similarly the generators of the system $\mathfrak{S}'$ correspond to the point-pairs in which the point on S_1' is kept fixed.

5. Linear spaces on a quadric, II. It is a simple matter to deduce from the results of § 4 the properties of the spaces of maximum dimension on a quadric Q_{n-1} of type ρ in S_n. If S_k is any k-space lying on Q_{n-1}, the line joining any point of the vertex of Q_{n-1} to any point of S_k not in the vertex lies on Q_{n-1}. Hence the join of S_k and the vertex (that is, the smallest space containing both) lies on Q_{n-1}. Therefore, in looking for the spaces of highest dimension on Q_{n-1}, we need only consider the spaces which pass through the vertex. Now if $S_{n-\rho}$ is an $(n-\rho)$-space not meeting the vertex, it meets Q_{n-1} in a non-singular quadric $Q_{n-1-\rho}$, and Q_{n-1} is the locus of points in the ρ-spaces joining the vertex to points of $Q_{n-1-\rho}$. It follows that the spaces of maximum dimension on Q_{n-1} are the $(\rho+m_0)$-spaces joining the vertex to the spaces of maximum dimension in $Q_{n-1-\rho}$, which are of dimension m_0, where m_0 is the integer part of $\frac{1}{2}(n-1-\rho)$. If $n-1-\rho$ is even there are two systems of m_0-spaces on $Q_{n-1-\rho}$, and hence two systems of $(\rho+m_0)$-spaces on Q_{n-1}. The intersection properties of these systems can be determined at once from the properties for non-singular quadrics proved in § 4.

With these preliminaries, we are able to investigate the d-spaces on a non-singular quadric Q_{n-1} for any value of d less than the maximum value. Let S_d be any d-space lying on Q_{n-1}, and let S_{n-d-1} be its polar with respect to Q_{n-1}. Then [p. 217] S_{n-d-1} touches Q_{n-1} along S_d, and cuts Q_{n-1} in a quadric Q_{n-d-2} which is of type $d+1$ having S_d as vertex.

Case (*i*) $n = 2m+1$. Then $n-d-2 = 2m-d-1$. Now a quadric of dimension $2m-d-1$ of type $d+1$ is the join of a d-space (the vertex) to a non-singular quadric of dimension $2m-2d-2$. Hence the spaces of maximum dimension on Q_{n-d-2} are of dimension $m-d-1+d+1 = m$, and form two systems, each of dimension $\frac{1}{2}(m-d)(m-d-1)$. By considering the intersection properties of the m-spaces on Q_{n-d-1}, we see that the two systems belong to opposite systems on Q_{n-1}. Moreover, any m-space of Q_{n-1} which passes through S_d lies in the tangent space at every point of S_d, and hence lies in S_{n-d-1}, as well as on Q_{n-1}. Hence it lies on Q_{n-d-1}. Thus each d-space on Q_{n-1} lies in an m-space of each system of Q_{n-1}, and the m-spaces through any d-space of Q_{n-1} form two irreducible systems of dimension $\frac{1}{2}(m-d)(m-d-1)$.

Now let ξ be a generic m-space of either system $\mathfrak{S}$ of m-spaces on Q_{n-1}, and let η be a generic d-space contained in ξ. We consider the correspondence whose generic point is (ξ, η). Since $K(\xi)$ is of dimension $\frac{1}{2}m(m+1)$, the object-variety is of this dimension. To the generic point ξ of the object-variety there corresponds a variety of dimension $(d+1)(m-d)$, since the system of d-spaces in an m-space is of this dimension. Hence the dimension of the correspondence is $\frac{1}{2}m(m+1)+(d+1)(m-d)$. The image-variety of the correspondence has generic point η. Let its dimension be ρ_d. We first show that any d-space on Q_{n-1} is a proper specialisation of η. If y' is a d-space on Q_{n-1}, we have seen that through y' there passes an irreducible system of dimension $\frac{1}{2}(m-d)(m-d-1)$ of m-spaces belonging to $\mathfrak{S}$. Let x' be any m-space of $\mathfrak{S}$ on Q_{n-1}, through y'. Then x' is a proper specialisation of ξ, and since y' is contained in x', (x', y') is a proper specialisation of (ξ, η), and hence y' is a proper specialisation of η. Since η is contained in ξ, and hence lies on Q_{n-1}, any proper specialisation of η is a d-space on Q_{n-1}. Thus the d-spaces of Q_{n-1} form a single irreducible system of which η is a generic member. The dimension of the system of d-spaces on Q_{n-1} is thus ρ_d. Now, as we have seen, the m-spaces of $\mathfrak{S}$ through η form a system of dimension $\frac{1}{2}(m-d)(m-d-1)$. Hence, by the Principle of Counting Constants [XI, § 4],

$$\rho_d+\tfrac{1}{2}(m-d)(m-d-1) = \tfrac{1}{2}m(m+1)+(d+1)(m-d),$$

that is,

$$\rho_d = \tfrac{1}{2}(d+1)(4m-3d) = \tfrac{1}{2}(d+1)(2n-2-3d).$$

Case (ii) $n = 2m$. The procedure in this case is similar, and we need only summarise the results. If S_d lies on Q_{n-1}, the section of Q_{n-1} by its polar space cuts Q_{n-1} in a quadric of dimension $2m-d-2$ and of type $d+1$. In this case it contains a single system of $(m-1)$-spaces of dimension $\frac{1}{2}(m-d-1)(m-d)$, and these are all the m-spaces of Q_{n-1} through S_d.

Considering the correspondence between the $(m-1)$-spaces on Q_{n-1} and the d-spaces contained in them, we find, as before, the dimension ρ_d of the system of d-spaces on Q_{n-1}, obtaining the formula

$$\rho_d = \tfrac{1}{2}(d+1)(4m-3d-2) = \tfrac{1}{2}(d+1)(2n-2-3d).$$

Hence in either case we have

THEOREM I. *If Q_{n-1} is a non-singular quadric, and d is less than the maximum dimension of linear spaces contained in Q_{n-1}, the d-spaces of Q_{n-1} form a single irreducible system of dimension $\frac{1}{2}(d+1)(2n-2-3d)$. Every d-space on Q_{n-1} is contained in a space of higher dimension contained in Q_{n-1}.*

From this result we can immediately obtain similar properties for d-spaces, which contain the vertex, of a quadric Q_{n-1} of type ρ. To extend this to d-spaces not containing the vertex we show again that these are all contained in the spaces of maximum dimension on Q_{n-1}, but if the spaces of maximum dimension fall into two irreducible systems the d-spaces may not all belong to a single irreducible system. The details are easily worked out, and we do not give them here.

6. The subvarieties of a quadric. In Chapter XII, § 11, we introduced the idea of a base for the subvarieties of dimension r on a variety U of dimension d. A base is a set of multiplicative varieties $V_1, \ldots, V_k$ of dimension r such that, given any multiplicative variety V of dimension r on U, there exists a virtual equivalence

$$aV \sim a_1 V_1 + \ldots + a_k V_k \quad (a \neq 0).$$

We did not prove the existence of a base for varieties of dimension r on a general variety U, but we proved that if $U = S_n$, then a base for the varieties of dimension r on U consists of an r-space; we also found a base on the Segre varieties. In this section we find a base for the varieties of dimension d on a quadric Q_{n-1}. We shall confine ourselves to non-singular quadrics; the methods used are applicable to singular quadrics, but in applications the singularities introduce difficulties, and the general theory of the base is of most significance in the case of non-singular varieties. We shall prove

THEOREM I. *On a non-singular quadric Q_{n-1} there exists a base for the varieties of dimension d $(0 \leqslant d \leqslant n-2)$. If n is odd and $d > \frac{1}{2}(n-1)$ a section of Q_{n-1} by a linear space of dimension $d+1$ constitutes a base; if $d = \frac{1}{2}(n-1)$ a base is constituted by a $\frac{1}{2}(n-1)$-space of each of the two systems of spaces of this dimension on Q_{n-1}; and if $d < \frac{1}{2}(n-1)$ a base is constituted by a d-space on Q_{n-1}. If n is even, and $d > \frac{1}{2}n-1$, a section of Q_{n-1} by a linear space of dimension $d+1$ constitutes a base, and if $d \leqslant \frac{1}{2}n-1$, a d-space on Q_{n-1} constitutes a base.*

The argument is by induction on the dimension of Q_{n-1}, the theorem being proved for Q_{n-1} on the assumption that it is true for Q_{n-3}. There is nothing to prove in the case of a conic Q_1, since any subvariety of Q_1 is already a set of points. The theorem has been proved for Q_2 [Ch. XII, § 11], since Q_2 is a Segre variety. The method which we shall follow will, however, enable us to deduce the theorem for Q_2 from a theorem for Q_0. It is true that Q_0 is reducible, but the theorem remains true (but trivial) in this case.

Let us consider any irreducible variety V of dimension d and order g on Q_{n-1}. Since there is nothing to prove if $d = 0$, we shall assume $0 < d < n-1$. We select a point A on Q_{n-1}, but not on V, such that the tangent prime to Q_{n-1} at A does not contain V, and also a point B not conjugate to A, at which the tangent prime to Q_{n-1} does not contain V; and we choose the coordinate system so that A is the point $(1, 0, \ldots, 0)$, B is $(0, 0, \ldots, 0, 1)$, and Q_{n-1} has the equation

$$\omega \equiv x_0 x_n - x_1^2 - \ldots - x_{n-1}^2 = 0. \tag{1}$$

Let $f(x_0, \ldots, x_n)$ be any homogeneous polynomial in $x_0, \ldots, x_n$, and regard it as a polynomial in x_0, x_n with coefficients in $K[x_1, \ldots, x_{n-1}]$, so that

$$f(x) = \sum_{i,j} a_{ij}(x_1, \ldots, x_{n-1})\, x_0^i x_n^j.$$

Consider any term in this in which i, j are both greater than zero. If $i \leqslant j$, we write

$$x_0^i x_n^j = x_n^{j-i}(\omega + x_1^2 + \ldots + x_{n-1}^2)^i,$$

and if $i > j$ we write

$$x_0^i x_n^j = x_0^{i-j}(\omega + x_1^2 + \ldots + x_{n-1}^2)^j.$$

It follows that we can write

$$f(x) = a(x_1, \ldots, x_{n-1}) + \sum_{i>0} b_i(x_1, \ldots, x_{n-1})\, x_n^i$$
$$+ \sum_{i>0} c_i(x_1, \ldots, x_{n-1})\, x_0^i + d(x_0, \ldots, x_n)\, \omega.$$

In what follows we shall assume that all the polynomials considered are written in this form.

Using this result, we write a basis for the equations of V in the form

$$\omega = 0, \tag{1}$$

$$a_i(x_1, \ldots, x_{n-1}) + \sum_{j>0} b_{ij}(x_1, \ldots, x_{n-1})\, x_n^j = 0 \quad (i = 1, \ldots, r), \tag{2}$$

$$a_i(x_1, \ldots, x_{n-1}) + \sum_{j>0} b_{ij}(x_1, \ldots, x_{n-1})\, x_n^j + \sum_{j>0} c_{ij}(x_1, \ldots, x_{n-1})\, x_0^j = 0$$
$$(i = r+1, \ldots, s), \tag{3}$$

where no combination of the equations (3) is independent of x_0.

We note that since V does not pass through $(1, 0, \dots, 0)$, there exists at least one form f of degree m:

$$f(x_0, \dots, x_n) = x_0^m + x_0^{m-1} f_1(x_1, \dots, x_{n-1}) + \dots,$$

which vanishes on V. Hence, among the equations (3) there exists at least one in which the highest power of x_0 has a non-zero element of K as coefficient. We suppose that the last equation of the set has this property.

We may assume that in equations (2) every $a_i(x_1, \dots, x_{n-1})$ is different from zero. For if, say, $a_1(x_1, \dots, x_{n-1}) = 0$, then

$$\sum_{j>0} b_{1j}(x_1, \dots, x_{n-1}) x_n^j = 0.$$

Since $x_n = 0$ does not contain any of the irreducible components of V, it follows that

$$\sum_{j>0} b_{1j}(x_1, \dots, x_{n-1}) x_n^{j-1} = 0$$

on V, and we can replace the first equation (2) by one of lower degree. Proceeding in this way until we obtain a form which does not vanish when we put $x_n = 0$, we see that our assumption can be justified.

By our choice of the equations (1), (2), (3), the resultant system of the equations obtained by eliminating x_0 from the equations of V is contained in (2). Since $(1, 0, \dots, 0)$ is not on V, it follows that the equations (2) have the property that $(\bar{x}_1, \dots, \bar{x}_n)$ satisfies them if and only if we can find $\bar{x}_0$ such that $(\bar{x}_0, \bar{x}_1, \dots, \bar{x}_n)$ lies on V.

Let $\xi = (\xi_0, \dots, \xi_n)$ be a generic point of V. Since V is not in $x_n = 0$, we may assume that ξ is normalised so that $\xi_n = 1$. Then since V satisfies $\omega = 0$,

$$\xi_0 = \xi_1^2 + \dots + \xi_{n-1}^2,$$

and therefore $K(\xi_1, \dots, \xi_{n-1}) = K(\xi_0, \dots, \xi_{n-1})$ is the function field of V, and hence it is of dimension d over K. Let η_0 be an indeterminate over $K(\xi_1, \dots, \xi_{n-1})$, and let $\eta = (\eta_0, \xi_1, \dots, \xi_{n-1}, 1)$. η is then the generic point of a variety of dimension $d+1$. Since ξ satisfies the equations (2), so does η. Conversely, if

$$g(x_0, \dots, x_n) = \sum_{i=0}^{t} g_i(x_1, \dots, x_n) x_0^i$$

is any equation satisfied by η,

$$\Sigma g_i(\xi_1, \dots, \xi_{n-1}, 1) \eta_0^i = 0,$$

and therefore $$g_i(\xi_1, \dots, \xi_{n-1}, 1) = 0.$$

$g_i(x_1, \ldots, x_n)$ therefore vanishes at every point of V, and since it is independent of x_0 it therefore vanishes at every solution of (2). Hence $g(x_0, \ldots, x_n)$ vanishes for all solutions of (2). η is therefore a generic point of the variety C defined by (2). C is thus irreducible and of dimension $d+1$. It is, in fact, the cone joining A to V.

The section of C by $x_n = 0$ is therefore an unmixed variety Γ of dimension d (since V is not contained in $x_n = 0$). Γ has the equations

$$\left.\begin{aligned} a_i(x_1, \ldots, x_{n-1}) = 0 \quad (i = 1, \ldots, r), \\ x_n = 0, \end{aligned}\right\}$$

and since these are independent of x_0, Γ is a cone with vertex A, containing the section of V by $x_n = 0$. Any line through A meeting V in $x_n = 0$ lies in the tangent space to Q_{n-1} at A and contains a point of Q_{n-1} different from A; hence it lies on Q_{n-1}. Thus Γ lies on Q_{n-1}.

The collineation

$$\left.\begin{aligned} x'_0 &= x_n, \\ x'_i &= x_i \quad (i = 1, \ldots, n-1), \\ x'_n &= x_0, \end{aligned}\right\}$$

transforms Γ into a cone Δ with vertex B; and since it transforms Q_{n-1} into itself, Δ lies on Q_{n-1}. Since Δ is a sum of cones each of dimension d, with vertex B, the section of Δ by $x_n = 0$ is an unmixed variety V^* of dimension $d-1$, lying on the non-singular quadric Q_{n-3} whose equations are

$$\left.\begin{aligned} x_1^2 + \ldots + x_{n-1}^2 = 0, \\ x_0 = x_n = 0. \end{aligned}\right\}$$

With these preliminaries, we denote by V^ϵ the variety obtained from V by the projective transformation

$$\left.\begin{aligned} x'_0 &= \epsilon^2 x_0, \\ x'_i &= \epsilon x_i \quad (i = 1, \ldots, n-1), \\ x'_n &= x_n, \end{aligned}\right\}$$

where ϵ is an indeterminate. V^ϵ is irreducible, of dimension d over $K(\epsilon)$, and of order g [cf. Ch. XII, § 2]. The equations of V^ϵ are obtained from the equations of V by replacing $x_0, \ldots, x_n$ in (1), (2), (3) by $x_0, \epsilon x_1, \ldots, \epsilon x_{n-1}, \epsilon^2 x_n$. Hence the equations of V^ϵ are

$$\omega = 0, \tag{1$^\epsilon$}$$

$$a_i(x_1, \ldots, x_{n-1}) + \sum_{j>0} \epsilon^j b_{ij}(x_1, \ldots, x_{n-1}) x_n^j = 0 \quad (i = 1, \ldots, r), \tag{2$^\epsilon$}$$

$$\epsilon^{m_i - \rho_i} a_i(x_1, \ldots, x_{n-1}) + \sum_{j>0} b_{ij}(x_1, \ldots, x_{n-1}) x_n^j \epsilon^{m_i + j - \rho_i}$$
$$+ \sum_{j>0} c_{ij}(x_1, \ldots, x_{n-1}) x_0^j \epsilon^{m_i - j - \rho_i} = 0$$
$$(i = r+1, \ldots, s), \tag{3$^\epsilon$}$$

where m_i is the degree of $a_i(x_1, \ldots, x_{n-1})$, $\rho_i = m_i - k_i$, k_i being the largest value of j for which $c_{ij}(x_1, \ldots, x_{n-1})$ is not zero. By our arrangements, $\rho_i < m_i$.

We consider the normal problem given by $(1)^\epsilon$, $(2)^\epsilon$, $(3)^\epsilon$. Corresponding to a generic ϵ, the solution is the variety V^ϵ. The object-variety is a line; hence if for any specialisation of ϵ the equations $(1)^\epsilon$, $(2)^\epsilon$, $(3)^\epsilon$ define a d-dimensional variety, the corresponding specialisation of V^ϵ is uniquely determined [XI, § 7, Th. I, Cor.]. We consider the specialisation $\epsilon \to 0$. We have the equations

$$\omega = 0, \tag{1^0}$$

$$a_i(x_1, \ldots, x_{n-1}) = 0 \quad (i = 1, \ldots, r), \tag{2^0}$$

$$c_{i\, m_i - \rho_i}(x_1, \ldots, x_{n-1}) x_0^{m_i - \rho_i} = 0 \quad (i = r+1, \ldots, s). \tag{3^0}$$

Since $\rho_i < m_i$, each of the equations $(3)^0$ contains x_0 as a factor, and by our previous assumption, the last equation is

$$x_0^{m_s} = 0.$$

Hence $(1)^0$, $(2)^0$, $(3)^0$ define a variety with the equations

$$\omega = 0.$$

$$a_i(x_1, \ldots, x_{n-1}) = 0 \quad (i = 1, \ldots, r),$$

$$x_0 = 0,$$

that is, the variety Δ. Since Δ is of dimension d, there is a unique specialisation V^0 of V^ϵ corresponding to the specialisation $\epsilon \to 0$, and $V^0 \subseteq \Delta$.

Now consider the variety V as a multiplicative variety $\mathbf{V}$. Then

$$\mathbf{V} \equiv \mathbf{V}^0.$$

V^0 is a cone contained in Δ, and we can write

$$\boldsymbol{V}^0 = \sum_{i=1}^{k} \lambda_i \boldsymbol{V}_i,$$

where $V_1, \ldots, V_k$ are the irreducible cones comprising Δ. V_i meets $x_n = 0$ in an irreducible variety U_i of dimension $d-1$, lying on the quadric Q_{n-3} which has the equations

$$x_1^2 + \ldots + x_{n-1}^2 = 0,$$

$$x_0 = x_n = 0.$$

As in XII, § 8, we can express the Cayley form of $\Sigma\lambda_i \boldsymbol{V}_i$ in terms of the Cayley form of $\Sigma\lambda_i U_i$, and we can show that if $\Sigma\lambda_i \boldsymbol{U}_i \equiv \Sigma\mu_i \boldsymbol{U}'_i$ on Q_{n-3}, then $\Sigma\lambda_i \boldsymbol{V}_i \equiv \Sigma\mu_i \boldsymbol{V}'_i$ on Q_{n-1}, where V'_i is the cone joining U'_i to B.

We can now apply the hypothesis of induction to Q_{n-3}. Consider the case in which n is odd. If $d-1 > \frac{1}{2}(n-3)$, $\boldsymbol{U}_i \equiv \rho_i \overline{\boldsymbol{U}}$, where $\overline{U}$ is the section of Q_{n-3} by a linear space of dimension d in $x_0 = x_n = 0$. The cone joining $\overline{U}$ to B is the section $\overline{V}$ of Q_{n-1} by a linear space of dimension $d+1$, and we have

$$\boldsymbol{V} \equiv \boldsymbol{V}^0 \equiv \rho \overline{\boldsymbol{V}} \quad (\rho = \Sigma\lambda_i \rho_i).$$

If $d-1 = \frac{1}{2}(n-3)$, $\boldsymbol{U}_i \equiv \alpha_i \boldsymbol{S} + \beta_i \boldsymbol{T}$, where S and T are linear spaces of dimension $d-1$ on Q_{n-3}, of opposite systems. The joins X and Y of S and T to B are linear spaces of opposite systems on Q_{n-1}, since $Q_{n-3} \subseteq Q_{n-1}$ and lies in the tangent space to Q_{n-3} at B, and we have

$$\boldsymbol{V} \equiv \boldsymbol{V}^0 \equiv \alpha \boldsymbol{X} + \beta \boldsymbol{Y} \quad (\alpha = \Sigma\lambda_i \alpha_i,\ \beta = \Sigma\lambda_i \beta_i).$$

If $d-1 < \frac{1}{2}(n-3)$, $\boldsymbol{U}_i \equiv \rho_i \boldsymbol{S}$, where S is a linear $(d-1)$-space on Q_{n-3}, and its join to B is a d space X on Q_{n-1}. We have $\boldsymbol{V} \equiv \boldsymbol{V}^0 \equiv \rho \boldsymbol{X}$. This proves Theorem I when n is odd. A similar argument proves the theorem when n is even.

Theorem I determines a base for the varieties of dimension d on Q_{n-1}. We now show how to express any given unmixed variety $\boldsymbol{V}$ of dimension d in terms of this base. Let $\boldsymbol{C}_d$ be a base variety on Q_{n-1} of dimension d ($d \neq \frac{1}{2}(n-1)$, when n is odd), and let $\boldsymbol{X}$ and $\boldsymbol{Y}$ be m-spaces of opposite systems on Q_{n-1} in the case $n = 2m+1$.

Then, clearly

$$(\boldsymbol{C}_d \cdot \boldsymbol{C}_{n-d-1}) = 1 \quad (d \neq \tfrac{1}{2}(n-1)).$$

When $n = 2m+1$,

$$\begin{pmatrix}(\boldsymbol{X}.\boldsymbol{X}) & (\boldsymbol{X}.\boldsymbol{Y})\\ (\boldsymbol{Y}.\boldsymbol{X}) & (\boldsymbol{Y}.\boldsymbol{Y})\end{pmatrix} = \begin{pmatrix}1 & 0\\ 0 & 1\end{pmatrix} \text{ or } \begin{pmatrix}0 & 1\\ 1 & 0\end{pmatrix},$$

according as m is even or odd.

(*a*) Suppose that $d \neq \frac{1}{2}(n-1)$. Then

$$\boldsymbol{V} \equiv \lambda \boldsymbol{C}_d.$$

Hence $$(\boldsymbol{V}.\boldsymbol{C}_{n-d-1}) = \lambda(\boldsymbol{C}_d.\boldsymbol{C}_{n-d-1}) = \lambda.$$

If $d < \frac{1}{2}(n-1)$, C_{n-d-1} is the section of Q_{n-1} by a linear space of dimension $n-d$, and hence λ is the order of V. If $d > \frac{1}{2}(n-1)$, C_d is a variety of order 2, hence V is of even order, and λ is half the order of V.

(*b*) If $d = m = \frac{1}{2}(n-1)$, then

$$\boldsymbol{V} = \lambda \boldsymbol{X} + \mu \boldsymbol{Y}.$$

Therefore $$\begin{pmatrix}(\boldsymbol{V}.\boldsymbol{X})\\ (\boldsymbol{V}.\boldsymbol{Y})\end{pmatrix} = \begin{pmatrix}(\boldsymbol{X}.\boldsymbol{X}) & (\boldsymbol{X}.\boldsymbol{Y})\\ (\boldsymbol{Y}.\boldsymbol{X}) & (\boldsymbol{Y}.\boldsymbol{Y})\end{pmatrix}\begin{pmatrix}\lambda\\ \mu\end{pmatrix}.$$

So, given $(\boldsymbol{V}.\boldsymbol{X})$ and $(\boldsymbol{V}.\boldsymbol{Y})$, λ, μ can be determined.

We conclude this section by proving a result of some importance concerning the varieties of dimension $n-2$ on a quadric Q_{n-1} in S_n.

THEOREM II. *If $n > 3$, and Q_{n-1} is a quadric of type ρ, where $\rho < n-3$, any unmixed variety V of dimension $n-2$ on Q_{n-1} is the complete intersection of Q_{n-1} and a primal.*

It is clearly sufficient to prove this theorem when V is irreducible. Let A be a simple point of Q_{n-1} not on V. Since the section of Q_{n-1} by the tangent prime at A is a quadric Q_{n-2} of dimension $n-2$ and type $\rho+1$ $(<n-2)$, it is irreducible; hence it contains only one variety of dimension $n-2$, namely Q_{n-2} itself, which passes through A. Therefore V does not lie in the tangent space to Q_{n-1} at A. As in § 2, we choose our coordinate system so that A is $(1, 0, \ldots, 0)$, and the equation of Q_{n-1} is

$$\omega \equiv x_0 x_n - x_1^2 - \ldots - x_{n-1-\rho}^2 = 0.$$

The tangent space to Q_{n-1} at A is $x_n = 0$. Let

$$\left.\begin{aligned}\omega &= 0,\\ f_i(x_0, \ldots, x_n) &= 0 \quad (i = 1, \ldots, r)\end{aligned}\right\}$$

be a basis for the equations of V. We can find an exponent m_i such that

$$x_n^{m_i} f_i(x_0, \ldots, x_n) = \sum_j a_{ij}(x_1, \ldots, x_n)\, x_0^j x_n^j, \tag{4}$$

and hence

$$x_n^{m_i} f_i(x_0, \ldots, x_n) = F_i(x_1, \ldots, x_n) + A_i(x_0, \ldots, x_n)\, \omega.$$

We consider the variety U in S_n given by the equations

$$x_n F_i(x_1, \ldots, x_n) = 0 \quad (i = 1, \ldots, r), \tag{5}$$

and prove that this is an unmixed variety of dimension $n-1$. Clearly, every point in the prime π whose equation is $x_n = 0$ lies on U. Let $x^* = (x_0^*, \ldots, x_n^*)$ be any point of U not in $x_n = 0$. We can then find x_0' uniquely so that

$$x_0' x_n^* = x_1^{*2} + \ldots + x_{n-1-\rho}^{*2}.$$

Then since

$$F_i(x_1^*, \ldots, x_n^*) = 0,$$

and $x' = (x_0', x_1^*, \ldots, x_n^*)$ lies on Q_{n-1},

$$x_n^{*m_i} f_i(x_0', x_1^*, \ldots, x_n^*) = 0,$$

and since $x_n^* \neq 0$, x' must lie on V. This implies that x^* is on the cone C joining A to V. Conversely, we see that if x^* is on the cone joining A to V, x^* satisfies equation (5), that is, it lies on U. Hence

$$U = \pi \dotplus C.$$

Further, by the proof given in Theorem I, C is of dimension $n-1$. Thus U is unmixed, and of dimension $n-1$. It is therefore given by a single equation

$$G(x_1, \ldots, x_n) = 0,$$

and we have

$$x_n F_i(x_1, \ldots, x_n) = B_i(x_1, \ldots, x_n)\, G(x_1, \ldots, x_n).$$

We consider the intersection of this primal with Q_{n-1};

$$U_\wedge Q_{n-1} = \pi_\wedge Q_{n-1} \dotplus C_\wedge Q_{n-1}.$$

Now $\pi_\wedge Q_{n-1}$ is the irreducible quadric Q_{n-2} in which Q_{n-1} is met by $x_n = 0$. If x^* is any point on $C_\wedge Q_{n-1}$ not in $x_n = 0$, it satisfies the equation

$$\omega = 0,$$

and (5). Hence, as above, it lies on V. Conversely, any point on V lies on C and on Q_{n-1}. Thus we see that the primal $G = 0$ meets Q_{n-1} in V, and in a variety which lies in $x_n = 0$.

G may contain x_n as a factor. If it contains it as a factor k times we write

$$G(x_1, \ldots, x_n) = x_n^k H(x_1, \ldots, x_n),$$

and we see immediately that $H = 0$ meets Q_{n-1} in V, and possibly in another variety lying in $x_n = 0$. The intersection of Q_{n-1} with

$$H(x_1, \ldots, x_n) = 0$$

is an unmixed variety of dimension $n-2$. Hence, since Q_{n-2} is irreducible, if the intersection contains any component other than V, the primal $H = 0$ contains the intersection of Q_{n-1} with $x_n = 0$. In this case, we must have

$$H(x_1, \ldots, x_{n-1}, 0) = C(x_1, \ldots, x_{n-1})(x_1^2 + \ldots + x_{n-1-\rho}^2),$$

and therefore

$$H(x_1, \ldots, x_{n-1}, x_n) = K(x_0, \ldots, x_n)\, x_n - C(x_1, \ldots, x_n)\, \omega.$$

In this case also we see that the primal

$$K(x_0, \ldots, x_n) = 0$$

meets Q_{n-1} in V, and possibly in another variety in $x_n = 0$. Since the intersection is unmixed and of dimension $n-2$, we see that if this second component exists, it is Q_{n-2}. As before, we deduce that

$$K(x_0, \ldots, x_n) = K_1(x_0, \ldots, x_n)\, x_n - D(x_0, \ldots, x_n)\, \omega,$$

where

$$K_1(x_0, \ldots, x_n) = 0$$

meets Q_{n-1} in a pure $(n-2)$-dimensional variety, of which one component is V, and the other, if it exists, is Q_{n-2}. Proceeding in this way, we must eventually arrive at a primal

$$K_r(x_0, \ldots, x_n) = 0,$$

whose total intersection with Q_{n-1} is V. This proves the theorem.

7. Stereographic projection. Let Q_{n-1} be a quadric in S_n, of type ρ $(\rho < n-1)$, and let A be any simple point on it. As we saw [§ 2], we can choose the coordinate system so that A is the point $(1, 0, \ldots, 0)$ and the equation of the quadric is

$$\omega \equiv x_0 x_n - x_1^2 - \ldots - x_{n-\rho-1}^2 = 0. \tag{1}$$

Let $t_1, \ldots, t_n$ be independent indeterminates over the ground field. Then

$$\frac{x_1}{t_1} = \ldots = \frac{x_n}{t_n} \tag{2}$$

is a generic line through A. This meets Q_{n-1} in A, and in the point

$\xi = (t_1^2 + \ldots + t_{n-\rho-1}^2, t_1 t_n, \ldots, t_{n-1}t_n, t_n^2)$, and [XI, § 10, Th. I] ξ is a generic point of Q_{n-1}. The line (2) meets the prime Π, whose equation is $x_0 = 0$, in the point $\eta = (0, t_1, \ldots, t_n)$, and η is obviously a generic point of Π. Consider the correspondence Γ between S_n and Π which has (ξ, η) as generic point. The object-variety is the variety in S_n whose generic point is ξ, that is, Q_{n-1}, and the image-variety is the variety whose generic point is η, that is, it is the prime Π. The dimension of the correspondence Γ is clearly $n-1$. This correspondence defines a mapping of Q_{n-1} on Π which is called a *stereographic projection* of Q_{n-1} (from A on Π). We give a brief account of its properties.

To any point $\bar{x}$ of Q_{n-1} there corresponds at least one point of Π, and since

$$\xi_i \eta_j - \xi_j \eta_i = 0 \quad (i, j = 1, \ldots, n), \tag{3}$$

it follows that if $\bar{x}$ is not A, then the corresponding point $\bar{y}$ of Π is determined uniquely by the equations

$$\bar{x}_i y_j - \bar{x}_j y_i = 0 \quad (i, j = 1, \ldots, n);$$

but if $\bar{x}$ is at A the specialisations derived from (3) do not tell us anything about $\bar{y}$. On the other hand, if $\bar{y}$ is any point of Π, we see from equation (3) that the corresponding points of Q_{n-1} satisfy the equations

$$x_i \bar{y}_j - x_j \bar{y}_i = 0 \quad (i, j = 1, \ldots, n),$$

and hence lie on the line joining $\bar{y}$ to A. In general, this line meets Q_{n-1} in A and in one other point. It is possible to show that A does not correspond to $\bar{y}$; but again the equations (3) give us little information if the line joining $\bar{y}$ to A lies on Q_{n-1}.

In order to determine the corresponding points of Q_{n-1} and Π in Γ in all cases, it is necessary to write down a basis for the equations of Γ. This can be done as follows. We use $(y_1, \ldots, y_n)$ to denote the coordinates of points in Π; these are the points whose coordinates in S_n are $(0, y_1, \ldots, y_n)$. The point (ξ, η) clearly satisfies the equations

$$\omega \equiv x_0 x_n - x_1^2 - \ldots - x_{n-\rho-1}^2 = 0, \tag{1}$$

$$f_{ij} \equiv x_i y_j - x_j y_i = 0 \quad (i, j = 1, \ldots, n), \tag{4}$$

$$g \equiv x_0 y_n - x_1 y_1 - \ldots - x_{n-\rho-1} y_{n-\rho-1} = 0. \tag{5}$$

We now show that if $F(x, y) = 0$ is any equation satisfied by (ξ, η), then

$$F(x, y) \equiv a(x, y)\,\omega + \Sigma b_{ij}(x, y) f_{ij} + c(x, y)\,g.$$

If $F(x,y)$ is a form in $(x_0, \ldots, x_n)$ which vanishes at a generic point of Q_{n-1}, it vanishes on Q_{n-1}. It therefore contains ω as a factor, and our statement follows in this case. Hence we need only consider forms in which y appears. Consider any form $F(x,y)$ of degree r in $(x_0, \ldots, x_n)$ and of degree s in $(y_1, \ldots, y_n)$, where $s>0$. Let $px_0^{\lambda_0} \ldots x_n^{\lambda_n} y_1^{\mu_1} \ldots y_n^{\mu_n}$ be any term of it. If λ_0 and λ_n are both positive we can express it, *modulo* ω, as a sum of power-products in each of which x_0 appears with exponent less than λ_0. Again, if λ_0 and μ_n are both positive, we can express it, *modulo* g, as a sum of power-products in each of which x_0 has exponent less than λ_0. If we can find integers i, j $(i>j \geqslant 1)$ such that $\mu_i > 0$, $\lambda_j > 0$, we can reduce the power-product, *modulo* f_{ij}, to a similar power-product in which λ_j, μ_i are replaced by $\lambda_j - 1$, $\mu_i - 1$. Using these three methods of reducing the terms of $F(x,y)$, we obtain an identity

$$F(x,y) \equiv F^*(x,y) + a(x,y)\,\omega + \sum_{i,j} b_{ij}(x,y) f_{ij} + c(x,y)\,g,$$

where

$$F^*(x,y) = \Sigma k^*_{\lambda,\mu} y_1^{\mu_1} \ldots y_p^{\mu_p} x_p^{\lambda_p} x_{p+1}^{\lambda_{p+1}} \ldots x_n^{\lambda_n} + \Sigma m^*_{\lambda,\mu} x_0^{\lambda} y_1^{\mu_1} \ldots y_p^{\mu_p} x_p^{\lambda_p} \ldots x_{n-1}^{\lambda_{n-1}},$$

where in the first sum p may be any integer $(1 \leqslant p \leqslant n)$, and in the second p may be any integer $(1 \leqslant p \leqslant n-1)$, and λ is a positive integer.

Now suppose that $F(\xi,\eta) = 0$. Then we have $F^*(\xi,\eta) = 0$. We show that this implies that $F^*(x,y)$ is identically zero. When we replace (x,y) by ξ,η in $y_1^{\mu_1} \ldots y_p^{\mu_p} x_p^{\lambda_p} x_{p+1}^{\lambda_{p+1}} \ldots x_n^{\lambda_n}$ we obtain

$$t_n^{\Sigma\lambda_i} t_1^{\mu_1} \ldots t_{p-1}^{\mu_{p-1}} t_p^{\mu_p+\lambda_p} t_{p+1}^{\lambda_{p+1}} \ldots t_n^{\lambda_n};$$

and, since $\Sigma\lambda_i = r$, each term of the first sum yields a product of the t_i having t_n to the power $r+\lambda_n$ as a factor. On the other hand, $x_0^{\lambda} y_1^{\mu_1} \ldots y_p^{\mu_p} x_p^{\lambda_p} \ldots x_{n-1}^{\lambda_{n-1}}$ yields a polynomial in the t_i in which t_n appears with the exponent $r-\lambda$. Since $t_1, \ldots, t_n$ are independent indeterminates, we deduce that if

$$F^*(x,y) = \sum_{\nu=0}^{r} x_n^{\nu} F^*_{\nu}(x,y) + \sum_{\lambda=1}^{r} x_0^{\lambda} G^*_{\lambda}(x,y),$$

where $$F^*_{\nu}(x,y) = \Sigma c^*_{\lambda,\mu} y_1^{\mu_1} \ldots y_p^{\mu_p} x_p^{\lambda_p} \ldots x_{n-1}^{\lambda_{n-1}},$$

and $$G^*_{\lambda}(x,y) = \Sigma d^*_{\lambda,\mu} y_1^{\mu_1} \ldots y_p^{\mu_p} x_p^{\lambda_p} \ldots x_{n-1}^{\lambda_{n-1}},$$

then the equation $F(\xi,\eta)=0$ implies that the forms $F^*_i(x,y)$, $G^*_i(x,y)$ vanish at $(x,y) = (\xi,\eta)$. But

$$F^*_{\nu}(\xi,\eta) = t_n^{r-\nu}\, \Sigma c^*_{\lambda,\mu} t_1^{\mu_1} \ldots t_p^{\mu_p+\lambda_p} t_{p+1}^{\lambda_{p+1}} \ldots t_{n-1}^{\lambda_{n-1}}$$

A given power-product $t_n^{r-\nu} t_1^{\rho_1} \dots t_{n-1}^{\rho_{n-1}}$ on the right-hand side cannot arise from two distinct terms of $F_\nu^*(x, y)$. Indeed, consider the smallest integer k such that $\rho_1 + \dots + \rho_k \geqslant s$. Then $t_n^{r-\nu} t_1^{\rho_1} \dots t_{n-1}^{\rho_{n-1}}$ must arise from the product $y_1^{\rho_1} \dots y_{k-1}^{\rho_{k-1}} y_k^{\alpha} x_k^{\beta} x_{k+1}^{\rho_{k+1}} \dots x_{n-1}^{\rho_{n-1}}$, where $\alpha + \rho_1 + \dots + \rho_{k-1} = s$, $\beta = \rho_k - \alpha$. Hence if $F_\nu^*(\xi, \eta) = 0$, each coefficient $c_{\lambda,\mu}^*$ is zero and hence $F_\nu^*(x, y)$ is identically zero. Similarly, if $G_\lambda^*(\xi, \eta)$ is zero, $G_\lambda^*(x, y)$ is identically zero. Thus if $F^*(\xi, \eta) = 0$, $F^*(x, y)$ is zero, and therefore

$$F(x, y) = a(x, y)\,\omega + \Sigma b_{ij}(x, y) f_{ij} + c(x, y)\,g.$$

Thus the equations (1), (4), (5) form a basis for the equations of the correspondence Γ. We can now find the locus in Π corresponding to any point of Q_{n-1}, and the locus on Q_{n-1} corresponding to any point of Π.

(i) (*a*) If $\bar{x}$ is not A, the unique point of Π corresponding to it is $(\bar{x}_1, \dots, \bar{x}_n)$;

(*b*) If $\bar{x}$ is at A, the corresponding locus of Γ is given by $y_n = 0$ [from (5)].

(ii) If $\bar{y}$ is any point of Π, equation (4) tells us that the points x corresponding to it are of the form $(\rho, \sigma\bar{y}_1, \dots, \sigma\bar{y}_n)$, where ρ and σ are arbitrary (but not both zero).

From (5), we obtain the equation

$$\rho\bar{y}_n = \sigma(\bar{y}_1^2 + \dots + \bar{y}_{n-\rho-1}^2).$$

(*a*) If $\bar{y}_n \neq 0$, σ cannot be zero, since this would imply $\rho = 0$. Hence to $\bar{y}$ there corresponds a unique point of Q_{n-1}, different from A.

(*b*) If $\bar{y}_n = 0$, $\bar{y}_1^2 + \dots + \bar{y}_{n-\rho-1}^2 \neq 0$, $\sigma = 0$, and hence the only point on Q_{n-1} corresponding to $\bar{y}$ is A.

(*c*) If $\bar{y}_1^2 + \dots + \bar{y}_{n-\rho-1}^2 = 0$, $\bar{y}_n = 0$, ρ and σ are indeterminates, and to $\bar{y}$ there corresponds any point on the line joining $\bar{y}$ to A.

A is the only point on Q_{n-1} to which there corresponds a locus of dimension greater than zero (if $n > 2$); it is called the *fundamental point* of Γ on Q_{n-1}. The points on the intersection of Q_{n-1} with the tangent prime at A are the only points on Q_{n-1} which are not uniquely determined by the corresponding points of Π; these points (except A itself) are called the *irregular points* of the correspondence Γ. All the other points on Q_{n-1} are said to be *regular* for Γ. This nomenclature is in conformity with that to be introduced in Vol. III in connection with the general theory of birational correspondences. In the same way the fundamental points of Γ in Π are the points of the quadric Ω whose equations are

$$y_n = 0, \quad y_1^2 + \ldots + y_{n-\rho-1}^2 = 0, \tag{6}$$

and the irregular points of the correspondence are the points of $y_n = y_0 = 0$, other than those of Ω.

The stereographic projection of Q_{n-1} from A on Π is often used to investigate the properties of subvarieties of Q_{n-1}; but care must be exercised in passing from Q_{n-1} to Π or from Π to Q_{n-1} because of the presence of fundamental and irregular elements of the correspondence. On the other hand, it should be pointed out that the actual prime Π on which Q_{n-1} is projected from A is unimportant (provided it does not pass through A). If Π' is any other prime of S_n not through A, we can establish a correspondence Γ' between Q_{n-1} and Π' in which to a generic point ξ of Q_{n-1} there corresponds the unique point in which the line $A\xi$ meets Π'. The two correspondences Γ and Γ' can easily be seen to establish a one-to-one correspondence between Π and Π', points y and y' of Π and Π' corresponding if they correspond to the same locus on Q_{n-1} in Γ and Γ'. The correspondence between Π and Π' is such that points y and y' correspond if and only if A, y, y' are collinear. The correspondence between Π and Π' is therefore a projective one.

We now outline the principles to be followed in considering the correspondence between varieties on Q_{n-1} and varieties of Π. Let V be an irreducible variety on Q_{n-1}, not consisting solely of the point A, let d be its dimension and $\xi' = (\xi'_0, \ldots, \xi'_n)$ a generic point of it. Since V is not the point A, some ξ'_i $(i > 0)$ is different from zero. Hence we can determine the unique point η' in Π corresponding to ξ', namely $\eta' = (\xi'_1, \ldots, \xi'_n)$. Let $\bar{V}$ be the variety in Π whose generic point is η'. $\bar{V}$ is defined to be the variety in Π which corresponds to V. Since (ξ', η') is a proper specialisation of (ξ, η), every pair of points of V and $\bar{V}$ which correspond in the correspondence with generic point (ξ', η') correspond in the stereographic projection. Every point on V other than A has a unique corresponding point on $\bar{V}$, but if A lies on V there may be more than one corresponding point on $\bar{V}$; indeed if $\bar{y}$ is any point of $\bar{V}$ in $y_n = 0$ but not on the quadric Ω given by (6), the only point of Q_{n-1} corresponding to $\bar{y}$ in the stereographic projection is A, and hence A is the only point on V corresponding to $\bar{y}$. Thus, corresponding to the point A of V

there corresponds any point of $\bar{V}$ in $y_n = 0$ but not on Ω; the relation of points on Ω requires special investigation.

At a later stage we shall find it desirable at times to modify the definition of the variety in Π corresponding to a variety V on Q_{n-1} which contains A by adding certain varieties of Π which lie in $y_n = 0$. These varieties are to be regarded as all corresponding to the point A of V.

We first consider the case in which V does not lie in $x_n = 0$. Then $\xi'_n \neq 0$, and we may normalise the coordinates of ξ' so that $\xi'_n = 1$. Since ξ' lies on Q_{n-1},

$$\xi'_0 = \sum_1^{n-\rho-1} \xi'^2_i.$$

Hence $K(\xi'_0, \ldots, \xi'_{n-1}) = K(\xi'_1, \ldots, \xi'_{n-1})$, so that $\bar{V}$ is of the same dimension as V. If, on the other hand, V lies in $x_n = 0$, $\xi'_n = 0$. We normalise with respect to some ξ'_i $(1 \leqslant i \leqslant n-1)$. Then $K(\xi'_1, \ldots, \xi'_{n-1})$ may be of degree of transcendency d or $d-1$, according as ξ'_0 is algebraic or transcendental over $K(\xi'_1, \ldots, \xi'_{n-1})$. In the latter case V is a cone on Q_{n-1} with vertex A, in the former it is not. In either case, since V lies on Q_{n-1} and is in $x_n = 0$,

$$\xi'^2_1 + \ldots + \xi'^2_{n-\rho-1} = 0,$$

and hence $\bar{V}$ lies on Ω, and is the projection of V from A.

Now consider once more the case in which V is not in $x_n = 0$. It then meets $x_n = 0$ in a variety U of dimension $d-1$. If this is not a cone with vertex A, it corresponds to a variety $\bar{U}$ of dimension $d-1$ on Ω, and it is easily seen that $\bar{U}$ is on $\bar{V}$. Hence $\bar{V}$ meets Ω in a variety of dimension $d-1$. If, however, U is a cone with vertex A, the intersection of $\bar{V}$ and Ω is of dimension $d-2$.

Conversely, let us consider an irreducible variety $\bar{V}$ of Π, of dimension d and with the normalised generic point $\eta' = (\eta'_1, \ldots, \eta'_n)$. If $\bar{V}$ is not in $y_n = 0$, there is a unique point $\xi' = \left(\sum_1^{n-\rho-1} \eta'^2_i, \eta'_n\eta'_1, \ldots, \eta'^2_n\right)$ of Q_{n-1} corresponding to η', and this is a generic point of a variety V on Q_{n-1}. $\bar{V}$ is the variety in Π which corresponds, according to the definition given above, to V, and V is, indeed, the only variety whose corresponding variety in Π is $\bar{V}$. But if $\eta'_n = 0$, the situation is different. If $\bar{V}$ is not contained in Ω, η' is not on Ω, and hence A is the only point of Q_{n-1} which corresponds to η'. Hence there is no variety V on Q_{n-1} which determines $\bar{V}$ in Π. But if $\bar{V} \subseteq \Omega$,

$$\eta'^2_1 + \ldots + \eta'^2_{n-\rho-1} = 0,$$

and the $(d+1)$-dimensional variety V whose generic point is $\xi' = (\xi_0', \eta_1', \ldots, \eta_{n-1}', 0)$, where ξ_0' is transcendental over $K(\eta_1', \ldots, \eta_{n-1}')$, is on Q_{n-1}, and $\bar{V}$ is the variety in Π defined by V. If

$$\xi'' = (\xi_0'', \eta_1', \ldots, \eta_{n-1}', 0),$$

where ξ'' is algebraic over $K(\eta_1', \ldots, \eta_{n-1}')$, the variety V'' whose generic point is ξ'' also determines $\bar{V}$.

To illustrate these principles, we consider certain elementary loci on Q_{n-1}. Let Q_d, supposed irreducible, be the section of Q_{n-1} by a $(d+1)$-space S_{d+1}. Suppose first that Q_d does not lie in $x_n = 0$, or pass through A. Then the equations of Q_d can be written in the form

$$\left.\begin{aligned} \omega \equiv x_0 x_n - x_1^2 - \ldots - x_{n-\rho-1}^2 &= 0, \\ x_0 - \sum_{j=1}^{n} a_j x_j &= 0, \\ \sum_{j=1}^{n} b_{ij} x_j = 0 \quad (i &= 1, \ldots, n-d-2). \end{aligned}\right\}$$

If $\xi' = (\xi_0', \ldots, \xi_{n-1}', 1)$ is a generic point of Q_d, ξ' satisfies all these equations. Hence $\eta' = (\xi_1', \ldots, \xi_{n-1}', 1)$ satisfies the equations

$$\left.\begin{aligned} y_n \sum_{j=1}^{n} a_j y_j &= y_1^2 + \ldots + y_{n-\rho-1}^2, \\ \sum_{j=1}^{n} b_{ij} y_j = 0 \quad (i &= 1, \ldots, n-d-2). \end{aligned}\right\}$$

This is the equation of a quadric $\bar{Q}_d$ in Π, and it can be verified at once that this is irreducible if Q_d is irreducible. But the variety in Π which corresponds to Q_d is irreducible and of dimension d, and has generic point η'; hence it must be $\bar{Q}_d$. It is seen that $\bar{Q}_d$ meets Ω in the quadric

$$\left.\begin{aligned} y_1^2 + \ldots + y_{n-\rho-1}^2 &= 0, \\ y_n &= 0, \\ \sum_{j=1}^{n} b_{ij} y_j = 0 \quad (i &= 1, \ldots, n-d-2), \end{aligned}\right\}$$

which is of dimension $d-1$.

If Q_d passes through A, but does not lie in $x_n = 0$, its equation is of the form

$$\left.\begin{aligned} \omega &= 0, \\ \sum_{j=1}^{n} b_{ij} x_j = 0 \quad (i &= 1, \ldots, n-d-1). \end{aligned}\right\}$$

The corresponding variety $\bar{V}$ in Π then satisfies the equations

$$\sum_{j=1}^{n} b_{ij} y_j = 0 \quad (i = 1, \ldots, n-d-1),$$

and as this is a d-space, $\bar{V}$ is a linear space of d dimensions. $\bar{V}$ in this case meets Ω in a variety of dimension $d-2$, in general. This corresponds to the fact that the intersection of Q_d with $x_n = 0$ is a cone with vertex A.

If Q_d lies in $x_n = 0$, it lies on the cone in which this prime meets Q_{n-1}, and the corresponding variety in Π is obtained by ordinary projection in $x_n = 0$.

The fact that the variety in Π corresponding to a quadric on Q_{n-1} can be of different order (and dimension), according to the position of the quadric relative to A, raises the question of the system described in Π by the varieties $\bar{V}$ corresponding to the varieties V of an algebraic system on Q_{n-1}. Let $\{Q_d\}$ be the system of equivalence containing all quadrics of dimension d on Q_{n-1}, and let X be the point on the Cayley image of quadrics in S_n which corresponds to a generic member of $\{Q_d\}$. To $Q_d(X)$ there corresponds a quadric $\bar{Q}_d(Y)$ in Π, where Y is the Cayley image of $\bar{Q}_d(Y)$. Let ξ' be a generic point of $Q_d(X)$ and let $\eta' = (\xi'_1, \ldots, \xi'_n)$; η' is a generic point of $\bar{Q}_d(Y)$. Now specialise $X \to X'$, where $Q_d(X')$ is a quadric through A. Corresponding to this specialisation, there is at least one specialisation Y' of Y. We thus obtain a quadric $\bar{Q}_d(Y')$. If y' is any point of this, (X', Y', y') is a proper specialisation of (X, Y, η'). Suppose y' is not in $y_n = 0$. Then since η' has the property that $(\eta_1'^2 + \ldots + \eta_{n-\rho-1}'^2, \eta'_1\eta'_n, \ldots, \eta_n'^2)$ satisfies the equations of $Q_d(X)$, $(y_1'^2 + \ldots + y_{n-\rho-1}'^2, y'_1 y'_n, \ldots, y_n'^2)$ lies on $Q_d(X')$. Hence y' lies in the d-space of Π corresponding to $Q_d(X')$. Conversely, any point of this d-space lies on $\bar{Q}_d(Y')$. Hence $\bar{Q}_d(Y')$ consists of this d-space, $S_d(X')$, say, and another d-space $S'_d(X')$ which lies in $y_n = 0$. Moreover, since $\bar{Q}_d(Y)$ lies in a $(d+1)$-space, so does $\bar{Q}_d(Y')$; hence $S_d(X')$ and $S'_d(X')$ meet in a $(d-1)$-space. Therefore $S'_d(X')$ passes through the quadric in which $S_d(X')$ meets Ω.

If $Q_d(X')$ lies in $x_n = 0$, $\bar{Q}_d(Y')$ lies entirely in $y_n = 0$.

We conclude by showing how stereographic projection can be used to investigate properties of d-spaces lying on Q_{n-1}. We consider only the most important case, in which $n = 2m+1, d = m$, and Q_{n-1} is non-singular. Then Ω is non-singular. Our general theory tells us that an m-space S_m on Q_{n-1}, not through A, corresponds to an m-space $\bar{S}_m$ in Π, not in $y_n = 0$ and meeting Ω in an $(m-1)$-space. The theory of m-spaces on Q_{n-1} is derived from this fact, assuming as hypothesis of induction that we know the properties of $(m-1)$-spaces on Ω. Two m-spaces S_m and S'_m of Q_{n-1} are defined to be of the same system if the corresponding spaces in Π meet Ω in spaces of the same system on Ω (or if S_m passes through A, if the $(m-1)$-space of Π corresponding to it is of the same system as the intersection of Ω with the m-space of Π which corresponds to S'_m). Two independent generic m-spaces of the same system in Q_{n-1} correspond to generic m-spaces $\bar{S}_m$, $\bar{S}'_m$ of Π through generic $(m-1)$-spaces of the same system on Ω. If m is odd, $m-1$ is even, and the linear spaces on Ω have a point in common. This is the only point common to $\bar{S}_m$ and $\bar{S}'_m$, and does not arise from an intersection of S_m and S'_m. Hence we conclude that S_m and S'_m do not meet. If m is even, the two $(m-1)$-spaces on Ω do not meet. But $\bar{S}_m$ and $\bar{S}'_m$, lying in Π, have a point in common, and since this does not lie in $y_n = 0$, it corresponds to a true intersection of S_m and S'_m. A similar argument applies when S_m and S'_m are of opposite systems.

The reader will realise that this method of dealing with the m-spaces on a quadric Q_{2m} requires very careful scrutiny. In actual fact, it is possible to develop the theory of m-spaces on a quadric Q_{2m} in a rigorous manner on these lines, but the method used in § 5 is more powerful, and we do not give the details of the method of stereographic projection here. Nevertheless, the method of stereographic projection for the investigation of properties of varieties lying on a quadric is an important one. One need only recall the properties of quadric surfaces which are obtained in this way in elementary geometry to see this.

The matters dealt with in this section will also prove useful as illustrations of the general theory of birational transformations which will be given in Vol. III.

8. The projective transformations of a quadric into itself.

Let
$$x'bx = 0$$
be the equations of a quadric Q_{n-1} in S_n, and consider the collineation C in S_n which takes the point x into the point y, where
$$y = px.$$

The point y lies on Q_{n-1} if and only if

$$y'by = x'p'bpx = 0.$$

Hence the relation

$$p'bp = \lambda b, \tag{1}$$

where λ is a scalar, is a necessary and sufficient condition that the transform by C of any point x of Q_{n-1} either does not exist ($px = 0$) or lies on Q_{n-1}. In this and the following section we are concerned with properties of the matrices p which satisfy a condition such as (1) for a given matrix b.

Let

$$x'cx = 0$$

be any other quadric Q'_{n-1} of the same type as Q_{n-1}; then b and c are of the same rank, and [§ 1] there exists a non-singular matrix t such that

$$b = t'ct.$$

Then

$$\bar{t}p't'ctpt^{-1} = \lambda c \quad (\bar{t} = (t')^{-1}),$$

and

$$y = tpt^{-1}x$$

is a collineation which transforms Q'_{n-1} into itself. Thus if we know the collineation which transforms a quadric Q_{n-1} into itself we can deduce the collineation which transforms any quadric of the same type as Q_{n-1} into itself. Suppose that Q_{n-1} is of type ρ; then the quadric Q'_{n-1} whose equation is

$$x_0^2 + \ldots + x_{n-\rho}^2 = 0,$$

that is,

$$x'\begin{pmatrix} I & 0 \\ 0 & 0 \end{pmatrix} x = 0,$$

where I is the unit matrix of $n-\rho+1$ rows and columns, is of the same type as Q_{n-1}, and we may consider the collineation which transforms this into itself. We write

$$p = \begin{pmatrix} q & r \\ s & t \end{pmatrix},$$

where q, r, s, t are matrices of $n-\rho+1$, $n-\rho+1$, ρ, ρ rows and $n-\rho+1$, ρ, $n-\rho+1$, ρ columns. Then if

$$p'\begin{pmatrix} I & 0 \\ 0 & 0 \end{pmatrix} p = \lambda \begin{pmatrix} I & 0 \\ 0 & 0 \end{pmatrix},$$

we have

$$q'q = \lambda I,$$

$$q'r = 0, \quad r'q = 0, \quad r'r = 0.$$

If $\lambda \neq 0$, it follows that qq' is non-singular, and hence q' is non-singular; and from the second equation we deduce that r is zero. Conversely, if $q'q$ is a non-zero multiple λ of the unit matrix I, and s, t are arbitrary,

$$p = \begin{pmatrix} q & 0 \\ s & t \end{pmatrix}$$

satisfies the equation

$$p'\begin{pmatrix} I & 0 \\ 0 & 0 \end{pmatrix}p = \lambda\begin{pmatrix} I & 0 \\ 0 & 0 \end{pmatrix}.$$

Thus our search for matrices p which satisfy an equation of the type (1) with λ different from zero reduces to the case in which b is non-singular.

It is this case, with the additional condition that λ is not zero, with which we shall be concerned, but before leaving the case in which λ is zero, we indicate how this case can be dealt with.

Let p be any $(n+1)\times(n+1)$ matrix such that

$$p'bp = 0,$$

and let σ be its rank. Let $\omega = (\omega_{ij})$ $(i = 0, \ldots, n, j = 1, \ldots, \sigma)$ be a matrix such that the columns of p are linearly dependent on the columns of ω. Then

$$p = \omega\nu,$$

where ν is a $\sigma \times (n+1)$ matrix of rank σ. Since ν is of rank σ, and $\omega' b\omega$ is a $\sigma\times\sigma$ matrix, the equation

$$p'bp = \nu'\omega' b\omega\nu = 0$$

implies the equation $\omega' b\omega = 0$.

If ω_i is the point whose coordinates are the elements of the ith column of ω, ω_i and ω_j $(i, j = 1, \ldots, \sigma)$ are conjugate with regard to the quadric

$$x'bx = 0,$$

and hence the $(\sigma-1)$-space determined by the points $\omega_1, \ldots, \omega_\sigma$ lies on this quadric. Conversely, if we take any σ points $\omega_1, \ldots, \omega_\sigma$ which determine a $(\sigma-1)$-space on the quadric, and ω is the matrix whose columns are coordinates of the points, we have

$$\omega' b\omega = 0,$$

and therefore if ν is any $\sigma\times(n+1)$ matrix and $p = \omega\nu$, we have

$$p'bp = 0.$$

Thus the determination of matrices p such that

$$p'bp = 0$$

reduces to the determination of linear spaces on the quadric

$$x'bx = 0.$$

It should be noted that a matrix p which satisfies the equation

$$p'bp = 0$$

cannot always be obtained by specialisation from a matrix q such that

$$q'bq = \lambda b,$$

where λ is different from zero. To see this we consider the case $n = 1$, where b is the unit matrix. If λ is not zero, q is non-singular, and hence the equation

$$q'q = \lambda I$$

implies $$qq' = q\lambda Iq^{-1} = \lambda I.$$

Hence $$q'q = qq' = \lambda I.$$

If p is any specialisation of q such that $p'p = 0$, then we must have

$$pp' = 0.$$

But for any α and β the matrix

$$P = \begin{pmatrix} \alpha & \beta \\ i\alpha & i\beta \end{pmatrix}$$

satisfies the equation $$P'P = 0,$$

but not the equation $$PP' = 0,$$

unless $\alpha^2 + \beta^2$ is zero. Thus P cannot be a proper specialisation of q.

We now return to the case in which Q_{n-1} is non-singular, and λ is not zero. We may take the equation of Q_{n-1} in the form

$$x_0^2 + \ldots + x_n^2 = 0.$$

Then if the collineation $y = px$ transforms Q_{n-1} into itself, we have

$$p'p = \lambda I,$$

I being the unit matrix of order $n+1$. If, as we assume, $\lambda \neq 0$, it follows that p is non-singular. If we write $q = \mu p$, where μ is scalar, we have

$$q'q = \mu^2 \lambda I.$$

We choose μ so that $\mu^2\lambda = 1$; then $q'q$ is the unit matrix. We then say that q is an *orthogonal matrix*; this generalises the usual definition of an orthogonal matrix found in works on algebra, since the elements of q are not restricted to be real numbers, but belong to the ground field K.

Since
$$q'q = I,$$
$$\det q = \pm 1.$$

If $\det q = -1$, and n is even,
$$(-q)'(-q) = I,$$
and
$$\det(-q) = 1,$$
so that, in the case in which n is even, we can effectively confine our attention to orthogonal matrices whose determinant is equal to plus one. But if n is odd there are two types of orthogonal matrix, those of determinant plus one, and those of determinant minus one. Later, we shall see that this is related to the fact that if n is odd there are two systems of $\frac{1}{2}(n-1)$-dimensional linear spaces on Q_{n-1}; indeed, we shall see that the orthogonal matrices of determinant plus one transform any $\frac{1}{2}(n-1)$-dimensional space of Q_{n-1} into a space of the same system, while those of determinant minus one interchange the systems.

We now consider some purely algebraic properties of orthogonal matrices.

Let q and r be two orthogonal matrices of $n+1$ rows and columns. Then
$$(qr)'(qr) = r'q'qr = r'r = I.$$

Hence qr is also orthogonal. Again, if q is orthogonal,
$$(q^{-1})'q^{-1} = (q')'q^{-1} = qq^{-1} = I,$$
so that the inverse of an orthogonal matrix is orthogonal. Again, if $\det q = 1$ and $\det r = 1$, then $\det qr = \det q \det r = 1$. Thus we have

THEOREM I. *The orthogonal matrices of $n+1$ rows and columns form a group G under multiplication. The orthogonal matrices of $n+1$ rows and columns having determinant plus one form a subgroup G_1 of G.*

We also note that if

$$F = \begin{pmatrix} 1 & 0 & . & 0 & 0 \\ 0 & 1 & . & 0 & 0 \\ . & . & . & . & . \\ 0 & 0 & . & 1 & 0 \\ 0 & 0 & . & 0 & -1 \end{pmatrix},$$

F is orthogonal and is of determinant minus one. If q is any orthogonal matrix of determinant minus one, $r = qF$ is in G_1, and $q = rF$. Hence the orthogonal matrices of determinant minus one are obtained as products rF, where r ranges over G_1.

Let q be an orthogonal matrix of $n+1$ rows and columns, and suppose that $q+I$ is non-singular. Let

$$\alpha = (I-q)(I+q)^{-1}. \tag{2}$$

Then

$$\begin{aligned} \alpha' &= (I+q')^{-1}(I-q') \\ &= (I+q^{-1})^{-1}(I-q^{-1}) \\ &= (q+I)^{-1}qq^{-1}(q-I) \\ &= -(I+q)^{-1}(I-q); \end{aligned}$$

and, since

$$(I+q)(I-q) = I-q^2 = (I-q)(I+q),$$

we have

$$\alpha' = -\alpha.$$

Hence α is skew, and we find on solving equation (2) that

$$(I+\alpha)q = I-\alpha.$$

Now

$$\begin{aligned} I+\alpha &= (I+q+I-q)(I+q)^{-1} \\ &= 2I(I+q)^{-1}, \end{aligned}$$

and hence $I+\alpha$ is non-singular. Therefore

$$q = (I+\alpha)^{-1}(I-\alpha).$$

Conversely, we show that if α is a skew matrix such that $I+\alpha$ is non-singular, then $q = (I+\alpha)^{-1}(I-\alpha)$ is orthogonal. For, since

$$(I+\alpha)(I-\alpha) = I-\alpha^2 = (I-\alpha)(I+\alpha),$$

$$(I+\alpha)^{-1}(I-\alpha) = q = (I-\alpha)(I+\alpha)^{-1},$$

and hence

$$\begin{aligned} q' &= (I+\alpha')^{-1}(I-\alpha') \\ &= (I-\alpha)^{-1}(I+\alpha). \end{aligned}$$

Therefore $q'q = (I-\alpha)^{-1}(I+\alpha)(I+\alpha)^{-1}(I-\alpha) = I.$
Thus we have

THEOREM II. *If q is an orthogonal matrix such that $I+q$ is non-singular, then we can find a skew matrix α such that*

$$(I+\alpha)^{-1}(I-\alpha) = q = (I-\alpha)(I+\alpha)^{-1}.$$

Conversely, if α is a skew matrix such that $I+\alpha$ is non-singular,

$$q = (I+\alpha)^{-1}(I-\alpha) = (I-\alpha)(I+\alpha)^{-1}$$

is orthogonal.

We note that if α is skew and such that $I+\alpha$ is non-singular,

$$\det(I-\alpha) = \det(I-\alpha)' = \det(I-\alpha') = \det(I+\alpha),$$

hence $$\det q = \det[(I+\alpha)^{-1}(I-\alpha)] = 1.$$

Hence all the orthogonal matrices q which can be written in the form $(I+\alpha)^{-1}(I-\alpha)$, where α is skew, have determinant equal to one. In other words, if q has determinant equal to minus one, $q+I$ is singular.

We complete Theorem II by proving

THEOREM III. *Let q be an r-rowed matrix of determinant one such that $q'q = I_r$, where I_r is the unit matrix of r rows and columns. Then we can find ρ_r skew-symmetric matrices of order r, $\alpha_1, \ldots, \alpha_{\rho_r}$, such that $I_r+\alpha_i$ is non-singular and*

$$q = \prod_{i=1}^{\rho_r}(I_r+\alpha_i)^{-1}(I_r-\alpha_i),$$

where ρ_r is a number depending only on r.

The actual value of ρ_r is not important; in the proof which follows we shall find it convenient to take

$$\rho_{2m} = 3m-1, \quad \rho_{2m+1} = 3m+1.$$

In the course of the proof, which will be by induction on r, we shall also have to prove

THEOREM IV. *If q is an orthogonal $r\times r$ matrix of determinant minus one, then we can find σ_r skew-symmetric matrices $\alpha_1, \ldots, \alpha_\sigma$ of r rows and columns such that $I_r+\alpha_i$ is non-singular and*

$$q = \left[\prod_{i=1}^{\sigma_r}(I_r+\alpha_i)^{-1}(I_r-\alpha_i)\right]X_j,$$

where X_j is a diagonal matrix obtained from I_r by changing the jth diagonal element to -1, and

$$\sigma_{2m} = 3m-2, \quad \sigma_{2m+1} = 3m-1.$$

We first establish a Lemma.

Lemma. *If* $x = \begin{pmatrix} x_1 \\ x_2 \\ \vdots \\ x_r \end{pmatrix}$ *and* $e = \begin{pmatrix} 1 \\ 0 \\ \vdots \\ 0 \end{pmatrix}$ *are two $r \times 1$ matrices,*

and $$x_1^2 + \ldots + x_r^2 = 1, \quad x_1 \neq -1,$$

we can find a skew-symmetric matrix α such that $I_r + \alpha$ is non-singular and

$$(I_r+\alpha)^{-1}(I_r-\alpha)x = e.$$

In fact, we take

$$\alpha = \begin{pmatrix} 0 & -\frac{x_2}{x_1+1} & . & -\frac{x_r}{x_1+1} \\ \frac{x_2}{x_1+1} & 0 & . & 0 \\ . & . & . & . \\ \frac{x_r}{x_1+1} & 0 & . & 0 \end{pmatrix}.$$

Then $$\det(I_r+\alpha) = 1 + \frac{\sum_2^r x_i^2}{(1+x_1)^2} = 1 + \frac{1-x_1^2}{(1+x_1)^2} = \frac{2}{1+x_1} \neq 0.$$

Also $$\alpha(x+e) = \begin{pmatrix} -\frac{\sum_2^r x_i^2}{x_1+1} \\ x_2 \\ \vdots \\ x_r \end{pmatrix} = \begin{pmatrix} x_1 - 1 \\ x_2 \\ \vdots \\ x_r \end{pmatrix} = x - e,$$

that is, $$(I_r+\alpha)e = (I_r-\alpha)x.$$

We begin our induction by considering the case $r = 2$. Let

$$q = \begin{pmatrix} a & b \\ c & d \end{pmatrix}$$

be a matrix such that $q'q = qq' = I_2$, and let Δ be the determinant of q. From the equations

$$qq' = q'q = I_2,$$

we see that

$$a^2+b^2 = a^2+c^2 = c^2+d^2 = b^2+d^2 = 1.$$

Suppose, first, that $a^2 = 1$. Then $b = c = 0$, $d^2 = 1$. Then

$$q = \begin{pmatrix} a & 0 \\ 0 & d \end{pmatrix},$$

where $ad = \Delta$, and $a = \pm 1$, $d = \pm 1$.

In the case of Theorem IV, $\Delta = -1$, and hence either a or d is equal to plus one, and the other is minus one. Since $\sigma_2 = 1$, we see that the theorem is satisfied by taking $\alpha_1 = 0$.

In the case of Theorem III, a and d are of like sign, $\rho_2 = 2$. If

$$a = d = 1,$$

we obtain the theorem by taking $\alpha_1 = \alpha_2 = 0$. If, however,

$$a = d = -1,$$

we have to take $\alpha_1 = \alpha_2 = \begin{pmatrix} 0 & 1 \\ -1 & 0 \end{pmatrix}$; we can then verify that

$$(I_2+\alpha_1)^{-1}(I_2-\alpha_1)(I_2+\alpha_2)^{-1}(I_2-\alpha_2) = \begin{pmatrix} -1 & 0 \\ 0 & -1 \end{pmatrix}.$$

Now suppose that $a \neq \pm 1$. Then, by the Lemma, we can find a skew matrix α such that

$$(I_2+\alpha)^{-1}(I_2-\alpha)\begin{pmatrix} a \\ c \end{pmatrix} = \begin{pmatrix} 1 \\ 0 \end{pmatrix}.$$

Then

$$(I_2+\alpha)^{-1}(I_2-\alpha)\,q = \begin{pmatrix} 1 & b_1 \\ 0 & d_1 \end{pmatrix}.$$

Since $(I_2+\alpha)^{-1}(I_2-\alpha)$ and q are orthogonal, so is their product [Theorem I]. Hence, as above, $b_1 = 0$ and $d_1 = \Delta$.

If $\Delta = 1$ and $\alpha_1 = -\alpha$, $\alpha_2 = 0$,

$$q = (I_2+\alpha_1)^{-1}(I_2-\alpha_1)(I_2+\alpha_2)^{-1}(I_2-\alpha_2);$$

and if $\Delta = -1$, we take $\alpha_1 = -\alpha$, and have

$$q = (I_2+\alpha_1)^{-1}(I_2-\alpha_1)\,X_2.$$

Now assume the truth of Theorems III and IV when $r = n$. Let $q = (q_{ij})$ $(i,j = 0, \dots, n)$ be an orthogonal matrix of determinant Δ. We have

$$q_{00}^2 + q_{10}^2 + \dots + q_{n0}^2 = 1.$$

If $q_{00} \neq -1$ we take x in the Lemma to be the first column of q, and if $q_{00} = -1$ we take $-x$ to be the first column of q. We thus see that we can find a skew $(n+1) \times (n+1)$ matrix α such that $(I_{n+1} + \alpha)$ is non-singular and such that

$$q_1 = (I_{n+1} + \alpha)^{-1} (I_{n+1} - \alpha) q = \begin{pmatrix} \epsilon & p_1 & \cdot & p_n \\ 0 & r_{11} & \cdot & r_{1n} \\ \cdot & \cdot & \cdot & \cdot \\ 0 & r_{n1} & \cdot & r_{nn} \end{pmatrix},$$

where $\epsilon = \pm 1$. Since $(I_{n+1} + \alpha)^{-1} (I_{n+1} - \alpha)$ and q are orthogonal, so is q_1. From the equation $q_1' q_1 = I_{n+1}$, we obtain $p_1 = \dots = p_n = 0$ and

$$r'r = I_n.$$

Also, $\Delta_1 = \det r = \epsilon \Delta$. We consider four cases.

(i) $\Delta = \epsilon = 1$. Then $\Delta_1 = 1$. By the hypothesis of induction, we can find ρ_n skew $n \times n$ matrices $\beta_1, \dots, \beta_{\rho_n}$ such that $I_n + \beta_i$ is non-singular and

$$r = \prod_{i=1}^{\rho_n} (I_n + \beta_i)^{-1} (I_n - \beta_i).$$

Let $\alpha_1 = -\alpha$, and let α_i be the matrix obtained from β_{i-1} by adding a first row and first column of zeros $(i = 2, \dots, \rho_n + 1)$; also take $\alpha_{\rho_n+2} = 0$, if $\rho_{n+1} = \rho_n + 2$. Then we have

$$q = \prod_{i=1}^{\rho_{n+1}} (I_{n+1} + \alpha_i)^{-1} (I_{n+1} - \alpha_i).$$

(ii) $\Delta = 1, \epsilon = -1$. Then $\Delta_1 = -1$. By the hypothesis of induction, we have, in the same way,

$$r = \left[\prod_{i=1}^{\sigma_n} (I_n + \beta_i)^{-1} (I_n - \beta_i) \right] X_j.$$

Taking $\alpha_1 = -\alpha$, and constructing $\alpha_2, \dots, \alpha_{\sigma_n+1}$ by bordering $\beta_1, \dots, \beta_{\sigma_n}$ as in (i), we obtain

$$q = \left[\prod_{i=1}^{\sigma_n+1} (I_{n+1} + \alpha_i)^{-1} (I_{n+1} - \alpha_i) \right] Y,$$

where Y is the matrix obtained by changing the sign of the first and $(j+1)$th elements of I_{n+1}. Let $\alpha_{\sigma_n+2} = \alpha_{\sigma_n+3}$ be the $(n+1)\times(n+1)$ skew matrix whose elements above the principal diagonal are zero except that in the first row and $(j+1)$th column, which is one. We then have

$$q = \prod_{i=1}^{\rho_{n+1}} (I_{n+1}+\alpha_i)^{-1}(I_{n+1}-\alpha_i),$$

since $\rho_{n+1} = \sigma_n + 3$.

(iii) $\Delta = -1$, $\epsilon = 1$. Then $\Delta_1 = -1$, and, as above,

$$r = \left[\prod_{i=1}^{\sigma_n} (I_n+\beta_i)^{-1}(I_n-\beta_i)\right] X_j.$$

Defining $\alpha_1, \ldots, \alpha_{\sigma_n+1}$ as above, we obtain

$$q = \left[\prod_{i=1}^{\sigma_{n+1}} (I_{n+1}+\alpha_i)^{-1}(I_{n+1}-\alpha_i)\right] X_j,$$

where $\alpha_{\sigma_{n+1}} = 0$, if $\sigma_{n+1}-\sigma_n = 2$.

(iv) $\Delta = -1$, $\epsilon = -1$. Then $\Delta_1 = 1$, and we have, in this case,

$$r = \prod_{i=1}^{\rho_n} (I_n+\beta_i)^{-1}(I_n-\beta_i),$$

and forming $\alpha_1, \ldots, \alpha_{\rho_n}$ from $\beta_1, \ldots, \beta_{\rho_n}$, as above, we have

$$q = \left[\prod_{i=1}^{\sigma_{n+1}} (I_{n+1}+\alpha_i)^{-1}(I_{n+1}-\alpha_i)\right] X_1,$$

since $\sigma_{n+1} = \rho_n$.

This completes the proof of Theorems III and IV. We are mainly concerned with Theorem III. Let $\gamma_1, \ldots, \gamma_{\rho_{n+1}}$ be ρ_{n+1} skew-symmetric matrices of $n+1$ rows and columns whose elements above the principal diagonal are all independent indeterminates. Then

$$Q = \prod_{i=1}^{\rho_{n+1}} (I_{n+1}+\gamma_i)^{-1}(I_{n+1}-\gamma_i)$$

is an orthogonal matrix whose determinant is plus one. Theorem III tells us that any orthogonal matrix of order $n+1$ whose determinant is plus one can be obtained from Q by proper specialisation of the γ_i. The orthogonal matrices of order $n+1$ and determinant plus one form an irreducible system, in the following sense. We represent the $(n+1)\times(n+1)$ matrix (a_{ij}) $(i, j = 0, \ldots, n)$ by the point with coordinates $(a_{00}, a_{01}, \ldots, a_{0n}, a_{10}, \ldots, a_{nn})$ of an $(n+1)^2$-dimensional space in which non-homogeneous coordinates are

$(X_{00}, \ldots, X_{nn})$. Then the orthogonal matrices of determinant plus one are represented by just those points which satisfy the equations

$$\left.\begin{aligned} \sum_{i=0}^{n} X_{ih} X_{ik} &= \delta_{hk} \quad (h, k = 0, \ldots, n), \\ \det(X_{ij}) &= 1, \end{aligned}\right\} \tag{3}$$

and do not lie in the prime at infinity. The matrix Q is a point Ξ of this locus, and the point which represents any orthogonal matrix of determinant plus one is a proper specialisation of Ξ. Hence Ξ is a generic point of the locus M given by (3), which is therefore irreducible.

Not all orthogonal matrices q of determinant plus one satisfy the condition

$$\det(I_{n+1} + q) = 0.$$

Hence the generic orthogonal matrix Q cannot satisfy this equation. It follows from Theorem II that we can find a skew-symmetric matrix A such that

$$Q = (I_{n+1} + A)^{-1}(I_{n+1} - A), \quad A = (I_{n+1} - Q)(I_{n+1} + Q)^{-1}.$$

The dimension of M is equal to the dimension of the field

$$K(Q) = K(\ldots, Q_{ij}, \ldots)$$

over K. But

$$K(Q) = K(A),$$

and hence the dimension of $K(Q)$ cannot exceed $\frac{1}{2}n(n+1)$. Now let B be a skew-symmetric matrix of order $n+1$ whose elements above the principal diagonal are independent indeterminates, and let

$$R = (I_{n+1} + B)^{-1}(I_{n+1} - B), \quad B = (I_{n+1} - R)(I_{n+1} + R)^{-1}.$$

R is an orthogonal matrix of determinant plus one, hence it is a proper specialisation of Q. But

$$K(R) = K(B)$$

is of dimension $\frac{1}{2}n(n+1)$ over K; hence the dimension of $K(Q)$, which is not less than the dimension of $K(R)$, is $\frac{1}{2}n(n+1)$ at least. Hence $K(Q)$ must be of dimension $\frac{1}{2}n(n+1)$ exactly. Thus we have

THEOREM V. *The orthogonal $(n+1) \times (n+1)$ matrices of determinant plus one form an irreducible system of dimension $\frac{1}{2}n(n+1)$.*

We now return to the problem of the transformation of the quadric Q_{n-1}:

$$x_0^2 + \ldots + x_n^2 = 0$$

into itself. We saw that if n is even, this reduces to the problem of finding the collineation

$$y = qx,$$

where q is an orthogonal matrix of determinant plus one. This collineation transforms linear spaces into linear spaces, and in particular it transforms any r-space of Q_{n-1} into an r-space of Q_{n-1}.

If n is odd, we have two types of collineation which transform Q_{n-1} into itself, namely, those which can be written in the form $y = qx$, where q is orthogonal and of determinant plus one, and those which can be written in the form $y = qx$, where q is orthogonal and of determinant minus one. Both types again transform r-spaces on Q_{n-1} into r-spaces of Q_{n-1}, but we now have to consider the case in which $r = m = \frac{1}{2}(n-1)$, since there are two systems of linear spaces of dimension m [§ 4] on the Q_{n-1} in this case.

We consider first the case in which $\det q = 1$. If Q is a generic orthogonal matrix of determinant plus one, the collineation $y = Qx$ transforms any m-space π of Q_{n-1} into an m-space π_Q. π_Q is a generic member of an irreducible system σ of m-spaces on Q_{n-1}. When $Q \rightarrow q$, $\pi_Q \rightarrow \pi_q$, where π_q is the m-space into which π is transformed by the collineation q. In particular, if q is the unit matrix, $\pi_q = \pi$. Hence the irreducible system σ contains π. It follows that the system σ is contained in the system $\mathfrak{S}$ of m-spaces on Q_{n-1} which contains π. Hence the collineations given by the orthogonal matrices of determinant plus one transform any m-space of Q_{n-1} into an m-space of the same system.

Now suppose that $\det q = -1$. Then we can write

$$q = rF,$$

where F is obtained from the unit matrix by changing the sign of the bottom right-hand element, and r is an orthogonal matrix with determinant plus one. Let the collineation

$$y = Fx$$

transform the m-space π into the m-space π'. Then the collineation

$$y = rx$$

transforms π' into an m-space π_1 of Q_{n-1}, which is the transform of π under

$$y = qx.$$

π_1 and π' are therefore in the same system of m-spaces on Q_{n-1}. We therefore study the effect of the collineation Γ with equations

$$y = Fx$$

on the m-spaces of Q_{n-1}. Writing the equation of Q_{n-1} in the form

$$(x_0+ix_1)(x_0-ix_1)+\ldots+(x_{n-1}+ix_n)(x_{n-1}-ix_n) = 0,$$

it follows that any m-space π on Q_{n-1} is given by an equation of the form

$$\begin{pmatrix} x_{i_0}+ix_{i_0+1} \\ \cdot \\ x_{i_k}+ix_{i_k+1} \\ x_{i_{k+1}}-ix_{i_{k+1}+1} \\ \cdot \\ x_{i_m}-ix_{i_m+1} \end{pmatrix} = P \begin{pmatrix} x_{i_0}-ix_{i_0+1} \\ \cdot \\ x_{i_k}-ix_{i_k+1} \\ x_{i_{k+1}}+ix_{i_{k+1}+1} \\ \cdot \\ x_{i_m}+ix_{i_m+1} \end{pmatrix},$$

where $i_0, i_1, \ldots, i_m$ is a derangement of $0, 2, \ldots, n-1$, and P is a skew-symmetric matrix. The collineation Γ has the effect of changing the sign of x_n in these equations. The equations so derived for π' are again in the standard form, but in this case we have either k or $k+2$ equations in which $i = \sqrt{(-1)}$ appears with positive sign on the left. Hence π' belongs to the opposite system of m-spaces on Q_{n-1} to π. Hence π and π_1 are of opposite systems. Thus we have

THEOREM VI. *The collineations $y = qx$, where q is orthogonal, transforming the quadric Q_{n-1},*

$$x_0^2+\ldots+x_n^2 = 0$$

into itself transform any r-space on Q_{n-1} into an r-space on Q_{n-1}. If $n = 2m+1$, the m-spaces of either system on Q_{n-1} are transformed into m-spaces of the same system if $\det q = 1$, *and into m-spaces of the opposite system if* $\det q = -1$.

9. The elementary divisors of orthogonal matrices. In this section we consider the elementary divisors of an orthogonal matrix p of order $n+1$, and the relation between the united spaces of the collineation

$$y = px$$

and the quadric Q_{n-1} whose equation is

$$x_0^2 + \ldots + x_n^2 = 0.$$

THEOREM I. *If a and b are orthogonal matrices of order $n+1$, a necessary and sufficient condition that there exists an orthogonal matrix c such that $a = c'bc$ is that the matrices $a - \lambda I_{n+1}$ and $b - \lambda I_{n+1}$ have the same elementary divisors.*

The condition is obviously necessary, since if

$$a = c'bc,$$

where c is orthogonal, then $c' = c^{-1}$ so that

$$a = c^{-1}bc,$$

and therefore $$a - \lambda I_{n+1} = c^{-1}(b - \lambda I_{n+1})c.$$

On the other hand, if $a - \lambda I_{n+1}$ and $b - \lambda I_{n+1}$ have the same elementary divisors, we can find a non-singular matrix t so that [VIII, § 2, Th. I]

$$a = t^{-1}bt.$$

Then, since $\tilde{a} = (a')^{-1} = a$, $\tilde{b} = b$, we have

$$t^{-1}bt = a = \tilde{a} = (\tilde{t}^{-1})\tilde{b}\tilde{t} = t'b\tilde{t}.$$

Therefore $$tt'b = btt'.$$

From this we deduce that

$$(tt')^2 b = tt'btt' = b(tt')^2,$$

and, generally, $$(tt')^r b = b(tt')^r.$$

Therefore, if $f(x)$ is any polynomial with coefficients in the ground field K,

$$f(tt')b = bf(tt').$$

Since tt' is non-singular, and since K is algebraically closed, we can [II, § 10, Th. II] choose $f(x)$ so that, if $s = f(tt')$,

$$s^2 = tt'.$$

Then we have $$sb = bs.$$

Also, since $\quad (tt')' = tt', \quad s' = f((tt')') = f(tt') = s.$

Now
$$\begin{aligned} a &= t^{-1}bt \\ &= t^{-1}btt'\tilde{t} \\ &= t^{-1}bs^2\tilde{t} \\ &= t^{-1}sbs\tilde{t}. \end{aligned}$$

Let $c = s\tilde{t}$. Then $c' = t^{-1}s$, and
$$c'c = t^{-1}s^2\tilde{t} = t^{-1}tt'\tilde{t} = I_{n+1}.$$

Thus
$$a = c'bc,$$

where c is orthogonal. This completes the proof of Theorem I.

Now let q be any orthogonal matrix of order $n+1$. Since $q'q = I_{n+1}$,
$$q - \lambda I_{n+1} = \tilde{q} - \lambda I_{n+1}.$$

Since $q^{-1} - \lambda I_{n+1}$ and $\tilde{q} - \lambda I_{n+1}$ have the same elementary divisors, we deduce that $q - \lambda I_{n+1}$ and $q^{-1} - \lambda I_{n+1}$ have the same elementary divisors. If the elementary divisors of $q - \lambda I_{n+1}$ are
$$(\lambda - \alpha_1)^{e_1}, \quad (\lambda - \alpha_2)^{e_2}, \quad \ldots, \quad (\lambda - \alpha_k)^{e_k},$$

the elementary divisors of $q^{-1} - \lambda I_{n+1}$ are [VIII, § 3, p. 337]
$$(\lambda - \alpha^{-1})^{e_1}, \quad (\lambda - \alpha^{-1})^{e_2}, \quad \ldots, \quad (\lambda - \alpha_k^{-1})^{e_k}.$$

Hence we deduce that if q is orthogonal its elementary divisors are of the form
$$\left.\begin{array}{lll} (\lambda - \beta_i)^{e_i}, & (\lambda - \beta_i^{-1})^{e_i} & (i = 1, 2, \ldots, t)\ (\beta_i \neq \pm 1), \\ (\lambda - 1)^{\mu_i} & & (i = 1, 2, \ldots, a), \\ (\lambda + 1)^{\nu_i} & & (i = 1, 2, \ldots, b). \end{array}\right\} \tag{1}$$

We have to determine whether there exists an orthogonal matrix of order $n+1$ with these elementary divisors, where $\beta_1, \ldots, \beta_t$, $t, a, b, e_1, e_2, \ldots, \mu_1, \mu_2, \ldots, \nu_1, \nu_2, \ldots$ can take any values, subject to the obvious condition $2\Sigma e_i + \Sigma \mu_i + \Sigma \nu_i = n+1$, or whether further restrictions are imposed on the elementary divisors of orthogonal matrices.

Let $C_e(\alpha)$ be the $e \times e$ matrix
$$C_e(\alpha) = \begin{pmatrix} \alpha & 1 & 0 & . & 0 \\ 0 & \alpha & 1 & . & 0 \\ . & . & . & . & . \\ . & . & . & \alpha & 1 \\ 0 & 0 & 0 & . & \alpha \end{pmatrix},$$

and write $C_e(1) = E_e$, $C_e(-1) = F_e$. Let

$$A = \begin{pmatrix} C_{e_1}(\beta_1) & 0 & & & & & & & & & \\ 0 & C_{e_1}(\beta_1^{-1}) & & & & & & & & & \\ & & \ddots & & & & & & & & \\ & & & C_{e_t}(\beta_t) & & & & & & & \\ & & & & C_{e_t}(\beta_t^{-1}) & & & & & & \\ & & & & & E_{\mu_1} & & & & & \\ & & & & & & \ddots & & & & \\ & & & & & & & E_{\mu_a} & & & \\ & & & & & & & & F_{\nu_1} & & \\ & & & & & & & & & \ddots & 0 \\ & & & & & & & & & 0 & F_{\nu_b} \end{pmatrix}, \quad (2)$$

where we may assume $\mu_1 \geqslant \mu_2 \geqslant \ldots \geqslant \mu_a$, $\nu_1 \geqslant \nu_2 \geqslant \ldots \geqslant \nu_b$. If $q - \lambda I_{n+1}$ has the invariant factors (1), there exists a non-singular matrix t such that [VIII, § 2, Th. I]

$$q = t^{-1}At,$$

and since $q'q = I_{n+1}$,

$$t'A'\tilde{t}t^{-1}At = I_{n+1},$$

and hence

$$A'(\tilde{t}t^{-1})A = \tilde{t}t^{-1}.$$

We note that $\tilde{t}t^{-1}$ is non-singular and symmetric.

Conversely, suppose there exists a non-singular symmetric matrix B such that

$$A'BA = B.$$

Since B is a symmetric and non-singular matrix, and the ground field K is algebraically closed, we can find a non-singular matrix s such that $s's = B$. Then

$$A's'sA = s's,$$

and hence sAs^{-1} is an orthogonal matrix. The elementary divisors of $sAs^{-1} - \lambda I_{n+1}$ are given by (1). Hence we see that *a necessary and sufficient condition that there should exist an orthogonal matrix q such that $q - \lambda I_{n+1}$ has the elementary divisors* (1) *is that there should exist a non-singular symmetric matrix B such that $A'BA = B$, where A is given by* (2).

We use this fact to determine what restrictions, if any, must be imposed on the elementary divisors (1) in order that they should be elementary divisors of $q - \lambda I_{n+1}$, where q is orthogonal. Let B be an $(n+1)\times(n+1)$ matrix, and divide it into blocks

$$B = \begin{pmatrix} B_{11} & B_{12} & . & B_{1\,2l+a+b} \\ B_{21} & . & . & . \\ . & . & . & . \\ B_{2l+a+b\,1} & B_{2l+a+b\,2} & . & B_{2l+a+b\,2l+a+b} \end{pmatrix},$$

corresponding to the division of A into blocks in (2). If

$$A'BA = B, \tag{3}$$

we obtain equations of the form

$$C'_r(\alpha)\, B_{lm} C_s(\beta) = B_{lm}, \tag{4}$$

where B_{lm} is an $r \times s$ submatrix of B. We first show that if $\alpha\beta \neq 1$, this implies that $B_{lm} = 0$. Let

$$B_{lm} = \begin{pmatrix} b_{11} & . & b_{1s} \\ . & . & . \\ b_{r1} & . & b_{rs} \end{pmatrix}.$$

Then (4) becomes

$$(\alpha\beta - 1)\, b_{ij} + \alpha b_{ij-1} + \beta b_{i-1\,j} + b_{i-1\,j-1} = 0$$
$$(i = 1, \ldots, r;\ j = 1, \ldots, s), \tag{5}$$

where $$b_{0j} = 0, \quad b_{i0} = 0.$$

Put $j = 1$, and consider these equations for $i = 1, \ldots, r$ successively. We get, in turn,

$$b_{11} = 0, \quad b_{21} = 0, \quad \ldots, \quad b_{r1} = 0.$$

Similarly, putting $i = 1$, and then $j = 1, \ldots, r$ in turn, we get

$$b_{11} = 0, \quad b_{12} = 0, \quad \ldots, \quad b_{1r} = 0.$$

Equations (5) then become the same equations but with $i = 2, \ldots, r$, $j = 2, \ldots, s$, and with

$$b_{1j} = 0, \quad b_{i1} = 0,$$

and by an elementary induction we deduce that each b_{ij} is zero. Hence $B_{lm} = 0$. It follows that if B satisfies (3) it is of the form

$$B = \begin{pmatrix} Y & 0 & 0 \\ 0 & Z & 0 \\ 0 & 0 & T \end{pmatrix},$$

where Y is a $2\Sigma e_i \times 2\Sigma e_i$ matrix, Z is a $\Sigma\mu_i \times \Sigma\mu_i$ matrix and T is a $\Sigma\nu_i \times \Sigma\nu_i$ matrix. If B is to be non-singular, Y, Z, T must be non-singular. If

$$A = \begin{pmatrix} A_1 & 0 & 0 \\ 0 & A_2 & 0 \\ 0 & 0 & A_3 \end{pmatrix}$$

is the corresponding division of A into blocks we must have

$$A_1' Y A_1 = Y, \quad A_2' Z A_2 = Z, \quad A_3' T A_3 = T.$$

The existence of a symmetric matrix Y satisfying this condition is easily established. $C_{e_i}(\beta_i^{-1}) - \lambda I_{e_i}$ and $\tilde{C}_{e_i}(\beta_i) - \lambda I_{e_i}$ have the same elementary divisors. Hence there exists a non-singular matrix P_i such that

$$P_i^{-1} \tilde{C}_{e_i}(\beta_i) P_i = C_{e_i}(\beta_i^{-1}).$$

Then if

$$Y_1 = \begin{pmatrix} 0 & P_1 & 0 & 0 & . & . & 0 \\ P_1' & 0 & 0 & 0 & . & . & 0 \\ 0 & 0 & 0 & P_2 & . & . & . \\ 0 & 0 & P_2' & . & . & . & . \\ . & . & . & . & . & . & . \\ 0 & . & . & . & . & 0 & P_t \\ 0 & . & . & . & . & P_t' & 0 \end{pmatrix},$$

Y_1 is symmetric and non-singular, and

$$A_1' Y_1 A_1 = Y_1.$$

Hence there exists a non-singular symmetric matrix Y_1 satisfying the required condition. Y_1 is a particular form of Y, and we shall continue to denote by Y *any* matrix satisfying the requirement $A_1' Y A_1 = Y$,

We next try to find a matrix Z such that

$$A_2' Z A_2 = Z. \tag{6}$$

In this case it will be necessary to find the most general form of Z satisfying this condition. We write Z in blocks:

$$Z = \begin{pmatrix} Z^{(11)} & . & Z^{(1a)} \\ Z^{(21)} & . & Z^{(2a)} \\ . & . & . \\ Z^{(a1)} & . & Z^{(aa)} \end{pmatrix},$$

where $Z^{(ij)}$ is a $\mu_i \times \mu_j$ matrix. Then equation (6) leads to the equation

$$E'_{\mu_i} Z^{(ij)} E_{\mu_j} = Z^{(ij)},$$

and this is equivalent to

$$z^{(ij)}_{p\,q-1} + z^{(ij)}_{p-1\,q} + z^{(ij)}_{p-1\,q-1} = 0 \quad (p = 1, \ldots, \mu_i;\ q = 1, \ldots, \mu_j),$$

where $Z^{(ij)} = (z^{(ij)}_{pq})$ and $z^{(ij)}_{po} = z^{(ij)}_{oq} = 0$.

These equations can be solved by recurrence methods just as the equations (5) for B_{lm} above. We find that if $\mu_i > \mu_j$, $z^{(ij)}_{pq} = 0$ if $p+q \leqslant \mu_i$, and finally we obtain

$$Z^{(ij)} = \begin{pmatrix} 0 & \cdot & \cdot & 0 \\ 0 & \cdot & \cdot & 0 \\ & & & \theta_{ij} \\ & & \cdots & \\ 0 & \pm\theta_{ij} & & \\ \mp\theta_{ij} & \cdot & & \cdot \end{pmatrix},$$

where θ_{ij} is arbitrary, and the elements above the diagonal in which the $\pm\theta_{ij}$ appears are zero; the elements in the last row are arbitrary, and the form of the remaining elements below the diagonal is easily calculated if required. Similarly, if $\mu_i \leqslant \mu_j$, we find that

$$Z^{(ij)} = \begin{pmatrix} 0 & \cdot & \cdot & \cdot & 0 & \theta_{ij} \\ & & & 0 & -\theta_{ij} & \cdot \\ & & \cdots & & & \\ 0 & \cdot & 0 & \pm\theta_{ij} & \cdot & \cdot \end{pmatrix},$$

where the elements above the diagonal containing $\pm\theta_{ij}$ are zero, the elements in the last column are arbitrary, and the remaining elements are easily calculated when required.

We now have to consider the conditions for $Z = (Z^{(ij)})$ to be symmetric and non-singular. Let us suppose that

$$\mu_1 = \mu_2 = \ldots = \mu_{a_1} \quad = \rho_1,$$

$$\mu_{a_1+1} = \mu_{a_1+2} = \ldots = \mu_{a_1+a_2} = \rho_2 < \rho_1,$$

$$\cdot \quad \cdot \quad \cdot \quad \cdot$$

$$\mu_{a-a_k+1} = \ldots = \mu_a \quad = \rho_k < \rho_{k-1},$$

where $a_1 + \ldots + a_k = a$. We then have the following form for Z (taking, as an example, the case $\mu_1 = \mu_2 > \mu_3$):

$$
Z = \begin{array}{cccccccccccccccc}
0 & . & . & . & 0 & \theta_{11} & 0 & . & . & . & 0 & \theta_{12} & 0 & . & . & 0 \\
 & & & & -\theta_{11} & * & & & & & -\theta_{12} & * & & & & \\
 & & & . & & & & & & . & & & & & & \theta_{13} \\
 & & . & & & & & & . & & & & & & . & * \\
 & . & & & & & & . & & & & & & . & & \\
\pm\theta_{11} & * & & & & & \pm\theta_{12} & * & & & & & \pm\theta_{13} & * & & \\
0 & . & . & . & 0 & \theta_{21} & 0 & . & . & . & 0 & \theta_{22} & 0 & . & . & 0 \\
 & & & & -\theta_{21} & * & & & & & -\theta_{22} & * & & & & \\
 & & & . & & & & & & . & & & & & & \theta_{23} \\
 & & . & & & & & & . & & & & & & . & * \\
 & . & & & & & & . & & & & & & . & & \\
\pm\theta_{21} & * & & & & & \pm\theta_{22} & * & & & & & \pm\theta_{23} & * & & \\
0 & . & . & . & 0 & \theta_{31} & 0 & . & . & . & 0 & \theta_{32} & 0 & . & 0 & \theta_{33} \\
 & & & & -\theta_{31} & * & & & & & -\theta_{32} & * & & & -\theta_{33} & * \\
 & & & . & & & & & & . & & & & . & & \\
0 & & \pm\theta_{31} & * & & & 0 & & \pm\theta_{32} & * & & & \pm\theta_{33} & * & & \\
. & . & . & . & . & . & . & . & . & . & . & . & . & . & . & .
\end{array} \tag{7}
$$

If Z is symmetrical, we must have $\theta_{ij} = (-1)^{\rho_1 - 1}\theta_{ji}\ (i,j = 1, \ldots, a_1)$, $\theta_{ij} = (-1)^{\rho_2 - 1}\theta_{ji}\ (i,j = a_1+1, \ldots, a_1+a_2)$, etc. We expand the determinant of Z in terms of the first, (μ_1+1)th, $(\mu_1+\mu_2+1)$th, ... rows, obtaining

$$\det Z = \Theta \det Z^{(1)},$$

where

$$
\Theta = \det\begin{pmatrix}
\theta_{11} & . & \theta_{1a_1} & 0 & . & 0 & . \\
. & . & . & . & . & . & . \\
\theta_{a_1 1} & . & \theta_{a_1 a_1} & 0 & . & 0 & . \\
\theta_{a_1+1\,1} & . & \theta_{a_1+1\,a_1} & \theta_{a_1+1\,a_1+1} & . & . & . \\
. & . & . & . & . & . & . \\
\theta_{a_1+a_2\,1} & . & \theta_{a_1+a_2\,a_1} & . & . & \theta_{a_1+a_2\,a_1+a_2} & . \\
. & . & . & . & . & . & .
\end{pmatrix}
= \prod_{i=1}^{k} \Phi_i,
$$

where $\Phi_i = \det(\theta_{ab})\quad (a,b = a_1 + \ldots + a_{i-1} + 1, \ldots, a_1 + \ldots + a_i)$.

Expanding $\det Z^{(1)}$ in terms of the first, μ_1th, $(\mu_1+\mu_2-1)$th, ... columns, we obtain

$$\det Z^{(1)} = \pm\, \Theta_1 \det Z^*,$$

where Θ_1 differs from Θ only by the omission of the factor Φ_k in the case $\mu_a = 1$. Z^* is the matrix obtained from Z by striking out the first and last rows and columns of each submatrix $Z^{(ij)}$. Proceeding to expand Z^* similarly, and so on, we finally obtain

$$\det Z = \pm \prod_{i=1}^{k} \Phi_i^{\mu_i}.$$

In order that Z be non-singular, it is necessary and sufficient that Φ_i should be different from zero, for $i = 1, \ldots, k$. Now if ρ_i is even, Φ_i is the determinant of a skew-symmetric matrix of order a_i; hence it can only be non-singular if a_i is even. On the other hand, if a_i is even whenever μ_i is even, it is easily seen that we can choose the θ_{ij} so that $\det Z$ is different from zero. Thus a necessary and sufficient condition that there exists a symmetric matrix Z satisfying equation (6) is that the number of elementary divisors of A of the form $(\lambda-1)^{2m}$ should be even.

When this condition is satisfied we shall find it convenient to take Z in the general form represented by (7). A special form Z_1 of Z can, however, be constructed as follows. We pair off the elementary divisors $(\lambda-1)^e$ (each pair having the same exponent) as far as possible. Then for any pair $(\lambda-1)^e$, $(\lambda-1)^e$; corresponding to the ith and $(i+1)$th blocks of A_2, we find P_i such that

$$P_i^{-1}\bar{C}_e(1)\,P_i = C_e(1),$$

and put

$$Z_1^{(i,i+1)} = P_i, \quad Z_1^{(i+1,i)} = P_i'.$$

The only elementary divisors not paired are of the form $(\lambda-1)^{2m+1}$, no two such divisors having the same exponent. Suppose $(\lambda-1)^{2m+1}$ is the elementary divisor corresponding to the jth block of A_2. Let

$$W_{2m+1} = \begin{pmatrix} 0 & 0 & 0 & . & 0 & 1 \\ 0 & 0 & 0 & . & -1 & 0 \\ . & . & . & . & . & . \\ . & . & 1 & . & . & . \\ 0 & -1 & -\binom{2m-2}{1} & . & -1 & 0 \\ 1 & \binom{2m-1}{1} & \binom{2m-1}{2} & . & 1 & 0 \end{pmatrix},$$

in which non-zero terms in the $(r+2)$th row are

$$(-1)^{r-1}\left[1, \binom{r}{1}, \binom{r}{2}, \ldots, 1\right].$$

Then $$E'_{2m+1}W_{2m+1}E_{2m+1} = W_{2m+1}.$$

If $$Z_1^{(jj)} = W_{2m+1} + W'_{2m+1},$$

$Z_1^{(jj)}$ is non-singular and symmetric, and

$$E'_{2m+1}Z_1^{(jj)}E_{2m+1} = Z^{(jj)}.$$

We take this for the block $Z_1^{(jj)}$ of Z_1. Put all the remaining $Z_1^{(ij)}$ equal to zero. Then Z_1 is symmetrical and non-singular, and satisfies the equation (6).

Finally, we have to determine whether there exists a non-singular symmetric matrix T such that

$$A'_3TA_3 = T.$$

This case is treated in exactly the same way as equation (6), and we find that T is necessarily of the form (7) and can be chosen to be non-singular if and only if the elementary divisors of the form $(\lambda+1)^{2m}$ are even in number. Thus we have

THEOREM II. *If q is an orthogonal $(n+1)\times(n+1)$ matrix, $q-\lambda I_{n+1}$ has elementary divisors of the form*

$$(\lambda-\beta_i)^{e_i}, \quad (\lambda-\beta_i^{-1})^{e_i} \quad (i = 1, \ldots, t),$$

$$(\lambda-1)^{\mu_i} \quad (i = 1, \ldots, a),$$

where if μ_i is even there is an even number with exponent μ_i;

$$(\lambda+1)^{\nu_i} \quad (i = 1, \ldots, b),$$

where if ν_i is even there is an even number with exponent ν_i; and

$$2\Sigma e_i + \Sigma\mu_i + \Sigma\nu_i = n+1;$$

conversely, if $\beta_i, e_i, \mu_i, \nu_i$ are any numbers satisfying these conditions, we can find an orthogonal $(n+1)\times(n+1)$ matrix q so that $q-\lambda I_{n+1}$ has these elementary divisors.

Now let B be any symmetric matrix such that

$$A'BA = B.$$

The collineation $$y = Ax \tag{8}$$

transforms the quadric $$x'Bx = 0 \tag{9}$$

into itself. We study the relation of the united spaces of the collineation to this quadric. If β is any latent root of A, the points of the corresponding united space satisfy the condition

$$\beta x = Ax.$$

Hence if x is in the united space we have

$$\beta^2 x'Bx = x'A'BAx = x'Bx.$$

If $\beta \neq \pm 1$, it follows that $x'Bx = 0$; hence the united space of (8) corresponding to any latent root $\beta \neq \pm 1$, lies on the quadric (9).

Now consider a latent root of A equal to 1. The corresponding united space is obtained by putting each x_i equal to zero except those x_i which correspond to the

$$\text{1st}, \quad (\mu_1 + 1)\text{th}, \quad \ldots, \quad (\mu_1 + \ldots + \mu_{a-1} + 1)\text{th}$$

rows and columns of Z, given by (7). When we substitute in (9) the resulting equation is

$$\Sigma B_{ij} x_i x_j = 0, \tag{10}$$

where the only terms present come from the top left-hand elements of the matrices $Z^{(ij)}$. Unless $\mu_i = \mu_j = 1$ this element is zero; hence if $\rho_k > 1$, this is an identity; the united space therefore lies on the quadric. If $\rho_k \doteq 1$, however, (10) is not identically zero, and the determinant of the form is $\Phi_k \neq 0$. Hence the united space meets the quadric (9) in a cone of type $a_1 + \ldots + a_{k-1}$.

Similar reasoning applies to the case of a latent root of A equal to -1, and we obtain

THEOREM III. *If q is an orthogonal matrix, the united space of the collineation*

$$y = qx$$

corresponding to a latent root of q not equal to plus or minus one lies on the quadric

$$x_0^2 + \ldots + x_n^2 = 0;$$

the united space corresponding to the latent root $+1$ (-1) lies on the quadric if $q - \lambda I_{n+1}$ has no linear elementary divisor $\lambda - 1$ $(\lambda + 1)$, and if it has such an elementary divisor the united space corresponding to the latent root $+1$ (-1) meets the quadric in a cone of type σ, where σ is the number of non-linear elementary divisors $(\lambda - 1)^e$ $((\lambda + 1)^e)$.

10. Pairs of quadrics. As a preliminary to the study of the intersection of two quadrics, this section is devoted to the problem of finding a canonical form for the equations of two quadrics in S_n, assumed distinct. Let the equations of the quadrics be

$$x'ax = 0, \quad x'bx = 0,$$

referred to the coordinate system $(x_0, \ldots, x_n)$. If

$$x = py,$$

where p is a non-singular $(n+1) \times (n+1)$ matrix with elements in K, is any allowable transformation of coordinates, the equations of the quadrics in the coordinate system $(y_0, \ldots, y_n)$ are

$$y'p'apy = 0, \quad y'p'bpy = 0,$$

and our problem is to find a canonical form to which the equations of the quadrics can be reduced by a suitable choice of the matrix p. The problem is very similar to that investigated in Chapter IX, where we sought a canonical form for a pair of bilinear forms, one of which was symmetric and the other skew-symmetric. We shall follow a similar method to that used in Chapter IX, and where the arguments are the same as, or differ only trivially from, those used earlier, we shall save space by not giving them in full.

Our first result deals with the case in which the matrix b is non-singular.

THEOREM I. *If $x'ax$ and $x'bx$ are two quadratic forms in $(x_0, \ldots, x_n)$, and b is a non-singular matrix, a necessary and sufficient condition that there should exist an allowable transformation of coordinates $x = py$ which transforms the quadratic forms to $y'cy$ and $y'dy$ is that the λ-matrices $a - \lambda b$ and $c - \lambda d$ have the same elementary divisors.*

The condition is clearly necessary. For if there exists a non-singular matrix p over K such that

$$p'ap = c, \quad p'bp = d,$$

then
$$p'(a-\lambda b)p = c - \lambda d,$$

and [II, § 9, Th. III] $a - \lambda b$ and $c - \lambda d$ have the same elementary divisors.

Conversely, if $a - \lambda b$ and $c - \lambda d$ have the same elementary divisors, and if b is non-singular, it follows from Chapter II, § 9, Th. V that there exist non-singular matrices m and n over K such that

$$m(c - \lambda d)\, n = a - \lambda b,$$

that is $$mcn = a, \quad mdn = b.$$

Now a, b, c, d are all symmetric, hence we have

$$n'cm' = n'c'm' = a' = a,$$

$$n'dm' = n'd'm' = b' = b;$$

therefore $$n'm^{-1}an^{-1}m' = a,$$

and $$n'm^{-1}bn^{-1}m' = b.$$

We write these equations in the form

$$n'm^{-1}a = a(n'm^{-1})', \quad n'm^{-1}b = b(n'm^{-1})'.$$

If $t = n'm^{-1}$, t is non-singular, since m^{-1} and n are non-singular. We have

$$ta = at', \qquad tb = bt',$$

$$t^2a = tat' = a(t')^2, \quad t^2b = b(t')^2,$$

$$\cdots\cdots\cdots\cdots\cdots, \qquad \cdots\cdots\cdots\cdots,$$

$$t^ra = a(t')^r, \quad t^rb = b(t')^r;$$

and, more generally, if $f(x)$ is any polynomial over K,

$$f(t)\,a = af(t') = a[f(t)]',$$

$$f(t)\,b = bf(t') = b[f(t)]'.$$

Since t is non-singular, we can choose $f(x)$ [II, § 10, Th. II] so that if $s = f(t)$, then $s^2 = t$.

With this choice of s, we have

$$sa = as', \quad sb = bs'.$$

Now $$\begin{aligned} c = m^{-1}an^{-1} &= (n^{-1})'\,tan^{-1} \\ &= (n^{-1})'\,s^2an^{-1} \\ &= (n^{-1})'\,sas'n^{-1} \\ &= (s'n^{-1})'\,as'n^{-1}, \end{aligned}$$

and similarly $$d = (s'n^{-1})'\,bs'n^{-1}.$$

Now s' and n^{-1} are non-singular, and therefore $p = s'n^{-1}$ is non-singular. We have thus found a non-singular matrix p over K such that

$$c = p'ap, \quad d = p'bp,$$

and thus Theorem I is proved.

We use Theorem I to construct a canonical form for the pair of quadratic forms $x'ax$ and $x'bx$ in the case in which b is non-singular. Suppose that the elementary divisors of $a-\lambda b$ are

$$(\lambda-\alpha_1)^{e_1}, \quad (\lambda-\alpha_2)^{e_2}, \quad \ldots, \quad (\lambda-\alpha_r)^{e_r},$$

where
$$e_1+e_2+\ldots+e_r = n+1.$$

Let $P_e(\alpha)$ be the $e \times e$ symmetric matrix

$$P_e(\alpha) = \begin{pmatrix} 0 & . & . & . & 0 & 0 & \alpha \\ 0 & . & . & . & 0 & \alpha & 1 \\ 0 & . & . & . & \alpha & 1 & 0 \\ . & . & . & . & . & . & . \\ \alpha & 1 & 0 & . & 0 & 0 & 0 \end{pmatrix},$$

and let Q_e be the $e \times e$ symmetric matrix

$$Q_e = \begin{pmatrix} 0 & . & 0 & 0 & 1 \\ 0 & . & 0 & 1 & 0 \\ 0 & . & 1 & 0 & 0 \\ . & . & . & . & . \\ 1 & . & 0 & 0 & 0 \end{pmatrix}.$$

If
$$p = \begin{pmatrix} P_{e_1}(\alpha_1) & 0 & . & . \\ 0 & P_{e_2}(\alpha_2) & . & . \\ . & . & . & . \\ . & . & . & P_{e_r}(\alpha_r) \end{pmatrix}$$

and
$$q = \begin{pmatrix} Q_{e_1} & 0 & . & . \\ 0 & Q_{e_2} & . & . \\ . & . & . & . \\ . & . & . & Q_{e_r} \end{pmatrix},$$

an elementary calculation shows that the elementary divisors of $p-\lambda q$ are
$$(\lambda-\alpha_1)^{e_1}, \quad (\lambda-\alpha_2)^{e_2}, \quad \ldots, \quad (\lambda-\alpha_r)^{e_r}.$$

Hence, by Theorem I, the quadratic forms $x'ax$, $x'bx$ can be transformed to $y'py$, $y'qy$. Since these quadratic forms are uniquely determined by the elementary divisors of $a-\lambda b$, they may be taken as a canonical form for the pairs of forms $x'ax$, $x'bx$ with the given elementary divisors.

For notational convenience in what follows, we state the result just proved in a different form. Let

$$\theta(\alpha, e) = \alpha \sum_{i=1}^{e} X_i X_{e+1-i} + \sum_{i=1}^{e-1} X_{i+1} X_{e+1-i},$$

$$\phi(\alpha, e) = \sum_{i=1}^{e} X_i X_{e+1-i}.$$

We then state the result proved above as

THEOREM II. *If $x'ax$, $x'bx$ are two quadratic forms, of which the second is non-singular, then by an allowable transformation they can be reduced to the form*

$$x'ax = \sum_{i=1}^{r} \theta(\alpha_i, e_i),$$

$$x'bx = \sum_{i=1}^{r} \phi(\alpha_i, e_i),$$

where $(\lambda - \alpha_1)^{e_1}$, ..., $(\lambda - \alpha_r)^{e_r}$ are the elementary divisors of $a - \lambda b$.

When we use this notation we assume that the forms $\theta(\alpha_i, e_i)$ and $\phi(\alpha_i, e_i)$ are paired and are forms in the same set of coordinates, and two distinct pairs involve non-overlapping sets of coordinates [cf. Vol. I, p. 396].

In order to deal with the case in which b is singular, we have to introduce four new types of pairs of forms [cf. Vol. I, p. 397]:

(i) $\theta_1(e) = 2X_0X_1 + 2X_2X_3 + \ldots + 2X_{2e-2}X_{2e-1} + X_{2e}^2,$
$\phi_1(e) = 2X_1X_2 + 2X_3X_4 + \ldots + 2X_{2e-1}X_{2e};$

(ii) $\theta_2(e) = 2X_0X_1 + 2X_2X_3 + \ldots + 2X_{2e}X_{2e+1},$
$\phi_2(e) = 2X_1X_2 + 2X_3X_4 + \ldots + 2X_{2e-1}X_{2e};$

(iii) $\theta_3(e) = 2X_0X_1 + 2X_2X_3 + \ldots + 2X_{2e-2}X_{2e-1},$
$\phi_3(e) = 2X_1X_2 + 2X_3X_4 + \ldots + 2X_{2e-3}X_{2e-2} + X_{2e-1}^2;$

(iv) $\theta_4(e) = 2X_0X_1 + 2X_2X_3 + \ldots + 2X_{2e-2}X_{2e-1},$
$\phi_4(e) = 2X_1X_2 + 2X_3X_4 + \ldots + 2X_{2e-1}X_{2e}.$

We now prove

THEOREM III. *If $x'ax$ and $x'bx$ are two quadratic forms in $x_0, \ldots, x_n$, then they can be simultaneously reduced to the forms*

$$x'ax = \sum_i \theta(\alpha_i, e_i) + \sum_{i=1}^{4} \sum_j \theta_i(e_j),$$

$$x'bx = \sum_i \phi(\alpha_i, e_i) + \sum_{i=1}^{4} \sum_j \phi_i(e_j).$$

As in Theorem II, the forms $\theta(\alpha_i, e_i)$ and $\phi(\alpha_i, e_i)$ are paired, as are the forms $\theta_i(e_j)$ and $\phi_i(e_j)$. Pairs of forms involve the same sets of coordinates, and distinct pairs involve non-overlapping sets of coordinates.

The proof is by induction on n. We can begin the induction with the case $n = 0$, when $x'ax = Ax^2$, $x'bx = Bx^2$. In this case there is nothing to prove. We therefore assume that the theorem is true for pairs of quadratic forms involving less than $n+1$ indeterminates. We follow the method used in Chapter IX, § 5, very closely; the quadratic forms $x'bx$, $x'ax$ taking the place of the bilinear forms $y'Bx$, $y'Cx$.

Suppose that a is of rank $n+1-r_2$, and that b is of rank $n+1-r_1$, while $\begin{pmatrix} a \\ b \end{pmatrix}$ is of rank $n+1-r_0$. By a preliminary transformation of coordinates we arrange that the space S_{r_2-1} defined by $ax = 0$ is

$$x_0 = 0, \quad \ldots, \quad x_{n-r_1-r_2+r_0} = 0, \quad x_{n-r_1+1} = 0, \quad \ldots, \quad x_{n-r_0} = 0,$$

and that the space S_{r_1-1} defined by $bx = 0$ is

$$x_0 = 0, \quad \ldots, \quad x_{n-r_1} = 0;$$

S_{r_1-1} and S_{r_2-1} intersect in the (r_0-1)-space S_{r_0-1} whose equations are

$$x_0 = 0, \quad \ldots, \quad x_{n-r_0} = 0.$$

Having done this, we confine our choice of transformations to the *permissible* ones, which are of the form

$$y = Px,$$

where

$$P = \begin{pmatrix} D & 0 & 0 & 0 \\ E & F & 0 & 0 \\ G & 0 & H & 0 \\ L & M & N & K \end{pmatrix},$$

in which the matrices in the first, second, third, fourth rows (columns) have $n-r_1-r_2+r_0+1$, r_2-r_0, r_1-r_0, r_0 rows (columns), respectively. These transformations preserve the form of the equations of S_{r_0-1}, S_{r_1-1}, S_{r_2-1} [cf. IX, § 5].

After the preliminary transformation, the forms $x'ax$, $x'bx$ can be written

$$x'ax = \sum_{i,j=0}^{n-r_0}{}' a_{ij}x_ix_j,$$

$$x'bx = \sum_{i,j=0}^{n-r_1} b_{ij}x_ix_j,$$

where the summation Σ' is over the values

$$0,\quad 1,\quad \ldots,\quad n-r_1-r_2+r_0,\quad n-r_1+1,\quad \ldots,\quad n-r_0$$

of the suffixes i, j. If $r_0 > 0$, the two quadratic forms involve less than $n+1$ of the coordinates, and Theorem III follows at once from the hypothesis of induction. We need therefore only consider the case $r_0 = 0$.

If $r_1 = 0$, b is non-singular, and Theorem III follows from Theorem II.

Suppose now that $r_1 > 0$. Then a_{in} $(i = 0, \ldots, n)$ are not all zero, since $r_0 = 0$. Let k be the largest value of i such that $a_{in} \neq 0$. Then [Chap. IX, § 2], we can find a permissible transformation such that

$$x'ax = \sum_0^{n-1}{}' a'_{ij} y_i y_j + 2y_k y_n \quad \left(\text{or } \sum_0^{n-1}{}' a'_{ij} y_i y_j + y_n^2\right),$$

$$x'bx = \sum_0^{n-r_1}{}' b'_{ij} y_i y_j,$$

where $a'_{kj} = 0$ if $k < n$.

(ia) $k = n$ (in this case we have the second form for $x'ax$). By the hypothesis of induction a permissible transformation, derived from a permissible transformation on $y_0, \ldots, y_{n-1}$ only, can be found which gives

$$\sum_0^{n-1} a'_{ij} y_i y_j = \Sigma\, \theta(\alpha_i, e_i) + \sum_{i=1}^{4} \sum_j \theta_i(e_j),$$

$$\sum_0^{n-1} b'_{ij} y_i y_j = \Sigma\, \phi(\alpha_i, e_i) + \sum_{i=1}^{4} \sum_j \phi_i(e_j).$$

Hence
$$x'ax = \Sigma\, \theta(\alpha_i, e_i) + \sum_{i=1}^{4} \sum_j \theta_i(e_j) + \theta_1(0),$$

$$x'bx = \Sigma\, \phi(\alpha_1, e_i) + \sum_{i=1}^{4} \sum_j \phi_i(e_j) + \phi_1(0),$$

and so the necessary reduction is obtained.

(ib) $n - r_1 < k < n$ (in this case we use the first form of $x'ax$). An exactly similar argument is applicable in this case, the only difference being that instead of a pair of terms $\theta_1(0)$, $\phi_1(0)$ we have the pair $\theta_2(0)$, $\phi_2(0)$.

(ii) $k \leqslant n - r_1$. Then $k \leqslant n - r_1 - r_2$. This case is treated in a manner very similar to the treatment of the corresponding case considered in Chapter IX, § 5. We need only give the conclusions.

(ii$a$)
$$x'ax = 2z_0 z_n + \sum_1^{n-1}{}' a_{ij}^* z_i z_j,$$
$$x'bx = z_0^2 + \sum_1^{n-r_1} b_{ij}^* z_i z_j.$$

By the hypothesis of induction we can write

$$\sum_1^{n-1}{}' a_{ij}^* z_i z_j = \sum \theta(\alpha_i, e_i) + \sum_{i=1}^{4} \sum_j \theta_i(e_j),$$
$$\sum_1^{n-r_1} b_{ij}^* z_i z_j = \sum \phi(\alpha_i, e_i) + \sum_{i=1}^{4} \sum_j \phi_i(e_j).$$

Then
$$x'ax = \sum \theta(\alpha_i, e_i) + \sum_{i=1}^{4} \sum_j \theta_i(e_j) + \theta_3(1),$$
$$x'bx = \sum \phi(\alpha_i, e_i) + \sum_{i=1}^{4} \sum_j \phi_i(e_j) + \phi_3(1).$$

(iib_1)
$$x'ax = 2z_0 z_n + \sum_1^{n-1}{}^* a_{ij}^* z_i z_j,$$
$$x'bx = 2z_0 z_h + \sum_1^{n-r_1} b_{ij}^* z_i z_j \quad (h > n - r_1 - r_2),$$

when the summation Σ^* omits the value h. By the hypothesis of induction, we can write

$$\sum_1^{n-1}{}^* a_{ij}^* z_i z_j = \sum \theta(\alpha_i, e_i) + \sum_{i=1}^{4} \sum_j \theta_i(e_j),$$
$$\sum_1^{n-1} b_{ij}^* z_i z_j = \sum \phi(\alpha_i, e_i) + \sum_{i=1}^{4} \sum_j \phi_i(e_j),$$

and hence we obtain the reduction

$$x'ax = \sum \theta(\alpha_i, e_i) + \sum_{i=1}^{4} \sum_j \theta_i(e_j) + \theta_4(1),$$
$$x'bx = \sum \phi(\alpha_i, e_i) + \sum_{i=1}^{4} \sum_j \phi_i(e_j) + \phi_4(1).$$

(iib_2). In this case we obtain [cf. Vol. I, pp. 408, 409]

$$x'ax = 2z_0 z_n + \sum_1^{n-1}{}' a_{ij}^* z_i z_j,$$
$$x'bx = 2z_0 z_1 + \sum_1^{n-r_1} b_{ij}^* z_i z_j,$$

where $\sum_1^{n-1}{}' a_{ij}^* z_i z_j$ and $\sum_1^{n-r_1} b_{ij}^* z_i z_j$ are already expressed in reduced form. In this case, the terms $2z_0 z_n$ and $2z_0 z_1$ have to be added to a pair of corresponding terms in the expressions for $\Sigma' a^*_{ij} z_i z_j$ and $\Sigma b^*_{ij} z_i z_j$, and change the e in these terms to $e+1$. Thus the proof of Theorem III is complete.

Let us now consider a pair of corresponding forms $\theta_i(e)$, $\phi_i(e)$. If $i = 1, 2, 3$, the matrix of $\theta_i(e)$ is non-singular, and we can consider the elementary divisors of $\lambda\theta_i(e) - \phi_i(e)$. Direct calculation shows that if $i = 1$, there is one elementary divisor λ^{2e+1}; if $i = 2$, there are two elementary divisors λ^{e+1}, λ^{e+1}, and if $i = 3$, there is one elementary divisor λ^{2e}. If e is even, say $e = 2f$, the elementary divisors of $\lambda\theta_2(e) - \phi_2(e)$ are the same as the elementary divisors of

$$\lambda\theta_1(f) + \phi_1(f) + \lambda\theta_1'(f) + \phi_1'(f),$$

where $\theta_1(f)$, $\phi_1(f)$ and $\theta_1'(f)$, $\phi_1'(f)$ are pairs of forms involving independent sets of indeterminates. Hence, by Theorem I, we can obtain an allowable transformation of coordinates so that

$$\theta_2(e) = \theta_1''(f) + \theta_1'''(f), \quad \phi_2(e) = \phi_1''(f) + \phi_1'''(f).$$

Similar reasoning shows that if e is odd, $e = 2f-1$, we can find a transformation which reduces

$$\theta_2(e) \quad \text{to} \quad \theta_3''(f) + \theta_3'''(f),$$

$$\phi_2(e) \quad \text{to} \quad \phi_3''(f) + \phi_3'''(f).$$

Hence we can find a reduction of $x'ax$, $x'bx$ to the form given in Theorem III in which there are no terms $\theta_2(e_j)$, $\phi_2(e_j)$.

With this additional reduction, we can show that the reduced form given by Theorem III is uniquely determined by the matrices a and b.

It is clear that the number r_0 is determined by these matrices. The proof given in Vol. I, pp. 416–17, is immediately applicable and shows that the pairs of forms $\theta_4(e_j)$, $\phi_4(e_j)$ are also determined by the matrices a, b. Then if there are k such pairs of forms, the rank of $\lambda a + \mu b$ is $n+1-r_0-k$, and as in Chapter IX, § 5, we see that if the elementary divisors of this matrix are

$$(\alpha_i \lambda + \mu)^{e_i} \quad (i = 1, \ldots, r),$$

$$\lambda^{2f_i+1} \quad (i = 1, \ldots, s),$$

$$\lambda^{2e_i} \quad (i = 1, \ldots, t),$$

the only reduced pair of forms equivalent to $x'ax$, $x'bx$ are

$$\sum_{i=1}^{r}\theta(\alpha_i, e_i)+\sum_{1}^{s}\theta_1(f_i)+\sum_{i=1}^{t}\theta_3(g_i)+\sum\theta_4(\rho_i),$$

$$\sum_{i=1}^{r}\phi(\alpha_i, e_i)+\sum_{1}^{s}\phi_1(f_i)+\sum_{i=1}^{t}\phi_3(g_i)+\sum\phi_4(\rho_i).$$

Thus the reduced forms given by Theorem III, in the case where there are no pairs of terms $\theta_2(e)$, $\phi_2(e)$, form canonical forms for pairs of quadratic forms.

11. The intersection of two quadrics in S_n. In this section we study the intersection of two quadrics Q, Q' in S_n, whose equations are

$$x'ax = 0, \quad x'bx = 0.$$

We assume that the quadrics are distinct. The intersection of two primals in S_n is an unmixed variety of dimension $n-2$ [XII, § 4, Th. I], except in the case in which they have a variety of dimension $n-1$ in common. This can only happen when Q and Q' are both reducible, and have a prime in common. This case is trivial, and we shall omit it, though in fact some of our general theorems may include it as a special case.

The system of quadrics $Q_{\lambda\mu}$, where $Q_{\lambda\mu}$ has equation

$$\lambda x'ax+\mu x'bx = 0,$$

and λ, μ are variable parameters, is called a *pencil* of quadrics. Two quadrics Q_{pq} and Q_{rs} of the pencil are distinct if and only if $ps-qr \neq 0$. It is easily seen that if Q_{pq} and Q_{rs} are distinct quadrics of the pencil, the intersection $Q_{pq} \wedge Q_{rs}$ coincides with the intersection $Q \wedge Q'$. For the equations

$$px'ax+qx'bx = 0,$$

$$rx'ax+sx'bx = 0,$$

imply the equations

$$x'ax = 0, \quad x'bx = 0,$$

and conversely. We wish to show that a similar result holds when we consider the varieties multiplicatively. If Q is irreducible (that is, if its type ρ is less than $n-1$), $\mathbf{Q}$ is the multiplicative variety obtained by attaching unit multiplicity to Q; if Q is of type $n-1$ it consists of two distinct primes, and $\mathbf{Q}$ is the variety obtained by attaching unit multiplicity to each of these primes. If Q is of type n it coincides, as a point-set, with a prime; $\mathbf{Q}$ is then obtained by attaching multiplicity two to this.

Now let $x'Ax$, $x'Bx$ be two quadratic forms in $x_1, \ldots, x_n$ whose $(n+1)(n+2)$ coefficients are independent indeterminates over K, and let $\mathfrak{S}_{\lambda\mu}$ be the multiplicative variety obtained by attaching unit multiplicity to the quadric whose equation is

$$\lambda x'Ax + \mu x'Bx = 0.$$

We denote $\mathfrak{S}_{10}$ by $\mathfrak{S}$, $\mathfrak{S}_{01}$ by $\mathfrak{S}'$. If $ps - qr \neq 0$, the intersection $\mathfrak{S}_{pq} \wedge \mathfrak{S}_{\lambda\mu}$ coincides with $\mathfrak{S} \wedge \mathfrak{S}'$. But $\mathfrak{S}_{\lambda\mu}$ is of order two, hence [XII, § 4, Th. II] $\mathfrak{S}_{pq} . \mathfrak{S}_{rs}$ and $\mathfrak{S} . \mathfrak{S}'$ are of order four. Since $\mathfrak{S}$, $\mathfrak{S}'$ are generic quadrics which are independent, the intersection $\mathfrak{S} \wedge \mathfrak{S}'$ is simple, and hence the sum of the orders of the components of $\mathfrak{S} \wedge \mathfrak{S}'$ is four. Since each of these components counts at least once in $\mathfrak{S}_{pq} . \mathfrak{S}_{rs}$, it follows that

$$\mathfrak{S}_{pq} . \mathfrak{S}_{rs} = \mathfrak{S} . \mathfrak{S}'.$$

Now consider the specialisation $(\mathfrak{S}, \mathfrak{S}') \to (\mathfrak{Q}, \mathfrak{Q}')$. Since $Q \wedge Q'$ is, by hypothesis, of dimension $n-2$, and the object-variety of the normal problem is a 2-way space, we deduce that the specialisations of $\mathfrak{S}_{pq} . \mathfrak{S}_{rs}$ and $\mathfrak{S} . \mathfrak{S}'$ are uniquely determined. This specialisation then gives

$$\mathfrak{Q}_{pq} . \mathfrak{Q}_{rs} = \mathfrak{Q} . \mathfrak{Q}'.$$

Thus the properties of the intersection of two quadrics, both in the point-set sense and in the multiplicative sense, belong to the pencil determined by Q and Q' rather than to the individual quadrics. We use this fact to simplify the canonical form of the equations of the intersection of the two quadrics.

Using Theorem III of § 10, we can choose the coordinate system so that

$$x'ax = \Sigma\theta(\alpha_i, e_i) + \Sigma\theta_1(f_i) + \Sigma\theta_3(g_i) + \Sigma\theta_4(\rho_i),$$

$$x'bx = \Sigma\phi(\alpha_i, e_i) + \Sigma\phi_1(f_i) + \Sigma\phi_3(g_i) + \Sigma\phi_4(\rho_i).$$

Then,

$$px'ax + qx'bx = \Sigma[p\theta(\alpha_i, e_i) + q\phi(\alpha_i, e_i)] + \Sigma[p\theta_1(f_i) + q\phi_1(f_i)] + \Sigma[p\theta_3(g_i) + q\phi_3(g_i)] + \Sigma[p\theta_4(\rho_i) + q\phi_4(\rho_i)],$$

and we express $rx'ax + sx'bx$ similarly. We may choose p, q, r, s so that $ps - qr \neq 0$, and $r \neq 0$, $r\alpha_i + s \neq 0$ (any i).

The various terms in these summations arising from paired forms involve independent sets of coordinates, so we can treat them separately.

(i) The matrix of the quadratic form

$$[p\theta(\alpha_i, e_i) + q\phi(\alpha_i, e_i)] - \lambda[r\theta(\alpha_i, e_i) + s\phi(\alpha_i, e_i)]$$

has the single elementary divisor $(\lambda-\beta_i)^{e_i}$, where

$$\beta_i = \frac{p\alpha_i+q}{r\alpha_i+s}.$$

Hence [§ 10, Th. II] we can make an allowable transformation of the coordinates involved so that

$$p\theta(\alpha_i, e_i)+q\phi(\alpha_i, e_i) \quad \text{and} \quad r\theta(\alpha_i, e_i)+s\phi(\alpha_i, e_i)$$

become, respectively, $\theta(\beta_i, e_i)$, $\phi(\beta_i, e_i)$.

(ii) The matrix of the quadratic form

$$[p\theta_i(k)+q\phi_i(k)]-\lambda[r\theta_i(k)+s\phi_i(k)]$$

has the unique elementary divisor $(\lambda-p/r)^{2k+1}$ if $i = 1$, and $(\lambda-p/r)^{2k}$ if $i = 3$. Hence, by an allowable transformation of the coordinates involved, we can reduce

$$p\theta_i(k)+q\phi_i(k) \quad \text{and} \quad r\theta_i(k)+s\phi_i(k)$$

to $\qquad \theta(p/r, 2k+1), \quad \phi(p/r, 2k+1) \quad \text{if } i = 1,$

and to $\qquad \theta(p/r, 2k), \quad \phi(p/r, 2k) \quad \text{if} \quad i = 3.$

(iii) The matrix of the quadratic form

$$\mu[p\theta_4(\rho)+q\phi_4(\rho)]-\lambda[r\theta_4(\rho)+s\phi_4(\rho)]$$

is a $(2\rho+1)\times(2\rho+1)$ matrix whose rank is 2ρ for all values of λ, μ not both zero.

This is, however, a characteristic property of the matrices of two forms $x'Ax$, $x'Bx$ in $2\rho+1$ variables which can be simultaneously reduced to forms $\theta_4(\rho)$, $\phi_4(\rho)$.

It follows that we can choose the coordinate system so that

$$px'ax+qx'bx = \sum_i \theta(\beta'_i, e'_i)+\sum_i \theta_4(f'_i),$$

$$rx'ax+sx'bx = \sum_i \phi(\beta'_i, e'_i)+\sum_i \phi_4(f'_i).$$

It will be observed that what we have done is to choose r, s so that the rank of the matrix $ra+sb$ is the same as the rank of $a-\lambda b$ for indeterminate λ.

In studying the intersections of the two quadrics Q and Q' we may replace them by Q_{pq} and Q_{rs}. In other words, we may assume that Q and Q' are such that the canonical forms for their equations given by Theorem III of § 10 do not involve pairs of forms $\theta_i(e_j)$, $\phi_i(e_j)$ $(i = 1, 2, 3)$. With a simple change of notation we can then write

$$\left.\begin{aligned} x'ax \equiv\, & 2x_0x_1 + 2x_2x_3 + \ldots + 2x_{2r_1-2}x_{2r_1-1} \\ & + 2y_0y_1 + 2y_2y_3 + \ldots + 2y_{2r_2-2}y_{2r_2-1} \\ & + \ldots \\ & + 2z_0z_1 + 2z_2z_3 \ + \ldots + 2z_{2r_k-2}z_{2r_k-1} \\ & + t'At, \end{aligned}\right\} \tag{1}$$

$$\left.\begin{aligned} x'bx \equiv\, & 2x_1x_2 + 2x_3x_4 + \ldots + 2x_{2r_1-1}x_{2r_1} \\ & + 2y_1y_2 + 2y_3y_4 + \ldots + 2y_{2r_2-1}y_{2r_2} \\ & + \ldots \\ & + 2z_1z_2 + 2z_3z_4 \ + \ldots + 2z_{2r_k-1}z_{2r_k} \\ & + t'Bt \end{aligned}\right\}, \tag{2}$$

where B is a non-singular matrix of $n+1-2\sum_{i=1}^{k} r_i - k - r_0$ rows and columns, and $t'At$ and $t'Bt$ are in the reduced form given by Theorem II of § 10. The total number of coordinates involved is $n+1-r_0$. If we wish to give a name to the coordinates which do not appear in (1) and (2) we shall denote them by $u_1, \ldots, u_{r_0}$.

Let the elementary divisors of $A - \lambda B$ be

$$(\lambda-\alpha_1)^{e_1}, \quad (\lambda-\alpha_2)^{e_2}, \quad \ldots, \quad (\lambda-\alpha_s)^{e_s}.$$

It will appear that the geometrical properties of the pencil $Q_{\lambda\mu}$ with which we are concerned depend on the numbers $e_1, e_2, \ldots, e_s$, $r_1, \ldots, r_k, r_0$, and also on equalities between the α_i, but not on their actual values. We therefore introduce a symbol to represent the different kinds of pencil which can arise in S_n. This is analogous to the Segre symbol in Chapter VIII for collineations, and is called *the Segre symbol of the pencil*. We form it by writing down the numbers $e_1, \ldots, e_s$ in any order, subject only to the condition that the values of e_i which correspond to elementary divisors with the same α_i come together, and to indicate the equality of these α_i we enclose the corresponding e_i in round brackets. Having written down the e_i, we put a semicolon, after which we write the numbers $r_1, \ldots, r_k$ in any order. Then we write another semicolon and finally r_0. Enclosing the whole in square brackets, we have the Segre symbol of the pencil. If we are given the space S_n in which the quadrics lie, the number r_0 is determined by the other numbers in the Segre symbol, since

$$\Sigma e_i = n+1-2\Sigma r_i - k - r_0,$$

and giving r_0 is equivalent to giving the dimension of the space in which the quadrics lie. When $r_0 = 0$, it is usual to omit it.

Let $\bar{Q}$ and $\bar{Q}'$ be the sections of Q and Q' by the linear space

$$u_1 = u_2 = \ldots = u_{r_0} = 0,$$

and let S_{ρ_0-1} be the linear space given by

$$\begin{aligned} x_i &= 0 \quad (i = 0, \ldots, 2r_1),\\ y_i &= 0 \quad (i = 0, \ldots, 2r_2),\\ &\ldots\ldots\ldots\ldots\ldots\ldots\ldots\ldots\\ z_i &= 0 \quad (i = 0, \ldots, 2r_k),\\ t_i &= 0 \quad (i = 0, \ldots, n-2\Sigma r_i - k - r_0). \end{aligned}$$

Then Q and Q' are the joins of $\bar{Q}$ and $\bar{Q}'$ to S_{ρ_0-1}, and the intersection (either in the point-set sense or the multiplicative sense) of Q and Q' is the join of the intersection of $\bar{Q}$ and $\bar{Q}'$ to S_{ρ_0-1}. It follows that we can obtain all the intersection properties of Q and Q' from those of $\bar{Q}$ and $\bar{Q}'$, and it is usually more convenient to consider the latter quadrics. In the $(n-r_0)$-spaces containing them, the pencil $\bar{Q}_{\lambda\mu}$ has the same Segre symbol as $Q_{\lambda\mu}$, save that r_0 is replaced by zero. We therefore confine our attention to pencils whose Segre symbol has $r_0 = 0$, and for the remainder of this chapter we shall assume that $r_0 = 0$. It should be noted that if r_0 is not zero the statement of some of the results to be given later may require minor modifications.

The vertex of the quadric $Q_{\lambda\mu}$ is given by the equations

$$\begin{aligned} \lambda x_1 &= 0, & \lambda y_{2r_2-2}+\mu y_{2r_2} &= 0,\\ \lambda x_0+\mu x_2 &= 0, & \mu y_{2r_2-1} &= 0,\\ \mu x_1+\lambda x_3 &= 0, & &\ldots\ldots\ldots\ldots\\ &\ldots\ldots\ldots\ldots & \lambda z_1 &= 0,\\ \lambda x_{2r_1-2}+\mu x_{2r_1} &= 0, & \lambda z_0+\mu z_2 &= 0,\\ \mu x_{2r_1-1} &= 0, & &\ldots\ldots\ldots\ldots\\ \lambda y_1 &= 0, & \lambda z_{2r_k-2}+\mu z_{2r_k} &= 0,\\ \lambda y_0+\mu y_2 &= 0, & \mu z_{2r_k-1} &= 0,\\ &\ldots\ldots\ldots\ldots & (\lambda A+\mu B)t &= 0. \end{aligned}$$

Let $S_{k-1}^{(\lambda,\mu)}$ be the space spanned by the points

$$A_1:\quad x_{2i} = (-1)^i \lambda^i \mu^{r_1-i},\ x_{2i+1} = 0,\ y_i = 0,\ \ldots,\ z_i = 0,\ t_i = 0,$$
$$A_2:\quad x_i = 0;\ y_{2i} = (-1)^i \lambda^i \mu^{r_2-1},\ y_{2i+1} = 0,\ \ldots,\ z_i = 0,\ t_i = 0;\ \ldots$$
$$A_k:\quad x_i = 0;\ y_i = 0,\ \ldots,\ z_{2i} = (-1)^i \lambda^i \mu^{r_k-i},\ z_{2i+1} = 0;\ t_i = 0.$$

Every point of $S_{k-1}^{(\lambda,\mu)}$ is in the vertex of $Q_{\lambda\mu}$, and indeed $S_{k-1}^{(\lambda,\mu)}$ is the vertex of the quadric, unless $\det(\lambda A + \mu B) = 0$. Thus the general quadric of the pencil given by

$$x'ax = 0, \quad x'bx = 0,$$

where $x'ax$ and $x'bx$ are given by (1) and (2), is of type k.

The matrix $pA + qB$ is singular if and only if $-q/p$ is a latent root of $A - \lambda B$. Suppose that σ of the elementary divisors of type $(\lambda - \alpha_i)^{e_i}$ of $A - \lambda B$ have $\alpha_i = -q/p$. Then the equations

$$x_i = 0, \quad y_i = 0, \quad \ldots, \quad z_i = 0, \quad (pA + qB)t = 0$$

determine a linear space $S_{\sigma-1}$, and the vertex of Q_{pq} is the $(k+\sigma-1)$-space joining this $S_{\sigma-1}$ to $S_{k-1}^{(p,q)}$. The relation between the spaces $S_{\sigma-1}$ obtained by making $-q/p$ equal to the different latent roots of $A - \lambda B$ is easily discussed by a method similar to that used in Chapter VIII to discuss the united spaces of a collineation. We may leave it to the reader to prove that if the Segre symbol of the pencil is

$$[(e_1, \ldots, e_{\sigma_1})(e_{\sigma_1+1}, \ldots, e_{\sigma_1+\sigma_2}) \ldots (e_{\sigma_1+\ldots+\sigma_{f-1}+1}, \ldots, e_{\sigma_1+\ldots+\sigma_f}); r_1, \ldots, r_k],$$

the spaces $S_{\sigma_1-1}, \ldots, S_{\sigma\ -1}$ in

$$x_i = 0, \quad y_i = 0, \quad \ldots, \quad z_i = 0,$$

are independent, their join being a space of dimension $\sum_{i=1}^{f} \sigma_i - 1$.

We return to the spaces $S_{k-1}^{(\lambda,\mu)}$. It is easily verified that if $ps - qr \neq 0$, $S_{k-1}^{(p,q)}$ and $S_{k-1}^{(r,s)}$ do not intersect. The spaces $S_{k-1}^{(\lambda,\mu)}$ are all contained in the space Π:

$$x_1 = x_3 = \ldots = x_{2r_1-1} = y_1 = y_2 = \ldots = y_{2r_2-1} = \ldots$$
$$= z_1 = z_3 = \ldots = z_{2r_k-1} = t_1 = t_2 = \ldots = t_m = 0$$

$(m = n + 1 - 2\Sigma r_i - k)$, which is of dimension $\Sigma r_i + k - 1$. By referring to the equations of Q and Q', we see that Π lies on both quadrics, and hence on every quadric of the pencil determined by them. The tangent prime to $Q_{\lambda\mu}$ at the point X of Π whose coordinates are

$$(\xi_0, 0, \xi_2, 0, \ldots, \xi_{2r}; \eta_0, 0, \eta_2, 0, \ldots, \eta_{2r_2}; \ldots; \zeta_1, 0, \zeta_2, 0 \ldots \zeta_{2r_k}; 0, 0, \ldots, 0)$$

is

$$(\lambda\xi_0+\mu\xi_2)\,x_1+(\lambda\xi_2+\mu\xi_4)\,x_3+\ldots+(\lambda\xi_{2r_1-2}+\mu\xi_{2r_1})\,x_{2r_1-1}$$
$$+(\lambda\eta_0+\mu\eta_2)\,y_1+(\lambda\eta_2+\lambda\eta_4)\,y_3+\ldots+(\lambda\eta_{2r_2-2}+\mu\eta_{2r_2})\,y_{2r_2-1}$$
$$+\ldots$$
$$+(\lambda\zeta_0+\mu\zeta_2)\,z_1+(\lambda\zeta_2+\mu\zeta_4)\,z_3+\ldots+(\lambda\zeta_{2r_k-2}+\mu\zeta_{2r_k})\,z_{2r_k-1}=0.$$

It is uniquely determined unless X lies on $S_{k-1}^{(\lambda,\mu)}$. It is independent of λ, μ if and only if

$$\xi_0:\xi_2:\ldots:\xi_{2r_1-2}:\eta_0:\eta_2:\ldots:\eta_{2r_2-2}:\ldots:\zeta_{2r_k-2}$$
$$=\xi_2:\xi_4:\ldots:\xi_{2r_1}:\eta_2:\eta_4:\ldots:\eta_{2r_2}:\ldots:\zeta_{2r_k},$$

that is, if and only if X lies in $S_{k-1}^{(p,q)}$ for some choice of p, q. If X lies in $S_{k-1}^{(p,q)}$, then the tangent prime of $Q_{\lambda\mu}$ at X, where $\lambda q-\mu p\neq 0$, is independent of λ, μ, that is, the quadrics of the pencil touch each other at X. The locus of points X in $S_{k-1}^{(p,q)}$ for any choice of p, q is therefore given by equating to zero all the 2-rowed minors of the matrix:

$$\begin{pmatrix} x_0\,x_2\ldots x_{2r_1-2} & y_0\,y_2\ldots y_{2r_2-2} & \ldots z_{2r_k-2} \\ x_2\,x_4\ldots x_{2r_1} & y_2\,y_4\ldots y_{2r_2} & \ldots z_{2r_k} \end{pmatrix}.$$

It is easily verified that $\rho_1A_1+\rho_2A_2+\ldots+\rho_kA_k$, where $\rho_1,\ldots,\rho_k$ are independent indeterminates over $K(\lambda,\mu)$, is a generic point of this locus, and from this we deduce that the locus is of dimension k.

We now consider the intersection of two quadrics Q and Q' in S_n, assuming that $Q\wedge Q'$ is of dimension $n-2$, and determine conditions for this to be reducible. If $n=2$, the intersection consists of a finite set of points, in which case it is always reducible. We shall later enumerate the intersections of quadrics of all possible Segre symbols in the case $n=2$. We shall also enumerate all possible cases when $n=3$, so for the present we shall assume $n>3$. We also assume [cf. p. 290] that it is not possible to express the quadrics simultaneously by equations involving less than $n+1$ of the coordinates. With these assumptions we prove

THEOREM I. *The intersection of two irreducible quadrics in S_n $(n>3)$ contains an $(n-2)$-space as a component if and only if the Segre symbol of the pencil defined by the quadrics is*

$$[(1,1);1]\quad\text{or}\quad[2;1]\quad\text{or}\quad[;2]\quad\text{or}\quad[;1,1].$$

We have seen [p. 236] that a quadric in S_n of type ρ does not contain a linear space of dimension greater than $\frac{1}{2}(n+\rho-1)$. If

$Q \wedge Q'$ contains an $(n-2)$-space, it follows that $Q_{\lambda\mu}$ is of type $\rho \geqslant n-3$. If $\rho \geqslant n-1$, $Q_{\lambda\mu}$ is reducible, and we have excepted this case. Hence $\rho = n-2$ or $n-3$. If the Segre symbol of the pencil is

$$[(e_1, \ldots, e_\sigma) \ldots (e_{\nu+1}, \ldots, e_\tau); r_1, \ldots, r_k],$$

$Q_{\lambda\mu}$ (for λ, μ independent indeterminates) is of type k. Hence

$$k = n-2 \quad \text{or} \quad k = n-3. \tag{3}$$

Again
$$2\sum_{i=1}^{k} r_i + k \leqslant n+1, \tag{4}$$

and therefore
$$\sum_{i=1}^{k} r_i \leqslant 2. \tag{5}$$

Since, by hypothesis, $n > 3$, we find that the only solutions of (3), (4), (5) are

(i) $n = 4, \quad k = 1, \quad r_1 = 1;$

(ii) $n = 4, \quad k = 1, \quad r_1 = 2;$

(iii) $n = 5, \quad k = 2, \quad r_1 = r_2 = 1.$

(i) The equations of Q and Q' can be written in the form

$$2x_0x_1 + t'At = 0,$$

$$2x_1x_2 + t'Bt = 0,$$

where A and B are 2×2 matrices, and we may assume that B is non-singular. There are two cases to consider: (ia) when the elementary divisors of $A - \lambda B$ are $(\lambda - \alpha), (\lambda - \beta)$, and (i$b$) when there is one elementary divisor $(\lambda - \alpha)^2$.

(ia) The equations of Q and Q' can be written in the form

$$2x_0x_1 + \alpha x_3^2 + \beta x_4^2 = 0,$$

$$2x_1x_2 + x_3^2 + x_4^2 = 0.$$

Any plane on Q passes through $(0, 0, 1, 0, 0)$, and any plane on Q' passes through $(1, 0, 0, 0, 0)$. Hence the equation of any plane on their intersection can be written in the form

$$\frac{x_1}{\lambda} = \frac{x_3}{\mu} = \frac{x_4}{\nu}.$$

This lies on Q if and only if $\lambda = 0$, $\alpha\mu^2 + \beta\nu^2 = 0$, and it lies on Q' if and only if $\lambda = 0$, $\mu^2 + \nu^2 = 0$. Hence in order to have a plane common to the two quadrics, we must have $\alpha = \beta$. When this condition is fulfilled, the pencil defined by Q and Q' has the Segre symbol $[(1, 1); 1]$, and conversely if the Segre symbol has this form, the equations of Q and Q' can be taken as

$$2x_0x_1 + \alpha x_3^2 + \alpha x_4^2 = 0,$$
$$2x_1x_2 + x_3^2 + x_4^2 = 0.$$

The intersection of the quadrics consists of the two planes π, π' whose equations are

$$x_1 = 0, \quad x_3 \pm ix_4 = 0,$$

and the quadric q whose equations are

$$x_0 = \alpha x_2, \quad 2x_1x_2 + x_3^2 + x_4^2 = 0.$$

Since $\boldsymbol{Q}.\boldsymbol{Q}'$ is of order four, and each of the components of $Q_\wedge Q'$ must have positive multiplicity in $\boldsymbol{Q}.\boldsymbol{Q}'$, we have

$$\boldsymbol{Q}.\boldsymbol{Q}' = \boldsymbol{\pi} + \boldsymbol{\pi}' + \boldsymbol{q}.$$

(i*b*) The equations of Q and Q' can be written in the form

$$2x_0x_1 + 2\alpha x_3x_4 + x_4^2 = 0,$$
$$2x_1x_2 + 2x_3x_4 = 0.$$

The Segre symbol of the pencil in this case is $[2; 1]$.

The plane π whose equations are

$$x_1 = x_4 = 0$$

is common to Q and Q'. It can then be verified that the remaining component of $Q_\wedge Q'$ is the surface F having the generic point $(-\nu(\alpha\mu + \frac{1}{2}\nu),\ \lambda^2,\ -\mu\nu,\ \lambda\mu,\ \lambda\nu)$. It can then be shown that this surface has three points in common with a generic plane of S_4, and hence that it is of order three. We thus have

$$\boldsymbol{Q}.\boldsymbol{Q}' = \boldsymbol{\pi} + \boldsymbol{F}.$$

(ii) This corresponds to the Segre symbol $[; 2]$. The equations of Q and Q' can be written in the form

$$2x_0x_1 + 2x_2x_3 = 0,$$
$$2x_1x_2 + 2x_3x_4 = 0.$$

The intersection $Q_\wedge Q'$ consists of the plane π whose equation is

$$x_1 = x_3 = 0,$$

and the surface F whose generic point is $(\lambda^2, \mu\nu, -\lambda\nu, \lambda\mu, \nu^2)$, which can be shown to be of order three. Hence

$$\mathbf{Q}.\mathbf{Q}' = \pi + \mathbf{F}.$$

(iii) In this case the Segre symbol is [; 1, 1]. We can take the equations of Q and Q' in the form

$$2x_0x_1 + 2x_3x_4 = 0,$$

$$2x_1x_2 + 2x_4x_5 = 0.$$

The 3-space π given by the equation

$$x_1 = x_4 = 0$$

is common to Q and Q'. The remaining intersection is a V_3 with generic point $(\lambda\mu, -\rho^2, \mu\nu, \rho\lambda, \rho\mu, \rho\nu)$, which can again be shown to be of order 3. Hence

$$\mathbf{Q}.\mathbf{Q}' = \pi + \mathbf{V}_3.$$

This completes the proof of Theorem I. It is worth noting that the cases (i*b*) and (ii) are distinguished by the fact that in (i*b*) π and F meet in the lines $x_1 = x_3 = x_4 = 0$ and $x_1 = x_4 = x_0 - \alpha x_2 = 0$ while in (ii) π and F meet in the conic $x_1 = x_3 = x_0x_4 - x_2^2 = 0$.

Since the intersection $\mathbf{Q}.\mathbf{Q}'$ of two quadrics in S_n is of order four, this intersection is either an irreducible variety of order four, or two quadrics of dimension $n-2$ (possibly coincident), or else the intersection contains an $(n-2)$-space. We therefore consider when the intersection can become a pair of quadrics of dimension $n-2$.

Let Q_{n-2} be a quadric common to Q and Q'. We choose the coordinate system so that Q_{n-2} has the equations

$$x_0 = 0, \quad x_1^2 + \ldots + x_{n-\rho}^2 = 0.$$

The equations of Q and Q' are then of the form

$$x_0(a_0x_0 + \ldots + a_nx_n) + x_1^2 + \ldots + x_{n-\rho}^2 = 0,$$

$$x_0(b_0x_0 + \ldots + b_nx_n) + x_1^2 + \ldots + x_{n-\rho}^2 = 0.$$

The intersection of Q and Q' is equally well defined by the quadrics

$$x_0(c_0x_0 + \ldots + c_nx_n) = 0, \tag{6}$$

$$x_0(b_0x_0 + \ldots + b_nx_n) + x_1^2 + \ldots + x_{n-\rho}^2 = 0, \tag{7}$$

where $c_i = a_i - b_i$, and it will be more convenient to define Q and Q' by the equations (6) and (7). Not all c_i can be zero, since the original quadrics are distinct. There are several cases to be considered.

We first consider certain transformations designed to reduce the number of coordinates which appear in the equation of Q. We have four cases to consider:

I. Suppose that in (6), $c_1 = c_2 = \ldots = c_n = 0$. Then $c_0 \neq 0$. The equations of Q and Q' are

$$x_0^2 = 0,$$

$$x_0(b_0 x_0 + \ldots + b_n x_n) + x_1^2 + \ldots + x_{n-\rho}^2 = 0.$$

II. Suppose that in (6), $c_{n-\rho+1} = \ldots = c_n = 0$, and that $c_k \neq 0$, for some value of k satisfying the inequality $0 < k \leqslant n-\rho$. In the space $S_{n-\rho-1}$ given by the equations

$$x_0 = 0, \quad x_{n-\rho+1} = 0, \quad \ldots, \quad x_n = 0,$$

the equation

$$x_1^2 + \ldots + x_{n-\rho}^2 = 0$$

defines a non-singular quadric $Q_{n-\rho-2}$, and the equation

$$c_1 x_1 + \ldots + c_{n-\rho} x_{n-\rho} = 0$$

defines a linear space $S_{n-\rho-2}$. In the present case we assume that $S_{n-\rho-2}$ does not touch $Q_{n-\rho-2}$. Then there exists an allowable transformation

$$x_i' = \sum_{j=1}^{n-\rho} c_{ij} x_j \quad (i = 1, \ldots, n-\rho),$$

which transforms $\sum_1^{n-\rho} c_i x_i$ and $\sum_1^{n-\rho} x_i^2$ into x_1' and $\sum_1^{n-\rho} x_i'^2$, respectively. If we write

$$y_0 = x_0,$$

$$y_1 = \sum_{i=1}^{n-\rho} c_{1i} x_i + c_0 x_0,$$

$$y_i = \sum_{j=1}^{n-\rho} c_{ij} x_j \quad (i = 2, \ldots, n-\rho),$$

$$y_i = x_i \qquad (i = n-\rho+1, \ldots, n),$$

the equations of Q and Q' become

$$y_0 y_1 = 0,$$

$$y_0(b_0' y_0 + \ldots + b_n' y_n) + y_1^2 + \ldots + y_{n-\rho}^2 = 0.$$

III. This is the same as II, except that $S_{n-\rho-2}$ touches $Q_{n-\rho-2}$. We can in this case construct an allowable transformation

$$x_i' = \sum_{j=1}^{n-\rho} c_{ij} x_j \quad (i = 1, \ldots, n-\rho),$$

which transforms $\sum_{i=1}^{n-\rho} c_i x_i$ and $\sum_{i=1}^{n-\rho} x_i^2$ into x_1' and $2x_1'x_2'+x_3'^2+\ldots+x_{n-\rho}'^2$.

Putting
$$\begin{aligned} y_0 &= x_0, \\ y_1 &= \Sigma c_{1j}x_j + c_0x_0, \\ y_i &= \Sigma c_{ij}x_j \quad (i = 2, \ldots, n-\rho), \\ y_i &= x_i \qquad (i = n-\rho+1, \ldots, n), \end{aligned}$$
the equations of Q and Q' become
$$y_0y_1 = 0,$$
$$y_0(b_0'y_0+\ldots+b_n'y_n)+2y_1y_2+y_3^2+\ldots+y_{n-\rho}^2 = 0.$$

IV. Suppose that there exists a value of k satisfying $n-\rho<k\leqslant n$ for which $c_k \neq 0$. The transformation
$$\begin{aligned} y_n &= c_0x_0+\ldots+c_nx_n, \\ y_k &= x_n, \\ y_i &= x_i \quad (i \neq k, n), \end{aligned}$$
transforms the equations of Q, Q' into
$$y_0y_n = 0,$$
$$y_0(b_0'y_0+\ldots+b_n'y_n)+y_1^2+\ldots+y_{n-\rho}^2 = 0.$$

We consider the cases I, ..., IV in turn. In the transformations which follow, the form of the equation of Q is unaltered, but the equation of Q' will be simplified. It will sometimes be convenient to replace Q' by another quadric of the pencil defined by Q and Q', so that Q' is non-singular. There is one case, however, in which this is not possible, since every quadric of the pencil $Q_{\lambda\mu}$ is singular.

I. We take the equations of Q and Q' in the form
$$x_0^2 = 0,$$
$$x_0(b_0x_0+\ldots+b_nx_n)+x_1^2+\ldots+x_{n-\rho}^2 = 0.$$

(i) If $b_i = 0$, for all $i>n-\rho$, we put
$$y_i = x_i+\tfrac{1}{2}b_ix_0 \quad (i = 1, \ldots, n-\rho).$$
The equation of Q' becomes
$$by_0^2+y_1^2+\ldots+y_{n-\rho}^2 = 0.$$

By replacing Q' by another quadric of the pencil, we can arrange that $b \neq 0$; the transformation

$$z_0 = b^{\frac{1}{2}} y_i, \quad z_1 = y_1, \quad \ldots, \quad z_n = y_n$$

then reduces the equations of Q and Q' to the form

$$z_0^2 = 0,$$

$$z_0^2 + \ldots + z_{n-\rho}^2 = 0.$$

If we assume, as usual, that every z_i is present in these equations, we must have $\rho = 0$. $Q_\wedge Q'$ is therefore the intersection of the non-singular quadric Q' with a prime not tangent to it. The Segre symbol is $[(1, 1, \ldots, 1), 1\,;]$.

(ii) If there exists a value of k greater than $n-\rho$ for which $b_k \neq 0$, the transformation

$$y_i = x_i \quad (i \neq k, n),$$

$$y_k = x_n,$$

$$2y_n = b_0 x_0 + \ldots + b_n x_n,$$

reduces the equations to $\quad y_0^2 = 0,$

$$2y_0 y_n + y_1^2 + \ldots + y_{n-\rho}^2 = 0,$$

and if all the y_i are present, $\rho = 1$. $Q_\wedge Q'$ is the intersection of the non-singular quadric Q' and a tangent prime. The Segre symbol is $[(1, 1, \ldots, 1, 2)\,;]$.

II. The equations of Q and Q' are of the form

$$x_0 x_1 = 0,$$

$$x_0(b_0 x_0 + \ldots + b_n x_n) + x_1^2 + \ldots + x_{n-\rho}^2 = 0.$$

(i) Suppose that $b_{n-\rho+1} = \ldots = b_n = 0$. Then if we write

$$y_i = x_i + \tfrac{1}{2} b_i x_0 \quad (i = 2, \ldots, n-\rho),$$

$$y_i = x_i \qquad \text{(all other } i\text{)},$$

the equations become $\quad y_0 y_1 = 0,$

$$y_0(by_0 + cy_1) + y_1^2 + \ldots + y_{n-\rho}^2 = 0,$$

and if all y_i are present, $\rho = 0$.

If $b \neq 0$, we replace Q' by another quadric of the pencil so that we may take $c = 0$, and then, by absorbing the multiplier b, we obtain the equations

$$y_0 y_1 = 0,$$

$$y_0^2 + y_1^2 + \ldots + y_n^2 = 0.$$

$Q_\wedge Q'$ is the section of the non-singular quadric Q' by two primes not touching Q'. The Segre symbol is $[(1, 1, \ldots, 1), 1, 1\,;]$.

If, on the other hand, $b = 0$, we can choose Q' in the pencil so that $c \neq 0$; and by absorbing a multiplier we obtain the equations

$$z_0 z_1 = 0,$$
$$2z_0 z_1 + z_1^2 + \ldots + z_n^2 = 0.$$

$Q_\wedge Q'$ is the intersection of the non-singular quadric Q' with a tangent prime $z_1 = 0$, and with a prime $z_0 = 0$ not tangent to it. The Segre symbol is $[(1, 1, \ldots, 1), 2\,;]$.

(ii) Suppose that there exists a $b_k \neq 0$, $k > n-\rho$. The transformation

$$2y_n = b_0 x_0 + \ldots + b_n x_n,$$
$$y_k = x_n,$$
$$y_i = x_i \quad (i \neq k, n),$$

transforms the equation to the form

$$y_0 y_1 = 0,$$
$$2y_0 y_n + y_1^2 + \ldots + y_{n-\rho}^2 = 0,$$

where $\rho = 1$ if all the variables are present. $Q_\wedge Q'$ is the section of the non-singular quadric Q' with a tangent prime $y_0 = 0$, and a prime through the point of contact. The Segre symbol is

$$[(1, 1, \ldots, 1, 3)\,;].$$

III. We take the equations of Q and Q' to be

$$x_0 x_1 = 0,$$
$$x_0(b_0 x_0 + \ldots + b_n x_n) + 2x_1 x_2 + x_3^2 + \ldots + x_{n-\rho}^2 = 0.$$

(i) Suppose that $b_{n-\rho+1} = \ldots = b_n = 0$. By writing

$$y_i = x_i + \tfrac{1}{2} b_i x_0 \quad (i = 3, \ldots, n-\rho),$$
$$y_i = x_i \quad \text{(all other } i),$$

these equations become

$$y_0 y_1 = 0,$$
$$y_0(by_0 + cy_1 + dy_2) + 2y_1 y_2 + y_3^2 + \ldots + y_{n-\rho}^2 = 0,$$

and if all y_i are to be present, $\rho = 0$. By replacing Q' by another member of the pencil, we can choose c as we wish. If d is zero, and b is not zero, we take $c = 0$, and we see that by rearranging the coordinates we obtain the case considered in II (ii). If $d = 0$, $b = 0$, the equation can be written as

$$y_0 y_1 = 0,$$

$$2y_1 y_2 + y_3^2 + \ldots + y_n^2 = 0.$$

In this case, all the quadrics of the pencil are singular. Then $Q \wedge Q'$ is the intersection of a quadric of type one with a prime touching it along a generator, and with a second prime not through the vertex. The Segre symbol is $[(1, 1, \ldots, 1) ; 1]$.

We next consider the cases in which d is not zero. Put

$$z_i = y_i \quad (i \neq 2),$$

$$z_2 = y_2 + b y_0 / d.$$

Then the equations become

$$z_0 z_1 = 0,$$

$$z_0(c' z_1 + d z_2) + 2 z_1 z_2 + z_3^2 + \ldots + z_n^2 = 0.$$

We can if necessary replace Q' by another member of the pencil and arrange the multipliers so that $c' = d = 2$. Then the equations of Q and Q' become

$$z_0 z_1 = 0,$$

$$2(z_1 z_2 + z_2 z_0 + z_0 z_1) + z_3^2 + \ldots + z_n^2 = 0.$$

$Q \wedge Q'$ is the intersection of the non-singular quadric Q' with two primes, not tangent to Q', but meeting in an $(n-2)$-space tangent to Q'. The Segre symbol is $[(1, \ldots, 1, 2), 1 ;]$.

(ii) If $b_k \neq 0$ for some k greater than $n - \rho$, we make the transformation

$$2y_n = b_0 x_0 + \ldots + b_n x_n,$$

$$y_k = x_n,$$

$$y_i = x_i \quad (i \neq k, n),$$

and obtain the equations of Q and Q' in the form

$$y_0 y_1 = 0,$$

$$2y_0 y_n + 2y_1 y_2 + y_3^2 + \ldots + y_{n-1}^2 = 0.$$

$Q \wedge Q'$ is the intersection of Q' with two tangent primes to Q' whose points of contact lie in a line on Q'. The Segre symbol of the pencil is $[(1, 1, \ldots, 1, 2, 2) ;]$.

IV. We take the equations of Q and Q' in the form

$$x_0 x_n = 0,$$

$$x_0(b_0 x_0 + \ldots + b_n x_n) + x_1^2 + \ldots + x_{n-\rho}^2 = 0.$$

In this case $\rho \geqslant 1$.

(i) We suppose, first, that $b_{n-\rho+1}, \ldots, b_{n-1}$ are zero. Then we write

$$y_i = x_i + \tfrac{1}{2} b_i x_0 \quad (i = 1, \ldots, n-\rho),$$

$$y_i = x_i \qquad \text{(all other } i\text{)}.$$

The equations become $\quad y_0 y_n = 0,$

$$y_0(b y_0 + c y_n) + y_1^2 + \ldots + y_{n-\rho}^2 = 0,$$

where $\rho = 1$ if all y_i are present. If $b = 0$, we choose the quadric Q' in the pencil so that $c = 2$. Then $Q_\wedge Q'$ is the intersection of a non-singular quadric Q' with a pair of tangent primes whose points of contact are not conjugate with regard to Q'. The Segre symbol of the pencil is $[(1, 1, \ldots, 1), (1, 1);]$.

If $b \neq 0$, we can choose the quadric in the pencil so that $c = 2$, and arrange the multipliers so that $b = 1$. We thus obtain a case already considered in II (i).

(ii) If $b_k \neq 0$, for some k such that $n-\rho < k < n$, we write

$$2y_{n-1} = b_0 x_0 + \ldots + b_n x_n,$$

$$y_k = x_{n-1},$$

$$y_i = x_i \quad (i \neq k, n-1).$$

The equations then become

$$y_0 y_n = 0,$$

$$2y_0 y_{n-1} + y_1^2 + \ldots + y_{n-\rho}^2 = 0,$$

and in this case $\rho = 2$ if all y_i are present. This is the case of the pencil of singular quadrics considered in III (i).

We have thus obtained all cases in which the intersection of two quadrics reduces to a pair of quadrics. Recalling that our condition that all the variables are present in the equations of the quadrics is equivalent to the condition that the vertices of Q and Q' do not intersect, we can state

THEOREM II. *If Q and Q' are two quadrics whose vertices do not meet, their intersection reduces to a pair of quadrics if and only if the pencil of quadrics defined by them is of one of the following types*:

$$[(1,\ldots,1),1\,;],\quad [(1,1,\ldots,1,2)\,;],$$

$$[(1,1,\ldots,1),1,1\,;],\quad [(1,1,\ldots,1),2\,;],\quad [(1,1,\ldots,1,3)\,;],$$

$$[(1,\ldots,1,2),1\,;],\quad [(1,\ldots,1,2,2)\,;],\quad [(1,1,\ldots,1),(1,1)\,;],$$

$$[(1,1,\ldots,1)\,;1].$$

Theorems I and II and the enumeration of all possible pencils in S_n $(n \leqslant 3)$ given below enable us to determine effectively all cases in which the intersection of the quadrics of a pencil is reducible. We have deduced these cases from the cases in which the vertices of two quadrics of the pencil do not intersect. We observe that we have proved, incidentally, that if the intersection of $Q \wedge Q'$ is reducible, the function field of each component of the intersection is a pure transcendental extension of K, that is, each component is *rational*. We can go further and prove

THEOREM III. *If* $n \geqslant 4$, and Q, Q' *are two quadrics in* S_n *whose vertices do not meet, and whose intersection is irreducible, then their intersection is rational.*

We shall prove in the next Chapter [XIV, § 7] that two quadrics in S_4 have a line in common, and from this it follows that two quadrics in S_n $(n > 4)$ have a line in common. Choose the coordinate system so that Q and Q' have the line in common given by

$$x_1 = \ldots = x_{n-1} = 0.$$

Case (i). Suppose that the line lies in the vertex of a quadric of the pencil determined by Q and Q'. We may suppose that Q' is this quadric. Then we may suppose that Q has the equation

$$x_0 \sum_1^{n-1} a_i x_i + x_n \sum_1^{n-1} \alpha_i x_i + \sum_{i,j=1}^{n-1} A_{ij} x_i x_j = 0,$$

and that Q' has the equation

$$x_1 x_2 - x_3^2 - \ldots - x_{n-\rho}^2 = 0 \quad (\rho \geqslant 1).$$

Let $t_3, \ldots, t_{n-1}$ be $n-3$ independent indeterminates over K, and let $\xi_1 = \sum_i t_i^2$, $\xi_2 = 1$, $\xi_j = t_j$ $(j = 3, \ldots, n-1)$. Consider the equation

$$x_0 \sum_1^{n-1} a_i \xi_i + x_n \sum_1^{n-1} \alpha_i \xi_i + \sum_{i,j=1}^{n-1} A_{ij} \xi_i \xi_j = 0.$$

The generic solution of this is

$$\xi_0 = c + dt_1, \quad \xi_n = e + ft_1,$$

where c, d, e, f are in $K(t_3, \ldots, t_{n-1})$, and t_1 is indeterminate over this field. The point $\xi = (\xi_0, \ldots, \xi_n)$ lies on Q and Q', and, since

$$K(\xi_0, \ldots, \xi_n) = K(t_1, t_3, \ldots, t_{n-1})$$

is a transcendental extension of K of dimension $n-2$, ξ is a generic point of an irreducible variety V of dimension $n-2$ over K. But, by hypothesis, $Q_\wedge Q'$ is an irreducible variety of dimension $n-2$, and $V \subseteq Q_\wedge Q'$; hence $V = Q_\wedge Q'$. Since $K(\xi_1, \ldots, \xi_n)$ is a pure transcendental extension of K, V is rational, by the definition of a rational variety.

Case (ii). If the common line is not in the vertex of any quadric of the pencil determined by Q and Q', we can take the equations of Q and Q' to be

$$x_0 \sum_1^{n-1} a_i x_i + x_n \sum_1^{n-1} \alpha_i x_i + \sum_{i,j=1}^{n-1} A_{ij} x_i x_j = 0,$$

and
$$x_0 \sum_1^{n-1} b_i x_i + x_n \sum_1^{n-1} \beta_i x_i + \sum_{i,j=1}^{n-1} B_{ij} x_i x_j = 0,$$

respectively, where $\sum_1^{n-1} a_i x_i \sum_1^{n-1} \beta_i x_i - \sum_1^{n-1} b_i x_i \sum_1^{n-1} \alpha_i x_i$ is not identically zero. Let $t_1, \ldots, t_{n-2}$ be $n-2$ independent indeterminates over K, and let $t_{n-1} = 1$. If $\xi_i = t_i$ $(i = 1, \ldots, n-1)$, the equations

$$x_0 \sum_1^{n-1} a_i \xi_i + x_n \sum_1^{n-1} \alpha_i \xi_i + \sum_{i,j=1}^{n-1} A_{ij} \xi_i \xi_j = 0$$

and
$$x_0 \sum_1^{n-1} b_i \xi_i + x_n \sum_1^{n-1} \beta_i \xi_i + \sum_{i,j=1}^{n-1} B_{ij} \xi_i \xi_j = 0$$

have a unique solution ξ_0, ξ_n in $K(t_1, \ldots, t_{n-2})$. As in Case (i), we show that $\xi = (\xi_0, \ldots, \xi_n)$ is a generic point of $V = Q_\wedge Q'$, and since $K(\xi_0, \ldots, \xi_n) = K(t_1, \ldots, t_{n-2})$ is a pure transcendental extension of K, it follows that $Q_\wedge Q'$ is rational.

We conclude this chapter by giving a complete list of the possible types of pencils of quadrics in S_n, for $n = 1, 2, 3$. The methods used in § 10 are applied. The proofs are omitted. Statements about multiplicities can easily be verified.

If $n = 1$, any two binary quadratic forms $x'ax$, $x'bx$ can be simultaneously reduced to one of the following forms:

(i) $$y_0^2, \quad ay_0^2;$$

(ii) $$y_0^2+y_1^2, \quad a(y_0^2+y_1^2);$$

(iii) $$2y_0y_1, \quad 2by_0y_1+y_1^2;$$

(iv) $$y_0^2, \quad 2y_0y_1;$$

(v) $$y_0^2, \quad by_0^2+y_1^2;$$

(vi) $$y_0^2+y_1^2, \quad ay_0^2+by_1^2 \quad (a \neq b,\ a \neq 0).$$

When we are concerned with pencils of quadrics, (i) and (ii) can be rejected, since they do not define pencils, and (iii) and (iv) are equivalent, defining a pencil of type [2;]. Cases (v) and (vi) define a pencil of type [1, 1].

When we pass to pencils of quadrics in S_2, we first have two cases, obtained from the case $n = 1$ by joining the points of a pencil of quadrics on a line to a point outside the line. The quadrics in S_2 are then pairs of lines through a point. We omit these cases.

From our general theory, we see that there only remains one case in which every quadric of the pencil is reducible, and that in this case the coordinates can be chosen so that the equations of the two quadrics are

$$2x_1x_2 = 0,$$

$$2x_0x_1 = 0.$$

The pencil then consists of conics containing the line $x_1 = 0$ as a fixed line and a variable line of a pencil of lines. This is the case [; 1].

To consider other pencils, we may suppose that the two quadrics $S = 0$ and $S' = 0$ are non-singular, and we have only to consider the pairs of quadrics corresponding to the different types with $r_1 = \dots = r_k = r_0 = 0$.

(i) [1, 1, 1]. $$S \equiv ax_0^2+bx_1^2+cx_2^2 = 0,$$ $$S' \equiv x_0^2+x_1^2+x_2^2 = 0$$

(a, b, c all different). The quadrics meet in four distinct points.

(ii) [(1, 1), 1]. $$S \equiv a(x_0^2+x_1^2)+cx_2^2 = 0,$$ $$S' \equiv x_0^2+x_1^2+x_2^2 = 0.$$

The quadrics touch the lines $x_0^2+x_1^2 = 0$ where they meet $x_2 = 0$.

(iii) [(1, 1, 1)]. This is the identity case, which we omit.

(iv) [2, 1].
$$S \equiv 2ax_0x_1 + x_1^2 + cx_2^2 = 0,$$
$$S' \equiv 2x_0x_1 + x_2^2 = 0.$$

The quadrics touch $x_1 = 0$ at $(1, 0, 0)$, and meet in two other points.

(v) [(2, 1)].
$$S \equiv 2ax_0x_1 + x_1^2 + ax_2^2 = 0,$$
$$S' \equiv 2x_0x_1 + x_2^2 = 0.$$

The conics have 'four-point' contact at $(1, 0, 0)$, which is their only intersection.

(vi) [3].
$$S \equiv 2ax_0x_2 + ax_1^2 + 2x_1x_2 = 0,$$
$$S' \equiv 2x_0x_2 + x_1^2 = 0.$$

The conics meet in $(0, 0, 1)$, and in $(1, 0, 0)$ which counts three times.

When we pass to $n = 3$, we again omit the cases in which all the quadrics of the pencil have a common vertex, since these can be obtained by joining one of the cases in $n = 2$ to a point. We therefore assume that all the coordinates are necessarily present. In this case, there is again only one case of a pencil in which every quadric is singular, namely the case [1 ; 1]. The equations can be taken in the form
$$S \equiv 2x_1x_2 + ax_3^2 = 0,$$
$$S' \equiv 2x_0x_1 + x_3^2 = 0.$$

These are two point-cones having the line $x_1 = x_3 = 0$ in common, and touching $x_1 = 0$ at every point of this line. The remaining intersection is the quadric
$$ax_0 - x_2 = 0, \quad 2x_0x_1 + x_3^2 = 0.$$

We classify the remaining types by their Segre symbols; a, b, ... represent different elements of K.

(i) [1, 1, 1, 1].
$$S \equiv ax_0^2 + bx_1^2 + cx_2^2 + dx_3^2 = 0,$$
$$S' \equiv x_0^2 + x_1^2 + x_2^2 + x_3^2 = 0.$$

The intersection is an irreducible quartic curve.

(ii) [(1, 1), 1, 1].
$$S \equiv ax_0^2 + ax_1^2 + cx_2^2 + dx_3^2 = 0,$$
$$S' \equiv x_0^2 + x_1^2 + x_2^2 + x_3^2 = 0.$$

The intersection is the section of $S = 0$ by the two planes

$$(c-a)x_2^2+(d-a)x_3^2 = 0.$$

These are two 'general' sections of $S = 0$.

(iii) $[(1,1),(1,1)]$.

$$S \equiv ax_0^2+ax_1^2+cx_2^2+cx_3^2 = 0,$$

$$S' \equiv x_0^2+x_1^2+x_2^2+x_3^2 = 0.$$

The intersection is given by the two plane-pairs

$$x_0^2+x_1^2 = 0, \quad x_2^2+x_3^2 = 0,$$

and consists of four lines.

(iv) $[(1,1,1),1]$ $S \equiv ax_0^2+ax_1^2+ax_2^2+dx_3^2 = 0,$

$$S' \equiv x_0^2+x_1^2+x_2^2+x_3^2 = 0.$$

The intersection is the quadric Q_2 given by $x_0^2+x_1^2+x_2^2 = 0$, $x_3 = 0$, counted twice.

(v) $[(1,1,1,1)]$. This is the identical case, which must be omitted.

(vi) $[2,1,1]$. $S \equiv 2ax_0x_1+x_1^2+cx_2^2+dx_3^2 = 0,$

$$S' \equiv 2x_0x_1+x_2^2+x_3^2 = 0.$$

The two quadrics have the same tangent plane $x_1 = 0$ at $(1,0,0,0)$. If we eliminate x_0, we obtain the equation

$$x_1^2+(c-a)x_2^2+(d-a)x_3^2 = 0.$$

Let (ξ_1, ξ_2, ξ_3) be a generic point of this locus, regarded as a conic in the (x_1, x_2, x_3)-plane and let $\xi_0 = -(\xi_2^2+\xi_3^2)/2\xi_1$. Then $(\xi_0, \xi_1, \xi_2, \xi_3)$ is a point common to S and S', and is a generic point of a curve common to the quadrics. It can be verified that this curve is of order four; hence the intersection is a quartic curve. We can also verify that $(1,0,0,0)$ is a multiple point of this.

(vii) $[(2,1),1]$. $S \equiv 2ax_0x_1+x_1^2+ax_2^2+dx_3^2 = 0,$

$$S' \equiv 2x_0x_1+x_2^2+x_3^2 = 0.$$

The intersection is given by $S' = 0$, $x_1^2+(d-a)x_3^2 = 0$. We verify at once that this is a pair of quadrics, but in this case their intersection, given by $x_1 = x_3 = 0$, $S' = 0$, is a pair of coincident points.

(viii) $[2,(1,1)]$. $S \equiv 2ax_0x_1+x_1^2+cx_2^2+cx_3^2 = 0,$

$$S' \equiv 2x_0x_1+x_2^2+x_3^2 = 0.$$

In this case the intersection consists of the section of $S = 0$ by the tangent plane $x_1 = 0$ at $(1, 0, 0, 0)$, which is therefore a pair of lines, and the section by the plane $2(a-c)x_0+x_1 = 0$, which is an irreducible quadric not containing $(1, 0, 0, 0)$.

(ix) $[(2, 1, 1)]$. $$S \equiv 2ax_0x_1+x_1^2+ax_2^2+ax_3^2 = 0,$$ $$S' \equiv 2x_0x_1+x_2^2+x_3^2 = 0.$$

The intersection is the section of the non-singular quadric S' with the plane $x_1 = 0$ counted twice. The plane touches the quadric, so the intersection is two lines, each counted twice.

(x) $[2, 2]$. $$S \equiv 2ax_0x_1+x_1^2+2bx_2x_3+x_3^2 = 0,$$ $$S' \equiv 2x_0x_1+2x_2x_3 = 0.$$

The intersection contains the line $x_1 = x_3 = 0$. The remaining intersection can be shown to be a cubic curve, which meets the line in the points $(1, 0, 0, 0)$ and $(0, 0, 1, 0)$.

(xi) $[(2, 2)]$. $$S \equiv 2ax_0x_1+x_1^2+2ax_2x_3+x_3^2 = 0.$$ $$S' \equiv 2x_0x_1+2x_2x_3 = 0.$$

The intersection is given by $S' = 0, x_1^2+x_3^2 = 0$. The second equation represents a pair of planes which meet in $x_1 = x_3 = 0$, which is contained in the intersection. Each plane therefore meets S' in $x_1 = x_3 = 0$ and another line. The intersection is thus three lines, of which one is counted twice.

(xii) $[3, 1]$. $$S \equiv 2ax_0x_2+ax_1^2+2x_1x_2+dx_3^2 = 0,$$ $$S' \equiv 2x_0x_2+x_1^2+x_3^2 = 0.$$

In this case it can be shown, as in case (vi), that the intersection is a quartic curve, having $(1, 0, 0, 0)$ as a singular point. The difference here is that if λ is the tangent line to the quartic at a generic point ξ, λ has a unique specialisation as $\xi \to (1, 0, 0, 0)$, whereas in Case (vi) it has two specialisations. The quartic is said to be *cuspidal* in this case.

(xiii) $[(3, 1)]$. $$S \equiv 2ax_0x_2+ax_1^2+2x_1x_2+ax_3^2 = 0,$$ $$S' \equiv 2x_0x_2+x_1^2+x_3^2 = 0.$$

The intersection is the section of $S' = 0$ by $x_1 = 0$ and $x_2 = 0$. The former gives a non-singular quadric Q_1, while the latter is the tangent plane at $(1, 0, 0, 0)$, a point of Q_1. The intersection is thus Q_1 and a pair of lines meeting on Q_1.

(xiv) [4].
$$S \equiv 2ax_0x_3 + 2ax_1x_2 + 2x_1x_3 + x_2^2 = 0,$$
$$S' \equiv 2x_0x_3 + 2x_1x_2 = 0.$$

The intersection contains the line $x_2 = x_3 = 0$. The residual intersection is a cubic curve having this line as tangent line at $(1, 0, 0, 0)$.

CHAPTER XIV

GRASSMANN VARIETIES

WE saw in Chapter X that an unmixed algebraic variety of order g and dimension d can be represented by its Cayley form $F(u_0, \ldots, u_d)$, this form being homogeneous of degree g in each set of indeterminates

$$u_i = (u_{i0}, \ldots, u_{in}) \quad (i = 0, \ldots, d). \tag{1}$$

Conversely, a form which is homogeneous of degree g in the $d+1$ sets of indeterminates (1) is the Cayley form of an algebraic variety of order g and dimension d if and only if the coefficients of the form satisfy certain algebraic relations [X, § 8, Th. II]. This theorem suggested the definition of *algebraic systems* of varieties of a given order and dimension.

The investigation of the properties of an algebraic system of varieties of a given order and dimension is usually a difficult task, but there is one case, of considerable importance in algebraic geometry, in which a number of interesting results can be obtained. This is the case when $g = 1$, in which case the varieties in question are linear spaces of d dimensions [X, § 7, p. 48]. This chapter is devoted to the study of algebraic system of d-spaces in S_n. We shall make considerable use of the Grassmann coordinates $(\ldots, p_{i_0 \ldots i_d}, \ldots)$ of these spaces. The properties of these coordinates were examined in Chapter VII.

1. Grassmann varieties. Let a given S_d in S_n have the Grassmann coordinates $(\ldots, p_{i_0 \ldots i_d}, \ldots)$. Its Cayley form is

$$F(u_0, \ldots, u_d) = \sum_{i_0=0}^{n} \cdots \sum_{i_d=0}^{n} p_{i_0 \ldots i_d} u_{0i_0} \cdots u_{di_d}. \tag{2}$$

This was proved at the end of X, § 7. It follows immediately that if

$$\sum_{i_0=0}^{n} \cdots \sum_{i_d=0}^{n} a_{i_0 \ldots i_d} u_{0i_0} \cdots u_{di_d}$$

is the Cayley form of an S_d, the $a_{i_0 \ldots i_d}$ are skew-symmetric in their suffixes, and [VII, § 6, (2)]

$$\sum_{\lambda=0}^{d+1} (-1)^{\lambda} a_{i_1 \ldots i_d j_\lambda} a_{j_0 \ldots j_{\lambda-1} j_{\lambda+1} \ldots j_{d+1}} = 0.$$

Conversely, if the $a_{i_0\dots i_d}$ satisfy these conditions they are the Grassmann coordinates of a unique S_d [VII, § 6, Th. II], and, by the above reasoning,

$$\sum_{i_0=0}^{n} \dots \sum_{i_d=0}^{n} a_{i_0\dots i_d} u_{0i_0} \dots u_{di_d}$$

is the Cayley form of this S_d.

We have therefore found a set of algebraic relations which are satisfied by the coefficients of any form, homogeneous of degree 1 in the $d+1$ sets of indeterminates (1), if and only if the given form is the Cayley form of an S_d. The representation of the S_d of S_n is an immediate deduction from this.

Let $(\dots, X_{i_0\dots i_d}, \dots)$ be a system of homogeneous coordinates in S_N, where

$$N+1 = \binom{n+1}{d+1},$$

the range of $i_0, \dots, i_d$ being $0, \dots, n$, and the coordinates being skew-symmetric in the suffixes. We consider the variety $\Omega(d, n)$ in S_N given by the equations [VII, § 6, (2)]

$$F_{i_1\dots i_d, j_0\dots j_{d+1}}(X) = \sum_{\lambda=0}^{d+1} (-1)^{\lambda} X_{i_1\dots i_d j_\lambda} X_{j_0\dots j_{\lambda-1} j_{\lambda+1}\dots j_{d+1}} = 0, \quad (3)$$

where $i_1, \dots, i_d$ are d distinct numbers chosen from the set $0, \dots, n$, and $j_0, \dots, j_{d+1}$ are $d+2$ distinct numbers chosen from the same set. By VII, § 7, Th. I, the set of equations (3) forms a basis for all the equations satisfied by $\Omega(d, n)$. There is a one-to-one correspondence without exception between the points of $\Omega(d, n)$ and the S_d of S_n. The variety $\Omega(d, n)$ is called the *Grassmann variety* or *Grassmannian* of the d-spaces of S_n.

An allowable transformation of coordinates in S_n

$$x_i^* = \sum_{j=0}^{n} \tau_{ij} x_j \quad (i = 0, \dots, n)$$

induces the transformation

$$p^*_{i_0\dots i_d} = \sum_{h_0, \dots, h_d} \begin{vmatrix} \tau_{i_0 h_0} & \cdot & \tau_{i_d h_0} \\ \cdot & \cdot & \cdot \\ \tau_{i_0 h_d} & \cdot & \tau_{i_d h_d} \end{vmatrix} p_{h_0\dots h_d} \quad (4)$$

on the Grassmann coordinates of S_d [VII, § 2, (2)], and therefore the transformation

$$X^*_{i_0 \ldots i_d} = \sum_{h_0, \ldots, h_d} \begin{vmatrix} \tau_{i_0 h_0} & \cdot & \tau_{i_d h_0} \\ \cdot & \cdot & \cdot \\ \tau_{i_0 h_d} & \cdot & \tau_{i_d h_d} \end{vmatrix} X_{h_0 \ldots h_d} \tag{5}$$

in S_N. This transformation may also be written in the form

$$X^*_{i_0 \ldots i_d} = \sum_{h_0=0}^{n} \ldots \sum_{h_d=0}^{n} \tau_{i_0 h_0} \ldots \tau_{i_d h_d} X_{h_0 \ldots h_d}.$$

The equations of $\Omega(d, n)$ in the new coordinate system are the same as in the original coordinate system, for

$$\begin{aligned} F^*_{i_1 \ldots i_d, j_0 \ldots j_{d+1}} &= \sum_{\lambda=0}^{d} (-1)^\lambda X^*_{i_1 \ldots i_d j_\lambda} X^*_{j_0 \ldots j_{\lambda-1} j_{\lambda+1} \ldots j_{d+1}} \\ &= \sum_{\lambda=0}^{d+1} (-1)^\lambda \sum_{h_1=0}^{n} \ldots \sum_{h_d=0}^{n} \sum_{k_\lambda=0}^{n} \tau_{i_1 h_1} \ldots \tau_{i_d h_d} \tau_{j_\lambda k_\lambda} X_{h_1 \ldots h_d k_\lambda} \\ &\qquad \times \sum_{k_0=0}^{n} \ldots \sum_{k_{d+1}=0}^{n} \tau_{j_0 k_0} \ldots \tau_{j_{d+1} k_{d+1}} X_{k_0 \ldots k_{\lambda-1} k_{\lambda+1} \ldots k_{d+1}} \\ &= \sum_{h_1=0}^{n} \ldots \sum_{h_d=0}^{n} \sum_{k_0=0}^{n} \ldots \sum_{k_{d+1}=0}^{n} \tau_{i_1 h_1} \ldots \tau_{i_d h_d} \tau_{j_0 k_0} \ldots \tau_{j_{d+1} k_{d+1}} F_{h_1 \ldots h_d, k_0 \ldots k_{d+1}}. \end{aligned}$$

Hence the set of equations $F_{i_1 \ldots i_d, j_0 \ldots j_{d+1}} = 0$ becomes the set of equations $F^*_{i_1 \ldots i_d, j_0 \ldots j_{d+1}} = 0$. That is, $\Omega(d, n)$ is transformed into itself by the transformation (5).

We could, of course, regard the equations (5) as defining a *collineation* in S_n, in which case S_d is transformed into S^*_d, and the point $(\ldots, X_{i_0 \ldots i_d}, \ldots)$ on $\Omega(d, n)$ into the point $(\ldots, X^*_{i_0 \ldots i_d}, \ldots)$, which is also on $\Omega(d, n)$. Since a collineation in S_n can always be found which transforms any given S_d into any other given S^*_d, there is a projective transformation in S_N which transforms $\Omega(d, n)$ into itself and any given point on $\Omega(d, n)$ into any other given point. The set of projective transformations in a linear space S_N which transforms a given algebraic variety into itself evidently forms a group, and the property just proved tells us that when the variety is $\Omega(d, n)$ the group is a transitive group, that is, a group which contains a transformation taking any point A of $\Omega(d, n)$ into any other assigned point B of $\Omega(d, n)$.

By the Principle of Duality, properties of S_d in S_n indicate properties of S_{n-d-1} in S_n. More precisely, *the variety* $\Omega(d,n)$ *is projectively equivalent to the variety* $\Omega(n-d-1,n)$. In fact, if the coordinates of an S_d are $(\ldots, p_{i_0\ldots i_d}, \ldots)$, then these same coordinates are the dual coordinates of an S_{n-d-1}; and the equations of $\Omega(n-d-1,n)$ in a space S'_N can be taken as

$$\sum_{\lambda=0}^{d+1} (-1)^\lambda\, Y^{i_1\ldots i_d j_\lambda} Y^{j_0\ldots j_{\lambda-1} j_{\lambda+1}\ldots j_{d+1}} = 0,$$

where $Y^{i_0\ldots i_d} = Y_{i_{d+1}\ldots i_n}$ are the coordinates in S'_N, $(i_0, \ldots, i_n)$ being an even derangement of $(0, \ldots, n)$. The equations

$$X_{i_0\ldots i_d} = Y^{i_0\ldots i_d} \tag{6}$$

determine a projective correspondence between $\Omega(d,n)$ and $\Omega(n-d-1,n)$.

We now establish some fundamental properties of $\Omega(d,n)$. Let ξ^i_j $(i = 0, \ldots, d;\ j = 0, \ldots, n)$ be $(n+1)(d+1)$ independent indeterminates, and let

$$\Xi_{i_0\ldots i_d} = \begin{vmatrix} \xi^0_{i_0} & \cdot & \xi^0_{i_d} \\ \cdot & \cdot & \cdot \\ \xi^d_{i_0} & \cdot & \xi^d_{i_d} \end{vmatrix}. \tag{7}$$

The point $(\ldots, \Xi_{i_0\ldots i_d}, \ldots)$ corresponds to a generic S_d of S_n and lies on $\Omega(d,n)$. If $(\ldots, X'_{i_0\ldots i_d}, \ldots)$ is any point on $\Omega(d,n)$, there is a d-space which has $(\ldots, X'_{i_0\ldots i_d}, \ldots)$ as its Grassmann coordinates [VII, § 6, Th. II], and we can therefore find $(n+1)(d+1)$ elements x'^i_j $(i = 0, \ldots, d;\ j = 0, \ldots, n)$ in an extension of the ground field such that

$$X'_{i_0\ldots i_d} = \begin{vmatrix} x'^0_{i_0} & \cdot & x'^0_{i_d} \\ \cdot & \cdot & \cdot \\ x'^d_{i_0} & \cdot & x'^d_{i_d} \end{vmatrix}.$$

The point $(\ldots, X'_{i_0\ldots i_d}, \ldots)$ is therefore a proper specialisation of the point $(\ldots, \Xi_{i_0\ldots i_d}, \ldots)$. This proves that $\Omega(d,n)$ *is an irreducible variety* and that $(\ldots, \Xi_{i_0\ldots i_d}, \ldots)$ *is a generic point.*

The possible singular points on $\Omega(d,n)$ fill a proper subvariety [X, § 14]. Hence, since the projective transformations of S_N which leave $\Omega(d,n)$ invariant form a transitive group, it follows that *all points on* $\Omega(d,n)$ *are simple points.*

To find the dimension of $\Omega(d, n)$ we first normalise the coordinates of a generic point. From (7) none of the $\Xi_{i_0 \ldots i_d}$ is zero. We normalise the coordinates so that $\Xi_{0 \ldots d} = 1$. If we define $d+1$ points $B^i = (b_0^i, \ldots, b_n^i)$ $(i = 0, \ldots, d)$, where

$$b_j^i = \Xi_{0 \ldots i-1\, j\, i+1 \ldots d} \quad (i = 0, \ldots, d;\ j = 0, \ldots, n),$$

these $d+1$ points determine a basis for the S_d with Grassmann coordinates $(\ldots, \Xi_{i_0 \ldots i_d}, \ldots)$ [VII, § 4, p. 300]. The point B^i is the intersection of this S_d with the S_{n-d}^i whose equations are

$$x_j = 0 \quad (j = 0, 1, \ldots, i-1, i+1, \ldots, d).$$

Since $B^0, B^1, \ldots, B^d$ form a basis for the generic S_d, and

$$b_j^i = \Xi_{0 \ldots i-1\, j\, i+1 \ldots d},$$

all Grassmann coordinates of this S_d can be expressed integrally in terms of the coordinates $\Xi_{0 \ldots i-1\, j\, i+1 \ldots d}$ $(i = 0, \ldots, d;\ j = 0, \ldots, n)$, where $\Xi_{0 \ldots d} = 1$. When $j = 0, 1, \ldots, i-1, i+1, \ldots, d$, the corresponding coordinate is zero. Hence the Grassmann coordinates of a generic S_d can be expressed integrally in terms of $(d+1)(n-d)$ coordinates

$$\Xi_{0 \ldots i-1\, j\, i+1 \ldots d} \quad (i = 0, \ldots, d;\ j = d+1, \ldots, n).$$

These coordinates are independent indeterminates. This follows from the fact that there exists an S_d^* determined by the $d+1$ points

$$\left.\begin{aligned} B^0 &= (1, 0, \ldots, 0, b_{d+1}^0, \ldots, b_n^0), \\ B^1 &= (0, 1, \ldots, 0, b_{d+1}^1, \ldots, b_n^1), \\ &\cdot \quad \cdot \quad \cdot \quad \cdot \quad \cdot \quad \cdot \\ B_d &= (0, 0, \ldots, 1, b_{d+1}^d, \ldots, b_n^d), \end{aligned}\right\} \tag{8}$$

where the b_j^i which appear in (8) are independent indeterminates. The Grassmann coordinates of this S_d^* are given by the equations

$$\Xi_{0 \ldots i-1\, j\, i+1 \ldots d}^* = b_j^i,$$

and since we have proved above that the dimension of a generic point on $\Omega(d, n)$ is at most $(d+1)(n-d)$, the construction of a point on $\Omega(d, n)$ with dimension $(d+1)(n-d)$ over K proves that the dimension of a generic point is $(d+1)(n-d)$.

We have proved not only that $\Omega(d, n)$ is of dimension $(d+1)(n-d)$, but that $\Omega(d, n)$ is a *rational* variety, the coordinates of a generic point being expressible rationally in terms of $(d+1)(n-d)$

independent indeterminates which are themselves rational functions of the coordinates of the generic point.

We remark that $\Omega(d,n)$ lies in S_N and not in a space of lower dimension. This follows from VII, § 6, Th. I, where it is proved that there is no non-trivial relation of the form

$$\sum_{i_0,\ldots,i_d} u_{i_0\ldots i_d} p_{i_0\ldots i_d} = 0$$

connecting the Grassmann coordinates of a generic S_d. Hence $\Omega(d,n)$ is not contained in any prime

$$\sum_{i_0,\ldots,i_d} u_{i_0\ldots i_d} X_{i_0\ldots i_d} = 0.$$

We also note that if $d = 0$, $\Omega(d,n) = S_n$, and, by duality,

$$\Omega(n-1,n) = S_n.$$

For convenience we shall sometimes omit these cases.

There is a geometrical property of $\Omega(d,n)$ which it is convenient to insert here. This connects points P and Q on $\Omega(d,n)$ whose representative S_d meet in an S_{d-1}. We prove that the tangent space to $\Omega(d,n)$ at either point contains the other point.

If two S_d meet in an S_{d-1} their join is an S_{d+1}, and there is a pencil of S_d in this S_{d+1} through the S_{d-1}. If the coordinates of P and Q are $(\ldots, p_{i_0\ldots i_d}, \ldots)$ and $(\ldots, q_{i_0\ldots i_d}, \ldots)$ respectively, the pencil of S_d is given by $(\ldots, p_{i_0\ldots i_d} + \lambda q_{i_0\ldots i_d}, \ldots)$. Hence this point satisfies the equations (3), that is, the line PQ lies on $\Omega(d,n)$. Let

$$F(\ldots, X_{i_0\ldots i_d}, \ldots) = 0$$

be any one of these equations which represents a quadric through $\Omega(d,n)$ with a simple point at P. Then

$$\begin{aligned} F(\ldots, p_{i_0\ldots i_d} + \mu q_{i_0\ldots i_d}, \ldots) &= F(\ldots, p_{i_0\ldots i_d}, \ldots) + \mu \sum_i q_{i_0\ldots i_d} \frac{\partial F(\ldots, p_{i_0\ldots i_d}, \ldots)}{\partial p_{i_0\ldots i_d}} \\ &\quad + \mu^2 F(\ldots, q_{i_0\ldots i_d}, \ldots) \\ &= \mu \sum_i q_{i_0\ldots i_d} \frac{\partial F(\ldots, p_{i_0\ldots i_d}, \ldots)}{\partial p_{i_0\ldots i_d}} = 0. \end{aligned}$$

Hence the tangent prime to this quadric at P contains Q. Since the tangent space to $\Omega(d,n)$ at P is the intersection of the tangent primes at P to all quadrics given by (3) which have simple points at P, the tangent space to $\Omega(d,n)$ at P contains Q. Similarly, the tangent space to $\Omega(d,n)$ at Q contains P.

Conversely, let P and Q be two points in $\Omega(d,n)$ such that the tangent space at P contains Q. If $F(\ldots, X_{i_0\ldots i_d}, \ldots) = 0$ is one of the quadratic equations given by (3), which form a basis for $\Omega(d,n)$, we have

$$\sum_i q_{i_0\ldots i_d} \frac{\partial F(\ldots, p_{i_0\ldots i_d}, \ldots)}{\partial p_{i_0\ldots i_d}} = 0, \tag{9}$$

and since $F(\ldots, p_{i_0\ldots i_d}, \ldots) = 0$ and $F(\ldots, q_{i_0\ldots i_d}, \ldots) = 0$, and F is of degree two, it follows that $F(\ldots, p_{i_0\ldots i_d} + \mu q_{i_0\ldots i_d}, \ldots) = 0$, that is $(\ldots, p_{i_0\ldots i_d} + \mu q_{i_0\ldots i_d}, \ldots)$ are the coordinates of a d-space for all values of μ. Hence the line PQ lies on $\Omega(d,n)$.

We now show that this condition implies that the spaces $p = (\ldots, p_{i_0\ldots i_d}, \ldots)$ and $q = (\ldots, q_{i_0\ldots i_d}, \ldots)$ meet in a $(d-1)$-space. Let

$$\sum_{j=0}^{n} u_{ij} x_j = 0 \quad (i = 1, \ldots, d-1)$$

be the equations of a generic $(n-d+1)$-space S_{n-d+1} of S_n. This meets p in the line p' whose coordinates are $(\ldots, p'_{ij}, \ldots)$, where

$$p'_{ij} = \sum_{k_1} \ldots \sum_{k_{d-1}} u_{1k_1} \ldots u_{d-1\,k_{d-1}} p_{k_1\ldots k_{d-1}\,ij},$$

and q in the line q' whose coordinates $(\ldots, q'_{ij}, \ldots)$ are similarly defined. Equation (9) can be written in the form

$$\sum_{\lambda} (-1)^{\lambda} p_{i_1\ldots i_d j_\lambda} q_{j_0\ldots j_{\lambda-1} j_{\lambda+1}\ldots j_{d+1}} + \sum_{\lambda} (-1)^{\lambda} q_{i_1\ldots i_d j_\lambda} p_{j_0\ldots j_{\lambda-1} j_{\lambda+1}\ldots j_{d+1}} = 0, \tag{10}$$

for all choices of $i_1, \ldots, i_d$, and of $j_0, \ldots, j_{d+1}$. Multiply (10) by

$$u_{1i_1} \ldots u_{d-1\,i_{d-1}} u_{1j_1} \ldots u_{d-1\,j_{d-1}},$$

and sum over all possible values of $i_1, \ldots, i_{d-1}, j_1, \ldots, j_{d-1}$. We obtain the equation $0 = p'_{i_d j_0} q'_{j_d j_{d+1}} + p'_{i_d j_d} q'_{j_{d+1} j_0} + p'_{i_d j_{d+1}} q'_{j_0 j_d} +$ terms in which p' and q' are interchanged. Since this relation holds for all choices of i_d, j_0, j_d, j_{d+1}, we conclude that if $(a_0, \ldots, a_n)$, $(\alpha_0, \ldots, \alpha_n)$ are two points on p', and $(b_0, \ldots, b_n)$, $(\beta_0, \ldots, \beta_n)$ are two points on q', then the matrix

$$\begin{pmatrix} a_0 & . & . & a_n \\ \alpha_0 & . & . & \alpha_n \\ b_0 & . & . & b_n \\ \beta_0 & . & . & \beta_n \end{pmatrix}$$

is of rank less than four. Hence p' and q' have a point in common. It follows that S_{n-d+1} meets the intersection of p and q. Since S_{n-d+1} is generic, p and q must have in common a space of dimension $d-1$ at least, and since p and q are distinct they must meet in a space of precisely this dimension. Hence we have

THEOREM I. *A necessary and sufficient condition that two d-spaces in S_n meet in a $(d-1)$-space is that they be represented on the Grassmannian $\Omega(d,n)$ by two points whose join lies on $\Omega(d,n)$.*

2. Schubert varieties. An algebraic subvariety on $\Omega(d,n)$ defines an algebraic system of S_d in S_n. Of particular importance are the *Schubert systems*. A Schubert system is that system of d-spaces which satisfies a *Schubert condition* $(a_0, a_1, \ldots, a_d)$. This condition is now defined.

Let $A_0, A_1, \ldots, A_d$ be $d+1$ linear spaces such that

$$A_0 \subset A_1 \subset \ldots \subset A_d.$$

If their respective dimensions are $a_0, a_1, \ldots, a_d$, we have the inequalities

$$0 \leqslant a_0 < a_1 < \ldots < a_d \leqslant n.$$

The d-spaces satisfying the Schubert condition $(a_0, a_1, \ldots, a_d)$ are those which meet A_0 in a point (at least), A_1 in a line (at least), ..., A_{d-1} in a $[d-1]$ (at least), and lie in A_d.

The subvariety on $\Omega(d,n)$ which represents these d-spaces is called a *Schubert variety* and is denoted by $\Omega_{a_0 a_1 \ldots a_d}$.

When there is no risk of confusion we shall use the symbol $\Omega_{a_0 a_1 \ldots a_d}$ both for the subvariety on $\Omega(d,n)$ and for the system of d-spaces represented by $\Omega(d,n)$.

THEOREM I. *$\Omega_{a_0 a_1 \ldots a_d}$ is irreducible and of dimension*

$$\sum_{i=0}^{d+1} a_i - \tfrac{1}{2}d(d+1).$$

Let $\xi^{(k)}$ be a generic point of A_k $(k = 0, 1, \ldots, d)$. Since $A_{k+i} \nsubseteq A_k$, $\xi^{(k+i)}$ is not in A_k $(i > 0)$. On the other hand, $\xi^{(k-i)}$ lies in A_k. Then $\xi^{(0)}, \xi^{(1)}, \ldots, \xi^{(d)}$ are linearly independent. For, suppose that $\xi^{(0)}, \xi^{(1)}, \ldots, \xi^{(k)}$ are linearly independent, but $\xi^{(0)}, \xi^{(1)}, \ldots, \xi^{(k)}, \xi^{(k+1)}$ are linearly dependent. Then $\xi^{(k+1)}$ is dependent on $\xi^{(0)}, \ldots, \xi^{(k)}$, and therefore lies in A_k. We thus have a contradiction.

Let the S_k determined by $\xi^{(0)}, \ldots, \xi^{(k)}$ be denoted by η_k $(k = 0, \ldots, d)$ and let $\eta = \eta_d$. Clearly $\eta_{k+1} \wedge A_k = \eta_k$. It is also evident that $\eta \wedge A_k \supseteq \eta_k$. If this intersection contains a point P which is not in η_k, P cannot be in η_{k+1}, and since P is in $\eta \wedge A_{k+1}$, it follows that $\eta \wedge A_{k+1} \supset \eta_{k+1}$. Proceeding in this way, we see that the assumption $\eta \wedge A_k \supset \eta_k$ leads to the conclusion that η and A_{d-1} meet in a d-space; that is, $\eta \subseteq A_{d-1}$. But η contains $\xi^{(d)}$, which is not in A_{d-1}, and we have a contradiction. Hence $\eta \wedge A_k = \eta_k$.

Consider the correspondence with generic point $(\xi^{(0)}, \ldots, \xi^{(d)}, \eta)$. We may regard this as a correspondence between the $(d+1)$-way space $S_{a_0 \ldots a_d}$ and $\Omega(d, n)$. Among the equations of the correspondence are:

(a) the quadratic equations which express the fact that η is a d-space;

(b) the equations which express the fact that $\xi^{(k)}$ lies in η and in A_k;

(c) the equations which express the fact that η and A_k meet in a k-space.

From (a) and (c) we deduce that any proper specialisation of η is a d-space which meets A_k in a k-space, *at least*.

Conversely, let $\bar{y}$ be a d-space which meets A_k in a k-space at least $(k = 0, \ldots, d)$. We can find a point $\bar{x}^{(0)}$ in $\bar{y}$ and in A_0. Since $\bar{y}$ meets A_1 in a line, at least, we can find $\bar{x}^{(1)} \neq \bar{x}^{(0)}$ in $\bar{y}$ and in A_1. Proceeding in this way, we can find $\bar{x}^{(0)}, \bar{x}^{(1)}, \ldots, \bar{x}^{(d)}$, where $\bar{x}^{(k)}$ is in A_k, and the $d+1$ points are linearly independent.

Since $\xi^{(0)}, \ldots, \xi^{(d)}$ are independent generic points, $(\bar{x}^{(0)}, \ldots, \bar{x}^{(d)})$ is a proper specialisation of $(\xi^{(0)}, \ldots, \xi^{(d)})$. Corresponding to this, there exists at least one proper specialisation y' of η such that $(\bar{x}^{(0)}, \ldots, \bar{x}^{(d)}, y')$ is a proper specialisation of $(\xi^{(0)}, \ldots, \xi^{(d)}, \eta)$. From the properties of the correspondence we see that y' is a d-space containing $\bar{x}^{(k)}$ $(k = 0, \ldots, d)$. Since $\bar{y}$ is the only d-space which contains these points, we have $y' = \bar{y}$. Hence any point of $\Omega_{a_0 a_1 \ldots a_d}$ is a proper specialisation of η. It follows that η is a generic point of $\Omega_{a_0 a_1 \ldots a_d}$, which is therefore irreducible.

Since $\eta \wedge A_k = \eta_k$, the specialisations of $\xi^{(k)}$ which correspond to η lie in η_k. Conversely, if $\bar{\xi}^{(k)}$ is a generic point of η_k, $(\xi^{(0)}, \ldots, \bar{\xi}^{(k)}, \ldots, \xi^{(d)}, \eta)$ is a proper specialisation of $(\xi^{(0)}, \ldots, \xi^{(k)}, \ldots, \xi^{(d)}, \eta)$. Hence there corresponds to η, in the $(d+1)$-way space $S_{a_0 \ldots a_d}$, a variety of dimension $0 + 1 + \ldots + d = \frac{1}{2}d(d+1)$. It follows that if $\Omega_{a_0 \ldots a_d}$ is of dimension ρ_a, the correspondence we have been considering is of dimension

$$\rho_a + \tfrac{1}{2}d(d+1).$$

On the other hand, it is clear that the object-variety in $S_{a_0 \dots a_d}$ is the whole space, and to a generic point of it, $(\xi^{(0)}, \dots, \xi^{(d)})$, there corresponds a unique point η of $\Omega_{a_0 \dots a_d}$. Hence the dimension of the correspondence is also $\sum_{i=0}^{d+1} a_i$. It follows that

$$\rho_a = \sum_{i=0}^{d+1} a_i - \tfrac{1}{2}d(d+1).$$

Any two Schubert varieties corresponding to the same set of dimension numbers $(a_0, a_1, \dots, a_d)$ are projectively equivalent in S_N. Let $A_0 \subset A_1 \subset \dots \subset A_d$ and $B_0 \subset B_1 \subset \dots \subset B_d$ be two sets of defining spaces, where $\dim A_i = \dim B_i = a_i$. We can evidently find a projective transformation of S_n which transforms A_i into B_i $(i = 0, \dots, d)$. This transformation maps an S_d satisfying $(a_0, a_1, \dots, a_d)$ for one set of defining spaces on an S_d satisfying $(a_0, a_1, \dots, a_d)$ for the other set. The projective transformation induced in S_N by the projective transformation in S_n therefore transforms one variety $\Omega_{a_0 a_1 \dots a_d}$ into the other.

It follows that all Schubert varieties corresponding to the same set of dimension numbers $(a_0, a_1, \dots, a_d)$ belong to a continuous algebraic system.

Some special Schubert systems are of frequent occurrence. The system $\Omega_{n-d\, n-d+1 \dots n}$ coincides with $\Omega(d, n)$, since every S_d meets an S_{n-d} in a point (at least), an S_{n-d+1} in a line (at least), ..., and lies in S_n. A system $\Omega_{01 \dots d}$ consists of a unique S_d. A system $\Omega_{m-d\, m-d+1 \dots m}$ consists of all the S_d which lie in a fixed $[m]$, since all such S_d, and only these, satisfy the Schubert condition. All the S_d of S_n which meet a given $[l]$ in a space of i dimensions, at least, are given by a system

$$\Omega_{l-i\, l-i+1 \dots l-1\, l\, n-d+i+1\, n-d+i+2 \dots n}.$$

In particular, all the S_d which pass through a given $[l]$ and lie in a given $[m]$ are given by a system

$$\Omega_{01 \dots l\, m-d+l+1\, m-d+l+2 \dots m}.$$

Again, all the S_d which meet a given $[l]$ are given by a system

$$\Omega_{l\, n-d+1\, n-d+2 \dots n}.$$

Schubert varieties may have singular points. For example, the lines in S_3 which meet a given line satisfy the Schubert condition

(1, 3). If the line has coordinates $(\ldots, \pi_{ij}, \ldots)$, the Schubert variety Ω_{13} is given as the intersection of the Grassmann variety

$$X_{01}X_{23}+X_{02}X_{31}+X_{03}X_{12}=0$$

with the S_4

$$\pi_{01}X_{23}+\pi_{23}X_{01}+\pi_{02}X_{31}+\pi_{31}X_{02}+\pi_{03}X_{12}+\pi_{12}X_{03}=0.$$

This S_4 touches the quadric $\Omega(1,3)$ at the point $(\ldots, \pi_{ij}, \ldots)$, and this point is a singular point on the variety Ω_{13}.

We have seen that any two Schubert varieties $\Omega_{a_0 a_1 \ldots a_d}$ with the same dimension numbers $a_0, a_1, \ldots, a_d$ are projectively equivalent. In discussing the properties of a Schubert variety it is therefore permissible to consider one which corresponds to conveniently chosen defining spaces $A_0 \subset A_1 \subset \ldots \subset A_d$. This is what we shall do in the next section.

We conclude this section by finding the dual of a Schubert condition. We prove

THEOREM II. *The dual of a Schubert condition is a Schubert condition.*

We consider the Schubert variety $\Omega_{a_0 a_1 \ldots a_d}$, which consists of the d-spaces in S_n which satisfy the Schubert condition $(a_0, a_1, \ldots, a_d)$ by meeting A_0 in a point (at least), ..., A_k in a k-space (at least), ..., and lie in A_d, where

$$A_0 \subset A_1 \subset \ldots \subset A_d$$

are assigned spaces of dimensions $a_0, a_1, \ldots, a_d$. We consider the dual of the whole configuration.

Write $q = n-d-1$. The dual of an S_d is an S_q. The dual of A_k is a space B_k of dimension $\beta_k = n-1-a_k$, and

$$B_d \subset B_{d-1} \subset \ldots \subset B_0.$$

Since an S_d satisfying $(a_0, a_1, \ldots, a_d)$ meets A_k in a $[k]$ (at least), the dual S_q is such that its join to B_k is an $[n-1-k]$ (at most). A necessary and sufficient condition for this is that S_q and B_k intersect in a $[\beta_k+k-d]$ (at least). If we write

$$\gamma_k = \beta_k + k - d,$$

then $$\gamma_{k-1}-\gamma_k = \beta_{k-1}-\beta_k-1 = a_k-a_{k-1}-1 \geqslant 0.$$

Hence $$\gamma_d \leqslant \gamma_{d-1} \leqslant \ldots \leqslant \gamma_0.$$

Also $$\gamma_k = n-1-d+k-a_k,$$

and since $$a_0 < a_1 < \ldots < a_d \leqslant n,$$
it follows that $$a_k \leqslant n + k - d,$$
so that $$\gamma_k \geqslant -1.$$

The q-spaces which are the duals of the d-spaces satisfying the Schubert condition $(a_0, a_1, \ldots, a_d)$ therefore meet B_i in a γ_i-space $(i = 0, \ldots, d)$. We show that these are just the q-spaces which satisfy a certain Schubert condition.

Suppose that $\gamma_d = \ldots = \gamma_{s+1} = -1$, $\gamma_s \geqslant 0$. Then there is no condition imposed on the S_q to meet $B_d, \ldots, B_{s+1}$. The S_q must meet B_s in a γ_s-space, and we express this as a Schubert condition. We choose in B_s spaces $C_0, C_1, \ldots, C_{\gamma_s - 1}$, where

$$C_0 \subset C_1 \subset \ldots \subset C_{\gamma_s} = B_s,$$

the dimension of C_i being $\dim B_s - (\gamma_s - i) = \beta_s - \gamma_s + i = d - s + i$. Any q-space meeting C_i in an i-space $(i = 0, \ldots, \gamma_s)$ meets B_s in a γ_s-space, and conversely, any q-space meeting B_s in a γ_s-space meets C_i in an i-space $(i = 0, \ldots, \gamma_s - 1)$.

Now suppose that we have a sequence of spaces

$$C_0 \subset C_1 \subset \ldots \subset C_t \quad (t = \gamma_{k+1})$$

such that the q-spaces which meet C_i in an i-space $(i = 0, \ldots, t)$ are just those which meet B_j in a γ_j-space $(j = d, \ldots, k+1)$. If $\gamma_k = \gamma_{k+1}$, each S_q which meets B_{k+1} in a γ_{k+1}-space meets B_k in a γ_k-space, since $B_{k+1} \subset B_k$. Hence the q-spaces which meet C_i in an i-space $(i = 0, \ldots, t)$ meet B_j in a γ_j-space $(j = d, \ldots, k)$. On the other hand, if $\gamma_k > \gamma_{k+1}$, we can interpose spaces $C_{t+1}, \ldots, C_{t+\gamma_k-\gamma_{k+1}}$, where

$$C_t \subset \ldots \subset C_{t+\gamma_k-\gamma_{k+1}} = B_k,$$

the dimension of $C_{t+\gamma_k-\gamma_{k+1}-i}$ being $\beta_k - i$. We see, as above, that the q-spaces which meet C_i in an i-space $(i = 0, \ldots, t+\gamma_k-\gamma_{k+1})$ are just those which meet B_j in a γ_j-space $(j = d, \ldots, k)$.

We proceed to construct spaces C_i in this way, finally considering the case $k = -1$, when we take $\gamma_{-1} = q$, $\beta_{-1} = n$. We thus find a sequence of spaces

$$C_0 \subset C_1 \subset \ldots \subset C_q$$

of dimensions $\rho_0, \rho_1, \ldots, \rho_q$, where

$$0 \leqslant \rho_0 < \rho_1 < \ldots < \rho_q \leqslant n,$$

such that the q-spaces which satisfy the Schubert condition $(\rho_0, \ldots, \rho_q)$ defined by the set of spaces $C_0, \ldots, C_q$ are the duals of the d-spaces which satisfy the Schubert condition $(a_0, a_1, \ldots, a_d)$. We have therefore proved Theorem II.

A rule for finding the dual of a Schubert condition $(a_0, \ldots, a_d)$ may now be stated. *The dual is* $(\rho_0, \rho_1, \ldots, \rho_{n-d-1})$, *where the* $\rho_0, \rho_1, \ldots, \rho_{n-d-1}$ *are merely the set of integers* $0, 1, \ldots, n$, *with the integers* $n-a_d, n-a_{d-1}, \ldots, n-a_0$ *deleted.* Indeed, since

$$\gamma_d = \gamma_{d-1} = \ldots = \gamma_{s+1} = -1,$$

we have
$$n-a_k = d-k \quad (k > s),$$

and since $\rho_0 = \dim C_0 = d-s < \rho_i$ $(i \geqslant 1)$, the numbers $n-a_k$ $(k > s)$ are missing from the sequence $\rho_0, \ldots, \rho_q$. Similarly, if we have $\gamma_{k+s+1} > \gamma_{k+s} = \ldots = \gamma_{k+1} > \gamma_k$, when we pass from the sequence of spaces

$$C_0 \subset C_1 \subset \ldots \subset C_{\gamma_{k+s}}$$

to the sequence
$$C_0 \subset C_1 \subset \ldots \subset C_{\gamma_k},$$

we see by our formulae that

$$\rho_t = \dim C_t \quad = \beta_{k+s},$$

and
$$\rho_{t+1} = \dim C_{t+1} = \beta_k - \gamma_k + \gamma_{k+1} + 1,$$

and we see that $n-a_{k+s}, \ldots, n-a_{k+1}$ are omitted. In this way we can see that the rule for determining $\rho_0, \ldots, \rho_q$ is justified.

3. Equations of a Schubert variety. A Schubert condition $(a_0, \ldots, a_d)$ in S_n is given by a sequence of spaces

$$A_0 \subset A_1 \subset \ldots \subset A_d.$$

Given the equations of these spaces in any coordinate system in S_n, we can obtain the equations of the corresponding Schubert variety on $\Omega(d, n)$ from them. In this section we shall consider two methods [Representations A and B] of choosing the coordinate system in S_n which lead to particularly simple forms for the equations of the Schubert variety. Since any two Schubert varieties $\Omega_{a_0 \ldots a_d}$ with the same dimension numbers $a_0, \ldots, a_d$ are projectively equivalent, it is possible to choose the coordinate system so that any given variety $\Omega_{a_0 \ldots a_d}$ has its equation in one or other of these forms. The purpose of having two representations is to enable us to write down in convenient form the equations of two Schubert varieties $\Omega_{a_0 \ldots a_d}$ and $\Omega_{b_0 \ldots b_d}$ simultaneously; but it is important to observe that it is not implied that, given any two Schubert varieties $\Omega_{a_0\ldots a_d}$ and $\Omega_{b_0\ldots b_d}$ on $\Omega(d, n)$, we can choose a coordinate system in S_n so that one is given in this coordinate system by Representation A, and the other by Representation B. It will appear, however, that in the cases considered in later sections a Schubert variety $\mathbf{\Omega}_{a_0\ldots a_d}$ given by Representation A and a Schubert variety $\mathbf{\Omega}_{b_0\ldots b_d}$ given by Representation B in the same coordinate system are sufficiently general members of the systems $\{\mathbf{\Omega}_{a_0\ldots a_d}\}$ and $\{\mathbf{\Omega}_{b_0\ldots b_d}\}$ to enable us to deduce results of significance for generic members of these systems.

In both representations, we shall find it more convenient to denote the coordinates in S_n by $(x_0, \ldots, x_d, y_0, \ldots, y_q)$ $(q = n-d-1)$, rather than by $(x_0, \ldots, x_n)$.

Representation A. We choose the coordinates in S_n so that the a_d-space A_d has the equation

$$y_{m_0} = y_{m_0+1} = \ldots = y_q = 0 \quad (a_d = m_0+d).$$

We have still a choice of the spaces to be represented by $x_i = 0$ $(i = 0, \ldots, d)$, $y_j = 0$ $(j < m_0)$, and we can choose these so that A_{d-1}, which lies in A_d, is given by the equations

$$x_0 = 0, \quad y_{m_1} = y_{m_1+1} = \ldots = y_q = 0 \quad (a_{d-1} = m_1+d-1).$$

Since $A_{d-2} \subset A_{d-1}$, we may further choose the coordinates so that A_{d-2} is given by

$$x_0 = x_1 = 0, \quad y_{m_2} = y_{m_2+1} = \ldots = y_q = 0 \quad (a_{d-2} = m_2+d-2).$$

Proceeding in this way, we can choose the coordinate system so that the a_{d-r}-space A_{d-r} has the equations

$$x_0 = \ldots = x_{r-1} = 0, \quad y_{m_r} = \ldots = y_q = 0 \quad (a_{d-r} = m_r+d-r).$$

We observe that $\quad m_0 \geqslant m_1 \geqslant \ldots \geqslant m_d.$

A generic point of A_{d-r} is

$$(0, \ldots, 0, \alpha_{rr}, \ldots, \alpha_{rd}, \beta_{r0}, \ldots, \beta_{r\,m_r-1}, 0, \ldots, 0),$$

where the α_{rj} $(j \geqslant r)$ and β_{rj} $(j < m_r)$ are independent indeterminates. $\Omega_{a_0\ldots a_d}$ is irreducible, and the d-space which contains a generic point of A_i $(i = 0, \ldots, d)$ is represented by a generic point of $\Omega_{a_0\ldots a_d}$. Since

$$\begin{vmatrix} \alpha_{00} & \alpha_{01} & . & \alpha_{0d} \\ 0 & \alpha_{11} & . & \alpha_{1d} \\ . & . & . & . \\ 0 & 0 & . & \alpha_{dd} \end{vmatrix} = \alpha_{00}\alpha_{11}\ldots\alpha_{dd}$$

is not zero when the α_{ij} are independent indeterminates, not all points of $\Omega_{a_0 \ldots a_d}$ lie in

$$X_{0\ldots d} = 0,$$

and in particular a generic point of it does not lie in this space. Any S_d represented by a point of $\Omega_{a_0 \ldots a_d}$ not in this space is given by a set of equations [VII, § 2, (2)]

$$y_i = \sum_{j=0}^{d} u_{ij} x_j \quad (i = 0, \ldots, q), \tag{1}$$

where

$$p_{0\ldots d} u_{ij} = p_{01\ldots j-1\, d+i+1\, j+1 \ldots d}. \tag{2}$$

We now find the conditions which must be satisfied by the S_d given by (1) to belong to $\Omega_{a_0 \ldots a_d}$. It must meet A_{d-r} in a space of at least $d-r$ dimensions, for $r = 0, 1, \ldots, d$. If in the equations of S_d given by (1) we put

$$x_0 = 0, \quad \ldots, \quad x_{r-1} = 0, \quad y_{m_r} = 0, \quad \ldots, \quad y_q = 0,$$

we obtain the equations

$$\left.\begin{aligned} y_0 &= \sum_{j=r}^{d} u_{0j} x_j, \\ &\vdots \\ y_\sigma &= \sum_{j=r}^{d} u_{\sigma j} x_j \quad (\sigma = m_r - 1), \\ 0 &= \sum_{j=r}^{d} u_{m_r j} x_j, \\ &\vdots \\ 0 &= \sum_{j=r}^{d} u_{qj} x_j. \end{aligned}\right\} \tag{3}$$

These may be interpreted as giving $q+1$ primes in the space A_{d-r} of a_{d-r} dimensions $(x_r, \ldots, x_d, y_0, \ldots, y_{m_r-1})$. For all values of the u_{ij} which appear in the above equations the first m_r equations are linearly independent, and therefore define a space of $a_{d-r} - m_r = d-r$ dimensions. The remaining equations, if non-trivial, are independent of the first m_r equations, and therefore the equations (3) define a space of dimension $d-r$ if and only if $u_{ij} = 0$ $(i \geqslant m_r, j \geqslant r)$.

Hence the conditions to be satisfied if the S_d given by (1) belongs to $\Omega_{a_0 \ldots a_d}$ are

$$u_{ij} = 0 \quad (i \geqslant m_j;\ j = 0, \ldots, d).$$

If these conditions are satisfied, and the remaining u_{ij} are taken as indeterminates over K, $K(u_{ij})$ is of dimension

$$\rho = \sum_{i=0}^{d} m_i = \sum_{i=0}^{d} a_i - \tfrac{1}{2}d(d+1).$$

Hence the coordinates of S_d define a generic point of a variety of dimension ρ on $\Omega(d,n)$, and this variety is contained in $\Omega_{a_0\ldots a_d}$. But $\Omega_{a_0\ldots a_d}$ is irreducible and of dimension ρ. It follows that S_d is a generic member of the Schubert system defined by the condition $(a_0, \ldots, a_d)$.

We note that a particular S_d satisfying the Schubert condition is obtained by taking *all* the u_{ij} appearing in (1) to be zero. We then obtain an S_d for which $p_{01\ldots d} = 1$, and all the other $p_{i_0\ldots i_d}$ are zero.

We put on record that among the equations satisfied by $\Omega_{a_0\ldots a_d}$ in this representation there are the linear equations

$$X_{0\ldots j-1\, d+i+1\, j+1\ldots d} = 0 \quad (i \geqslant m_j;\ j = 0, \ldots, d). \tag{4}$$

The point P, corresponding to the particular S_d for which all u_{ij} are zero, is a simple point on $\Omega_{a_0\ldots a_d}$. To prove this we first find the equations to the tangent space at P on $\Omega(d,n)$. This space is the intersection of the tangent primes at P to the primals

$$\sum_{\lambda=0}^{d+1} (-1)^{\lambda} X_{i_1\ldots i_d j_\lambda} X_{j_0\ldots j_{\lambda-1} j_{\lambda+1}\ldots j_{d+1}} = 0$$

which have a simple point at P [X, § 14, Th. II]. Such primals must contain $X_{01\ldots d}$ in their equation and can therefore be written in the form [VII, § 6, (4)]

$$X_{01\ldots d} X_{j_0\ldots j_d} = \sum_{\lambda=0}^{d} X_{0\ldots s-1\, j_\lambda s+1\ldots d} X_{j_0\ldots j_{\lambda-1} s j_{\lambda+1}\ldots j_d}.$$

This equation is non-trivial only if at least two of the integers $j_0, \ldots, j_d$ are distinct from $0, 1, \ldots, d$, that is, greater than d. We can then choose s so that s is not in the set $j_0, \ldots, j_d$, and therefore the only term in the equation which contains $X_{01\ldots d}$ is the term $X_{01\ldots d} X_{j_0\ldots j_d}$. The tangent space at P to the primal we are considering is therefore

$$X_{j_0\ldots j_d} = 0.$$

Hence the tangent space to $\Omega(d,n)$ at P satisfies the equations

$$X_{j_0\ldots j_d} = 0 \tag{5}$$

for all $j_0, \ldots, j_d$ in which at least two of the integers in this set are greater than d. The equations (5) define a space of dimension $(d+1)(n-d)$, and therefore define the tangent space at P to $\Omega(d, n)$.

We now consider the intersection of the tangent spaces at P to primals through $\Omega_{a_0 \ldots a_d}$ which have P as a simple point. This intersection must satisfy the equations (4) and (5). But these equations define a space of dimension

$$(d+1)(n-d) - \sum_j (q+1-m_j) = \sum a_j - \tfrac{1}{2}d(d+1),$$

which is the dimension of $\Omega_{a_0 \ldots a_d}$. Hence P must be a simple point of $\Omega_{a_0 \ldots a_d}$, and the tangent space at P is given by (4) and (5).

Representation B. An argument similar to that used in connection with Representation A shows us that we can choose the coordinate system $(x_0, \ldots, x_d, y_0, \ldots, y_q)$ in S_n so that A_r satisfies the equations

$$x_{r+1} = \ldots = x_d = y_0 = \ldots = y_{q-m_{d-r}} = 0 \quad (m_{d-r} = a_r - r)$$

for $r = 0, \ldots, d$. Again, we see that not all d-spaces which meet A_r in an r-space $(r = 0, \ldots, d)$ satisfy the condition $p_{0 \ldots d} = 0$. Indeed, the space for which $p_{0 \ldots d} = 1$, while the remaining $p_{i_0 \ldots i_d}$ are zero, satisfies the Schubert condition. Any d-space for which $p_{0 \ldots d} \neq 0$, satisfying the Schubert condition, and in particular the generic d-space satisfying the condition, can be written in the form

$$y_i = \sum_{j=0}^{d} u_{ij} x_j \quad (i = 0, \ldots, q). \tag{1}$$

The S_d given by (1) meets A_r in a space which has the following equations:

$$\left.\begin{aligned}
0 &= \sum_{j=0}^{r} u_{0j} x_j, \\
\vdots \quad & \quad \vdots \\
0 &= \sum_{j=0}^{r} u_{\sigma j} x_j \quad (\sigma = q - m_{d-r}), \\
y_{\sigma+1} &= \sum_{j=0}^{r} u_{\sigma+1\, j} x_j, \\
\vdots \quad & \quad \vdots \\
y_q &= \sum_{j=0}^{r} u_{qj} x_j.
\end{aligned}\right\}$$

Here the final $q - \sigma = m_{d-r}$ equations define a space of $a_r - m_{d-r} = r$ dimensions in A_r, and for S_d to meet A_r in a space of r dimensions, the first $\sigma + 1$ equations must be trivial. Hence we must have

$$u_{ij} = 0 \quad (i \leqslant q - m_{d-r};\ j = 0, \ldots, r).$$

If S_d satisfies the Schubert condition $(a_0, \ldots, a_d)$, the whole set of equations is

$$u_{ij} = 0 \quad (i \leqslant q - m_{d-j};\ j = 0, \ldots, d).$$

Among the linear equations satisfied by $\Omega_{a_0 \ldots a_d}$ in this representation are therefore

$$X_{0 \ldots j-1\, d+i+1\, j+1 \ldots d} = 0 \quad (i \leqslant q - m_{d-j};\ j = 0, \ldots, d). \qquad (6)$$

There is, of course, a transformation of coordinates in S_n which gives us Representation A from Representation B, and conversely. The transformation is

$$\left.\begin{aligned} x_s &= x^*_{d-s} \quad (s = 0, \ldots, d), \\ y_r &= y^*_{q-r} \quad (r = 0, \ldots, q). \end{aligned}\right\} \qquad (7)$$

The special S_d for which $p_{01 \ldots d} = 1$ and all other $p_{i_0 \ldots i_d}$ are zero satisfies the condition $(a_0, \ldots, a_d)$ in this representation, and, as above, we can show that the point P on $\Omega(d, n)$ representing this special S_d is a simple point on $\Omega_{a_0 \ldots a_d}$.

4. Intersections of Schubert varieties: point-set properties. In this section we shall consider certain properties of the intersection of a number of Schubert varieties on $\Omega(d, n)$ which will be used later as the basis of a multiplicative theory. Our discussion will deal (i) with intersections $\Omega_{a_0 \ldots a_d} \wedge \Omega_{b_0 \ldots b_d}$, for any admissible values of $a_0, \ldots, a_d$ and $b_0, \ldots, b_d$, and (ii) with intersections $\Omega_{a_0 \ldots a_d} \wedge \Omega_{b_0 \ldots b_d} \wedge \sigma_h$, where $\sigma_h = \Omega_{n-d-h\, n-d+1 \ldots n}$, that is, σ_h is the Schubert variety which represents the d-spaces which meet a fixed $(n-d-h)$-space of S_n. In (ii) we are only concerned with the case in which

$$\dim \Omega_{a_0 \ldots a_d} + \dim \Omega_{b_0 \ldots b_d} + \dim \sigma_h = 2(d+1)(n-d) = 2 \dim \Omega(d, n),$$

when we should expect a finite number of intersections if the varieties $\Omega_{a_0 \ldots a_d}$, $\Omega_{b_0 \ldots b_d}$ and σ_h are in 'sufficiently general' position on $\Omega(d, n)$. Since $0 \leqslant h \leqslant n-d$, it is clear that this requirement on the dimensions puts a severe limitation on the possible choices of $a_0, \ldots, a_d, b_0, \ldots, b_d$.

We saw in § 3 [Representation A] that we may choose the coordinate system $(x_0, \ldots, x_d, y_0, \ldots, y_q)$ in S_n so that a given Schubert variety $\Omega'_{a_0 \ldots a_d}$ represents the d-spaces of S_n which meet A'_0 in a point, A'_1 in a line, ..., and lie in A'_d, where A'_{d-r} is defined by the equations

$$x_0 = \ldots = x_{r-1} = y_{m_r} = y_{m_r+1} = \ldots = y_q = 0 \quad (m_r = a_{d-r} - d + r).$$

Let $$f_i(\ldots, X_{i_0\ldots i_d}, \ldots) = 0 \quad (i = 1, \ldots, r'),$$

be the equations of $\Omega'_{a_0\ldots a_d}$. We have seen [§ 2, p. 318] that if $\Omega^t_{a_0\ldots a_d}$ is the Schubert variety of the same type as $\Omega'_{a_0\ldots a_d}$, defined by the sequence of spaces $A^t_0 \subset A^t_1 \subset \ldots \subset A^t_d$, it can be obtained from $\Omega'_{a_0\ldots a_d}$ by constructing a non-singular projective transformation $x = ty$ which transforms A'_i into A^t_i $(i = 0, \ldots, d)$; the equations of $\Omega^t_{a_0\ldots a_d}$ are then

$$f_i(\ldots, \sum_j t_{i_0 j_0} \ldots t_{i_d j_d} X_{j_0\ldots j_d}, \ldots) = 0 \quad (i = 1, \ldots, r').$$

Again [§ 3, Representation B], we can choose the coordinate system in S_n so that a given Schubert variety $\Omega'_{b_0\ldots b_d}$ is defined by a sequence of spaces $B'_0 \subset \ldots \subset B'_d$, where B'_r has the equation

$$x_{r+1} = \ldots = x_d = y_0 = \ldots = y_{q-n_{d-r}} = 0 \quad (n_{d-r} = b_r - r);$$

and we can define $\Omega^t_{b_0\ldots b_d}$ as the variety obtained from $\Omega'_{b_0\ldots b_d}$ by a projective transformation in S_n with matrix t; and we can choose t so that $\Omega^t_{b_0\ldots b_d}$ is any assigned Schubert variety of the same type as $\Omega'_{b_0\ldots b_d}$.

To consider the intersection $\Omega_{a_0\ldots a_d} \wedge \Omega_{b_0\ldots b_d}$, we choose the coordinate system so that $\Omega_{b_0\ldots b_d}$ is defined by the sequence of spaces $B_0 \subset B_1 \subset \ldots \subset B_d$, where B_r has the equations

$$x_{r+1} = \ldots = x_d = y_0 = \ldots = y_{q-n_{d-r}} = 0,$$

and take $\Omega_{a_0\ldots a_d}$ to be $\Omega^t_{a_0\ldots a_d}$, where $\Omega'_{a_0\ldots a_d}$ is the variety defined by Representation A for the same system of coordinates $(x_0, \ldots, x_d, y_0, \ldots, y_q)$. $\Omega^\tau_{a_0\ldots a_d}$ will denote the Schubert variety in which the matrix t is replaced by the matrix τ whose elements are independent indeterminates over K.

THEOREM I. *$\Omega^\tau_{a_0\ldots a_d}$ and $\Omega_{b_0\ldots b_d}$ intersect if and only if*

$$a_{d-i} + b_i \geqslant n \quad (i = 0, \ldots, d).$$

(i) If $\Omega^\tau_{a_0\ldots a_d}$ and $\Omega_{b_0\ldots b_d}$ have a point in common, there exists a d-space S_d meeting A^τ_{d-i} in a $(d-i)$-space and B_i in an i-space. These two spaces, of dimension $d-i$ and i respectively, lie in the d-space S_d, and hence meet in a point common to A^τ_{d-i} and B_i. Since A^τ_{d-i} is a generic $(d-i)$-space of S_n over K, it can only meet B_i if

$$a_{d-i} + b_i \geqslant n.$$

Since this holds for all values of i from 0 to d, the necessity of the condition is proved.

(ii) Suppose that $a_{d-i}+b_i \geqslant n \quad (i = 0, \ldots, d)$.

Then A^τ_{d-i} and B_i meet in a space of dimension $a_{d-i}+b_i-n$ (and not in a space of higher dimension, since A^τ_{d-i} is generic over K). Let P_i be a generic point of $A^\tau_{d-i} \wedge B_i$ over $K(\tau)$. We show that $P_0, \ldots, P_d$ are linearly independent. Suppose, indeed, that $P_0, \ldots, P_{k-1}$ are linearly independent, but that $P_0, \ldots, P_k$ are linearly dependent. We have

$$\begin{aligned} P_0 &\subseteq B_0 \subset B_1 \subset \ldots \subset B_{k-1}, \\ P_1 &\subset B_1 \subset \quad \ldots \subset B_{k-1}, \\ &\quad . \quad . \quad . \quad . \quad . \\ P_{k-1} &\subset B_{k-1}, \end{aligned}$$

and since P_k is linearly dependent on $P_0, \ldots, P_{k-1}$, it is contained in B_{k-1}. But P_k is a generic point of $A^\tau_{d-k} \wedge B_k$ over $K(\tau)$. Hence this implies that

$$A^\tau_{d-k} \wedge B_k \subseteq B_{k-1},$$

and therefore

$$A^\tau_{d-k} \wedge B_k \subseteq A^\tau_{d-k} \wedge B_{k-1}.$$

But since A^τ_{d-k} is generic over K, $A^\tau_{d-k} \wedge B_k$ is of dimension

$$a_{d-k}+b_k-n$$

and $A^\tau_{d-k} \wedge B_{k-1}$ is of dimension

$$a_{d-k}+b_{k-1}-n.$$

Hence $b_k \leqslant b_{k-1}$, which contradicts the fact that $b_k > b_{k-1}$.

It follows that $P_0, \ldots, P_d$ are linearly independent. These points therefore determine a d-space S_d. Since this contains the points $P_0, \ldots, P_i$, it meets B_i in an i-space, and since it contains $P_d, \ldots, P_{d-i}$ it meets A^τ_i in an i-space. Hence S_d is a d-space satisfying the two Schubert conditions defined by the sequences $A^\tau_0 \subset A^\tau_1 \subset \ldots \subset A^\tau_d$ and $B_0 \subset B_1 \subset \ldots \subset B_d$. The sufficiency of the condition is therefore proved.

If $a_{d-i}+b_i = n \; (i = 0, \ldots, d)$, A^τ_{d-i} and B_i meet in a unique point P_i. By the proof of (i) it follows that any d-space represented by a point common to $\Omega^\tau_{a_0 \ldots a_d}$ and $\Omega_{b_0 \ldots b_d}$ contains P_i. The d-space $P_0 \ldots P_d$ is therefore the only d-space which is represented by a point common to $\Omega^\tau_{a_0 \ldots a_d}$ and $\Omega_{b_0 \ldots b_d}$. Hence we have the

Corollary. There is a unique point common to the Schubert varieties $\Omega^\tau_{a_0 a_1 \ldots a_d}$ *and* $\Omega_{n-a_d\, n-a_{d-1} \ldots n-a_0}$.

If $\Omega'_{a_0 \ldots a_d}$ and $\Omega_{b_0 \ldots b_d}$ are Schubert varieties of respective dimensions $\rho_a = \Sigma a_i - \frac{1}{2}d(d+1)$, $\rho_b = \Sigma b_i - \frac{1}{2}d(d+1)$, and these dimensions satisfy the equation

$$\rho_a + \rho_b = (d+1)(n-d),$$

we say that they are of *complementary dimension*. Since this relation is equivalent to

$$\sum_{i=0}^{d} (a_i + b_{d-i}) = n(d+1), \tag{1}$$

and the two varieties $\Omega^{\tau}_{a_0 \ldots a_d}$ and $\Omega_{b_0 \ldots b_d}$ intersect if and only if

$$a_i + b_{d-i} \geqslant n \quad (i = 0, \ldots, d), \tag{2}$$

and since (1) and (2) imply

$$a_i + b_{d-i} = n \quad (i = 0, \ldots, d),$$

it follows that if $\Omega'_{a_0 \ldots a_d}$ and $\Omega_{b_0 \ldots b_d}$ are two Schubert varieties of complementary dimension, $\Omega^{\tau}_{a_0 \ldots a_d}$ and $\Omega_{b_0 \ldots b_d}$ either have no intersection, or else have a single point in common. We shall prove in Theorem II that in the latter case the Schubert varieties intersect simply in this point.

To obtain further information about the intersection of any two Schubert varieties $\Omega^{\tau}_{a_0 \ldots a_d}$ and $\Omega_{b_0 \ldots b_d}$, we use the methods of XII, § 3. Let the equations of $\Omega'_{a_0 \ldots a_d}$ be

$$f_i(\ldots, X_{i_0 \ldots i_d}, \ldots) = 0 \quad (i = 1, \ldots, r), \tag{3}$$

and let the equations of $\Omega_{b_0 \ldots b_d}$ be

$$g_i(\ldots, X_{i_0 \ldots i_d}, \ldots) = 0 \quad (i = 1, \ldots, s). \tag{4}$$

The dimension of $\Omega'_{a_0 \ldots a_d}$ is $\rho_a = \Sigma a_i - \frac{1}{2}d(d+1)$, and that of $\Omega_{b_0 \ldots b_d}$ is $\rho_b = \Sigma b_i - \frac{1}{2}d(d+1)$. If (τ_{ij}) is an $(n+1) \times (n+1)$ matrix of indeterminates, the equations

$$f_i(\ldots, \Sigma \tau_{i_0 j_0} \ldots \tau_{i_d j_d} X_{j_0 \ldots j_d}, \ldots) = 0 \quad (i = 1, \ldots, r) \tag{5}$$

define the Schubert variety $\Omega^{\tau}_{a_0 \ldots a_d}$.

The equations (4) and (5) determine a correspondence $\mathscr{C}$ between $\Omega(d, n)$ and the τ-space. As in XII, § 3 we find that this correspondence is unmixed. To a point t of the object-variety in the τ-space of a component of $\mathscr{C}$ there corresponds a component of the intersection $\Omega^{t}_{a_0 \ldots a_d} \wedge \Omega_{b_0 \ldots b_d}$.

To determine the dimension of $\mathscr{C}$, an argument similar to that used in XII, § 3 shows that it is $\rho_a + \rho_b + \sigma$, where σ is the dimension of the system of collineations in S_n which transform one given S_d into another given S_d. We find that

$$\sigma = n^2+2n-(d+1)(n-d),$$

so that the dimension of each irreducible component of $\mathscr{C}$ is

$$\rho_a+\rho_b+n^2+2n-(d+1)(n-d).$$

If $a_i+b_{d-i}<n$ for some value of i, Theorem I tells us that $\Omega^{\tau}_{a_0\ldots a_d}$ does not meet $\Omega_{b_0\ldots b_d}$, τ being generic. The object-variety of any component of $\mathscr{C}$ does not fill the τ-space in this case, and has dimension $\nu<n^2+2n$, the same for all components. To a generic point ζ of each object-variety there corresponds an irreducible component of $\Omega^{\zeta}_{a_0\ldots a_d}\wedge\Omega_{b_0\ldots b_d}$ of dimension

$$\rho_a+\rho_b+n^2+2n-(d+1)(n-d)-\nu>\rho_a+\rho_b-(d+1)(n-d),$$

and for any specialisation $\zeta\rightarrow t'$, the corresponding locus has every component of dimension greater than $\rho_a+\rho_b-(d+1)(n-d)$. If $\Omega_{a_0\ldots a_d}$ does not meet $\Omega_{b_0\ldots b_d}$ we can say no more, but if these varieties do meet, $t'=I$ is on the object-variety of one of the components of $\mathscr{C}$, and we can say that every component of

$$\Omega'_{a_0\ldots a_d}\wedge\Omega_{b_0\ldots b_d}$$

has dimension greater than $\rho_a+\rho_b-(d+1)(n-d)$.

We are more concerned, however, with the case

$$a_i+b_{d-i}\geqslant n \quad (i=0,\ldots,d).$$

By Theorem I the varieties $\Omega^{\tau}_{a_0\ldots a_d}$ and $\Omega_{b_0\ldots b_d}$ meet, and therefore the object-variety of any component of $\mathscr{C}$ fills the τ-space. It follows that $\Omega^{\tau}_{a_0\ldots a_d}\wedge\Omega_{b_0\ldots b_d}$ is unmixed of dimension

$$\rho_a+\rho_b-(d+1)(n-d),$$

and no component of $\Omega'_{a_0\ldots a_d}\wedge\Omega_{b_0\ldots b_d}$ is of lower dimension.

Later on we shall consider the multiplicative theory of the intersection of varieties on $\Omega(d,n)$, and we shall require the theorem that every component of $\Omega^{\tau}_{a_0\ldots a_d}\wedge\Omega_{b_0\ldots b_d}$ counts simply in the intersection $\mathbf{\Omega}^{\tau}_{a_0\ldots a_d}\cdot\mathbf{\Omega}_{b_0\ldots b_d}$. To show this it is necessary and sufficient to prove that there exists a point in the intersection at which the tangent spaces to $\Omega^{\tau}_{a_0\ldots a_d}$ and $\Omega_{b_0\ldots b_d}$ meet in a space of dimension $\rho_a+\rho_b-(d+1)(n-d)$. This in turn will follow if we can show that there is a non-singular specialisation t' of τ such that there is a point on $\Omega^{\tau'}_{a_0\ldots a_d} \wedge \Omega_{b_0\ldots b_d}$ at which the tangent spaces meet in a space of $\rho_a+\rho_b-(d+1)(n-d)$ dimensions. We therefore have merely to construct two Schubert varieties with this property.

This can be done at once by specialising $\Omega^{\tau}_{a_0\ldots a_d}$ to $\Omega'_{a_0\ldots a_d}$. Then the d-space

$$y_0 = \ldots = y_q = 0$$

is represented on $\Omega(d,n)$ by a point P common to $\Omega'_{a_0\ldots a_d}$ and $\Omega_{b_0\ldots b_d}$. The tangent space at P to $\Omega_{a_0\ldots a_d}$ is cut on the tangent space to $\Omega(d,n)$ by the primes [§ 3, (4)]

$$X_{0\ldots j-1\, d+i+1\, j+1\ldots d} = 0 \quad (i \geqslant m_j),$$

and similarly, the tangent space at P to $\Omega_{b_0\ldots b_d}$ is cut by the primes [§ 3, (6)]

$$X_{0\ldots j-1\, d+i+1\, j+1\ldots d} = 0 \quad (i \leqslant q-n_{d-j}).$$

These primes are certainly independent if

$$q-n_{d-j} < m_j \quad (j = 0, \ldots, d),$$

that is, if

$$a_{d-j}+b_j \geqslant n,$$

which is true, by hypothesis. Hence the tangent spaces to $\Omega_{a_0\ldots a_d}$ and $\Omega_{b_0\ldots b_d}$ at the simple point P intersect in a space of dimension $\rho_a+\rho_b-(d+1)(n-d)$, and we have proved

THEOREM II. *If $\Omega'_{a_0\ldots a_d}$ and $\Omega_{b_0\ldots b_d}$ are two Schubert varieties, and $a_{d-i}+b_i \geqslant n$ $(i = 0, \ldots, d)$, the intersection $\Omega^{\tau}_{a_0\ldots a_d} \wedge \Omega_{b_0\ldots b_d}$ is a simple intersection.*

It may happen that all the d-spaces represented by points of $\Omega^{\tau}_{a_0\ldots a_d} \wedge \Omega_{b_0\ldots b_d}$ lie in a subspace of S_n of dimension $n-k$ $(k>0)$. It is important for our future purposes to note a case in which this happens. We may assume that $a_{d-i}+b_i \geqslant n$ $(i = 0, \ldots, d)$, otherwise there are no points common to $\Omega^{\tau}_{a_0\ldots a_d}$ and $\Omega_{b_0\ldots b_d}$. If $a_d = n-\lambda_0$, all the d-spaces represented by points of $\Omega^{\tau}_{a_0\ldots a_d} \wedge \Omega_{b_0\ldots b_d}$ clearly lie in $S^{\tau}_{n-\lambda_0} = A^{\tau}_d$; and if $b_d = n-\lambda_{d+1}$, they all lie in $S_{n-\lambda_{d+1}} = B_d$. If, for $0 < i \leqslant d$, $\lambda_i = n-a_{d-i}-b_{i-1}-1 \geqslant 0$, the join of the spaces A^{τ}_{d-i} and B_{i-1} is of dimension $n-\lambda_i$ at most, and since A^{τ}_{d-i} is generic over K, A^{τ}_{d-i} and B_{i-1} do not meet when $\lambda_i \geqslant 0$, hence the join is a space $S^{\tau}_{n-\lambda_i}$ of dimension exactly $n-\lambda_i$. If

$$n-a_{d-i}-b_{i-1}-1 < 0,$$

we define $\lambda_i = 0$, and $S^{\tau}_{n-\lambda_i} = S_n$. Now any d-space S_d represented by a point of $\Omega^{\tau}_{a_0\ldots a_d} \wedge \Omega_{b_0\ldots b_d}$ meets A^{τ}_{d-i} in a $(d-i)$-space and

B_{i-1} in an $(i-1)$-space. If $\lambda_i > 0$, A^{τ}_{d-i} and B_{i-1} do not intersect; neither do these $(d-i)$- and $(i-1)$-spaces, and S_d is therefore their join; hence it lies in $S^{\tau}_{n-\lambda_i}$. Hence every d-space represented by a point of $\Omega^{\tau}_{a_0\ldots a_d} \wedge \Omega_{b_0\ldots b_d}$ lies in the intersection of the $d+2$ spaces $S^{\tau}_{n-\lambda_i}$ $(i = 0, \ldots, d+1)$. We show that this is an S_{n-k}, where $k = \sum_0^{d+1} \lambda_i$. If it were of dimension greater than $n-k$, it would clearly be so for all specialisations of τ, and in particular it would be of dimension greater than $n-k$ when $\Omega^{\tau}_{a_0\ldots a_d} \to \Omega'_{a_0\ldots a_d}$. For this specialisation, $S^{\tau}_{n-\lambda_i}$ becomes $S'_{n-\lambda_i}$, where

$$S'_{n-\lambda_0} \text{ has equations } \quad y_{m_0} = \ldots = y_q \quad = 0,$$

$$S_{n-\lambda_{d+1}} \text{ has equations } \quad y_0 = \ldots = y_{q-n_0} = 0,$$

and, for $0 < i \leqslant d$,

$$S'_{n-\lambda_i} \text{ has equations } y_{m_i} = \ldots = y_{q-n_{d-i+1}} = 0.$$

Since $a_{d-i} + b_i \geqslant n$, $q - n_{d-i} < m_i$, and hence these equations are independent. It follows that the intersection of the $S'_{n-\lambda_i}$ $(i = 0, \ldots, d+1)$ is of dimension $n - \sum_0^{d+1} \lambda_i$ exactly, and hence the intersection of the $S^{\tau}_{n-\lambda_i}$ $(i = 0, \ldots, d+1)$ is of this dimension.

This result plays an important role in the investigation of the intersection $\Omega^{\tau}_{a_0\ldots a_d} \wedge \Omega_{b_0\ldots b_d} \wedge \sigma_h$, where σ_h is the Schubert variety of d-spaces which meet a fixed $(n-d-h)$-space S_{n-d-h} which is generic with respect to $K(\tau)$. We are only concerned with the case

$$\dim \Omega_{a_0\ldots a_d} + \dim \Omega_{b_0\ldots b_d} + \dim \sigma_h = 2(d+1)(n-d),$$

that is, with

$$\begin{aligned} h &= \sum_0^d a_i + \sum_0^d b_i - n(d+1) \\ &= n - \lambda_0 + n - \lambda_{d+1} + \sum_1^d (a_{d-i} + b_{i-1} + 1) - d - n(d+1), \end{aligned}$$

where $\lambda_0, \ldots, \lambda_{d+1}$ are defined as above. Since

$$a_{d-i} + b_{i-1} + 1 = n - \lambda_i \quad (\text{if } \lambda_i > 0),$$

$$a_{d-i} + b_{i-1} + 1 \geqslant n - \lambda_i \quad (\lambda_i = 0),$$

we must have $\quad h \geqslant 2n - \sum_0^{d+1} \lambda_i + nd - d - n(d+1).$

that is, $$h+k \geqslant n-d,$$
where $k = \sum_0^{d+1} \lambda_i$. On the other hand, we know that all the d-spaces represented by points of $\Omega^\tau_{a_0\ldots a_d} \wedge \Omega_{b_0\ldots b_d}$ lie in the $(n-k)$-space S^τ_{n-k}. The spaces meet S_{n-d-h}, if at all, in the intersection of S^τ_{n-k} and S_{n-d-h}, and this intersection is empty if $h+k > n-d$. Hence if $\Omega^\tau_{a_0\ldots a_d}$, $\Omega_{b_0\ldots b_d}$ and σ_h have a point in common, we cannot have $h+k > n-d$. Thus we have, necessarily, $h+k = n-d$. This implies that
$$n-\lambda_i = a_{d-i}+b_{i-1}+1 \leqslant n \quad (i = 1, \ldots, d).$$

Since, by Theorem I, we must have $a_{d-i}+b_i \geqslant n$ in order that $\Omega^\tau_{a_0\ldots a_d}$, $\Omega_{b_0\ldots b_d}$ and σ_h have a point in common, we have established the first part of

THEOREM III. *If $\Omega^\tau_{a_0\ldots a_d}$, $\Omega_{b_0\ldots b_d}$ and σ_h, where*
$$h = \Sigma a_i + \Sigma b_i - n(d+1),$$
have a point in common, we must have
$$a_{d-i} < n - b_{i-1} \leqslant a_{d-i+1}$$
for all i. Conversely, if these inequalities are satisfied, $\Omega^\tau_{a_0\ldots a_d}$, $\Omega_{b_0\ldots b_d}$, σ_h intersect in a single point.

To prove the second part of this theorem, we proceed by induction on $k = \sum_0^{d+1} \lambda_i$.

Let us consider first the case $k = 0$. Then $h = n-d$, and hence S_{n-d-h} is a generic point P of S_n. Also $a_d = b_d = n$, and
$$a_{d-i}+b_{i-1}+1 = n \quad (i = 1, \ldots, d).$$
To find the number of points common to $\Omega^\tau_{a_0\ldots a_d}$, $\Omega_{b_0\ldots b_d}$, σ_h, we have to find the number of d-spaces in S_n which meet A^τ_i in an i-space $(i = 0, \ldots, d)$, B_j in a j-space $(j = 0, \ldots, d)$, and pass through P. Since A^τ_{d-i} and P are generic over K, and independent, the joins of P to A^τ_{d-i} and to B_{i-1} meet in a line l_i $(i = 1, \ldots, d)$. The line l_i meets B_{i-1} in the point Q_i where the join of P to A^τ_{d-i} meets it. This join is a generic $(a_{d-i}+1)$-space over K; hence it meets B_{i-1} in a generic point of B_{i-1}, that is, Q_i is a generic point of B_{i-1}. Then, as in the proof of Theorem I, we conclude that $Q_1, \ldots, Q_d$ are linearly independent. The lines $l_1, \ldots, l_d$ through P therefore lie in a unique d-space. Now consider any d-space through P which meets A^τ_{d-i} in a $(d-i)$-space and B_{i-1} in an $(i-1)$-space. It has a $(d-i+1)$-space

in common with the join of P to A^τ_{d-i}, and an i-space in common with the join of P to B_{i-1}. The $(d-i+1)$-space and the i-space, lying in a d-space, have a line at least in common, which is therefore common to the joins of P to A^τ_{d-i} and to B_{i-1}. Hence the d-space passes through l_i. Hence any d-space through P which meets A^τ_r in an r-space (at least), and B_s in an s-space at least, contains $P, Q_1, \ldots, Q_d$, and hence it must be the unique d-space containing these points. Conversely, the d-space containing $P, Q_1, \ldots, Q_d$ meets A^τ_r in an r-space and B_s in an s-space. Thus Theorem III is true when $k = 0$.

Now suppose that the second part of Theorem III is true when $k < k_1$ $(1 \leqslant k_1 \leqslant n-d)$, and consider the case $k = k_1$. Since $k > 0$, some $\lambda_i > 0$. Suppose, first, that $\lambda_0 > 0$. Then $a_d < n$. We take the space A^τ_d in place of S_n. Let $\Omega^*_{a_0 \ldots a_d}$ be the Schubert variety on $\Omega(d, a_d)$ representing d-spaces which meet A^τ_r in an r-space $(r = 0, \ldots, d)$, and let $\Omega^*_{b_0-\lambda_0 \ldots b_d-\lambda_0}$ be the Schubert variety on $\Omega(d, a_d)$ defined by the spaces

$$B_0 \wedge A^\tau_d \subset B_1 \wedge A^\tau_d \subset \ldots \subset B_d \wedge A^\tau_d.$$

We also denote by σ^*_h the variety on $\Omega(d, a_d)$ representing the d-spaces of A^τ_d which meet the $(n-\lambda_0-d-h)$-space $S_{n-d-h} \wedge A^\tau_d$. We consider the intersection $\Omega^\tau_{a_0 \ldots a_d} \wedge \Omega^*_{b_0-\lambda_0 \ldots b_d-\lambda_0} \wedge \sigma^*_h$. The join of A^τ_{d-i} and $B_{i-1} \wedge A^\tau_d$ is an $(n-\lambda_0-\lambda^*_i)$-space, where $\lambda^*_i = \lambda_i$ $(i = 1, \ldots, d)$, A^τ_d is an $(n-\lambda_0-\lambda^*_0)$-space $(\lambda^*_0 = 0)$, and $B_d \wedge A^\tau_d$ is an $(n-\lambda_0-\lambda^*_{d+1})$-space $(\lambda^*_{d+1} = \lambda_{d+1})$. Hence $\Sigma \lambda^*_i = k - \lambda_0 < k$. Also

$$a_{d-i} < (n-\lambda_0) - (b_{i-1} - \lambda_0) \leqslant a_{d-i+1}.$$

Hence, by the hypothesis of induction, $\Omega^\tau_{a_0 \ldots a_d} \wedge \Omega^*_{b_0-\lambda_0 \ldots b_d-\lambda_0} \wedge \sigma^*_h$ is a single point. This point represents the unique d-space which meets A^τ_r in an r-space $(r = 0, \ldots, d)$, B_s in an s-space $(s = 0, \ldots, d)$, and meets S_{n-d-h}, since any such space necessarily lies in A^τ_d. Hence $\Omega^\tau_{a_0 \ldots a_d} \wedge \Omega_{b_0 \ldots b_d} \wedge \sigma_h$ is a unique point. A similar argument applies if $\lambda_{d+1} > 0$.

Finally, suppose that $\lambda_0 = \lambda_{d+1} = 0$, $k = k_1 > 0$. Then for some i $(1 \leqslant i \leqslant d)$, $\lambda_i > 0$. For this value of i, we have

$$a_{d-i} + b_{i-1} + 1 < n.$$

Since

$$a_{d-i} + b_i \geqslant n, \quad a_{d-i+1} + b_{i-1} \geqslant n, \quad \text{and} \quad a_{d-i+1} > a_{d-i}, \quad b_i > b_{i-1},$$

we have
$$a_{d-i+1} \geqslant a_{d-i} + 2, \quad b_i \geqslant b_{i-1} + 2.$$

Let Π be a generic prime through $S^{\tau}_{n-\lambda_i}$. This contains A^{τ}_{d-j} $(j \geq i)$ and meets A^{τ}_{d-j} in an $(a_{d-j}-1)$-space A^{*}_{d-j} if $j<i$. Since $a_{d-i+1}-1 > a_{d-i}$, we have

$$A^{\tau}_0 \subset A^{\tau}_1 \subset \ldots \subset A^{\tau}_{d-i} \subset A^{*}_{d-i+1} \subset \ldots \subset A^{*}_d.$$

Similarly, Π contains B_j $(j<i)$ and meets B_j in a (b_j-1)-space B^{*}_j $(j \geq i)$, and we have

$$B_0 \subset B_1 \subset \ldots \subset B_{i-1} \subset B^{*}_i \subset \ldots \subset B^{*}_d.$$

Let $\Omega^{*}_{a_0 \ldots a_{d-i}\, a_{d-i+1}-1 \ldots a_d-1}$ be the Schubert variety defined by the first of these sets of spaces, and $\Omega^{*}_{b_0 \ldots b_{i-1} b_i-1 \ldots b_d-1}$ be the Schubert variety defined by the second set. Also let $S^{*}_{n-d-h-1}$ be the intersection of S_{n-d-h} and Π. Now any d-space which meets A^{τ}_r in an r-space $(r = 0, \ldots, d)$ and B_i in an s-space $(s = 0, \ldots, d)$ lies in $S^{\tau}_{n-\lambda_i}$, and hence in Π. The number of points common to $\Omega_{a_0 \ldots a_d}$, $\Omega_{b_0 \ldots b_d}$ and σ_h is therefore the number of d-spaces which meet $S^{*}_{n-d-h-1}$ and are represented by points of

$$\Omega^{*}_{a_0 \ldots a_{d-i}\, a_{d-i+1}-1 \ldots a_d-1} \wedge \Omega^{*}_{b_0 \ldots b_{i-1} b_i-1 \ldots b_d-1}.$$

Applying the hypothesis of induction, we deduce that

$$\Omega_{a_0 \ldots a_d} \wedge \Omega_{b_0 \ldots b_d} \wedge \sigma_h$$

is a single point. Thus Theorem III is proved.

Theorem III will play an important role when we come to consider the multiplicative theory of varieties on $\Omega(d, n)$. To use it, however, we want one more result. Suppose that the conditions stated in Theorem III in order that $\Omega^{\tau}_{a_0 \ldots a_d}$, $\Omega_{b_0 \ldots b_d}$, σ_h have an intersection are satisfied. The Schubert variety σ_h is determined by a generic S_{n-d-h} of S_n, that is, by a generic point ζ of $\Omega(n-d-h, n)$. The equations for $\Omega^{\tau}_{a_0 \ldots a_d} \wedge \Omega_{b_0 \ldots b_d} \wedge \sigma_h$ determine a correspondence between $\Omega(d, n)$ and the variety U which is the direct product of the τ-space whose points are the matrices t and $\Omega(n-d-h, n)$. Since both of these are irreducible and the τ-space is absolutely irreducible (as, indeed, is $\Omega(n-d-h, n)$), U is irreducible, and just as in the proof of Theorem III we show that for any specialisation of τ, σ_h such that $\Omega^{\bar{t}}_{a_0 \ldots a_d} \wedge \Omega_{b_0 \ldots b_d} \wedge \bar{\sigma}_h$ is of dimension zero, this intersection consists of a single point. If $\mathbf{\Omega}^{\bar{t}}_{a_0 \ldots a_d}$, $\mathbf{\Omega}_{b_0 \ldots b_d}$, $\bar{\boldsymbol{\sigma}}_h$ are the irreducible multiplicative varieties derived from $\Omega^{\bar{t}}_{a_0 \ldots a_d}$, $\Omega_{b_0 \ldots b_d}$, $\bar{\sigma}_h$, we are able to conclude that $\mathbf{\Omega}^{\bar{t}}_{a_0 \ldots a_d} \cdot \mathbf{\Omega}_{b_0 \ldots b_d} \cdot \bar{\boldsymbol{\sigma}}_h$ is a variety of dimension zero and order one provided that we can prove that $\mathbf{\Omega}^{\tau}_{a_0\ldots a_d}\cdot\mathbf{\Omega}_{b_0\ldots b_d}\cdot\boldsymbol{\sigma}_h$ is of order one, or, what is the same thing, that $\Omega^{\tau}_{a_0\ldots a_d}$, $\Omega_{b_0\ldots b_d}$, σ_h intersect simply at their one common point. This will follow if we can specialise $\Omega^{\tau}_{a_0\ldots a_d}$ and σ_h so that some point common to $\Omega'_{a_0\ldots a_d}$, $\Omega_{b_0\ldots b_d}$, $\overline{\sigma}_h$ is simple on each of these varieties, and so that the tangent spaces to them at the point have only the point in common.

It is sufficient to specialise $\Omega^{\tau}_{a_0\ldots a_d}$ to $\Omega'_{a_0\ldots a_d}$, and choose σ_h suitably so that the point $P[p_{0\ldots d}=1,\ p_{i_0\ldots i_d}=0$ if $(i_0,\ldots,i_d)\neq(0,\ldots,d)]$ has the required properties. We have already seen that the space S_{n-k} which contains all S_d represented by points of $\Omega'_{a_0\ldots a_d}\wedge\Omega_{b_0\ldots b_d}$ is the intersection of the k independent primes

$$y_i=0 \quad (i=m_j,\ldots,q-n_{d-j+1};\ j=0,\ldots,d+1).$$

Let y_{ρ_i} be the remaining $h=(q+1-k)$ y's, and let $c_0,\ldots,c_d$ be $d+1$ non-zero elements of K. Let $\overline{\sigma}_h$ be the Schubert variety of d-spaces which meet the $\overline{S}_{n-d-h}$ given by

$$\frac{x_0}{c_0}=\ldots=\frac{x_d}{c_d},\quad y_{\rho_i}=0 \quad (i=1,\ldots,h).$$

The d-space $$y_i=\sum_j v_{ij}x_j \quad (i=0,\ldots,q)$$

meets $\overline{S}_{n-d-h}$ if and only if

$$\sum_j v_{\rho_i j}c_j=0 \quad (i=1,\ldots,h).$$

These equations are satisfied by the space represented by the point P, which therefore lies on $\overline{\sigma}_h$. The tangent space to $\overline{\sigma}_h$ at P is given by the equation

$$\sum_j c_j X_{0\ldots j-1\,d+i+1\,j+1\ldots d}=0 \quad (i=\rho_1,\ldots,\rho_h).$$

The tangent spaces to $\Omega'_{a_0\ldots a_d}$, $\Omega_{b_0\ldots b_d}$ at P intersect where

$$X_{0\ldots j-1\,d+i+1\,j+1\ldots d}=0 \quad (i\leqslant q-n_{d-j} \text{ or } i\geqslant m_j;\ j=0,\ldots,d),$$

and we see at once that this meets the tangent space at P to $\overline{\sigma}_h$ only in P, provided $c_0,\ldots,c_d$ are sufficiently general. Hence we have

THEOREM IV. *If $a_0,\ldots,a_d$, $b_0,\ldots,b_d$, h satisfy the conditions of Theorem III, and if $\Omega_{a_0\ldots a_d}$, $\Omega_{b_0\ldots b_d}$, σ_h meet in a variety of dimension zero, then $\mathbf{\Omega}_{a_0\ldots a_d}\cdot\mathbf{\Omega}_{b_0\ldots b_d}\cdot\boldsymbol{\sigma}_h$ is a zero-dimensional variety of order one.*

5. The basis theorem. The importance of the Schubert varieties on the Grassmann variety $\Omega(d,n)$ lies in the fact that if V is any ρ-dimensional variety (in the multiplicative sense) on $\Omega(d,n)$, there exists an equivalence

$$V \sim \Sigma \alpha_{a_0 \ldots a_d} \mathbf{\Omega}_{a_0 \ldots a_d},$$

where $\mathbf{\Omega}_{a_0 \ldots a_d}$ is the multiplicative variety obtained by attaching multiplicity one to $\Omega_{a_0 \ldots a_d}$, $\alpha_{a_0 \ldots a_d}$ is an integer, and the summation is over all sets of integers $a_0, \ldots, a_d$ which satisfy the relations

$$0 \leqslant a_0 < a_1 < \ldots < a_d \leqslant n,$$

$$\sum_0^d a_i - \tfrac{1}{2}d(d+1) = \rho.$$

In other words, the Schubert varieties of dimension ρ form a base for the ρ-dimensional varieties on $\Omega(d,n)$ [XII, § 11]. In fact, as we shall see, they form a minimal base. The object of this section is to establish the truth of the fundamental property which we have stated. It is sufficient to consider an irreducible variety V, and to prove that V is equivalent to a sum of Schubert varieties.

Let

$$f_\lambda(\ldots, X_{i_0 \ldots i_d}, \ldots) = 0 \quad (\lambda = 1, 2, \ldots), \tag{1}$$

be a basis for the equations of V in S_N, let $\alpha = (\alpha_{ij})$ be a non-singular matrix of $n+1$ rows and columns, whose elements may lie in K or an extension of K, and let (A_{ij}) be the inverse of α. If $p = (\ldots, p_{i_0 \ldots i_d}, \ldots)$ is any d-space in S_n, $q = (\ldots, q_{i_0 \ldots i_d}, \ldots)$, where

$$q_{i_0 \ldots i_d} = \sum_j A_{i_0 j_0} \ldots A_{i_d j_d} p_{j_0 \ldots j_d}, \tag{2}$$

is the d-space obtained from p by the collineation Γ: $x \to y$,

$$y_i = \sum_j A_{ij} x_j.$$

p satisfies the equations of V if and only if q satisfies the equations

$$f_\lambda(\ldots, \sum_j \alpha_{i_0 j_0} \ldots \alpha_{i_d j_d} X_{j_0 \ldots j_d}, \ldots) = 0 \quad (\lambda = 1, 2, \ldots). \tag{3}$$

The equation (3) therefore defines a system of d-spaces which is the transform under Γ of the system defined by V. Moreover, the transformation $X \to Y$ in S_N given by

$$Y_{i_0 \ldots i_d} = \Sigma A_{i_0 j_0} \ldots A_{i_d j_d} X_{j_0 \ldots j_d} \tag{4}$$

transforms any point of $\Omega(d,n)$ into a point of $\Omega(d,n)$. Hence equations (3) represent the variety on $\Omega(d,n)$ obtained from V by the transformation (4). We denote it by V^α. The variety V^α is thus obtained from V by a projective transformation, as, in the particular case $V = \Omega_{a_0\ldots a_d}$, we obtained $\Omega^\tau_{a_0\ldots a_d}$ in § 4; and the method of passing from V to V^α is similar to that of passing from a variety V to a variety V^α in S_n, discussed in Chapter XII, § 2. In particular, if α is a matrix τ of independent indeterminates, we prove that V^τ is an irreducible variety of dimension ρ over $K(\tau)$, and is a generic member of a system of varieties of dimension ρ on $\Omega(d, n)$. The properties of this system can be worked out by methods exactly similar to those used to discuss the system of varieties obtained for a variety V in S_n by projective transformation, and it will not be necessary to repeat all the proofs. The main facts which we shall use are

(i) that if $m = n(n+2)$ and the matrix α is represented by the point $(\ldots, \alpha_{ij}, \ldots)$ in S_m, then in the irreducible correspondence between S_m and S_N in which to α there corresponds the points of the variety whose equations are (3), the object-variety is S_m; and if for a specialisation α of τ, the equations (3) define a variety of dimension ρ, there is a unique specialisation of the variety V^τ corresponding to $\tau \to \alpha$;

(ii) V^τ meets a fixed variety U on $\Omega(d, n)$ simply, if at all.

The method which we shall follow to prove the main result stated at the beginning of this section is similar to that used in Chapter XII, § 2, to degenerate an irreducible variety in S_n. We construct the variety V^τ on $\Omega(d, n)$, where τ is a matrix of indeterminates, and consider a specialisation of V^τ corresponding to a specialisation of τ to a suitably chosen singular matrix. But in order to simplify our work, we shall find it convenient to consider, not the complete system $\{V^\alpha\}$ of varieties of which V^τ is a generic member, but a one-dimensional subsystem of this.

Let a^* be some fixed $(n+1) \times (n+1)$ matrix whose elements are independent indeterminates over K. We write the equations of V^{a^*} in the form

$$F_\lambda(\ldots, X_{i_0\ldots i_d}, \ldots) \equiv f_\lambda(\ldots, \sum_j a^*_{i_0 j_0} \ldots a^*_{i_d j_d} X_{j_0\ldots j_d}, \ldots) = 0,$$

the coefficients of F_λ being in $K(a^*)$.

Let $Q = (b_0, \ldots, b_n)$ be a generic point of S_n, and let S_{n-1} be a generic prime of S_n with equation

$$B_0 x_0 + \ldots + B_n x_n = 0.$$

These are also to be kept fixed. Let $\Delta = b_0 B_0 + \dots + b_n B_n$. We shall denote the field

$$K(a^*, b_0, \dots, b_n, B_0, \dots, B_n)$$

by K^*.

Consider the variety $V(\epsilon)$ having the equations

$$\left.\begin{aligned} &F_\lambda(\dots, \sum_j b_{i_0 j_0} \dots b_{i_d j_d} X_{j_0 \dots j_d}, \dots) = 0 \quad (\lambda = 1, 2, \dots),\\ \text{where} \quad &b_{ij} = \delta_{ij} - \epsilon b_i B_j, \end{aligned}\right\} \tag{5}$$

δ_{ij} being a Kronecker delta, and ϵ being an indeterminate over K^*. Since $\det(b_{ij}) \neq 0$, $V(\epsilon)$ is contained in $\{V^\tau\}$, and for specialisations of ϵ in K^* defines a one-dimensional system over K^* contained in the larger system. For the specialisation $\epsilon \to 0$, we obtain V^{a*}. We write

$$c_{ij} = \delta_{ij} - \Delta^{-1} b_i B_j. \tag{6}$$

When $\epsilon \to \Delta^{-1}$ there may, *a priori*, be several specialisations of $V(\epsilon)$. Our immediate object is to obtain a set of equations which must be satisfied by any such specialisation, and which define a variety of dimension ρ. Since every point of the object-variety of the correspondence in which $V(\epsilon)$ corresponds to ϵ is simple, it will follow from this that the specialisation of $V(\epsilon)$ when $\epsilon \to \Delta^{-1}$ is unique. Let us call it V^0. From the equations satisfied by this specialisation we shall be able to make the deduction we require regarding V^0.

Among the equations satisfied by any specialisation of $V(\epsilon)$ corresponding to $\epsilon \to \Delta^{-1}$ are the equations

$$F_\lambda(\dots, \sum_j c_{i_0 j_0} \dots c_{i_d j_d} X_{j_0 \dots j_d}, \dots) = 0 \quad (\lambda = 1, 2, \dots), \tag{7}$$

obtained by putting $\epsilon = \Delta^{-1}$ in (5). But the variety defined by (7) may depend on the particular basis (5) which we choose for the equations of $V(\epsilon)$, and may be of dimension greater than ρ. Any polynomial $F(\dots, X_{i_0 \dots i_d}, \dots)$ which vanishes on $V(\epsilon)$ can be written in the form

$$F = \Sigma A_\lambda F_\lambda(\dots, \Sigma b_{i_0 j_0} \dots b_{i_d j_d} X_{j_0 \dots j_d}, \dots),$$

where A_λ is a polynomial in $X_{i_0 \dots i_d}$, with coefficients in $K^*(\epsilon)$. This relation may not persist, however, when we specialise $\epsilon \to \Delta^{-1}$. For the coefficients of the A_λ may contain $(1 - \epsilon\Delta)$ in the denominator, and we must multiply through by a suitable power of $(1 - \epsilon\Delta)$ before replacing ϵ by Δ^{-1}. When we have done this, and replaced ϵ by Δ^{-1}. we may obtain, not a relation expressing the specialised form of F in terms of $F_\lambda(\ldots, \Sigma c_{i_0 j_0} \ldots c_{i_d j_d} X_{j_0 \ldots j_d}, \ldots)$, but a relation connecting the equations(7). We overcome this difficulty by showing that from the equations (5) we can deduce a set of equations satisfied by V^0 which specialise to equations which, with (7), define a variety of dimension ρ. From this, we are able to deduce the uniqueness of V^0.

If we express $\sum_j b_{i_0 j_0} \ldots b_{i_d j_d} X_{j_0 \ldots j_d}$ in ascending powers of ϵ by means of the relation $b_{ij} = \delta_{ij} - \epsilon b_i B_j$, we see that all terms of degree greater than one vanish, since they are linear combinations of expressions such as

$$\Sigma B_j B_k X_{jkl_2 \ldots l_d},$$

which vanish because $X_{i_0 \ldots i_d}$ is skew-symmetric in its suffixes. Hence

$$\begin{aligned}\sum_j b_{i_0 j_0} \ldots b_{i_d j_d} X_{j_0 \ldots j_d} &= \sum_j (\delta_{i_0 j_0} - \epsilon b_{i_0} B_{j_0}) \ldots (\delta_{i_d j_d} - \epsilon b_{i_d} B_{j_d}) X_{j_0 \ldots j_d} \\ &= X_{i_0 \ldots i_d} - \epsilon \sum_j \sum_{k=0}^{d} b_{i_k} B_{j_k} \delta_{i_0 j_0} \ldots \delta_{i_{k-1} j_{k-1}} \delta_{i_{k+1} j_{k+1}} \ldots \delta_{i_d j_d} X_{j_0 \ldots j_d} \\ &= X_{i_0 \ldots i_d} - \epsilon \sum_{k=0}^{d} \sum_j b_{i_k} B_j X_{i_0 \ldots i_{k-1} j i_{k+1} \ldots i_d}.\end{aligned}$$

If we substitute $\epsilon = \Delta^{-1}$ in this identity we obtain our first result:

$$X_{i_0 \ldots i_d} = \sum_j c_{i_0 j_0} \ldots c_{i_d j_d} X_{j_0 \ldots j_d} + \Delta^{-1} \sum_{k=0}^{d} \sum_j b_{i_k} B_j X_{i_0 \ldots i_{k-1} j i_{k+1} \ldots i_d}. \quad (8)$$

We also have the equation

$$\begin{aligned}\sum_j c_{ij} b_{jk} &= \sum_j (\delta_{ij} - \Delta^{-1} b_i B_j)(\delta_{jk} - \epsilon b_j B_k) \\ &= \sum_j \delta_{ij} \delta_{jk} - \Delta^{-1} \sum_j b_i B_j \delta_{jk} - \epsilon \sum_j \delta_{ij} b_j B_k + \epsilon \Delta^{-1} \sum_j b_i B_j b_j B_k \\ &= \delta_{ik} - \Delta^{-1} b_i B_k - \epsilon b_i B_k + \epsilon \Delta^{-1} b_i B_k \Delta = \delta_{ik} - \Delta^{-1} b_i B_k = c_{ik}.\end{aligned}$$

It follows immediately that

$$\sum_j \sum_k c_{i_0 j_0} \ldots c_{i_d j_d} b_{j_0 k_0} \ldots b_{j_d k_d} X_{k_0 \ldots k_d} = \sum_j c_{i_0 j_0} \ldots c_{i_d j_d} X_{j_0 \ldots j_d}. \quad (9)$$

Finally, from the proof of (8),

$$\begin{aligned}&\sum_j \sum_k B_j b_{j k_0} b_{i_1 k_1} \ldots b_{i_d k_d} X_{k_0 \ldots k_d} \\ &= \sum_j B_j \left[X_{j i_1 \ldots i_d} - \epsilon \sum_l b_j B_l X_{l i_1 \ldots i_d} - \epsilon \sum_{k=1}^{d} \sum_l b_{i_k} B_l X_{j i_1 \ldots i_{k-1} l i_{k+1} \ldots i_d} \right] \\ &= \sum_j B_j X_{j i_1 \ldots i_d} - \epsilon \Delta \sum_j B_j X_{j i_1 \ldots i_d},\end{aligned}$$

the third term vanishing because of the skew-symmetry property of the suffixes of Grassmann coordinates. Hence

$$\sum_j \sum_k B_j b_{jk_0} b_{i_1 k_1} \dots b_{i_d k_d} X_{k_0 \dots k_d} = (1-\epsilon\Delta) \sum_j B_j X_{j i_1 \dots i_d}. \qquad (10)$$

Now let $F(\dots, X_{i_0 \dots i_d}, \dots) = 0$ be any equation satisfied by the fixed variety V^{a*}, and substitute for $X_{i_0 \dots i_d}$ from (8). The left-hand side of the resulting equation is a form in the variables $Y_{i_0 \dots i_d}$ and $Z_{i_1 \dots i_d}$, where

$$Y_{i_0 \dots i_d} = \sum_j c_{i_0 j_0} \dots c_{i_d j_d} X_{j_0 \dots j_d},$$

and

$$Z_{i_1 \dots i_d} = \sum_j B_j X_{j i_1 \dots i_d}.$$

Let us suppose that it is possible to extract from every term of this form r factors $Z_{i_1 \dots i_d}$, and that for at least one term we cannot extract more. We express this by writing the equation as

$$\Sigma g_\lambda\left(\sum_j B_j X_{j i_1 \dots i_d}\right) h_\lambda\left(\sum_j c_{i_0 j_0} \dots c_{i_d j_d} X_{j_0 \dots j_d}, \sum_j B_j X_{j i_1 \dots i_d}\right) = 0,$$

where each g_λ is homogeneous of degree r in the $Z_{i_1 \dots i_d}$ and has coefficients in K^*, and for at least one value of λ,

$$h_\lambda\left(\sum_j c_{i_0 j_0} \dots c_{i_d j_d} X_{j_0 \dots j_d}, 0\right) \neq 0.$$

The variety $V(\epsilon)$ satisfies the equation

$$F\left(\dots, \sum_j b_{i_0 j_0} \dots b_{i_d j_d} X_{j_0 \dots j_d}, \dots\right) = 0,$$

which, using (8), becomes

$$\Sigma g_\lambda\left(\sum_j \sum_k B_j b_{jk_0} b_{i_1 k_1} \dots b_{i_d k_d} X_{k_0 \dots k_d}\right)$$
$$\times h_\lambda\left(\sum_j \sum_k c_{i_0 j_0} \dots c_{i_d j_d} b_{j_0 k_0} \dots b_{j_d k_d} X_{k_0 \dots k_d}, \right.$$
$$\left. \sum_j \sum_k B_j b_{jk_0} b_{i_1 k_1} \dots b_{i_d k_d} X_{k_0 \dots k_d}\right) = 0,$$

and by (9) and (10) this simplifies to

$$(1-\epsilon\Delta)^r \Sigma g_\lambda\left(\sum_j B_j X_{j i_1 \dots i_d}\right)$$
$$\times h_\lambda\left(\sum_j c_{i_0 j_0} \dots c_{i_d j_d} X_{j_0 \dots j_d}, (1-\epsilon\Delta) \sum_j B_j X_{j i_1 \dots i_d}\right) = 0,$$

that is, to

$$\Sigma g_\lambda\Big(\sum_j B_j X_{j i_1 \dots i_d}\Big)$$
$$\times h_\lambda\Big(\sum_j c_{i_0 j_0} \dots c_{i_d j_d} X_{j_0 \dots j_d}, (1-\epsilon\Delta)\sum_j B_j X_{j i_1 \dots i_d}\Big) = 0.$$

All equations satisfied by $V(\epsilon)$ can be obtained in this way, since V^{a*} is obtained from $V(\epsilon)$ by the non-singular transformation in S_n,

$$x_i' = \sum_j b_{ij} x_j.$$

We now specialise $\epsilon \to \Delta^{-1}$, and obtain the equation

$$\Sigma g_\lambda\Big(\sum_j B_j X_{j i_1 \dots i_d}\Big) h_\lambda\Big(\sum_j c_{i_0 j_0} \dots c_{i_d j_d} X_{j_0 \dots j_d}, 0\Big) = 0. \tag{11}$$

Any specialisation V^0 of $V(\epsilon)$ corresponding to the specialisation $\epsilon \to \Delta^{-1}$ satisfies the equations obtained in this way. In particular these include, besides the equations (7),

$$\sum_\lambda g_{\mu\lambda}\Big(\sum_j B_j X_{j i_1 \dots i_d}\Big)$$
$$\times h_{\mu\lambda}\Big(\sum_j c_{i_0 j_0} \dots c_{i_d j_d} X_{j_0 \dots j_d}, 0\Big) = 0 \quad (\mu = 1, 2, \dots), \tag{12}$$

$$G_\nu\Big(\dots, \sum_j B_j X_{j i_1 \dots i_d}, \dots\Big) = 0. \tag{13}$$

Equations (12) are those in which the forms $h_{\mu\lambda}$ are of degree greater than zero, while equations (13) involve only the expressions $\sum_j B_j X_{j i_1 \dots i_d}$. From the manner in which they are derived we see that equations (13) are satisfied by V^{a*} as well as by any specialisation V^0 of $V(\epsilon)$ corresponding to $\epsilon \to \Delta^{-1}$; and that every equation in $\sum_j B_j X_{j i_1 \dots i_d}$ only, satisfied by the points of V^{a*}, is included in (13).

The variety V^0 satisfies the equations (7), (12) and (13). We proceed to interpret these equations, showing that equations (7) and (13) and the equations

$$\sum_{\lambda=0}^{d+1} (-1)^\lambda X_{i_1 \dots i_d j_\lambda} X_{j_0 \dots j_{\lambda-1} j_{\lambda+1} \dots j_{d+1}} = 0, \tag{14}$$

which are contained in (12), determine an algebraic variety of dimension ρ. Since any specialisation of $V(\epsilon)$ as $\epsilon \to \Delta^{-1}$ satisfies these equations, we are able to deduce that $V(\epsilon)$ has a unique specialisation for $\epsilon \to \Delta^{-1}$. Moreover, it follows that any further equations, such as those given by (12), can at most result in the exclusion of certain of the components of the variety defined by (7) and (13).

We now give geometrical meanings to the expressions

$$\sum_j c_{i_0 j_0} \dots c_{i_d j_d} X_{j_0 \dots j_d} \quad \text{and} \quad \sum_j B_j X_{j i_1 \dots i_d}.$$

If $(x_0, \dots, x_n)$ is any point in S_n, its projection from $Q = (b_0, \dots, b_n)$ on to the S_{n-1} given by $B_0 x_0 + \dots + B_n x_n = 0$ is $(x_0 - \lambda b_0, \dots, x_n - \lambda b_n)$,

where
$$\sum_{j=0}^{n} B_j x_j - \lambda \sum_{j=0}^{n} b_j B_j = 0,$$

so that
$$\lambda = \Delta^{-1} \sum_{j=0}^{n} B_j x_j.$$

Hence, since $c_{ij} = \delta_{ij} - \Delta^{-1} b_i B_j$, the projection

$$\left(x_0 - \Delta^{-1} b_0 \sum_{j=0}^{n} B_j x_j, \dots, x_n - \Delta^{-1} b_n \sum_{j=0}^{n} B_j x_j\right)$$

of $(x_0, \dots, x_n)$ is the same point as

$$\left(\sum_{j=0}^{n} c_{0j} x_j, \dots, \sum_{j=0}^{n} c_{nj} x_j\right).$$

We consider the d-space in S_n spanned by the points $x^0, x^1, \dots, x^d$. If its Grassmann coordinates are $(\dots, p_{i_0 \dots i_d}, \dots)$, the coordinates of the space spanned by the projections of $x^0, x^1, \dots, x^d$ are

$$\left(\dots, \sum_j c_{i_0 j_0} \dots c_{i_d j_d} p_{j_0 \dots j_d}, \dots\right)$$

if the projected points are independent. If not,

$$\sum_j c_{i_0 j_0} \dots c_{i_d j_d} p_{j_0 \dots j_d} = 0$$

for all $i_0, \dots, i_d$. The projected points are dependent if and only if their joins to Q form an S_d, at most. The joins must, however, contain $x^0, x^1, \dots, x^d$, which span the original d-space. Hence the original d-space passes through Q.

To interpret the second expression, let $u^1, \dots, u^{n-d}$ be $n-d$ independent primes which pass through the S_d whose coordinates are $(\dots, p_{i_0 \dots i_d}, \dots)$. With a suitable factor of proportionality k we can write

$$k p_{i_0 \dots i_d} = \begin{vmatrix} u^1_{i_{d+1}} & . & u^1_{i_n} \\ . & . & . \\ u^{n-d}_{i_{d+1}} & . & u^{n-d}_{i_n} \end{vmatrix},$$

where $(i_0, \ldots, i_n)$ is an even permutation of $(0, \ldots, n)$. It follows that

$$(-1)^d k \sum_j B_j p_{ji_0 \ldots i_{d-1}} = \begin{vmatrix} B_{i_d} & . & B_{i_n} \\ u^1_{i_d} & . & u^1_{i_n} \\ . & . & . \\ u^{n-d}_{i_d} & . & u^{n-d}_{i_n} \end{vmatrix}.$$

Hence if $\sum_j B_j p_{ji_1 \ldots i_d} = 0$ for all $i_1, \ldots, i_d$, the primes $S_{n-1}, u^1, \ldots, u^{n-d}$ are linearly dependent, that is, the d-space lies in S_{n-1}. If this is not the case, $\Sigma B_j p_{j\,i_1 \ldots i_d}$ are the coordinates of the $(d-1)$-space in which the d-space meets S_{n-1}.

These facts lead to an immediate interpretation of the equations

$$F_\lambda\left(\ldots, \sum_j c_{i_0 j_0} \ldots c_{i_d j_d} X_{j_0 \ldots j_d}, \ldots\right) = 0 \quad (\lambda = 1, 2, \ldots), \tag{7}$$

$$\sum_{\lambda=0}^{d+1} (-1)^\lambda X_{i_0 \ldots i_d j_\lambda} X_{j_0 \ldots j_{\lambda-1} j_{\lambda+1} \ldots j_{d+1}} = 0. \tag{14}$$

Any solution of these equations represents a d-space

$$p = (\ldots, p_{i_0 \ldots i_d}, \ldots)$$

in S_n. If p is any d-space in S_n through Q,

$$\sum_j c_{i_0 j_0} \ldots c_{i_d j_d} p_{j_0 \ldots j_d} = 0 \quad (\text{all } i_0, \ldots, i_d).$$

Hence p satisfies the equations (7) and (14). If p does not pass through Q, a necessary and sufficient condition that p satisfy the equation (7) is that it projects from Q into a d-space in S_{n-1} belonging to the system represented by V^{a*}. Hence the solutions of (7) and (14) are the d-spaces which pass through Q or project from Q into d-spaces in S_{n-1} belonging to the system represented by V^{a*}.

In order to investigate the solutions of (13) and (14), we consider a three-way correspondence Γ between the space S'_n in which $(B_0, \ldots, B_n)$ are coordinates, the space S_N containing $\Omega(d, n)$, and the space $S_{N'}$ containing the Grassmann variety $\Omega(d-1, n)$ of $(d-1)$-spaces in S_n. We denote the coordinates of points on $\Omega(d-1, n)$ by $Z_{i_1 \ldots i_d}$. The correspondence Γ is defined by the equations

(i) $$F_\lambda(\ldots, X_{i_0 \ldots i_d}, \ldots) = 0 \quad (\lambda = 1, 2, \ldots), \tag{5}$$

the equations of V^{a*}, with coefficients in $K_1 = K(a^*, b_0 \ldots b_n)$, which is taken as the ground field;

$$\left.\begin{aligned}\text{(ii)}\qquad \sum_{\lambda=0}^{d+1}(-1)^{\lambda} Z_{i_1\ldots i_{d-1}j_\lambda} X_{j_0\ldots j_{\lambda-1}j_{\lambda+1}\ldots j_{d+1}} &= 0,\\ \sum_{\lambda=0}^{d}(-1)^{\lambda} Z_{i_1\ldots i_{d-1}j_\lambda} Z_{j_0\ldots j_{\lambda-1}j_{\lambda+1}\ldots j_d} &= 0,\\ \sum_h B_h Z_{h i_1\ldots i_{d-1}} &= 0.\end{aligned}\right\}\qquad(15)$$

Equations (15) simply imply that $(\ldots, Z_{i_1\ldots i_d}, \ldots)$ is a $(d-1)$-space in the d-space represented by $(\ldots, X_{i_0\ldots i_d}, \ldots)$ and that

$$B_0 x_0 + \ldots + B_n x_n = 0$$

is a prime through it. Then (B', X', Z') is a solution of these equations if and only if X' represents a d-space of the system defined by V^{a*}, Z' is any $(d-1)$-space in it, and B' is a prime through Z'.

We prove that Γ is irreducible by constructing a generic point of it. Let $\pi = (\ldots, \pi_{i_0\ldots i_d}, \ldots)$ be a generic point of V^{a*}, and let $u_0, \ldots, u_n, v_{i0}, \ldots, v_{in}$ $(i = 1, \ldots, n-d)$ be $(n+1)(n-d+1)$ independent indeterminates over $K_1(\pi)$. Let $\zeta'_{i_1\ldots i_d} = \sum_j u_j \pi_{j i_1\ldots i_d}$, and let

$$\beta'_i = \sum_j (-1)^i \delta^{j_1\ldots\ldots\ldots\ldots j_n}_{0\ldots i-1\, i+1\ldots n} \zeta'_{j_1\ldots j_d} v_{1 j_{d+1}} \ldots v_{n-d\, j_n}.$$

π can be interpreted as a d-space in S_n, ζ' as a $(d-1)$-space in it, and β' as the join of this $(d-1)$-space to a generic $(n-d-1)$-space of S_n, that is, as a generic prime through ζ'. Clearly (β', π, ζ') satisfies the equations (5) and (15). Now consider any equation over K_1 which is satisfied by (β', π, ζ'), say

$$f(B, X, Z) = 0.$$

We have
$$f(\beta', \pi, \zeta') = 0, \qquad (16)$$
and since the u_i, v_{ij} are independent over $K_1(\pi)$, the coefficient of each power-product of the u_i, v_{ij} must vanish, and this must remain true for any proper specialisation of π. Now let (B', X', Z') be any point of Γ. X' is a proper specialisation of π, and since Z' represents a $(d-1)$-space contained in the d-space represented by X', we can find a prime
$$u'_0 x_0 + \ldots + u'_n x_n = 0$$
which meets X' in Z'. We can then write $Z'_{i_1\ldots i_d} = \sum_j u'_j X'_{j i_1\ldots i_d}$. Since B' is a prime through Z', we can, similarly, find
$$v'_{ij} \quad (j = 0, \ldots, n;\ i = 1, \ldots, n-d)$$
so that
$$B'_i = \sum_j (-1)^i \delta^{j_0\ldots\ldots\ldots\ldots j_n}_{0\ldots i-1\, i+1\ldots n} Z'_{j_1\ldots j_d} v'_{1 j_{d+1}} \ldots v'_{n-d\, j_n}.$$

Since (16) is satisfied for any proper specialisation of π, it is satisfied when π is replaced by X', with u_i, v_{ij} independent indeterminates. The resulting identity remains true when we replace u_i, v_{ij} by u'_i, v'_{ij}. Hence

$$f(B', X', Z') = 0,$$

that is $f(B, X, Z)$ vanishes on Γ. It follows that (β', π, ζ') is a generic point of Γ, which is therefore irreducible.

While the method just used is the most convenient for proving the existence of a generic point of Γ, once we have established its existence it is more convenient to construct a generic point in a different way. We again take π to be a generic solution of (5), and let β be a generic point (over $K_1(\pi)$) of S'_n, that is, a generic prime of S_n. The d-space π and the independent generic prime β meet in a $(d-1)$-space ζ whose coordinates are given by

$$\zeta_{i_1 \ldots i_d} = \Sigma \beta_j \pi_{j i_1 \ldots i_d}.$$

Then it is clear that (β, π, ζ) is a generic point of Γ. Hence if we regard Γ as a correspondence between the two-way space S_{nN} and the one-way space $S_{N'}$, we see that the locus Π in S_{nN} which is the direct product of S'_n and V^{a*} is contained in the object-variety, since its generic point (β, π) is on the object-variety. On the other hand, it is clear that the object-variety is contained in Π; hence Π is the object-variety. We now construct the image of (β, π) in Γ. From (15) any corresponding point in $S_{N'}$ represents a $(d-1)$-space lying both in the prime β and in the d-space π. But there is only one such $(d-1)$-space, namely ζ. Since (β, π) is a generic point of the object-variety of Γ, and ζ is a generic point of the image of (β, π), it follows that (β, π, ζ) is a generic point of Γ. Since Π is of dimension $n+\rho$, and the image of (β, π) is of dimension zero, it follows that the dimension of Γ is $\rho+n$.

We now consider the correspondence Γ_N between S'_n and $S_{N'}$ whose generic point is (β, ζ). We first obtain the equations of this correspondence. Amongst the equations (5) are the equations (13), which we write in the form

$$G_\lambda(B, \ldots, \Sigma B_j X_{j i_1 \ldots i_d}, \ldots) = 0 \quad (\lambda = 1, 2, \ldots).$$

Then, clearly, (β, ζ) satisfies the equations

$$G_\lambda(B, \ldots, Z_{i_1 \ldots i_d}, \ldots) = 0 \quad (\lambda = 1, 2, \ldots), \tag{17}$$

$$\left.\begin{aligned} \sum_{\lambda=0}^{d} (-1)^\lambda Z_{i_1 \ldots i_{d-1} j_\lambda} Z_{j_0 \ldots j_{\lambda-1} j_{\lambda+1} \ldots j_d} &= 0, \\ \sum_j B_j Z_{j i_1 \ldots i_{d-1}} &= 0. \end{aligned}\right\} \tag{18}$$

Consider any equation over K_1 satisfied by (β, ζ), say,

$$g(B, \ldots, Z_{i_1 \ldots i_d}, \ldots) = 0.$$

Then
$$g(\beta, \ldots, \Sigma \beta_j \pi_{j i_1 \ldots i_d}, \ldots) = 0,$$

and since $\beta_0, \ldots, \beta_n$ are independent indeterminates over $K_1(\pi)$ it follows that

$$g(B, \ldots, \Sigma B_j X_{j i_1 \ldots i_d}, \ldots) = 0$$

on V^{a*}. But (13) is, by hypothesis, a basis for the equations of V^{a*} which involve only $\Sigma B_j X_{j i_1 \ldots i_d}$. Therefore we must have an identity

$$\phi(B)\, g(B, \ldots, \Sigma B_j X_{j i_1 \ldots i_d}, \ldots) = \sum_\lambda a_\lambda(B, \ldots, \Sigma B_j X_{j i_1 \ldots i_d}, \ldots)\, G_\lambda(B, \ldots, \Sigma B_j X_{j i_1 \ldots i_d}, \ldots),$$

where the forms are polynomials in $B_0, \ldots, B_n$ and $\Sigma B_j X_{j i_1 \ldots i_d}$, and $\phi(B) \neq 0$. Then

$$\phi(B)\, g(B, \ldots, Z_{i_1 \ldots i_d}, \ldots) - \Sigma a_\lambda(B, \ldots, Z_{i_1 \ldots i_d}, \ldots)\, G_\lambda(B, \ldots, Z_{i_1 \ldots i_d}, \ldots) \quad (19)$$

vanishes whenever $Z_{i_1 \ldots i_d}$ is replaced by $\sum_j B_j X_{j i_1 \ldots i_d}$. Since whenever $Z_{i_0 \ldots i_d}$ satisfies (18) we can find $X_{i_0 \ldots i_d}$ so that

$$Z_{i_1 \ldots i_d} = \sum_j B_j X_{j i_1 \ldots i_d},$$

it follows from Hilbert's zero-theorem that a power of (19) is expressible in terms of the left-hand sides of the equations in (18). Since $\phi(B) \neq 0$, we conclude that any solution of (17) and (18) satisfies

$$g(B, \ldots, Z_{i_1 \ldots i_d}, \ldots) = 0.$$

Thus (17) and (18) are the equations of the correspondence with generic point (β, ζ).

We have to determine certain dimension numbers related to the correspondences Γ_N, Γ. Let σ be the dimension of the variety in $S_{N'}$ which has generic point ζ. ζ is on the object-variety of Γ, regarded as a correspondence between $S_{N'}$ and $S_{n,N}$, and hence through it there passes an irreducible system of d-spaces of V^{a*}, of dimension σ_1, say. Any prime β' through ζ, and any d-space ξ of V^{a*} through ζ, and ζ itself constitute a point (β', ξ, ζ) of Γ. Hence, by counting constants, we see that the dimension of Γ is

$$\sigma + \sigma_1 + n - d = \rho + n.$$

Therefore
$$\sigma = \rho + d - \sigma_1 \leqslant \rho + d.$$

The dimension of Γ_N is similarly seen to be $\sigma+n-d = \rho+n-\sigma_1$. Moreover, if β is a generic prime of S_n it meets a generic d-space of the system represented by V^{a*} in a $(d-1)$-space ζ' which must correspond to β in Γ_N. Hence the object-variety in S'_n of Γ_N is S'_n itself, and to a generic prime of S_n there therefore corresponds a system of $(d-1)$-spaces of dimension $\tau = \rho+n-\sigma_1-n = \rho-\sigma_1$. Thus the equations (17) and (18), regarded as equations over $K^* = K_1(B_0, \ldots, B_n)$, determine an irreducible system of $(d-1)$-spaces of dimension $\tau \leqslant \rho$.

We are now in a position to determine the variety given by the equations (13) and (14). Any solution represents d-spaces of S_n. If $p = (\ldots, p_{i_0\ldots i_d}, \ldots)$ is a d-space in S_n lying in the S_{n-1}

$$B_0 x_0 + \ldots + B_n x_n = 0,$$

we have

$$\sum_j B_j p_{j\,i_1\ldots i_d} = 0,$$

and hence p satisfies (13) and (14). If p does not lie in S_{n-1} and satisfies equations (13) and (14), its intersection with S_{n-1} has coordinates $(\ldots, \Sigma B_j p_{j\,i_1\ldots i_d}, \ldots)$ and is a $(d-1)$-space p' satisfying equations (17) and (18). p' is therefore a proper specialisation of ζ, and hence through it there passes a d-space of the system represented by V^{a*}. Conversely, it is clear that any d-space in S_n which meets S_{n-1} in a $(d-1)$-space contained in a d-space of the system V^{a*} satisfies (13) and (14). Thus the d-spaces which satisfy (13) are

(i) the d-spaces in S_{n-1}, and

(ii) the d-spaces of S_n which meet S_{n-1} in a $(d-1)$-space lying in a d-space of the system represented by V^{a*}.

Combining our results for equations (7), (13), (14) we see that the solutions of (7), (13), (14) are

(*a*) the d-spaces which lie in the $(d+1)$-spaces which join Q to any d-space of S_{n-1} belonging to the system V^{a*};

(*b*) the d-spaces joining Q to the $(d-1)$-spaces of S_{n-1} which lie on d-spaces of the system V^{a*}.

We now examine these two systems of d-spaces.

(*a*) We consider the equations (5) and the equation

$$\sum_{j=0}^{n} t_j X_{j\,i_1\ldots i_d} = 0. \qquad (20)$$

These determine a correspondence between V^{a*} and the n-space $(t_0, \ldots, t_n)$. For any specialisation X' of π the coefficients of equation (20) form a matrix of rank $d+1$. Thus [XI, §5, Th. I] the correspondence is irreducible and of dimension $\rho+n-d-1$. We consider the image-variety of this correspondence in the t-space. If this is of dimension less than n the equations (5) and

$$\Sigma B_j X_{j i_1 \dots i_d} = 0,$$

where $B_0, \dots, B_n$ are independent indeterminates, have no solution. Hence there are no d-spaces of V^{a*} in S_{n-1}. The solutions (a) of (7), (13), (14) are then non-existent.

If, on the other hand, the image-variety of the correspondence is of dimension n, to the generic point $(B_0, \dots, B_n)$ of it there corresponds a variety of dimension $\rho-d-1$. Thus the d-spaces of V^{a*} which lie in S_{n-1} form an irreducible system S^* of dimension $\rho-d-1$. Let the equations of this system be

$$\phi_\lambda(\dots, X_{i_0 \dots i_d}, \dots) = 0 \quad (\lambda = 1, 2, \dots).$$

The $(d+1)$-space $(\dots, p_{i_0 \dots i_{d+1}}, \dots)$ joins a d-space of S^* to Q if and only if it satisfies the equations

$$\phi_\lambda(\dots, B_j Y_{j i_0 \dots i_d}, \dots) = 0 \quad (\lambda = 1, 2, \dots),$$

$$\sum_{\lambda=0}^{d+2} (-1)^\lambda b_{i_\lambda} Y_{i_0 \dots i_{\lambda-1} i_{\lambda+1} \dots i_{d+2}} = 0.$$

An elementary application of the principle of counting constants shows that this system S^{**} of $(d+1)$-spaces is irreducible and of dimension $\rho-d-1$. We construct for this system of $(d+1)$-spaces a correspondence similar to the correspondence Γ constructed for V^{a*}, and deduce that the d-spaces of S_n which lie in $(d+1)$-spaces of S_n through Q which meet S_{n-1} in d-spaces of V^{a*} form an irreducible system S of dimension ρ. It is further clear from the equations that as S^* varies in a continuous system of d-spaces in S_{n-1}, S varies in a continuous system of d-spaces in S_{n-1}.

(b) The $(d-1)$-spaces which lie in S_{n-1} and in d-spaces of the system represented by V^{a*} form an irreducible system $\mathfrak{S}^*$ of dimension τ $(\tau \leqslant \rho)$, given by (17) and (18). Their joins to Q form the irreducible τ-dimensional system $\mathfrak{S}$ given by replacing $Z_{i_1 \dots i_d}$ in (17) by $\Sigma B_h X_{h i_1 \dots i_d}$, and adding the equation

$$\Sigma(-1)^\lambda b_{i_\lambda} X_{i_0 \dots i_{\lambda-1} i_{\lambda+1} \dots i_{d+1}} = 0.$$

Hence as $\mathfrak{S}^*$ describes a continuous system, so does $\mathfrak{S}$.

Now return to the system $\boldsymbol{V}(\epsilon)$. When $\epsilon \to \Delta^{-1}$, any corresponding specialisation of $\boldsymbol{V}(\epsilon)$ satisfies the equations (7), (13). Since these define a variety of dimension ρ, at most, we conclude that the specialisation of $V(\epsilon)$ is unique, say V^0, and that as a point set it is contained in $S+\mathfrak{S}$. Hence

$$V^0 = \lambda S + \mu \mathfrak{S},$$

where $\lambda, \mu \geqslant 0$, and if S is vacuous we put $S = 0$, and if $\mathfrak{S}$ is of dimension less than ρ we put $\mathfrak{S} = 0$. Since $V^{a*} \sim V^0$, we have $V^{a*} \sim \lambda S + \mu \mathfrak{S}$.

We are now in a position to prove the main theorem of this section.

THEOREM I. *If V is any ρ-dimensional variety on $\Omega(d,n)$, there exists an equivalence*

$$V \sim \Sigma \alpha_{a_0 \ldots a_d} \Omega_{a_0 \ldots a_d},$$

where the summation is over all values of $a_0, \ldots, a_d$ satisfying the relations

$$0 \leqslant a_0 < a_1 < \ldots < a_d \leqslant n,$$
$$\Sigma a_j - \tfrac{1}{2}d(d+1) = \rho.$$

The theorem is proved by induction on n. If $n = 1$, the truth of the theorem is obvious. For in this case $d = 0$, or $d = 1$. If $d = 0$, $\Omega(d,n)$ is a line, and V is a set of points; and as any two points in S_n are equivalent any 0-variety is equivalent to a multiple of a point. If $d = 1$, $\Omega(d,n)$ is a point, and there is nothing to prove. We therefore assume the truth of the theorem for $\Omega(d, n-1)$, for all d satisfying $0 \leqslant d \leqslant n-1$. If $d = n$, $\Omega(d,n)$ is again a point, and there is nothing to prove. If $d = 0$, $\Omega(d,n)$ is an n-space, and the theorem follows from XII, § 2, Th. I. We therefore assume that $0 < d < n$.

We have obtained the equivalence

$$V \sim \lambda S + \mu \mathfrak{S},$$

and we apply the hypothesis of induction to obtain equivalences for S and $\mathfrak{S}$, but only in the cases in which these are ρ-dimensional. (If they are of lower dimension, we have agreed to treat them as zero varieties.) S^* and $\mathfrak{S}^*$ are defined as above.

S^* is a $(\rho-d-1)$-dimensional variety on $\Omega(d, n-1)$, the Grassmann variety of d-dimensional spaces in S_{n-1}. Hence, by the hypothesis of induction,

$$S^* \sim \Sigma \lambda_{i_0 \ldots i_d} \Omega^*_{i_0 \ldots i_d},$$

where the summation is over values of $i_0, \ldots, i_d$ such that

$$0 \leqslant i_0 < i_1 \ldots < i_d \leqslant n-1,$$
$$\sum_0^d i_j - \tfrac{1}{2}d(d+1) = \rho - d - 1.$$

The d-spaces which are represented by points of any $\Omega^*_{i_0\ldots i_d}$ of this summation are those which meet the space C_r in an r-space, where

$$C_0 \subset C_1 \subset \ldots \subset C_d,$$

and C_r is of dimension i_r and lies in S_{n-1}. Let D_r be the (i_r+1)-space which joins C_r to Q. A d-space of S_n lies in the join of Q to a d-space of the system defined by $\Omega^*_{i_0\ldots i_d}$ if and only if it meets D_r in an r-space $(r = 0, \ldots, d)$. Let $\Omega_{i_0+1\ldots i_d+1}$ be the Schubert variety on $\Omega(d, n)$ defined by the spaces $D_0, \ldots, D_d$. Then the equivalence

$$\mathbf{S}^* \sim \Sigma\lambda_{i_0\ldots i_d}\mathbf{\Omega}^*_{i_0\ldots i_d}$$

implies the equivalence

$$\mathbf{S} \sim \Sigma\lambda'_{i_0\ldots i_d}\mathbf{\Omega}_{i_0+1\ldots i_d+1}.$$

We have
$$0 < i_0+1 < i_1+1 < \ldots < i_d+1 \leqslant n,$$

$$\sum_0^d (i_j+1) - \tfrac{1}{2}d(d+1) = \rho.$$

$\mathfrak{S}^*$ is a ρ-dimensional variety on $\Omega(d-1, n-1)$, the Grassmann variety of $(d-1)$-spaces in S_{n-1}. Hence, by the hypothesis of induction,

$$\mathfrak{S}^* \sim \Sigma\mu_{i_1\ldots i_d}\mathbf{\Omega}^*_{i_1\ldots i_d},$$

where
$$0 \leqslant i_1 < i_2 < \ldots < i_d \leqslant n-1,$$

$$\sum_1^d i_j - \tfrac{1}{2}(d-1)d = \rho.$$

The $(d-1)$-spaces represented by points of any $\Omega^*_{i_1\ldots i_d}$ of this summation are those which meet the i_r-space E_r in an $(r-1)$-space $(r = 1, \ldots, d)$, where

$$E_1 \subset E_2 \subset \ldots \subset E_{d-1} \subset S_{n-1}.$$

The joins of these spaces to Q are simply the d-spaces which meet F_r in an r-space, where $F_0 = Q$, and F_r is the (i_r+1)-space joining E_r to Q $(r = 1, \ldots, d)$. Hence the d-spaces joining Q to the $(d-1)$-spaces represented by points of $\Omega^*_{i_1\ldots i_d}$ form the system represented on $\Omega(d, n)$ by the Schubert variety $\Omega_{0\, i_1+1\ldots i_d+1}$ defined by the sequence of spaces

$$F_0 \subset F_1 \subset \ldots \subset F_d.$$

Writing $i_0 + 1 = 0$, we have

$$\sum_0^d (i_j+1) - \tfrac{1}{2}d(d+1) = \rho.$$

The equation $\mathfrak{S}^* \sim \Sigma \mu_{i_1 \dots i_d} \mathbf{\Omega}^*_{i_1 \dots i_d}$
implies the equation

$$\mathfrak{S} \sim \Sigma \mu'_{i_1 \dots i_d} \mathbf{\Omega}_{0\, i_1+1 \dots i_d+1}.$$

Combining these results, we obtain Theorem I, for irreducible varieties. The extension to virtual varieties is immediate.

We now show that if we assume that in the summation $\Sigma \alpha_{a_0 \dots a_d} \mathbf{\Omega}_{a_0 \dots a_d}$ no two terms involve the same set of dimension numbers $a_0, \dots, a_d$, then the coefficients $\alpha_{a_0 \dots a_d}$ are uniquely determined by $\boldsymbol{V}$. We select any particular set of number $a_0, \dots, a_d$. Since

$$\boldsymbol{V} \sim \Sigma \alpha_{a_0 \dots a_d} \mathbf{\Omega}_{a_0 \dots a_d},$$

$$(\boldsymbol{V} . \mathbf{\Omega}^{\tau}_{n-b_d \dots n-b_0}) = \Sigma \alpha_{a_0 \dots a_d} (\mathbf{\Omega}_{a_0 \dots a_d} . \mathbf{\Omega}^{\tau}_{n-b_d \dots n-b_0}),$$

[Chap. XII, § 10], where $\Sigma a_i = \Sigma b_i$ and $\Omega_{n-b_d \dots n-b_0}$ is any Schubert variety with dimension numbers $n - b_d, \dots, n - b_0$, and $\Omega^{\tau}_{n-b_d \dots n-b_0}$ is obtained from it by a generic projective transformation. Since [§ 4, Th. I]

$$\Omega_{a_0 \dots a_d} \wedge \Omega^{\tau}_{n-b_d \dots n-b_0}$$

is vacuous if $a_i \neq b_i$, $(\mathbf{\Omega}_{a_0 \dots a_d} . \mathbf{\Omega}^{\tau}_{n-b_d \dots n-b_0}) = 0$ unless $a_i = b_i$, and since $\Omega_{b_0 \dots b_d}$ and $\Omega_{n-b_d \dots n-b_0}$ intersect simply in a point,

$$(\mathbf{\Omega}_{b_0 \dots b_d} . \mathbf{\Omega}^{\tau}_{n-b_d \dots n-b_0}) = 1.$$

Hence $\alpha_{b_0 \dots b_d} = (\boldsymbol{V} . \mathbf{\Omega}^{\tau}_{n-b_d \dots n-b_0}) = (\boldsymbol{V} . \mathbf{\Omega}_{n-b_d \dots n-b_0}),$

and is therefore determined uniquely by $\boldsymbol{V}$.

An immediate corollary is

THEOREM II. *The varieties $\mathbf{\Omega}_{a_0 \dots a_d}$, for all distinct choices of $a_0, \dots, a_d$ such that*

$$0 \leqslant a_0 < a_1 < \dots < a_d \leqslant n,$$

$$\Sigma a_j - \tfrac{1}{2} d(d+1) = \rho,$$

form a minimal base for the ρ-dimensional varieties on $\Omega(d, n)$.

If

$$\Sigma \alpha_{a_0 \dots a_d} \mathbf{\Omega}_{a_0 \dots a_d} \sim \mathbf{0},$$

we have $\alpha_{a_0 \dots a_d} = (\boldsymbol{O} . \mathbf{\Omega}_{n-a_d \dots n-a_0}) = 0.$

6. The intersection formulae. We now consider systems of equivalence [XII, § 10] on $\Omega(d, n)$. Let $\{\boldsymbol{V}_\rho\}$ be any system of equivalence of virtual varieties of dimension ρ on $\Omega(d, n)$, and let $\boldsymbol{V}_\rho$ be any member of it. By § 5, Th. I, we have an equivalence

$$\boldsymbol{V}_\rho \sim \Sigma \alpha_{i_0 \dots i_d} \mathbf{\Omega}_{i_0 \dots i_d}, \tag{1}$$

summed over all sets of integers $i_0, \ldots, i_d$ which satisfy the condition

$$0 \leqslant i_0 < i_1 < \ldots < i_d \leqslant n,$$

$$\sum_0^d i_j - \tfrac{1}{2}d(d+1) = \rho,$$

and we shall assume that the same set of integers does not occur twice in the summation. We then have

$$\{V_\rho\} = \Sigma \alpha_{i_0 \ldots i_d} \{\boldsymbol{\Omega}_{i_0 \ldots i_d}\}. \tag{2}$$

Since any two Schubert varieties with the same dimension numbers are equivalent, the system $\{V_\rho\}$ is determined uniquely by the numbers $\alpha_{i_0 \ldots i_d}$, and any ρ-dimensional system determines the numbers $\alpha_{i_0 \ldots i_d}$ uniquely, since the numbers $\alpha_{i_0 \ldots i_d}$ in (1) are uniquely determined. We can then specify any ρ-dimensional system of equivalence by giving the appropriate numbers $\alpha_{i_0 \ldots i_d}$.

Now consider a system of equivalence $\{V_\sigma\}$ of virtual varieties of dimension σ, and let the equation corresponding to (2) be

$$\{V_\sigma\} = \Sigma \beta_{j_0 \ldots j_d} \{\boldsymbol{\Omega}_{j_0 \ldots j_d}\}. \tag{3}$$

Here, again, the summation is over all values of the integers $j_0, \ldots, j_d$ which satisfy the conditions

$$0 \leqslant j_0 < j_1 < \ldots < j_d \leqslant n,$$

$$\sum_0^d j_k - \tfrac{1}{2}d(d+1) = \sigma,$$

and we assume that each set appears only once in the summation. If $\rho + \sigma \geqslant (d+1)(n-d)$, the systems $\{V_\rho\}$, $\{V_\sigma\}$ define a unique system of intersection $\{V_\rho\}.\{V_\sigma\}$, which can similarly be expressed as a sum of Schubert systems $\{\boldsymbol{\Omega}_{k_0 \ldots k_d}\}$ for which

$$\Sigma k_i - \tfrac{1}{2}d(d+1) = \rho + \sigma - (d+1)(n-d),$$

say,

$$\{V_\rho\}.\{V_\sigma\} = \Sigma \gamma_{k_0 \ldots k_d} \{\boldsymbol{\Omega}_{k_0 \ldots k_d}\}.$$

The main problem of the intersection theory on $\Omega(d, n)$ is to determine the numbers $\gamma_{k_0 \ldots k_d}$ in terms of the coefficients $\alpha_{i_0 \ldots i_d}$, $\beta_{i_0 \ldots i_d}$ in (2) and (3). Now we have

$$\{V_\rho\}.\{V_\sigma\} = \sum_i \sum_j \alpha_{i_0 \ldots i_d} \beta_{j_0 \ldots j_d} \{\boldsymbol{\Omega}_{i_0 \ldots i_d}\}.\{\boldsymbol{\Omega}_{j_0 \ldots j_d}\};$$

hence this problem will be completely solved if we can find the equivalences for $\{\Omega_{i_0\ldots i_d}\}.\{\Omega_{j_0\ldots j_d}\}$, for all values of $i_0, \ldots, i_d$, $j_0, \ldots, j_d$ such that

$$\Sigma i_t + \Sigma j_k - d(d+1) \geqslant (n-d)(d+1).$$

This section is devoted to the solution of this problem.

An equivalent problem is the following: if $\Omega_{i_0\ldots i_d}$, $\Omega_{j_0\ldots j_d}$ are two Schubert varieties, to find an equivalence for $\Omega_{i_0\ldots i_d}.\Omega_{j_0\ldots j_d}$ in terms of Schubert varieties. We first solve this problem in the case in which $\Omega_{j_0\ldots j_d}$ is a Schubert variety $\sigma_h = \Omega_{n-d-h\, n-d+1\ldots n}$, as defined in § 4.

By § 5, Th. I,

$$\Omega_{i_0\ldots i_d}.\sigma_h \sim \Sigma\lambda_{k_0\ldots k_d}\Omega_{k_0\ldots k_d},$$

where the summation is over distinct sets of integers $k_0, \ldots, k_d$ such that

$$0 \leqslant k_0 < \ldots < k_d \leqslant n,$$
$$\Sigma k_t = \Sigma i_t - h.$$

Selecting any of these sets $k_0, \ldots, k_d$, we have, as in § 5,

$$\lambda_{k_0\ldots k_d} = (\Omega_{i_0\ldots i_d}.\Omega_{n-k_d\ldots n-k_0}.\sigma_h).$$

By Theorems III and IV of § 4, we conclude that

$$\lambda_{k_0\ldots k_d} = 0$$

unless

$$i_{d-t} < n-(n-k_{d-t+1}) \leqslant i_{d-t+1}$$

for all t, and if these inequalities are satisfied we have

$$\lambda_{k_0\ldots k_d} = 1.$$

Hence

$$\Omega_{i_0\ldots i_d}.\sigma_h \sim \Sigma\Omega_{k_0\ldots k_d}, \tag{4}$$

where the summation is over all distinct sets $k_0, \ldots, k_d$, such that

$$0 \leqslant k_0 \leqslant i_0, \quad i_0 < k_1 \leqslant i_1, \quad \ldots, \quad i_{d-1} < k_d \leqslant i_d,$$
$$\Sigma k_t = \Sigma i_t - h.$$

The formula (4) is usually called *Pieri's formula*.

We use formula (4) to show that any Schubert variety is virtually equivalent to a variety which is the sum of intersections of varieties σ_h $(h = 0, \ldots, n-d)$. Consider the variety $\Omega_{a_0\ldots a_d}$. If $a_i = n-d+i$ $(i = 0, \ldots, d)$, $\Omega_{a_0\ldots a_d} = \Omega(d, n) = \sigma_0$, and there is nothing to prove. Any other Schubert variety can be written in the form

$$\Omega_{a_0\ldots a_r\, n-d+r+1\ldots n}, \quad \text{where} \quad 0 \leqslant r \leqslant d,$$

and

$$0 \leqslant a_0 < a_1 < \ldots < a_r < n-d+r.$$

For convenience, we define a_{-1} to be -1, and $n-d+r$ to be m. If $0 \leqslant m-a_j \leqslant n-d$, we can define σ_{m-a_j}. The first condition, $m \geqslant a_j$, is satisfied by hypothesis. The second condition requires that $a_j \geqslant r$. Since $a_j \geqslant j$, there is at least one value of j which satisfies this condition. For any j satisfying the condition, we have

$$\begin{aligned}
&\dim[\Omega_{a_0 \ldots a_{j-1} a_{j+1} \ldots a_r m \ldots n} \wedge \sigma_{m-a_j}] \\
&= \sum_0^r a_i + (m+1) + \ldots + n - \tfrac{1}{2}d(d+1) \\
&= \dim[\Omega_{a_0 \ldots a_r m+1 \ldots n}].
\end{aligned}$$

Let ρ ($\rho \geqslant -1$) be the integer such that $a_\rho < r$, $a_{\rho+1} \geqslant r$, and consider the virtual variety

$$\boldsymbol{F} = \sum_{j=\rho+1}^{r} (-1)^j \, \boldsymbol{\Omega}_{a_0 \ldots a_{j-1} a_{j+1} \ldots a_r m \ldots n} \cdot \boldsymbol{\sigma}_{m-a_j}. \tag{5}$$

Using formula (4), we obtain the equivalence

$$\boldsymbol{F} \sim \sum_{j=\rho+1}^{r} (-1)^j \sum_b \boldsymbol{\Omega}_{b_0^j \ldots b_r^j m+1 \ldots n}, \tag{6}$$

where, in the second summation, we include values of the b_i^j which satisfy the conditions

$$\left.\begin{aligned}
&0 \leqslant b_0^j \leqslant a_0, \quad a_0 < b_1^j \leqslant a_1, \quad \ldots, \quad a_{j-1} < b_j^j \leqslant a_{j+1}, \\
&a_{j+1} < b_{j+1}^j \leqslant a_{j+2}, \quad \ldots, \quad a_r < b_r^j \leqslant m. \\
&\qquad \sum_{i=0}^{r} b_i^j = \sum_{i=0}^{r} a_i.
\end{aligned}\right\} \tag{7}$$

We consider any Schubert variety which appears in any summation included in (6). It is of the form

$$\boldsymbol{\Omega}_{c_0 \ldots c_r m+1 \ldots n},$$

where
$$0 \leqslant c_0 \leqslant a_0, \quad \ldots, \quad a_{\rho-1} < c_\rho \leqslant a_\rho,$$

and $c_{\rho+1}, \ldots, c_r$ satisfy the inequalities

$$c_\rho < c_{\rho+1} < \ldots < c_r \leqslant m,$$

and are distributed so that at most one of them lies in each of the closed intervals

$$(a_\rho + 1, a_{\rho+1}), \quad (a_{\rho+1} + 1, a_{\rho+2}), \quad \ldots, \quad (a_r + 1, m).$$

Moreover
$$\sum_0^r c_i = \sum_0^r a_r. \tag{8}$$

There are $r-\rho$ integers in the series $c_{\rho+1}, \ldots, c_r$, and there are $r-\rho+1$ intervals. Hence there is only one interval which is unoccupied.

(a) Suppose that the interval $(a_\rho+1, a_{\rho+1})$ is unoccupied. Then $\Omega_{c_0 \ldots c_r m+1 \ldots n}$ can arise only from the term in (6) with $j = \rho+1$. We have
$$0 \leqslant c_0 \leqslant a_0, \quad \ldots, \quad a_{\rho-1}+1 \leqslant c_\rho \leqslant a_\rho,$$
$$a_{\rho+1}+1 \leqslant c_{\rho+1} \leqslant a_{\rho+2}, \quad \ldots, \quad a_r+1 \leqslant c_r \leqslant m.$$

Adding these inequalities, we obtain
$$\sum_{i=0}^r a_i + r - a_\rho \leqslant \sum_{i=0}^r c_i \leqslant \sum_{i=0}^r a_i + m - a_{\rho+1},$$
and hence, on account of (8),
$$r - a_\rho \leqslant 0 \leqslant m - a_{\rho+1}.$$

But by the definition of ρ, $a_\rho < r$, and we have a contradiction. Hence we conclude that there is no term $\Omega_{c_0 \ldots c_r m+1 \ldots n}$ in the summation (6) for which no c_i lies in the interval $(a_\rho+1, a_{\rho+1})$.

(b) Suppose that the interval (a_r+1, m) is unoccupied. Then
$$0 \leqslant c_0 \leqslant a_0, \quad a_0+1 \leqslant c_1 \leqslant a_1, \quad \ldots, \quad a_{r-1}+1 \leqslant c_r \leqslant a_r.$$

Adding these inequalities, we obtain
$$\sum_0^r c_i \leqslant \sum_0^r a_i,$$
and in view of (8), we must have $c_i = a_i$ $(i = 0, \ldots, r)$. Further, the term can only arise from the term in (6) with $j = r$. Thus the only term which leaves the interval (a_r+1, m) unoccupied is
$$(-1)^r \Omega_{a_0 \ldots a_r m+1 \ldots n}.$$

(c) If the interval (a_k+1, a_{k+1}) is unoccupied, the term arises both from the term in (6) with $j = k$, and from the term with $j = k+1$. As these two terms have opposite signs, the two terms $\Omega_{c_0 \ldots c_r m+1 \ldots n}$ must cancel out. Hence we conclude that $F \sim (-1)^r \Omega_{a_0 \ldots a_r m+1 \ldots n}$, that is
$$(-1)^r \Omega_{a_0 \ldots a_r m+1 \ldots n} \sim \sum_{\rho+1}^r (-1)^j \Omega_{a_0 \ldots a_{j-1} a_{j+1} \ldots a_r m \ldots n} \cdot \sigma_{m-a_j}.$$

In what follows we shall formally introduce varieties σ_h, in which h does not satisfy the conditions $0 \leqslant h \leqslant n-d$. We shall agree that σ_h is the zero-variety when h does not satisfy these inequalities. With this convention, the formula written above for $\Omega_{a_0 \ldots a_r m+1 \ldots n}$ can then be written

$$(-1)^r \Omega_{a_0 \ldots a_r n-d+r+1 \ldots n} \sim \sum_0^r (-1)^j \Omega_{a_0 \ldots a_{j-1} a_{j+1} \ldots a_r n-d+r \ldots n} \cdot \sigma_{n-d+r-a_j}. \tag{9}$$

We now use (9) to obtain the equivalence

$$\Omega_{a_0 \ldots a_r n-d+r+1 \ldots n} \sim \begin{vmatrix} \sigma_{h_0} & \cdot & \sigma_{h_0+r} \\ \sigma_{h_1-1} & \cdot & \sigma_{h_1+r-1} \\ \cdot & \cdot & \cdot \\ \sigma_{h_r-r} & \cdot & \sigma_{h_r} \end{vmatrix}, \tag{10}$$

where $h_i = n-d+i-a_i$ $(i = 0, \ldots, r)$ $(a_i < n-d+i, i \geqslant 0)$, and the right-hand side is to be evaluated as an algebraic determinant, products being interpreted as intersections.

We prove this result by induction on r. For $r = 0$,

$$\Omega_{a_0 n-d+1 \ldots n} = \sigma_{h_0},$$

by definition, hence there is nothing to prove. We therefore assume that formula (10) is valid when $r = s-1$, and prove it true for $r = s$ $(s \leqslant d)$. By (9), we have

$$(-1)^s \Omega_{a_0 \ldots a_s n-d+s+1 \ldots n} \sim \sum_0^s (-1)^j \Omega_{a_0 \ldots a_{j-1} a_{j+1} \ldots a_s n-d+s \ldots n} \cdot \sigma_{n-d+s-a_j},$$

and since, by the hypothesis of induction, (10) is true for $r = s-1$, we have

$$(-1)^s \Omega_{a_0 \ldots a_s n-d+s+1 \ldots n}$$

$$\sim \sum_{j=0}^s (-1)^j \begin{vmatrix} \sigma_{h_0} & \cdot & \sigma_{h_0+s-1} \\ \cdot & \cdot & \cdot \\ \sigma_{h_{j-1}-j+1} & \cdot & \sigma_{h_{j-1}+s-j} \\ \sigma_{h_{j+1}-j-1} & \cdot & \sigma_{h_{j+1}+s-j-2} \\ \cdot & \cdot & \cdot \\ \sigma_{h_s-s} & \cdot & \sigma_{h_s} \end{vmatrix} \cdot \sigma_{h_j+s-j} = (-1)^s \begin{vmatrix} \sigma_{h_0} & \cdot & \sigma_{h_0+s} \\ \cdot & \cdot & \cdot \\ \sigma_{h_s-s} & \cdot & \sigma_{h_s} \end{vmatrix},$$

and this establishes formula (10). We can write this formula in another way, without having to distinguish in $\Omega_{a_0 \ldots a_d}$ between the suffixes in which $a_i < n-d+i$ and those in which $a_i = n-d+i$. In fact, if we write

$$h_i = n-d+i-a_i \quad (i = 0, \ldots, d),$$

we have

$$\begin{vmatrix} \sigma_{h_0} & \cdot & \sigma_{h_0+d} \\ \sigma_{h_1-1} & \cdot & \sigma_{h_1+d-1} \\ \cdot & \cdot & \cdot \\ \sigma_{h_d-d} & \cdot & \sigma_{h_d} \end{vmatrix} = \begin{vmatrix} \sigma_{h_0} & \cdot & \sigma_{h_0+r} & \sigma_{h_0+r+1} & \cdot & \cdot & \sigma_{h_0+d} \\ \cdot & \cdot & \cdot & \cdot & \cdot & \cdot & \cdot \\ \sigma_{h_r-r} & \cdot & \sigma_{h_r} & \sigma_{h_r+1} & \cdot & \cdot & \sigma_{h_r+d-r} \\ 0 & \cdot & 0 & \sigma_0 & \cdot & \cdot & \cdot \\ 0 & \cdot & 0 & 0 & \sigma_0 & \cdot & \cdot \\ \cdot & \cdot & \cdot & \cdot & \cdot & \cdot & \cdot \\ 0 & \cdot & \cdot & \cdot & 0 & \cdot & \sigma_0 \end{vmatrix},$$

where r is the largest integer for which $a_i < n-d+i$, and since

$$\begin{vmatrix} \sigma_{h_0} & \cdot & \sigma_{h_0+d} \\ \cdot & \cdot & \cdot \\ \sigma_{h_d-d} & \cdot & \sigma_{h_d} \end{vmatrix} = \begin{vmatrix} \sigma_{h_0} & \cdot & \sigma_{h_0+r} \\ \cdot & \cdot & \cdot \\ \sigma_{h_r-r} & \cdot & \sigma_{h_r} \end{vmatrix} . (\sigma_0)^{d-r},$$

and $\sigma_0 = \Omega(d, n)$, we can omit the factor $(\sigma_0)^{d-r}$. Hence

$$\Omega_{a_0 \ldots a_d} \sim \begin{vmatrix} \sigma_{h_0} & \cdot & \sigma_{h_0+d} \\ \cdot & \cdot & \cdot \\ \sigma_{h_d-d} & \cdot & \sigma_{h_d} \end{vmatrix}. \tag{11}$$

Formula (11) or (10) enables us to express any Schubert variety in terms of intersections of varieties σ_h. This virtually solves the intersection problem. To calculate an equivalence for $\Omega_{a_0 \ldots a_d} . \Omega_{b_0 \ldots b_d}$, we express each of the varieties in terms of intersections of varieties σ_h. We then obtain an equivalence for the intersection $\Omega_{a_0 \ldots a_d} . \Omega_{b_0 \ldots b_d}$ in terms of intersections of varieties σ_h. We express this intersection in terms of Schubert varieties $\Omega_{c_0 \ldots c_d}$, where

$$\Sigma c_i = \Sigma a_i + \Sigma b_i - (n-d)(d+1) - \tfrac{1}{2}d(d+1).$$

Now any term $\sigma_{i_1} \ldots \sigma_{i_k}$ can be expressed in terms of the Schubert varieties $\Omega_{c_0 \ldots c_d}$ by repeated application of (9), since

$$\sigma_h = \Omega_{n-d-h\, n-d+1 \ldots n}.$$

The formula is rather complicated, and we do not give it. It is simpler to apply the process outlined above in particular cases when required, than to use a general formula.

It should be pointed out that it is sometimes more convenient to use the varieties $\sigma_{i_1} \dots \sigma_{i_k}$, where the integers $i_1, \dots, i_k$ take all values subject to the conditions

$$1 \leqslant i_j \leqslant n-d, \quad \Sigma i_j = (n-d)(d+1)-\rho,$$

as a base for the ρ-dimensional varieties on $\Omega(d, n)$. For, using (11), we can express any Schubert variety $\mathbf{\Omega}_{a_0 \dots a_d}$ of dimension ρ in terms of these. In terms of this base, the intersection problem is trivial. But this basis has the disadvantage that it is not minimal. For instance, on $\Omega(1, 3)$, $\sigma_1^3 \sim 2\sigma_1 . \sigma_2$.

7. Applications to enumerative geometry. The results proved so far in this chapter are not only of interest as properties of Grassmann varieties, but are also of importance because of their applications to the enumerative geometry of systems of d-spaces in S_n. Any algebraic system $\mathfrak{S}$ of d-spaces of S_n consists of the d-spaces of S_n which are represented by the points of an algebraic variety V of $\Omega(d, n)$. $\mathfrak{S}$ is irreducible if and only if V is irreducible, and the dimension of $\mathfrak{S}$ is, by definition, the dimension of V. If there are x d-spaces of $\mathfrak{S}$ which have a property ϕ, x is called the enumerative character of $\mathfrak{S}$ determined by ϕ. The properties ϕ considered are the properties of belonging to given algebraic systems. Hence the d-spaces of S_n which have an assigned property ϕ are represented by the points of $\Omega(d, n)$ which lie on some sub-variety U, whose properties are assumed to be known. The x d-spaces of $\mathfrak{S}$ with the property ϕ are just the d-spaces represented by the intersection $U_\wedge V$ on $\Omega(d, n)$. This point-set intersection may be very complicated, and to obtain a satisfactory mathematical theory we change our point of view.

Let the equations of V be

$$f_i(\dots, X_{i_0 \dots i_d}, \dots) = 0 \quad (i = 1, \dots, r).$$

Then if α_{ij} $(i, j = 0, \dots, n)$ be $(n+1)^2$ independent indeterminates over K, the equations

$$f_i(\dots, \Sigma \alpha_{i_0 j_0} \dots \alpha_{i_d j_d} X_{j_0 \dots j_d}, \dots) = 0 \quad (i = 1, \dots, r)$$

define an algebraic variety V^α 'belonging to the same system as V', and therefore define a system of d-spaces $\mathfrak{S}^\alpha$ equivalent to $\mathfrak{S}$ [XII, § 10]. The enumerative properties of $\mathfrak{S}$ which we now seek are not so much properties of $\mathfrak{S}$ itself, but of the system of systems of which $\mathfrak{S}^{\alpha}$ is a generic member. Having obtained the enumerative properties of $\mathfrak{S}^{\alpha}$ corresponding to the property ϕ, we can obtain the enumerative properties of $\mathfrak{S}$ corresponding to ϕ by proper specialisation. If $U_{\wedge} V$ is of dimension zero (or vacuous), we can attach a multiplicity to each of the intersections of U and V, that is, to each d-space of $\mathfrak{S}$ with the property ϕ, and it is the number of d-spaces of $\mathfrak{S}$ with the property ϕ, *each counted with its multiplicity*, when this is defined, which is the enumerative property sought. Thus the enumerative properties of $\mathfrak{S}$ are intersection properties on $\Omega(d,n)$. It is clear that if $\dim U + \dim V > (d+1)(n-d)$, the intersection $U_{\wedge} V^{\alpha}$ is either empty or of dimension greater than zero [XII, §6, Th. I], and that if $\dim U + \dim V < (d+1)(n-d)$ we have $U_{\wedge} V^{\alpha} = 0$. Hence we need only consider the case in which $\dim U + \dim V = (d+1)(n-d)$.

When this equality holds the intersection $U_{\wedge} V^{\alpha}$ is evidently unaltered if we omit components of U and of V whose dimensions are respectively less than $\dim U$ and $\dim V$. Hence we need only consider the case in which U and V are unmixed. Since U and V are of complementary dimension, it then follows that the intersections of U and V, when finite in number, all have positive multiplicity [XII, §§7, 8].

Hence $\mathfrak{S}$ is taken to be an unmixed system of d-spaces of dimension ρ, and the property ϕ is represented by an unmixed variety U of dimension $\sigma = (d+1)(n-d)-\rho$. To find the enumerative property of $\mathfrak{S}$ corresponding to the property ϕ we must find $x = (\boldsymbol{U}.\boldsymbol{V})$. Now ϕ is supposed to be known, that is, the properties of $\boldsymbol{U}$ are known. Therefore, in the equivalence [§5]

$$\boldsymbol{U} \sim \Sigma\lambda_{b_0\ldots b_d}\boldsymbol{\Omega}_{b_0\ldots b_d}$$

the numbers $\lambda_{b_0\ldots b_d}$ are known. Then x is given by the equation

$$x = (\boldsymbol{U}.\boldsymbol{V}) = \Sigma\lambda_{b_0\ldots b_d}(\boldsymbol{\Omega}_{b_0\ldots b_d}.\boldsymbol{V}),$$

and x is known if the numbers

$$\mu_{b_0\ldots b_d} = (\boldsymbol{\Omega}_{b_0\ldots b_d}.\boldsymbol{V})$$

are known for all sets $b_0, \ldots, b_d$ such that

$$0 \leqslant b_0 < b_1 < \ldots < b_d,$$

$$\sum_{i=0}^{d} b_i - \tfrac{1}{2}d(d+1) = \sigma.$$

Thus the enumerative characters of $\mathfrak{S}$ can be determined whenever we know the numbers $\mu_{b_0 \ldots b_d}$ of V (that is, of $\mathfrak{S}$). These numbers are called the *elementary enumerative characters* of $\mathfrak{S}$.

Systems of d-spaces arise in many problems in geometry. For instance, when a system of lines in S_n generates a variety V_d, the enumerative characters of $\mathfrak{S}$ determine important properties of V_d.

Again, if M is an irreducible variety of dimension e in S_n, we can associate with it an irreducible system of d-spaces for each value of d from 0 to e, as follows. Let ξ be a generic point of M, and let η be a generic d-space of the system of d-spaces which pass through ξ and lie in the tangent e-space to M at ξ. The properties of the system $\mathfrak{S}_d$ of d-spaces whose generic member is η are clearly properties of M. The elementary enumerative characters of the $\mathfrak{S}_d$ are called *projective characters of M*.

Consider, for instance, the case $n = 3$, $e = 2$. We have to consider systems $\mathfrak{S}_0$, $\mathfrak{S}_1$, $\mathfrak{S}_2$. The first system, $\mathfrak{S}_0$, is simply the system of points of M. Its only enumerative character is the number of points of $\mathfrak{S}_0$ on a line. The corresponding projective character of M is its *order*. $\mathfrak{S}_1$ is a system of lines in S_3 of dimension three, and the one enumerative character of $\mathfrak{S}_1$ is the number of lines in $\mathfrak{S}_1$ which pass through a generic point of S_3 and lie in a generic plane through the point. Hence the corresponding projective character of M is the number of tangents which can be drawn in a generic plane section through a generic point, or the *class* of a generic plane section. Finally, the only enumerative character of $\mathfrak{S}_2$ is the number of planes of the system through a generic line of S_3; that is, it is the number of tangent planes to M through a generic line of S_3—the *class* of M.

Returning to the general problem of the determination of the enumerative characters of an irreducible system $\mathfrak{S}$ of d-spaces, we have seen that these are determined if the elementary enumerative characters of $\mathfrak{S}$ are known, that is if the intersection numbers $(\boldsymbol{V}.\boldsymbol{\Omega}_{b_0 \ldots b_d})$ are known, where V is the image of $\mathfrak{S}$ on $\Omega(d, n)$. A method for determining these intersection numbers has been indicated in § 5.

There is one important type of problem in enumerative geometry in which the elementary enumerative characters can be computed at once. If $\mathfrak{S}$, $\mathfrak{S}'$ are two irreducible systems of d-spaces in general position (that is, the corresponding varieties V, V' on $\Omega(d, n)$ intersect simply) whose enumerative characters are known, we can find the enumerative characters of $V_{\wedge} V'$, the components of intersection being counted simply. For if

$$V \sim \Sigma \nu_{a_0 \ldots a_d} \Omega_{a_0 \ldots a_d}$$

and

$$V' \sim \Sigma \nu_{b_0 \ldots b_d} \Omega_{b_0 \ldots b_d}$$

where the $\nu_{a_0 \ldots a_d}$, $\nu_{b_0 \ldots b_d}$ are known,

$$V . V' \sim \Sigma \nu_{a_0 \ldots a_d} \nu_{b_0 \ldots b_d} \Omega_{a_0 \ldots a_d} . \Omega_{b_0 \ldots b_d},$$

and by the intersection formulae of § 6 we can determine the equivalences

$$\Omega_{a_0 \ldots a_d} . \Omega_{b_0 \ldots b_d} \sim \Sigma c_{a,b} \Omega_{c_0 \ldots c_d}.$$

We then have the equivalence

$$V . V' \sim \Sigma \nu_{a_0 \ldots a_d} \nu_{b_0 \ldots b_d} c_{a,b} \Omega_{c_0 \ldots c_d},$$

and the coefficients $\nu_{a_0 \ldots a_d} \nu_{b_0 \ldots b_d} c_{a,b}$ in this equivalence are the enumerative characters of $V . V'$.

We now give some examples of the application of the methods developed in this chapter, and compare the algebraic processes involved, whose correctness we have established, with the corresponding traditional geometrical methods of the Schubert calculus.

We determine the number of lines which meet four generic lines a, b, c, d in S_3. The lines meeting a given line determine an $\Omega_{1,3} = \sigma_1$. By Pieri's formula [§ 6, (4), p. 354],

$$\Omega_{1,3} . \sigma_1 \sim \Sigma \Omega_{b_0, b_1},$$

where $0 \leqslant b_0 \leqslant 1$, $1 < b_1 \leqslant 3$, $b_0 + b_1 = 1+3-1 = 3$, so that $b_0 = 0$, $b_1 = 3$ and $b_0 = 1$, $b_1 = 2$ are the solutions. Hence

$$\Omega_{1,3} . \sigma_1 \sim \Omega_{0,3} + \Omega_{1,2}.$$

The system of lines meeting two given lines a, b is therefore equivalent to the system of lines passing through a point together with the system of lines lying in a plane. By making the two given lines a, b meet, and thus degenerating the system of lines we are considering, this result would have been indicated, but we should have been uncertain of the correct multiplicities to attach to the component systems. Again

$$\Omega_{1,3} . \Omega_{1,3} . \sigma_1 \sim \Omega_{0,3} . \sigma_1 + \Omega_{1,2} . \sigma_1 \sim \Omega_{0,2} + \Omega_{0,2},$$

and

$$\Omega_{1,3} . \Omega_{1,3} . \Omega_{1,3} . \sigma_1 \sim 2\Omega_{0,2} . \sigma_1 \sim 2\Omega_{0,1}.$$

Hence there are two solutions to the problem, since $\Omega_{0,1}$ consists of one line. By the traditional Schubert methods this result would have been indicated, but not proved, by taking the line through the intersection of a, b which meets c, d, and the line joining the points in which c and d respectively meet the plane determined by a, b.

As another example let us determine the number of lines which meet six planes generally chosen in S_4. The system of lines meeting any plane in S_4 is a Schubert system $\Omega_{2,4} = \sigma_1$. We have therefore to determine the intersection number of six such varieties on $\Omega(1, 4)$. Using Pieri's formula,

$$\Omega_{2,4} \cdot \sigma_1 \sim \Omega_{1,4} + \Omega_{2,3}, \tag{1}$$

$$\Omega_{2,4} \cdot \Omega_{2,4} \cdot \sigma_1 \sim \Omega_{1,4} \cdot \sigma_1 + \Omega_{2,3} \cdot \sigma_1 \sim (\Omega_{0,4} + \Omega_{1,3}) + \Omega_{1,3}, \tag{2}$$

$$\Omega_{2,4} \cdot \Omega_{2,4} \cdot \Omega_{2,4} \cdot \sigma_1 \sim \Omega_{0,4} \cdot \sigma_1 + 2\Omega_{1,3} \cdot \sigma_1 \sim \Omega_{0,3} + 2(\Omega_{0,3} + \Omega_{1,2}), \tag{3}$$

$$\Omega_{2,4} \cdot \Omega_{2,4} \cdot \Omega_{2,4} \cdot \Omega_{2,4} \cdot \sigma_1 \sim 3\Omega_{0,3} \cdot \sigma_1 + 2\Omega_{1,2} \cdot \sigma_1 \sim 3\Omega_{0,2} + 2\Omega_{0,2}, \tag{4}$$

and finally

$$\Omega_{2,4} \cdot \Omega_{2,4} \cdot \Omega_{2,4} \cdot \Omega_{2,4} \cdot \Omega_{2,4} \cdot \sigma_1 \sim 5\Omega_{0,2} \cdot \sigma_1 \sim 5\Omega_{0,1}. \tag{5}$$

Hence the number of lines is 5.

Let us consider some of the steps of the above process geometrically.

(i) By replacing $\Omega_{2,4} \cdot \sigma_1$ by $\Omega_{1,4} + \Omega_{2,3}$ we say that the system of lines which meet two generic planes in S_4 is equivalent to the system of lines meeting a line plus the system lying in an S_3. This is precisely what arises when the two given planes are specialised to meet in a line l_1, and hence lie in an S_3. The lines meeting the two planes either meet l_1, or meet both planes and lie in the S_3 which they determine.

(ii) When we replace $\Omega_{1,4} \cdot \sigma_1$ by $\Omega_{0,4} + \Omega_{1,3}$ we are replacing the system of lines meeting l_1 and a plane by the system of lines through a point in S_4 and the system of lines in an S_3 which meet a line. This is the situation which arises when l_1 meets the plane in a point O, and hence determines, with the plane, an S'_3. The lines required either pass through O, or meet l_1 and lie in S'_3.

(iii) The system $\Omega_{1,3} \cdot \sigma_1$ represents the lines in an S_3 which meet a line l_1 in S_3 and a plane π of S_4; this system is replaced by $\Omega_{0,3}$, the system of lines in an S_3 which pass through a point, and $\Omega_{1,2}$, the system of lines in a plane. If π is allowed to meet l_1 the given degeneration takes place. In fact, the lines either pass through the point $l_1 \wedge \pi$ and lie in S_3, or meet l_1, and the intersection of π and S_3, which is now a line meeting l_1.

Pieri's formula can also be used to find the order of the Schubert variety $\Omega_{a_0 a_1 \ldots a_d}$. If the Grassmann variety $\Omega(d, n)$ lies in S_N, and π denotes a generic prime section of $\Omega(d, n)$, we assert that

$$\pi \equiv \sigma_1.$$

This follows from the fact that only one linear condition has to be satisfied for an S_d to meet a generic S_{n-d-1}, so that σ_1 is cut on $\Omega(d, n)$ by a prime section, and all prime sections of $\Omega(d, n)$ are algebraically equivalent. Hence

$$\begin{aligned}\Omega_{a_0 a_1 \ldots a_d} \cdot \pi &\sim \Omega_{a_0 a_1 \ldots a_d} \cdot \sigma_1 \\ &\sim \Sigma \Omega_{b_0 b_1 \ldots b_d},\end{aligned}$$

by Pieri's formula, where the summation is over values of $b_0, b_1, \ldots, b_d$ satisfying the inequalities

$$0 \leqslant b_0 \leqslant a_0, \quad a_0 < b_1 \leqslant a_1, \quad \ldots, \quad a_{d-1} < b_d \leqslant a_d,$$

$$\sum_{i=0}^{d} b_i = \sum_{i=0}^{d} a_i - 1.$$

Hence

$$\Omega_{a_0 \ldots a_d} \cdot \pi \sim \Omega_{a_0 - 1\, a_1 \ldots a_d} + \Omega_{a_0\, a_1 - 1 \ldots a_d} + \ldots + \Omega_{a_0 a_1 \ldots a_d - 1},$$

where we exclude any variety $\Omega_{c_0 c_1 \ldots c_d}$ appearing in this equivalence if the inequalities $0 \leqslant c_0 < c_1 < \ldots < c_d$ are not satisfied.

The order of $\Omega_{a_0 \ldots a_d}$ is equal to the number of points in which it is met by ρ generic primes of S_N, where $\rho = \sum_{i=0}^{d} a_i - \frac{1}{2}d(d+1)$ is the dimension of $\Omega_{a_0 \ldots a_d}$. Hence the order of $\Omega_{a_0 \ldots a_d}$ is equal to the order of the prime section $\Omega_{a_0 \ldots a_d} \wedge \pi$, and this in its turn is equal to the sum of the orders of the Schubert varieties which appear in the above equivalence. Hence if we denote the order of $\Omega_{a_0 \ldots a_d}$ by $g(a_0, a_1, \ldots, a_d)$, we have the recurrence relation

$$g(a_0, a_1, \ldots, a_d) = g(a_0 - 1, a_1, \ldots, a_d) + g(a_0, a_1 - 1, \ldots, a_d) + \ldots + g(a_0, \ldots, a_d - 1), \quad (6)$$

where $g(c_0, c_1, \ldots, c_d) = 0$ if the inequalities $c_0 < c_1 < \ldots < c_d$ are not satisfied.

We know that $g(0, 1, \ldots, d) = 1$, since $\Omega_{01 \ldots d}$ consists of a single S_d and is represented on $\Omega(d, n)$ by a single point. We now prove that

$$g(a_0, a_1, \ldots, a_d) = \frac{\rho!}{a_0! \ldots a_d!} \prod_{\lambda > \mu} (a_\lambda - a_\mu), \quad (7)$$

where, as above, $\rho = \sum_{i=0}^{d} a_i - \frac{1}{2}d(d+1)$ is the dimension of $\Omega_{a_0 \ldots a_d}$. This formula is valid when $a_0 = 0, a_1 = 1, \ldots, a_d = d$, with the usual convention that $0! = 1$. We also note that we obtain the value zero if any of the inequalities in $a_0 < a_1 < \ldots < a_d$ is replaced by an equality. We therefore suppose that it is true for all Schubert varieties $\Omega_{b_0 b_1 \ldots b_d}$ whenever $\sum_{i=0}^{d} b_i < \sum_{i=0}^{d} a_i$, and apply it to equation (6). We then find that

$$g(a_0, a_1, \ldots, a_d) = \frac{(\rho-1)!}{a_0! \ldots a_d!}\left[a_0 \prod_{\lambda>0}(a_\lambda - a_0 + 1) \prod_{\lambda>\mu>0}(a_\lambda - a_\mu) + \ldots\right]$$

$$= \frac{(\rho-1)!}{a_0! \ldots a_d!}\left[a_0 \begin{vmatrix} a_d^d & a_d^{d-1} & . & 1 \\ a_{d-1}^d & a_{d-1}^{d-1} & . & 1 \\ . & . & . & . \\ a_1^d & a_1^{d-1} & . & 1 \\ (a_0-1)^d & (a_0-1)^{d-1} & . & 1 \end{vmatrix} + a_1 \begin{vmatrix} a_d^d & a_d^{d-1} & . & 1 \\ a_{d-1}^d & a_{d-1}^{d-1} & . & 1 \\ . & . & . & . \\ (a_1-1)^d & (a_1-1)^{d-1} & . & 1 \\ a_0^d & a_0^{d-1} & . & 1 \end{vmatrix} + \ldots\right].$$

We consider the expression inside the brackets, and denote it by $f(a_0, a_1, \ldots, a_d)$. It is evident that

$$f(a_0, \ldots, a_{i-1}, a_j, a_{i+1}, \ldots, a_{j-1}, a_i, a_{j+1}, \ldots, a_d)$$
$$= -f(a_0, \ldots, a_{i-1}, a_i, a_{i+1}, \ldots, a_{j-1}, a_j, a_{j+1}, \ldots, a_d),$$

that is, interchanging a_i and a_j changes the sign of $f(a_0, a_1, \ldots, a_d)$. Hence $f(a_0, a_1, \ldots, a_d)$ contains $D = \prod_{\lambda>\mu}(a_\lambda - a_\mu)$ as a factor, and f/D is a symmetric polynomial in $a_0, a_1, \ldots, a_d$. By comparing the degrees of $f(a_0, a_1, \ldots, a_d)$ and D, it is clear that f/D is of the first degree in $a_0, a_1, \ldots, a_d$. Hence

$$f(a_0, a_1, \ldots, a_d)/D = \alpha \sum_{i=0}^{d} a_i + \beta, \tag{8}$$

where α and β are rational numbers. The coefficient of a_0^{d+1} in $f(a_0, a_1, \ldots, a_d)$ is

$$(-1)^d \begin{vmatrix} a_d^{d-1} & . & 1 \\ . & . & . \\ a_1^{d-1} & . & 1 \end{vmatrix},$$

and this is equal to the coefficient of a_0^d in D. Therefore $\alpha = 1$. Now put $a_i = i$ $(i = 0, \ldots, d)$ in (8). Then $f(0, 1, \ldots, d) = 0$, since in the determinantal expression for $f(a_0, a_1, \ldots, a_d)$ the first determinant is zero after multiplication by $a_0 = 0$, and in each of the other determinants two rows become identical after the substitutions have been made. It follows from (8) that

$$\alpha[\tfrac{1}{2}d(d+1)] + \beta = 0,$$

and so

$$\beta = -\tfrac{1}{2}d(d+1).$$

Hence

$$\alpha \sum_{i=0}^{d} a_i + \beta = \rho,$$

and

$$g(a_0, a_1, \ldots, a_d) = \frac{\rho!}{a_0! \ldots a_d!} \prod_{\lambda > \mu} (a_\lambda - a_\mu),$$

which proves (7).

In particular, since

$$\Omega_{n-d\, n-d+1, \ldots n} = \Omega(d, n),$$

and

$$\rho = (d+1)(n-d)$$

in this case, the order of $\Omega(d, n)$ is

$$\frac{1!\,2! \ldots d!\,\rho!}{(n-d)!\,(n-d+1)! \ldots n!} = \frac{1!\,2! \ldots d!\,[(d+1)(n-d)]!}{(n-d)!\,(n-d+1)! \ldots n!}. \tag{9}$$

Finally, we prove that *two quadrics in S_4 have, in general, sixteen lines in common*. We saw, in Chapter XIII, § 4, Th. III, that an irreducible quadric Q in S_4 contains an irreducible system of lines, of dimension three. If this system of lines be represented on $\Omega(1, 4)$ by the irreducible variety V, then

$$V \sim \Sigma \lambda_{a_0, a_1} \mathbf{\Omega}_{a_0, a_1},$$

where

$$a_0 + a_1 - 1 = 3,$$

so that

$$V \sim \lambda \mathbf{\Omega}_{0,4} + \mu \mathbf{\Omega}_{1,3}.$$

Now

$$(V . \mathbf{\Omega}_{0,4}) = \lambda(\mathbf{\Omega}_{0,4} . \mathbf{\Omega}_{0,4}) + \mu(\mathbf{\Omega}_{0,4} . \mathbf{\Omega}_{1,3}) = \lambda;$$

but no line of the quadric passes through a generic point of S_4. Therefore $\lambda = 0$. On the other hand,

$$(V . \mathbf{\Omega}_{1,3}) = \lambda(\mathbf{\Omega}_{0,4} . \mathbf{\Omega}_{1,3}) + \mu(\mathbf{\Omega}_{1,3} . \mathbf{\Omega}_{1,3}) = \mu.$$

A generic three-dimensional space intersects the quadric in a quadric lying in this 3-space, and four lines of this quadric in 3-space meet a generic line of this space, since the line cuts the quadric at two distinct points, and therefore meets a generator of each system at each point. Hence $\mu = 4$, and

$$V \sim 4\Omega_{1,3}.$$

Let Q' be any other quadric in S_4, and let V' be the variety on $\Omega(1,4)$ which represents the lines on Q'. We consider a 5×5 matrix α whose elements are independent indeterminates. The collineation $x \rightarrow y$, where

$$x = \alpha y,$$

transforms Q into a quadric Q^α, and induces a transformation of V into a variety V^α which represents the lines on Q^α; V^α is the variety obtained from V by projective transformation [§ 5], and V^α and V' intersect simply, if at all. Since

$$(V^\alpha . V') = (V . V') = (4\Omega_{1,3} . 4\Omega_{1,3}) = 16(\Omega_{1,3} . \Omega_{1,3}) = 16,$$

V^α and V' have sixteen points in common. When we specialise α to the unit matrix, we conclude

(i) that V and V' have at least one point in common, that is, Q and Q' have at least one line in common;

(ii) if the number of lines common to Q and Q' is finite, unique positive multiplicities can be attached to each, and the sum of these multiplicities is sixteen.

8. Varieties of dimension $(n-d)(d+1)-1$ on $\Omega(d,n)$. So far, the properties of $\Omega(d,n)$ which we have considered in this chapter are those which illustrate certain general properties of algebraic varieties which have been discussed in earlier chapters. We conclude this chapter with the consideration of one or two further properties of $\Omega(d,n)$ which do not fall into this category (though that considered in the next section is, in fact, related to an important general question not discussed in this book).

We have seen [X, § 5, Th. III] that if V is an unmixed variety of $n-1$ dimensions in S_n, V is defined by a single equation

$$f(x) \equiv f(x_0, \ldots, x_n) = 0.$$

A similar proof, which we do not give, enables us to prove that if V is an unmixed variety of dimension $\Sigma n_i - 1$ in $S_{n_1 \ldots n_k}$, V is given by a single equation

$$f(x^1, \ldots, x^k) = 0,$$

irreducible if V is irreducible.

On the other hand, there are many varieties of dimension d in S_n which contain unmixed varieties of dimension $d-1$ which are not the intersection of the variety with a primal. For example, we consider a non-singular quadric surface Q_2 in S_3, and a line l lying on it. Suppose there exists a primal V in S_3 such that

$$V_\wedge Q_2 = l.$$

Any generator l' of the same system on Q_2 as l does not meet l. Hence

$$0 = l_\wedge l' = V_\wedge Q_{2\wedge} l' = V_\wedge l'.$$

Thus l' does not meet V. But any line has at least one point in common with a primal; so we have a contradiction. It follows that there is no surface in S_3 whose total intersection with Q_2 is a line.

In this section we prove

THEOREM I. *Every unmixed variety V of dimension $\rho(d,n)-1$ on $\Omega(d,n)$ is the complete intersection of $\Omega(d,n)$ with a primal.*

Here $\rho(d,n)$ denotes the dimension $(n-d)(d+1)$ of $\Omega(d,n)$. We need only consider the case in which V is irreducible.

Let $$F_i(\ldots, X_{i_0\ldots i_d}, \ldots) = 0 \quad (i = 1, 2, \ldots) \tag{1}$$

be the equations of V, and let η be a generic point of it. We consider a correspondence Γ, between the $(d+1)$-way space $S_{n\ldots n}$ and the space S_N containing $\Omega(d,n)$, which is defined by the equations (1) and

$$\sum_{r=0}^{d+1} (-1)^r x_{i_r}^{(j)} X_{i_0\ldots i_{r-1} i_{r+1}\ldots i_{d+1}} = 0 \quad (\text{all } i_0, \ldots, i_{d+1};\, j = 0, \ldots, d), \tag{2}$$

where $(x^{(0)}, \ldots, x^{(d)})$ is the coordinate system in $S_{n\ldots n}$. For any specialisation of η, the equations deduced from (2) are of rank $n-d$ in $(x_0^{(j)}, \ldots, x_n^{(j)})$; and by an immediate extension of the argument of § 5 of Chapter XI, we conclude that Γ is irreducible, and of dimension $\rho(d,n)-1+d(d+1)$. From the equations of Γ we see

(i) that if $\bar{X}$ is any point of V, the corresponding points of $S_{n\ldots n}$ are all the points $(\bar{x}^0, \ldots, \bar{x}^d)$ such that $\bar{x}^i$ is a point of the d-space in S_n corresponding to $\bar{X}$ $(i = 0, \ldots, d)$;

(ii) that if $(\bar{x}^0, \ldots, \bar{x}^d)$ is any point of the object-variety M of Γ in $S_{n\ldots n}$, any point of S_N corresponding to it must represent a d-space through $\bar{x}^i$ $(i = 0, \ldots, d)$, and in particular if $\bar{x}^0, \ldots, \bar{x}^d$ are independent points of S_n the point of S_N corresponding to it is unique, since only one d-space contains $\bar{x}^0, \ldots, \bar{x}^d$.

Let $\xi^0, \ldots, \xi^d$ be $d+1$ independent generic solutions of the equations

$$\sum_{r=0}^{d+1} (-1)^r x_{i_r} \eta_{i_0 \ldots i_{r-1} i_{r+1} \ldots i_{d+1}} = 0,$$

and let $\xi = (\xi^0, \ldots, \xi^d)$. Then (ξ, η) is a generic point of Γ. ξ is a generic point of the object-variety M. Since $\xi^0, \ldots, \xi^d$ are independent points of S_n, it follows from (ii) that to ξ there corresponds a unique point in S_N. If M is of dimension σ, we therefore have, by the principle of counting constants,

$$\sigma + 0 = \dim \Gamma = \rho(d, n) - 1 + d(d+1),$$

that is

$$\sigma = n(d+1) - 1.$$

Since M is irreducible and of dimension $n(d+1)-1$ in $S_{n \ldots n}$, it follows from the result quoted above that it is given by a unique irreducible equation

$$f(x^0, \ldots, x^d) = 0.$$

If $\bar{x} = (\bar{x}^0, \ldots, \bar{x}^d)$ is any point of M, there exists a d-space containing $\bar{x}^0, \ldots, \bar{x}^d$ which corresponds to $\bar{x}$ in Γ. Since $\bar{x}^{i_0}, \ldots, \bar{x}^{i_d}$, where $(i_0, \ldots, i_d)$ is a derangement of $(0, \ldots, d)$, lie in a d-space represented by a point of V, $(\bar{x}^{i_0}, \ldots, \bar{x}^{i_d})$ lies on M. In particular, $(\xi^{i_0}, \ldots, \xi^{i_d})$ lies on M. If $\begin{pmatrix} 0 & \ldots & d \\ j_0 & \ldots & j_d \end{pmatrix}$ is the substitution inverse to $\begin{pmatrix} 0 & \ldots & d \\ i_0 & \ldots & i_d \end{pmatrix}$, we conclude that

$$f(x^{j_0}, \ldots, x^{j_d})$$

vanishes on M, and hence contains $f(x^0, \ldots, x^d)$ as a factor. It is trivial to see that this implies that $f(x^0, \ldots, x^d)$ is of the same degree g in $(x^i_0, \ldots, x^i_n)$ for $i = 0, \ldots, d$.

A similar argument shows that if λ is any new indeterminate, $(\xi^0, \ldots, \xi^{i-1}, \xi^i + \lambda\xi^j, \xi^{i+1}, \ldots, \xi^d)$ lies on M. Hence

$$f(\xi^0, \ldots, \xi^i + \lambda\xi^j, \ldots, \xi^d) = 0,$$

that is,

$$f(\xi^0, \ldots, \xi^d) + \lambda \sum_k \xi^j_k \frac{\partial f}{\partial \xi^i_k}(\xi^0, \ldots, \xi^d) + \ldots = 0,$$

and hence

$$\sum_k \xi^j_k \frac{\partial f}{\partial \xi^i_k}(\xi^0, \ldots, \xi^d) = 0,$$

that is,

$$\sum_k x^j_k \frac{\partial f}{\partial x^i_k}(x^0, \ldots, x^d)$$

vanishes on M. Hence it contains $f(x^0, \ldots, x^d)$ as a factor, or else is zero. If $j \neq i$, the degree in x^i is $g-1$, hence f cannot be a factor.

Thus $$\sum_k x_k^j \frac{\partial f}{\partial x_k^i}(x^0, \ldots, x^d) \equiv 0 \quad (j \neq i).$$

The argument of X, § 7, Th. IV, then enables us to deduce that

$$f(x^0, \ldots, x^d) \equiv F(\ldots, |\, x_{i_0}^0 \ldots x_{i_d}^d \,|, \ldots),$$

where F is of order g.

Now consider any polynomial $\Phi(\ldots, X_{i_0 \ldots i_d}, \ldots)$ which vanishes on V. Then

$$\Phi(\ldots, \eta_{i_0 \ldots i_d}, \ldots) = 0.$$

But $(\ldots, \eta_{i_0 \ldots i_d}, \ldots)$ is the unique d-space containing $\xi^0, \ldots, \xi^d$. Hence we may take

$$\eta_{i_0 \ldots i_d} = \begin{vmatrix} \xi_{i_0}^0 & . & \xi_{i_d}^0 \\ . & . & . \\ \xi_{i_0}^d & . & \xi_{i_d}^d \end{vmatrix},$$

and therefore $$\Phi(\ldots, |\, \xi_{i_0}^0 \ldots \xi_{i_d}^d \,|, \ldots) = 0.$$

Therefore $\Phi(\ldots, |\, x_{i_0}^0 \ldots x_{i_d}^d \,|, \ldots)$ vanishes on M (or else vanishes identically) and we can therefore write

$$\Phi(\ldots, |\, x_{i_0}^0 \ldots x_{i_d}^d \,|, \ldots) \equiv \psi(x^0, \ldots, x^d)\, f(x^0, \ldots, x^d).$$

Now $$\sum_k x_k^j \frac{\partial \Phi}{\partial x_k^i} \equiv 0 \ (j \neq i),$$

and therefore $$\sum_k x_k^j \frac{\partial \psi}{\partial x_k^i} f + \psi \sum_k x_k^j \frac{\partial f}{\partial x_k^i} \equiv 0.$$

But since $f \not\equiv 0$, $$\sum_k x_k^j \frac{\partial f}{\partial x_k^i} \equiv 0,$$

and so $$\sum_k x_k^j \frac{\partial \psi}{\partial x_k^i} \equiv 0.$$

Therefore we may write

$$\psi(x^0, \ldots, x^d) \equiv \Psi(\ldots, |\, x_{i_0}^0 \ldots x_{i_d}^d \,|, \ldots).$$

Now $\Phi(\ldots, X_{i_0 \ldots i_d}, \ldots) - \Psi(\ldots, X_{i_0 \ldots i_d}, \ldots)\, F(\ldots, X_{i_0 \ldots i_d}, \ldots)$

is a form in $X_{i_0 \ldots i_d}$ which vanishes when we put

$$X_{i_0 \ldots i_d} = \begin{vmatrix} x_{i_0}^0 & . & x_{i_d}^0 \\ . & . & . \\ x_{i_0}^d & . & x_{i_d}^d \end{vmatrix},$$

where the x_j^i are indeterminates. Hence it must vanish on $\Omega(d,n)$, and is therefore expressible in terms of the basis for the equations of $\Omega(d,n)$ [§ 1]. Since, conversely, $F(\ldots, X_{i_0\ldots i_d}, \ldots) = 0$ is satisfied by the generic point $(\ldots, |\xi^0_{i_0} \ldots, \xi^d_{i_d}|, \ldots)$ of V, it follows that V is determined by the equations of $\Omega(d,n)$ and the single equation

$$F(\ldots, X_{i_0\ldots i_d}, \ldots) = 0, \tag{3}$$

that is, V is the intersection of $\Omega(d,n)$ and the primal (3).

9. Postulation formula for $\Omega_{a_0 a_1\ldots a_d}$. Let V be any algebraic variety in S_n. If $f(x_0, \ldots, x_n)$ and $g(x_0, \ldots, x_n)$ are two homogeneous polynomials over K of degree m which vanish on V, then $af+bg$, where a and b are elements of K, is either the zero form or is a form of order m vanishing on V. Hence the forms of order m which vanish on V form a linear set over K. Let

$$f_i(x_0, \ldots, x_n) \quad (i = 0, \ldots, r_m)$$

be a K-basis for the forms of order m which vanish on V; we further suppose that the basis is independent over K. The forms of order m in $x_0, \ldots, x_n$ can be represented by the points of S_{N_m}, where

$$N_m + 1 = \binom{n+m}{n};$$

each form being represented by the point whose coordinates are its coefficients. The forms which vanish on V form a linear subspace of dimension r_m. The number $\rho_m = N_m - r_m$, which is the number of linearly independent linear conditions which the coefficients of a form f are subjected to in order that f should vanish on V is an important character of the variety V in S_n; it is called the *postulation* of V for primals of order m. From what has been said, it has a definite value for each m, but it is the formula expressing it as a function of m which has to be obtained. The problem is a difficult one, and in general all that can be said is that for V there exists a polynomial $k(x)$ with rational coefficients, of degree [dim V], with the property that $\rho_m = k(m)$ *provided that m is sufficiently large.*

In the case in which V is $\Omega(d,n)$, or a Schubert variety $\Omega_{a_0\ldots a_d}$, it can be shown that the postulation of V for primals of order m is expressible as a polynomial in m valid for *all positive integral values of m.* As this formula is of frequent application in the theory of systems of d-spaces in S_n, we conclude this chapter by obtaining the formula for the postulation of $\Omega_{a_0\ldots a_d}$. This of course includes the case $\Omega_{n-d\ldots n} = \Omega(d, n)$.

Let $\Omega_{a_0 a_1\ldots a_d}$ be the Schubert variety on $\Omega(d, n)$ defined by the spaces $A_0 \subset A_1 \subset \ldots \subset A_d$, where A_r is of dimension a_r. We choose the coordinate system so that A_r has the equations

$$x_0 = \ldots = x_{\alpha_r - 1} = 0 \quad (\alpha_r = n - a_r). \tag{1}$$

Let ξ_i^{d-r} ($i \geqslant \alpha_r$; $r = 0, \ldots, d$) be $\Sigma(n - \alpha_r + 1)$ independent indeterminates over K. Then the point

$$(0, \ldots, 0, \xi_{\alpha_r}^{d-r}, \ldots, \xi_n^{d-r})$$

is a generic point of A_r. If $p_{i_0\ldots i_d}$ denotes the determinant formed from the $(i_0 + 1)$th, ..., $(i_d + 1)$th columns of the matrix

$$\begin{pmatrix} 0 & . & 0 & . & \xi_{\alpha_d}^0 & . & . & . & . & . & . & \xi_n^0 \\ 0 & . & 0 & . & 0 & . & \xi_{\alpha_{d-1}}^1 & . & . & . & . & \xi_n^1 \\ . & . & . & . & . & . & . & . & . & . & . & . \\ 0 & . & 0 & . & . & . & . & 0 & . & \xi_{\alpha_0}^d & . & \xi_n^d \end{pmatrix}, \tag{2}$$

then $p = (\ldots, p_{i_0\ldots i_d}, \ldots)$ is a generic point of $\Omega_{a_0 a_1\ldots a_d}$. If $p_{j_0\ldots j_d} = 0$, $\Omega_{a_0 a_1\ldots a_d}$ satisfies the linear equation

$$X_{j_0\ldots j_d} = 0$$

in the related coordinate system in S_N. We find which of the $p_{i_0\ldots i_d}$ are not identically zero. We may assume that $i_0 < i_1 < \ldots < i_d$.

The determinant $p_{i_0\ldots i_d}$ is zero if, for any value of r in the range $0, \ldots, d$, we have $i_r < \alpha_{d-r}$. For if this is the case for some value of r, the element in the $(i+1)$th row and $(j+1)$th column of the determinant is zero whenever $i \geqslant r$, $j \leqslant t$, and the result follows by expanding the determinant in terms of the first $r+1$ columns. On the other hand, if $i_r \geqslant \alpha_{d-r}$ ($r = 0, \ldots, d$), we consider the specialisation of the determinant obtained by specialising

$$\xi_i^r \to 0 \quad (i \neq i_r),$$

$$\xi_{i_r}^r \to 1 \quad (r = 0, \ldots, d).$$

The specialised determinant is equal to 1, and therefore $p_{i_0\ldots i_d} \neq 0$. Thus $p_{i_0\ldots i_d}$ is not identically zero if and only if $i_r \geqslant \alpha_{d-r}$ $(r = 0, \ldots, d)$.

To calculate the number of non-zero determinants $p_{i_0\ldots i_d}$ we have merely to find the number of sets of integers $i_0, \ldots, i_d$ which satisfy the inequalities

$$0 \leqslant i_0 < i_1 < \ldots < i_d \leqslant n,$$

$$i_r \geqslant \alpha_{d-r}.$$

This number is clearly

$$\sum_{i_d=\alpha_0}^{n} \sum_{i_{d-1}=\alpha_1}^{i_d-1} \ldots \sum_{i_1=\alpha_{d-1}}^{i_2-1} \sum_{i_0=\alpha_d}^{i_1-1} 1.$$

We find this sum by an induction argument, assuming as our hypothesis of induction the formula

$$(-1)^k \sum_{i_{k-1}=\alpha_{d-k+1}}^{i_k-1} \ldots \sum_{i_0=\alpha_d}^{i_1-1} 1$$

$$= \begin{vmatrix} \binom{\alpha_d}{1} & 1 & 0 & . & 0 \\ \binom{\alpha_{d-1}}{2} & \binom{\alpha_{d-1}}{1} & 1 & . & 0 \\ . & . & . & . & . \\ \binom{\alpha_{d-k+2}}{k-1} & \binom{\alpha_{d-k+2}}{k-2} & . & . & 1 \\ \binom{\alpha_{d-k+1}}{k} & \binom{\alpha_{d-k+1}}{k-1} & . & . & \binom{\alpha_{d-k+1}}{1} \end{vmatrix}$$

$$- \begin{vmatrix} \binom{\alpha_d}{1} & 1 & 0 & . & 0 \\ \binom{\alpha_{d-1}}{2} & \binom{\alpha_{d-1}}{1} & 1 & . & 0 \\ . & . & . & . & . \\ \binom{\alpha_{d-k+2}}{k-1} & \binom{\alpha_{d-k+2}}{k-2} & . & . & 1 \\ \binom{i_k}{k} & \binom{i_k}{k-1} & . & . & \binom{i_k}{1} \end{vmatrix}. \qquad (3)$$

This formula is clearly true when $k = 1$. We now have

$$(-1)^{k+1} \sum_{i_k=\alpha_{d-k}}^{i_{k+1}-1} \sum_{i_{k-1}=\alpha_{d-k+1}}^{i_k-1} \cdots \sum_{i_0=\alpha_d}^{i_1-1} 1$$

$$= - \sum_{i_k=\alpha_{d-k}}^{i_{k+1}-1} \begin{vmatrix} \binom{\alpha_d}{1} & 1 & 0 & . & 0 \\ \binom{\alpha_{d-1}}{2} & \binom{\alpha_{d-1}}{1} & 1 & . & 0 \\ . & . & . & . & . \\ \binom{\alpha_{d-k+2}}{k-1} & \binom{\alpha_{d-k+2}}{k-2} & . & . & 1 \\ \binom{\alpha_{d-k+1}}{k} & \binom{\alpha_{d-k+1}}{k-1} & . & . & \binom{\alpha_{d-k+1}}{1} \end{vmatrix}$$

$$+ \sum_{i_k=\alpha_{d-k}}^{i_{k+1}-1} \begin{vmatrix} \binom{\alpha_d}{1} & 1 & 0 & . & 0 \\ . & . & . & . & . \\ \binom{i_k}{k} & \binom{i_k}{k-1} & . & . & \binom{i_k}{1} \end{vmatrix}.$$

The first summation comes to

$$-(i_{k+1} - \alpha_{d-k}) \begin{vmatrix} \binom{\alpha_d}{1} & 1 & 0 & . & 0 \\ \binom{\alpha_{d-1}}{2} & \binom{\alpha_{d-1}}{1} & 1 & . & 0 \\ . & . & . & . & . \\ \binom{\alpha_{d-k+1}}{k} & . & . & . & \binom{\alpha_{d-k+1}}{1} \end{vmatrix}.$$

In the second summation we obtain the determinant

$$\begin{vmatrix} \binom{\alpha_d}{1} & 1 & 0 & . & 0 \\ . & . & . & . & . \\ \binom{\alpha_{d-k+2}}{k-1} & \binom{\alpha_{d-k+2}}{k-2} & . & . & 1 \\ \Sigma\binom{i_k}{k} & \Sigma\binom{i_k}{k-1} & . & . & \Sigma\binom{i_k}{1} \end{vmatrix}.$$

Using the formula‡

$$\sum_{\lambda=a}^{b}\binom{\lambda+c}{d}=\binom{b+c+1}{d+1}-\binom{a+c}{d+1}, \tag{*}$$

the last row of this determinant becomes

$$\binom{i_{k+1}}{k+1}-\binom{\alpha_{d-k}}{k+1},\ \binom{i_{k+1}}{k}-\binom{\alpha_{d-k}}{k},\ \ldots,\ \binom{i_{k+1}}{2}-\binom{\alpha_{d-k}}{2}.$$

This determinant is therefore equal to

$$-\begin{vmatrix} \binom{\alpha_d}{1} & 1 & 0 & . & . \\ . & . & . & . & . \\ \binom{\alpha_{d-k+2}}{k-1} & . & . & . & 1 \\ \binom{\alpha_{d-k}}{k+1} & . & . & . & \binom{\alpha_{d-k}}{2} \end{vmatrix} + \begin{vmatrix} \binom{\alpha_d}{1} & 1 & 0 & . & 0 \\ . & . & . & . & . \\ \binom{\alpha_{d-k+2}}{k-1} & . & . & . & 1 \\ \binom{i_{k+1}}{k+1} & . & . & . & \binom{i_{k+1}}{2} \end{vmatrix},$$

and these are easily combined with the two determinants resulting from the first summation to give two determinants of order $k+1$,

$$\begin{vmatrix} \binom{\alpha_d}{1} & 1 & . & . & 0 \\ . & . & . & . & . \\ \binom{\alpha_{d-k+1}}{k} & . & . & \binom{\alpha_{d-k+1}}{1} & 1 \\ \binom{\alpha_{d-k}}{k+1} & . & . & \binom{\alpha_{d-k}}{2} & \binom{\alpha_{d-k}}{1} \end{vmatrix}$$

$$-\begin{vmatrix} \binom{\alpha_d}{1} & 1 & 0 & . & . & 0 \\ . & . & . & . & . & . \\ \binom{\alpha_{d-k+1}}{k} & . & . & . & \binom{\alpha_{d-k+1}}{1} & 1 \\ \binom{i_{k+1}}{k+1} & . & . & . & \binom{i_{k+1}}{2} & \binom{i_{k+1}}{1} \end{vmatrix}.$$

‡ Derived immediately from $\binom{r+1}{s+1}=\binom{r}{s}+\binom{r}{s+1}$.

The formula (3) is therefore proved. If $k = d+1$, and $i_{d+1}-1 = n$, we obtain the number $\omega_{a_0 a_1 \dots a_d}$ of non-zero determinants $p_{i_0 \dots i_d}$ as the single determinant:

$$\omega_{a_0 a_1 \dots a_d} = (-1)^{d+1} \begin{vmatrix} \binom{\alpha_d}{1} & 1 & 0 & . & 0 \\ \binom{\alpha_{d-1}}{2} & \binom{\alpha_{d-1}}{1} & 1 & . & 0 \\ . & . & . & . & . \\ \binom{\alpha_1}{d} & \binom{\alpha_1}{d-1} & . & . & 1 \\ \binom{\alpha_0}{d+1}-\binom{n+1}{d+1} & \binom{\alpha_0}{d}-\binom{n+1}{d} & . & . & \alpha_0-(n+1) \end{vmatrix}.$$

To obtain this determinant in a more convenient form we form a new column 1 = column 1 − column 2 + column 3 − column 4 + ..., a new column 2 = column 2 − column 3 + column 4 − column 5 + ..., and so on, using the formula‡

$$\binom{r}{s}-\binom{r}{s-1}+\binom{r}{s-2}-\dots+(-1)^s = \binom{r-1}{s}. \qquad (\dagger)$$

We then find that

$$\omega_{a_0 a_1 \dots a_d} = (-1)^{d+1} \begin{vmatrix} \binom{\alpha_d-1}{1} & 1 & 0 & . & 0 \\ \binom{\alpha_{d-1}-1}{2} & \binom{\alpha_{d-1}-1}{1} & 1 & . & 0 \\ . & . & . & . & . \\ \binom{\alpha_1-1}{d} & \binom{\alpha_1-1}{d-1} & . & . & 1 \\ \binom{\alpha_0-1}{d+1}-\binom{n}{d+1} & \binom{\alpha_0-1}{d}-\binom{n}{d} & . & . & (\alpha_0-1)-n \end{vmatrix},$$

which is of the same form as the determinant above, but with $\alpha_0-1, \dots, \alpha_d-1, n-1$ for $\alpha_0, \dots, \alpha_d$, n respectively. We note that

‡ Obtained by comparing coefficients of x^s in $(1+x)^r(1+x)^{-1}$ and $(1+x)^{r-1}$.

$\binom{p}{q} = p(p-1)\ldots(p-q+1)/q!$, whatever the integer p, provided that $q>0$, and we define $\binom{p}{0} = 1$ and $\binom{p}{-k} = 0$ for use in future calculations. We now repeat the above transformation on the derived determinant and after n such steps we have

$$\omega_{a_0a_1\ldots a_d} = (-1)^{d+1}\begin{vmatrix} \binom{\alpha_d-n-1}{1} & 1 & 0 & . & 0 \\ \binom{\alpha_{d-1}-n-1}{2} & \binom{\alpha_{d-1}-n-1}{1} & 1 & . & 0 \\ . & . & . & . & . \\ \binom{\alpha_0-n-1}{d+1} & \binom{\alpha_0-n-1}{d} & . & . & \alpha_0-n-1 \end{vmatrix}. \tag{4}$$

It is this form of the determinant which we shall use as the basis of a further induction.

We now wish to calculate the number of power-products

$$X_{i_0\ldots i_d}X_{j_0\ldots j_d}\ldots X_{l_0\ldots l_d} \tag{5}$$

of degree m, no linear combination of which vanishes identically when the $X_{i_0\ldots i_d}$ are replaced by the determinantal expressions $p_{i_0\ldots i_d}$ obtained from (3). We do this by determining a certain set of power-products of degree m with the following properties:

(i) no linear combination of these power-products is identically zero after the determinantal substitution is made;

(ii) every power-product of degree m, and hence every form of degree m in the $X_{i_0\ldots i_d}$, can be expressed as a linear combination of the power-products in the set, *modulo* the equations of $\Omega_{a_0\ldots a_d}$.

This set of power-products will be called *standard* power-products of degree m. To define them we represent a power-product (5) by a *tableau*

$$\begin{matrix} i_0 & j_0 & . & l_0 \\ i_1 & j_1 & . & l_1 \\ . & . & . & . \\ i_d & j_d & . & l_d \end{matrix} \tag{6}$$

Each column contains the $d+1$ suffixes corresponding to a term in the power-product. Since the power-product is of degree m, there are m columns. We shall assume that the suffixes corresponding to any term are arranged so that the numbers increase as we go down any particular column. It may be possible to rearrange the order of the terms in (5) so that the numbers in any row do not decrease as we go along the row. If this is so the tableau is said to be a *standard* tableau, and the corresponding power-product is said to be a *standard* power-product. Hence any tableau of $d+1$ rows and m columns corresponds to a standard power-product in the $X_{i_0 \ldots i_d}$ of degree m if the numbers in any column increase (strictly) as we go down the column, and the numbers in any row do not decrease as we go along the row. In particular, any power-product $(X_{01\ldots d})^m$ is standard, and so is $(X_{i_0 i_1 \ldots i_d})^m$, where $i_0 < i_1 < \ldots < i_d$.

THEOREM I. *Any power-product of degree m can be expressed as a sum of standard power-products of degree m modulo the quadratic relations satisfied by Grassmann coordinates.*

We know that [VII, § 6]

$$\sum_{\rho=0}^{d+1} (-1)^\rho p_{i_0 \ldots i_{d-1} j_\rho} p_{j_0 \ldots j_{\rho-1} j_{\rho+1} \ldots j_{d+1}} = 0, \tag{7}$$

and that the quadratic relations may be written in the form

$$p_{i_0 \ldots i_d} p_{j_0 \ldots j_d} = \sum_{\rho=0}^{d} p_{i_0 \ldots i_{s-1} j_\rho i_{s+1} \ldots i_d} p_{j_0 \ldots j_{\rho-1} i_s j_{\rho+1} \ldots j_d}. \tag{8}$$

From (7) we deduce a more general form of relation

$$\sum_l \delta^{l_0 \ldots l_{d+1}}_{j_0 \ldots j_{d+1}} p_{i_1 \ldots i_\lambda l_0 \ldots l_{d-\lambda}} p_{l_{d-\lambda+1} \ldots l_{d+1} k_1 \ldots k_{d-\lambda}} = 0, \tag{9}$$

where the summation is over all possible values of $l_0, \ldots, l_{d+1}$, and $\delta^{l_0 \ldots l_{d+1}}_{j_0 \ldots j_{d+1}}$ is a Kronecker delta [II, § 8, p. 74]. To prove (9) we first note that when $\lambda = d$ we obtain (7). In fact, (7) is (in a slightly modified notation)

$$\sum_l \delta^{l_0 \ldots l_{d+1}}_{j_0 \ldots j_{d+1}} p_{i_1 \ldots i_d l_0} p_{l_1 \ldots l_{d+1}} = 0. \tag{10}$$

Now, using (8), we have the equation

$$\begin{aligned} p_{i_1 \ldots i_\lambda l_0 \ldots l_{d-\lambda}} & p_{l_{d-\lambda+1} \ldots l_{d+1} k_1 \ldots k_{d-\lambda}} \\ &= \sum_{\mu=1}^{\lambda+1} p_{i_1 \ldots i_\lambda l_0 \ldots l_{d-\lambda-1} l_{d-\lambda+\mu}} p_{l_{d-\lambda+1} \ldots l_{d-\lambda+\mu-1} l_{d-\lambda} l_{d-\lambda+\mu} \ldots l_{d+1} k_1 \ldots k_{d-\lambda}} \\ &+ \sum_{\mu=1}^{d-\lambda} p_{i_1 \ldots i_\lambda l_0 \ldots l_{d-\lambda-1} k_\mu} p_{l_{d-\lambda+1} \ldots l_{d+1} k_1 \ldots k_{\mu-1} l_{d-\lambda} k_{\mu-1} \ldots k_{d-\lambda}}. \end{aligned}$$

For a fixed value of μ

$$\sum_l \delta^{l_0 \ldots l_{d+1}}_{j_0 \ldots j_{d+1}} p_{i_1 \ldots i_\lambda l_0 \ldots l_{d-\lambda-1} l_{d-\lambda+\mu}} p_{l_{d-\lambda+1} \ldots l_{d-\lambda+\mu-1} l_{d-\lambda} l_{d-\lambda+\mu} \ldots l_{d+1} k_1 \ldots k_{d-\lambda}}$$
$$= -\sum_l \delta^{l_0 \ldots l_{d+1}}_{j_0 \ldots j_{d+1}} p_{i_1 \ldots i_\lambda l_0 \ldots l_{d-\lambda-1} l_{d-\lambda}} p_{l_{d-\lambda+1} \ldots l_{d+1} k_1 \ldots k_{d-\lambda}},$$

interchanging the summation suffixes $l_{d-\lambda+\mu}$ and $l_{d-\lambda}$. Also

$$\sum_l \delta^{l_0 \ldots l_{d+1}}_{j_0 \ldots j_{d+1}} p_{i_1 \ldots i_\lambda l_0 \ldots l_{d-\lambda-1} k_\mu} p_{l_{d-\lambda+1} \ldots l_{d+1} k_1 \ldots k_{\mu-1} l_{d-\lambda} k_{\mu+1} \ldots k_{d-\lambda}}$$
$$= (-1)^{d+\mu} \sum_l \delta^{l_0 \ldots l_{d+1}}_{j_0 \ldots j_{d+1}} p_{i_1 \ldots i_\lambda k_\mu l_0 \ldots l_{d-\lambda-1}} p_{l_{d-\lambda} \ldots l_{d+1} k_1 \ldots k_{\mu-1} k_{\mu+1} \ldots k_{d-\lambda}},$$

by the skew-symmetric property of Grassmann coordinates. Hence

$$\sum_l \delta^{l_0 \ldots l_{d+1}}_{j_0 \ldots j_{d+1}} p_{i_1 \ldots i_\lambda l_0 \ldots l_{d-\lambda}} p_{l_{d-\lambda+1} \ldots l_{d+1} k_1 \ldots k_{d-\lambda}}$$
$$= -(\lambda+1) \sum_l \delta^{l_0 \ldots l_{d+1}}_{j_0 \ldots j_{d+1}} p_{i_1 \ldots i_\lambda l_0 \ldots l_{d-\lambda-1} l_{d-\lambda}} p_{l_{d-\lambda+1} \ldots l_{d+1} k_1 \ldots k_{d-\lambda}}$$
$$+ \sum_{\mu=1}^{d-\lambda} (-1)^{d+\mu} \sum_l \delta^{l_0 \ldots l_{d+1}}_{j_0 \ldots j_{d+1}} p_{i_1 \ldots i_\lambda k_\mu l_0 \ldots l_{d-\lambda-1}} p_{l_{d-\lambda} \ldots l_{d+1} k_1 \ldots k_{\mu-1} k_{\mu+1} \ldots k_{d-\lambda}}.$$

If (9) is true when λ is replaced by $\lambda+1$, the second term of this last equation is zero, and we deduce that

$$(\lambda+2) \sum_l \delta^{l_0 \ldots l_{d+1}}_{j_0 \ldots j_{d+1}} p_{i_1 \ldots i_\lambda l_0 \ldots l_{d-\lambda}} p_{l_{d-\lambda+1} \ldots l_{d+1} k_1 \ldots k_{d-\lambda}} = 0,$$

that is, (9) is true. Since the equation (9) is true when $\lambda = d$, it is true when $\lambda = d-1, d-2, \ldots, 1, 0$.

Now consider any power-product of degree m in the $X_{i_0 \ldots i_d}$, and construct the corresponding tableau. Let N be any number greater than 1. We define the *measure* of the tableau to be N^m times the sum of the numbers in the first column plus N^{m-1} times the sum of the numbers in the second column plus, etc. We show that if the tableau is not standard we can replace the power-product of degree m by a sum of power-products of degree m, each of which has a tableau of smaller measure. Indeed, if the tableau is not standard there is a pair of adjacent columns, say the kth and $(k+1)$th, with the entries

$$\begin{matrix} i_0 & j_0 \\ \vdots & \vdots \\ i_d & j_d \end{matrix}$$

such that for some $\lambda \leqslant d$,

$$j_0 < j_1 < \ldots < j_\lambda < i_\lambda < \ldots < i_d.$$

We then consider the particular case of (9),

$$\sum_{l} \delta^{l_0 \ldots \ldots l_{d+1}}_{j_0 \ldots j_\lambda i_\lambda \ldots i_d} X_{j_{\lambda+1} \ldots j_d l_0 \ldots l_\lambda} X_{l_{\lambda+1} \ldots l_{d+1} i_0 \ldots i_{\lambda-1}} = 0. \tag{11}$$

By means of this equation we replace the factor

$$X_{i_0 \ldots i_d} X_{j_0 \ldots j_d}$$

in the power-product under consideration by a sum of quadratic terms, and the power-product is replaced by a sum of power-products, each of degree m, each of which can be represented by a tableau which differs from the original tableau only in the kth and $(k+1)$th columns; consideration of (11) shows that these columns have been altered by the interchange of certain of the numbers $j_0, \ldots, j_\lambda$ in the $(k+1)$th column with a corresponding set of *larger* numbers from the kth column, and then rearranging the numbers in each column so that they increase as we descend the column. If the sum of the numbers in the kth column has been reduced by σ, the sum of the numbers in the $(k+1)$th column is increased by σ, and the measure of the tableau is decreased by $\sigma(N^{m-k+1} - N^{m-k}) > 0$. Now the measure of any tableau is not negative; hence we cannot carry on this process indefinitely. We therefore reach a set of standard power-products of degree m after a finite number of operations. This completes the proof of Theorem I.

Now let $F(\ldots, X_{i_0 \ldots i_d}, \ldots)$ be any form of degree m in $X_{i_0 \ldots i_d}$, and let it be reduced, *modulo* the quadratic p-relations, to the sum of standard power-products of degree m

$$\Sigma a_{ij\ldots l} X_{i_0 \ldots i_d} X_{j_0 \ldots j_d} \ldots X_{l_0 \ldots l_d} + \Sigma b_{ij\ldots l} X_{i_0 \ldots i_d} X_{j_0 \ldots j_d} \ldots X_{l_0 \ldots l_d},$$

where the first summation involves power-products which become zero after the substitution $X_{i_0 \ldots i_d} = p_{i_0 \ldots i_d}$, and the second summation involves only those power-products which do not become zero after this substitution, $p_{i_0 \ldots i_d}$ being the determinant formed from the $(i_0 + 1)$th, ..., $(i_d + 1)$th columns of the matrix (2). A typical term $a_{ij\ldots l} X_{i_0 \ldots i_d} X_{j_0 \ldots j_d} \ldots X_{l_0 \ldots l_d}$ of the first summation cannot have all $i_r \geqslant \alpha_{d-r}$ $(r = 0, \ldots, d)$, for, being a standard power-product, we should have $j_r \geqslant \alpha_{d-r}$ $(r = 0, \ldots, d), \ldots, l_r \geqslant \alpha_{d-r}$ $(r = 0, \ldots, d)$, and the determinantal substitution would not make the term zero. Hence $i_r < \alpha_{d-r}$ for some value of r. In the second summation a typical term $b_{ij\dots l} X_{i_0\dots i_d} X_{j_0\dots j_d} \dots X_{l_0\dots l_d}$ has $i_r \geqslant \alpha_{d-r}$ $(r = 0, \dots, d)$. On $\Omega_{a_0 a_1 \dots a_d}$ we have

$$F(\dots, X_{i_0\dots i_d}, \dots) = \Sigma b_{ij\dots l} p_{i_0\dots i_d} p_{j_0\dots j_d} \dots p_{l_0\dots l_d}.$$

We wish to prove that if $F(\dots, X_{i_0\dots i_d}, \dots) = 0$ on $\Omega_{a_0 a_1 \dots a_d}$, all the coefficients $b_{ij\dots l}$ must be zero.

We do this by attaching a weight to every term in the expansion of the determinantal expression for $p_{i_0\dots i_d}$. We define $(d+1)(n+1)$ positive integers ρ_{ij} by the formula

$$\rho_{ij} = (n+1-i)(n+1-i+j) \quad (i = 0, \dots, d;\ j = 0, \dots, n).$$

If $k > j$,

$$\rho_{ik} - \rho_{ij} = (n+1-i)(k-j) > (n+1-i-1)(k-j) = \rho_{i+1\,k} - \rho_{i+1\,j},$$

so that
$$\rho_{ij} + \rho_{i+1\,k} < \rho_{ik} + \rho_{i+1\,j} \quad (j < k). \tag{12}$$

If we have a determinant whose elements are indeterminates,

$$\begin{vmatrix} y^0_{i_0} & \cdot & y^0_{i_d} \\ \cdot & \cdot & \cdot \\ y^d_{i_0} & \cdot & y^d_{i_d} \end{vmatrix} \quad (i_0 < i_1 < \dots < i_d),$$

we put
$$y^i_j = x^i_j t^{\rho_{ij}},$$

where the x^i_j and t are independent indeterminates. On expansion, the determinant is an expression of the form

$$\Sigma \pm x^0_{j_0} \dots x^d_{j_d} t^{\rho_{0j_0} + \dots + \rho_{dj_d}}.$$

We prove that the term of lowest degree in t in this polynomial is

$$x^0_{i_0} \dots x^d_{i_d} t^{\rho_{0i_0} + \dots + \rho_{di_d}}.$$

Consider any term

$$\pm x^0_{j_0} \dots x^d_{j_d} t^{\rho_{0j_0} + \dots + \rho_{dj_d}}.$$

The numbers $j_0, \dots, j_d$ are a permutation of the numbers $i_0, \dots, i_d$, and, by hypothesis, $i_0 < i_1 < \dots < i_d$. In the sum

$$\rho_{0j_0} + \rho_{1j_1} + \dots + \rho_{dj_d}$$

let us suppose that i_0 is not in its undisturbed position, and that $j_r = i_0$. Then $j_{r-1} > i_0$. If we interchange j_{r-1} and j_r, by (12) we see that

$$\rho_{r-1\,j_{r-1}} + \rho_{r\,j_r} > \rho_{r-1\,j_r} + \rho_{r\,j_{r-1}}.$$

Therefore

$$\rho_{0j_0} + \dots + \rho_{r-1\,j_{r-1}} + \rho_{r\,j_r} + \dots + \rho_{dj_d}$$
$$> \rho_{0j_0} + \dots + \rho_{r-1\,j_r} + \rho_{r\,j_{r-1}} + \dots + \rho_{dj_d}.$$

Continuing this process, we move i_0 to the first position, and then we move i_1 in to the second position, and so on, and finally we obtain the inequality

$$\rho_{0j_0}+\ldots+\rho_{dj_d}>\rho_{0i_0}+\ldots+\rho_{di_d}$$

whenever $(j_0, \ldots, j_d)$ is not the identical permutation of $(i_0, \ldots, i_d)$. With this result in mind, let us assume that

$$F(\ldots, X_{i_0\ldots i_d}, \ldots) = \Sigma b_{ij\ldots l} X_{i_0\ldots i_d} X_{j_0\ldots j_d} \ldots X_{l_0\ldots l_d} \qquad (13)$$

is a linear combination of standard products which is not zero, but which vanishes on $\Omega_{a_0\ldots a_d}$. We may assume that the symbol Σ means summation only over terms for which $b_{ij\ldots l}$ is not zero. Then

$$\Sigma b_{ij\ldots l} p_{i_0\ldots i_d} p_{j_0\ldots j_d} \ldots p_{l_0\ldots l_d} = 0, \qquad (14)$$

where the $p_{i_0\ldots i_d}$ are determinants formed from the columns of the matrix (2). We replace each ξ^i_j by $\eta^i_j t^{\rho_{ij}}$, and consider the resulting equation in t. The term of least degree in t on the left-hand side has coefficient

$$\Sigma' b_{ij\ldots l} \eta^0_{i_0} \ldots \eta^d_{i_d} \eta^0_{j_0} \ldots \eta^d_{j_d} \ldots \eta^0_{l_0} \ldots \eta^d_{l_d},$$

where the summation Σ' is over those terms of the summation Σ for which

$$\rho_{0i_0}+\ldots+\rho_{di_d}+\rho_{0j_0}+\ldots+\rho_{dj_d}+\ldots+\rho_{0l_0}+\ldots+\rho_{dl_d}$$

is least. Σ' is necessarily a summation over a non-zero number of terms, in each of which $b_{ij\ldots l}$ is not zero. Since t is an indeterminate over $K(\eta^i_j)$, (14) implies that

$$\Sigma' b_{ij\ldots l} \eta^0_{i_0} \ldots \eta^d_{i_d} \eta^0_{j_0} \ldots \eta^d_{j_d} \ldots \eta^0_{l_0} \ldots \eta^d_{l_d} = 0. \qquad (15)$$

It is clear, however, that any two distinct standard power-products in (13)

$$X_{i_0\ldots i_d} \ldots X_{l_0\ldots l_d}$$

and

$$X_{i'_0\ldots i'_d} \ldots X_{l'_0\ldots l'_d}$$

lead to *distinct* products $\eta^0_{i_0} \ldots \eta^d_{i_d} \ldots \eta^0_{l_0} \ldots \eta^d_{l_d}$ and $\eta^0_{i'_0} \ldots \eta^d_{i'_d} \ldots \eta^0_{l'_0} \ldots \eta^d_{l'_d}$. Hence from (15) we deduce that $b_{ij\ldots l} = 0$ for each term in (15). But by our assumption Σ' is a sum over terms with non-zero coefficients, and we have a contradiction. Hence if (13) vanishes on $\Omega_{a_0\ldots a_d}$, $F(\ldots, X_{i_0\ldots i_d}, \ldots)$ must be identically zero.

We have therefore proved

THEOREM II. *$F(\ldots, X_{i_0 \ldots i_d}, \ldots)$ vanishes on $\Omega_{a_0 a_1 \ldots a_d}$ if and only if*

$$F(\ldots, X_{i_0 \ldots i_d}, \ldots) \equiv \sum_{i,j} A_{i_1 \ldots i_d, j_0 \ldots j_{d+1}}(X)\, F_{i_1 \ldots i_d, j_0 \ldots j_{d+1}}(X) + \sum_{l} B_{l_0 \ldots l_d}(X)\, X_{l_0 \ldots l_d},$$

where $F_{i_1 \ldots i_d, j_0 \ldots j_{d+1}}(X) = \sum_{\lambda=0}^{d+1} (-1)^{\lambda} X_{i_1 \ldots i_d j_\lambda} X_{j_0 \ldots j_{\lambda-1} j_{\lambda+1} \ldots j_{d+1}}$, *and the second summation is over the sets $l_0 < l_1 < \ldots < l_d$ for which at least one $l_r < \alpha_{d-r}$.*

In other words, the quadratic p-relations and the equations

$$X_{l_0 \ldots l_d} = 0 \quad (l_r < \alpha_{d-r} \text{ for some } r)$$

form a basis for the equations of $\Omega_{a_0 a_1 \ldots a_d}$.

We note that in the particular case $a_r = n-d+r$, $\alpha_{d-r} = r$, $\Omega_{a_0 a_1 \ldots a_d} = \Omega(d, n)$, and Theorem II becomes the basis theorem of VII, § 7, Th. I.

We have seen that any form $F(\ldots, X_{i_0 \ldots i_d}, \ldots)$ of degree m can be written in the form

$$\Sigma A_{i_1 \ldots i_d, j_0 \ldots j_{d+1}}(X)\, F_{i_1 \ldots j_d, j_0 \ldots j_{d+1}}(X) + \sum_{l} B_{l_0 \ldots l_d}(X)\, X_{l_0 \ldots l_d} + F^*(\ldots, X_{i_0 \ldots i_d}, \ldots),$$

where the first two summations are as defined in Theorem II, and $F^*(\ldots, X_{i_0 \ldots i_d}, \ldots)$ is a sum of standard power-products of degree m. In order that F vanish on $\Omega_{a \ldots a_d}$, it is necessary and sufficient that F^* should be zero. The coefficients of F^* are linearly independent linear combinations of the coefficients of F, and hence if $\omega_{a_0 \ldots a_d}(m)$ is the number of distinct standard power-products of degree m, there are $\omega_{a_0 \ldots a_d}(m)$ linearly independent conditions to be satisfied by the coefficients of F in order that F should vanish on $\Omega_{a_0 \ldots a_d}$. Hence $\omega_{a_0 \ldots a_d}(m)$ is the postulation of $\Omega_{a_0 \ldots a_d}$ for forms of degree m. We now proceed to calculate it.

When $m = 1$, $\omega_{a_0 \ldots a_d}(1) = \omega_{a_0 \ldots a_d}$ is the number of $X_{i_0 \ldots i_d}$ $(i_0 < i_1 < \ldots < i_d)$ which do not vanish on $\Omega_{a_0 a_1 \ldots a_d}$. This is given by (4) as

$$\omega_{a_0 \ldots a_d} = (-1)^{d+1} \begin{vmatrix} \binom{\alpha_d - n - 1}{1} & 1 & 0 & . & 0 \\ \binom{\alpha_{d-1} - n - 1}{2} & \binom{\alpha_{d-1} - n - 1}{1} & 1 & . & 0 \\ . & . & . & . & . \\ \binom{\alpha_0 - n - 1}{d+1} & \binom{\alpha_0 - n - 1}{d} & . & . & \alpha_0 - n - 1 \end{vmatrix}.$$

We now prove that

$$\omega_{a_0\ldots a_d}(m) = (-1)^{m(d+1)} \begin{vmatrix} \binom{\alpha_d - n - 1}{m} & \binom{\alpha_d - n - 1}{m-1} & \cdots \\ \binom{\alpha_{d-1} - n - 1}{m+1} & \binom{\alpha_{d-1} - n - 1}{m} & \cdots \\ \cdot & \cdot & \cdots \\ \binom{\alpha_0 - n - 1}{m+d} & \binom{\alpha_0 - n - 1}{m+d-1} & \cdots \end{vmatrix}. \quad (16)$$

The proof is by induction, the formula having been established for $m = 1$. We shall assume that the formula is correct for standard power-products of degree $m-1$, and prove it for standard power-products of degree m.

We begin by determining the number of distinct standard tableaux of m columns for which the first column $\beta_d < \beta_{d-1} < \ldots < \beta_0$ is given. We have the inequalities

$$\beta_r \geqslant \alpha_r = n - a_r \quad (r = 0, \ldots, d).$$

The remaining $m-1$ columns form a standard tableau of the required type, corresponding to a non-vanishing power-product, and to form these standard tableaux we must see that the first column $\gamma_d < \gamma_{d-1} < \ldots < \gamma_0$ satisfies the inequalities

$$\gamma_d \geqslant \beta_d, \quad \gamma_{d-1} \geqslant \beta_{d-1}, \quad \ldots, \quad \gamma_0 \geqslant \beta_0.$$

These inequalities ensure the existence of the necessary inequalities

$$\gamma_r \geqslant n - a_r \quad (r = 0, \ldots, d).$$

The number of standard power-products of m columns with a given first column $\beta_d < \beta_{d-1} < \ldots < \beta_0$ is therefore

$$\omega_{n-\beta_0\, n-\beta_1 \ldots n-\beta_d}(m-1),$$

since they are obtained by constructing all the standard power-products of degree $m-1$ obtained from the matrix (2) in which $\alpha_0, \ldots, \alpha_d$ are replaced by $\beta_0, \ldots, \beta_d$, and then multiplying in front by $X_{\beta_d \ldots \beta_0}$.

If we now consider all possible values of $\beta_d < \beta_{d-1} < \ldots < \beta_0$ which can serve as the first column of a non-vanishing standard power-product, we have

$$\omega_{a_0 \ldots a_d}(m) = \sum_{\beta_0=\alpha_0}^{n} \cdots \sum_{\beta_{d-1}=\alpha_{d-1}}^{\beta_{d-2}-1} \sum_{\beta_d=\alpha_d}^{\beta_{d-1}-1} \omega_{n-\beta_0 \ldots n-\beta_d}(m-1)$$

$$= (-1)^{(m-1)(d+1)} \sum_{\beta_0=\alpha_0}^{n} \cdots \sum_{\beta_d=\alpha_d}^{\beta_{d-1}-1} \begin{vmatrix} \binom{\beta_d-n-1}{m-1} & \binom{\beta_d-n-1}{m-2} & \cdots \\ \cdot & \cdot & \cdots \\ \binom{\beta_0-n-1}{m-1+d} & \binom{\beta_0-n-1}{m-2+d} & \cdots \end{vmatrix},$$

by the hypothesis of induction. Now, using the formula (*),

$$\sum_{\beta_d=\alpha_d}^{\beta_{d-1}-1} \begin{vmatrix} \binom{\beta_d-n-1}{m-1} & \binom{\beta_d-n-1}{m-2} & \cdots \\ \binom{\beta_{d-1}-n-1}{m} & \binom{\beta_{d-1}-n-1}{m-1} & \cdots \\ \cdot & \cdot & \cdots \\ \binom{\beta_0-n-1}{m-1+d} & \binom{\beta_0-n-1}{m-2+d} & \cdots \end{vmatrix}$$

$$= \begin{vmatrix} \binom{\beta_{d-1}-n-1}{m}-\binom{\alpha_d-n-1}{m} & \binom{\beta_{d-1}-n-1}{m-1}-\binom{\alpha_d-n-1}{m-1} & \cdots \\ \binom{\beta_{d-1}-n-1}{m} & \binom{\beta_{d-1}-n-1}{m-1} & \cdots \\ \cdot & \cdot & \cdots \\ \binom{\beta_0-n-1}{m-1+d} & \binom{\beta_0-n-1}{m-2+d} & \cdots \end{vmatrix}$$

$$= - \begin{vmatrix} \binom{\alpha_d-n-1}{m} & \binom{\alpha_d-n-1}{m-1} & \cdots \\ \binom{\beta_{d-1}-n-1}{m} & \binom{\beta_{d-1}-n-1}{m-1} & \cdots \\ \cdot & \cdot & \cdots \\ \binom{\beta_0-n-1}{m-1+d} & \binom{\beta_0-n-1}{m-2+d} & \cdots \end{vmatrix}.$$

When we carry out the next summation the second row is affected, and we obtain the determinant

$$(-1)^2 \begin{vmatrix} \binom{\alpha_d - n - 1}{m} & \binom{\alpha_d - n - 1}{m-1} & \cdots \\ \binom{\alpha_{d-1} - n - 1}{m+1} & \binom{\alpha_{d-1} - n - 1}{m} & \cdots \\ \cdot & \cdot & \cdots \\ \binom{\beta_0 - n - 1}{m-1+d} & \binom{\beta_0 - n - 1}{m-2+d} & \cdots \end{vmatrix}.$$

Finally, carrying out the last summation

$$(-1)^d \sum_{\beta_0 = \alpha_0}^{n} \begin{vmatrix} \binom{\alpha_d - n - 1}{m} & \binom{\alpha_d - n - 1}{m-1} & \cdots \\ \binom{\alpha_{d-1} - n - 1}{m+1} & \binom{\alpha_{d-1} - n - 1}{m} & \cdots \\ \cdot & \cdot & \cdots \\ \binom{\beta_0 - n - 1}{m-1+d} & \binom{\beta_0 - n - 1}{m-2+d} & \cdots \end{vmatrix}$$

$$= (-1)^d \begin{vmatrix} \binom{\alpha_d - n - 1}{m} & \binom{\alpha_d - n - 1}{m-1} & \cdots \\ \binom{\alpha_{d-1} - n - 1}{m+1} & \binom{\alpha_{d-1} - n - 1}{m} & \cdots \\ \cdot & \cdot & \cdots \\ -\binom{\alpha_0 - n - 1}{m+d} & -\binom{\alpha_0 - n - 1}{m+d-1} & \cdots \end{vmatrix}$$

$$= (-1)^{d+1} \begin{vmatrix} \binom{\alpha_d - n - 1}{m} & \binom{\alpha_d - n - 1}{m-1} & \cdots \\ \binom{\alpha_{d-1} - n - 1}{m+1} & \binom{\alpha_{d-1} - n - 1}{m} & \cdots \\ \cdot & \cdot & \cdots \\ \binom{\alpha_0 - n - 1}{m+d} & \binom{\alpha_0 - n - 1}{m+d-1} & \cdots \end{vmatrix}.$$

Hence

$$\omega_{a_0\dots a_d}(m) = (-1)^{m(d+1)} \begin{vmatrix} \binom{\alpha_d - n - 1}{m} & \binom{\alpha_d - n - 1}{m-1} & \dots \\ \binom{\alpha_{d-1} - n - 1}{m+1} & \binom{\alpha_{d-1} - n - 1}{m} & \dots \\ \cdot & \cdot & \dots \\ \binom{\alpha_0 - n - 1}{m+d} & \binom{\alpha_0 - n - 1}{m+d-1} & \dots \end{vmatrix}.$$

We have therefore proved formula (16). An alternative form in terms of $a_0, \dots, a_d$ may be obtained, since

$$\binom{\alpha_k - n - 1}{r} = (-1)^r \binom{n + r - \alpha_k}{r} = (-1)^r \binom{a_k + r}{r}.$$

Then

$$\omega_{a_0\dots a_d}(m)$$
$$= (-1)^{m(d+1)} \begin{vmatrix} (-1)^m \binom{a_d + m}{m} & (-1)^{m-1} \binom{a_d + m - 1}{m-1} & \dots \\ \cdot & \cdot & \dots \\ (-1)^{m+d} \binom{a_0 + m + d}{m+d} & (-1)^{m+d-1} \binom{a_0 + m + d - 1}{m+d-1} & \dots \end{vmatrix}$$

$$= \begin{vmatrix} \binom{a_d + m}{m} & \binom{a_d + m - 1}{m-1} & \dots & \binom{a_d + m - d}{m-d} \\ \binom{a_{d-1} + m + 1}{m+1} & \cdot & \dots & \binom{a_d + m - d + 1}{m - d + 1} \\ \cdot & \cdot & \dots & \cdot \\ \binom{a_0 + m + d}{m+d} & \cdot & \dots & \binom{a_0 + m}{m} \end{vmatrix}, \qquad (17)$$

as can be seen by removing factors $(-1)^m, (-1)^{m-1}, \dots, (-1)^{m-d}$ from the first, second, ... columns and $(-1)^0, (-1)^1, \dots, (-1)^d$ from the first, second, ... rows.

We thus have

THEOREM III. *The postulation $\omega_{a_0\dots a_d}(m)$ of the Schubert variety $\Omega_{a_0\dots a_d}$ for primals of order m is given by equation* (17) *for all values of m greater than zero.*

BIBLIOGRAPHICAL NOTES

BOOK III

The treatment of algebraic varieties given here is inspired by the ideas developed by van der Waerden, Zariski, Weil and others during the past two decades. The work of van der Waerden has been the chief inspiration for our study of varieties in projective space, although ideas due to Weil have enabled us to prove some important results. The ideas of Zariski have been profitable in certain places, but his work will figure far more in vol. III than it does in this volume.

Chapter X. The ideas of the first five sections derive from van der Waerden (**10**). The notion of a *generic point* was first explicitly introduced by van der Waerden under the title *allgemeine Nullpunkt*. This was first translated into English as generic point, but lately the term *general point* has taken its place, and the term generic point is used to denote a point of a variety which does not lie on a certain special set of subvarieties. We have preferred to retain the name 'generic point' for *allgemeine Nullpunkt*, on the grounds that the word 'general' in mathematics is often used in association with rather vague concepts, whereas the word 'generic' has not, so far, been thus spoiled, and appears to be very suitable for the precise concept to which it is attached. What we have called the Cayley form was introduced by van der Waerden and Chow in (**12**), under the name *zugeordnete Form*. It is now often called the Chow form. Regarding it as leading to the equation of the cone projecting a variety V_d from a generic $(n-d-2)$-space (Pedoe (**7**)) the idea can be traced back to Cayley (**1**). For this reason we call it the Cayley form. The use of the Cayley form to obtain the parametric equations of an algebraic variety (p. 62) is based on some lectures given by Zariski at Princeton in 1935, which in turn owe much to the work of Macaulay (**6**).

In the sections dealing with absolute and relative irreducibility we have made use of the ideas of many writers. Results based on the use of the Cayley form are mainly due to van der Waerden (**10**), in addition to the many who have developed the theory of algebraic fields. The characterisation of an absolutely irreducible variety, as well as Theorem I of § 13 are taken from Zariski (**14**).

Chapter XI. The general treatment given here of the theory of algebraic correspondences is based on the work of van der Waerden (**10**). The notion of multiplicative variety is closely related to that of algebraic cycle introduced by Weil in his book (**13**), which appeared when preparations for this volume had begun. The variety used in § 2 to represent an r-way space as a variety in 1-way space was studied extensively by C. Segre (**8**) and is used in the study of correspondences between linear spaces by the Italian school of geometers. The use of the direct product (p. 146) in the treatment of general algebraic correspondences between two varieties is due to Severi, and his methods have been adopted by all geometers, and penetrated other branches of mathematics, such as topology.

Advantage has been taken of the use of powerful arithmetic methods in algebraic geometry, due mainly to Zariski, to give more general conditions for the unique determination of a multiplicity than are given by van der Waerden. Theorem I, our main theorem in this connection, is given by Weil (**13**).

Chapter XII. The theory of intersections given in this chapter is based on van der Waerden (**11**), but in order to develop the theory of equivalence in sections 9–11, his work has been considerably expanded. Our aim has been to develop intersection theory just as far as is necessary to establish the results of classical algebraic geometry. More general algebraic theories have been developed by Weil, Zariski, and Chevalley (**2**).

The first study of non-linear systems of curves on an algebraic surface is due to Castelnuovo, Enriques, Severi, Humbert and many others. The notion of a base for curves on an algebraic surface first arose in Picard's work on double integrals of the second kind on an algebraic surface. The geometric theory of the base was first formulated by Severi. Important contributions have been made by many other Italian geometers, especially Albanese. In view of the many excellent bibliographies and works of reference in which the achievements of the Italian school are recorded, it is sufficient to refer to the article by G. Castelnuovo and F. Enriques, III C 5 in *Ency. der Math. Wissenschaften*. The topological theory of the base for curves on a surface is due to Lefschetz (**5**).

Systems of d-dimensional varieties on an n-dimensional variety have been extensively studied by Severi. The methods used in § 11 to obtain a base on the Segre variety, and later in XIII, § 6, and XIV, § 5, rely on the fact that for each type of variety considered there exists a transitive group of transformations of the variety into itself. On more general varieties it is not possible to find such groups of transformations, and no general method for finding a base on an algebraic variety is known.

BOOK IV

Chapter XIII. The theory of quadrics is discussed in so many geometrical works that nearly all the results of this chapter are familiar to students of geometry. Our object has been to illustrate how many of the standard properties can be established by means of the methods of Book III.

Chapter XIV. Our purpose in this chapter is two-fold; first, to give another illustration of the general theory of Book III, and in the second place to provide an adequate basis for the theory of systems of d-spaces in S_n, in particular for the applications of the Schubert calculus to such systems. The geometrical treatment of the propositions discussed in this chapter will be found in papers by Severi, particularly Severi (**9**). The methods we use, though differing in several major points, are based on Hodge (**4**). The treatment given in these papers owes much to a study of Grassmann varieties by Ehresmann (**3**).

BIBLIOGRAPHY

(1) Cayley, A. *Collected Papers*, **4**, 446.

(2) Chevalley, C. *Trans. American Math. Soc.*, **57** (1945), 1–85.

(3) Ehresmann, C. *Annals of Math.*, **35** (1934), 396–443.

(4) Hodge, W. V. D. *Journal London Math. Soc.*, **16** (1941), 245–255; **17** (1942), 48–64; *Proc. Camb. Phil. Soc.*, **39** (1943), 22–30.

(5) Lefschetz, S. *L'analysis Situs et La Géométrie Algébrique* (Paris, 1924).

(6) Macaulay, F. S. *Algebraic Theory of Modular Systems* (Cambridge, 1916).

(7) Pedoe, D. *Proc. Camb. Phil. Soc.*, **43** (1947), 455–458.

(8) Segre, C. *Rendic. Palermo*, **5** (1891), 192–204.

(9) Severi, F. *Annali di Mat.*, **24** (1915), 89–120; or *Memorie Scelte* (Bologna, 1950), 405–438.

(10) van der Waerden, B. L. *Einfuehrung in die Algebraische Geometrie* (Berlin, 1939).

(11) van der Waerden, B. L. *Math. Annalen*, **115** (1938), 619–642.

(12) van der Waerden, B. L. and Chow, W.-L. *Math. Annalen*, **113** (1937), 692–704.

(13) Weil, A. *Foundations of Algebraic Geometry* (New York, 1946).

(14) Zariski, O. *American J. of Math.*, **62** (1940), 187–221.

INDEX

(The numbers refer to the page where the term occurs for the first time or is defined)

Printed in the United States
By Bookmasters